Large White Butterfly
THE BIOLOGY, BIOCHEMISTRY AND PHYSIOLOGY
OF PIERIS BRASSICAE (LINNAEUS)

SERIES ENTOMOLOGICA

EDITORS

E. SCHIMITSCHEK & K. A. SPENCER

VOLUME 18

DR. W. JUNK PUBLISHERS THE HAGUE–BOSTON–LONDON

Large White Butterfly

THE BIOLOGY, BIOCHEMISTRY AND PHYSIOLOGY OF PIERIS BRASSICAE (LINNAEUS)

JOHN FELTWELL

Battle, Sussex, Great Britain

DR W. JUNK PUBLISHERS THE HAGUE–BOSTON–LONDON

Distributors:

for the United States and Canada

Kluwer Boston Inc.
190 Old Derby Street
Hingham, MA 02043
USA

for all other countries

Kluwer Academic Publishers Group
Distribution Center
P.O. Box 322
3300 AH Dordrecht
The Netherlands

Library of Congress Cataloging in Publication Data   CIP

Feltwell, John.
  Large white butterfly.

  (Series entomologica; v. 18)
  Includes bibliographical references and indexes.
  1. Pieris brassicae. I. Title. II. Series.
QL561.P5F44   595.78′9   81-4378
                                 AACR2

ISBN-13: 978-94-009-8640-4   e-ISBN-13: 978-94-009-8638-1
DOI: 10.1007/978-94-009-8638-1

Cover design: Max Velthuijs

Plate I Frontispiece (From M. Herold, 1815). *Entwicklungsgeschichte der Schmetterlinge, anatomisch und physiologisch bearbeitet. Atlas des Planten.* Krieger, Cassel und Marburg. VI + 118 pp. + XXXIV

Preface

The literature is still one of our biggest frustrations to-day. There is, in one sense, too much of it, and in another not enough – for there are insufficient and inadequate published guidelines through this jungle. Last year two excellent books for students of ecological chemistry were published, one in France and one in England. The concordance of the references was a mere overall 3% rising to 7% in the chapters on pheromones. Even in the computer age, the channel remains a formidable barrier to the rapid exchange of biological information. At the present time we are in urgent need of compilations similar to John Feltwell's "The Large White Butterfly"; since the literature has become virtually unmanageable. This insect is now a demonstration object in the sixth form schoolroom; an experimental "rabbit" in the University laboratory; a test animal in virus and bacterial research projects; a tool for the study of flight mechanisms, migration, plant biochemistry, hormones, genetics, allergies, pigments, mimicry, etc., etc.

John Feltwell has, by this massive compilation, rendered us a great service – in fact he has given us a present of 4,000 hours of library time spent in 50 different libraries in seven countries. In the process he has collected 8000 references to the Large White. Of these, 4000 have been selected, and we are given a brief indication of their contents.

He does not claim that such a compilation absolves us from consulting the original publications – but he outlines the work which has been done on this species, and tells us where to find the information. This is a priceless service for the beginner, for it provides him with an *überblick* of the scene and a framework in which he can view his own particular tiny corner of the field. There is nothing more disconcerting for a student embarking on his first piece of research than to feel the animal concerned is fluttering about in a cabbage patch, but beyond that there is nothing but a cloudy mass of inconclusive, undigested, inaccessible information. Nor are the benefits confined to the student, for such compilations also render a great service to the specialist. Thus, for example, I myself intended to use the White Butterfly to test the chemical differences between the Turkish and Mexican strain of *Cannabis sativa*. From John Feltwell's compilation I was able to ascertain that Russian workers at the turn of the century had already demonstrated the repellent quality of this plant to the ovipositing Large White. I could provide many similar examples.

John Feltwell has shown us the way, and put us on the right road – for the time has come when we are in desperate need of many similar compilations.

Miriam Rothschild

Foreword

The Large White Butterfly is one of the most widely known butterflies in western Europe, notorious not only for its former abundance, but for the voracious way in which the larvae defoliate cabbages. One of the first descriptions of *P. brassicae* was by the 17th century biologist Johan Jacob Swammerdam (1637–1680), who was the first to note the difference between complete and incomplete metamorphosis in insects. It was probably because *P. brassicae* was so readily available in the past that it gradually gained acceptance as a laboratory insect on which experiments and observations could be made.

There are, however, several other factors which influenced the adoption of *P. brassicae* as a research tool and object of investigation; it is easy to rear at home and in the laboratory, it can be fed on a variety of foodplants which are usually available all the year, and the fully grown larva is of a sufficiently large size for manipulation in experiments.

This book is designed for the research scientist as a source to relevant literature and as an introduction to all aspects of the biology of *P. brassicae*. The aim of the author has been to assemble as many references as possible, to abstract the facts, figures, ideas and methods, and fit them into a text which treats the biology of *P. brassicae* in a logical manner. The 4000[+] references include key papers on the insect, over 50 theses and diplomas from six European countries written specifically on *P. brassicae*, as well as notes and personal communications. Much information on *P. brassicae* is embodied in the 79 tables, 50 figures and 10 plates. A wide coverage of the literature pertaining to this insect has been achieved in this book, but it is appreciated that much will have escaped attention.

As far as possible the book has been entirely restricted to accounts of papers on *P. brassicae*, but where there have been obvious gaps in knowledge reference to other closely related pierids has been made. Comparisons with *Pieris rapae* (Linnaeus) have been kept to a minimum as this would have doubled the number of references. The papers for this book were collected over nine years by manually going through as many scientific journals from year of publication to present day as was physically possible, and from making use of the current agricultural, chemical, entomological and horticultural abstracting journals. No computerised retrieval system was used as these are not sufficiently retrospective for the needs of such a book.

In the past, *P. brassicae* has been studied more as a typical laboratory insect than as a pest, so that now more is known about, for example, the constituents of the haemolymph, or the structure of the brain, than about the bionomics of the insect in the field. This one-sided interest has necessarily left gaps in the understanding of *P. brassicae* as a whole animal. In such fields as basic anatomy and morphology or the biology of the parasites and hyperparasites of *P. brassicae* there is much more work to be done, and such simple descriptive studies of the legs of the larva or the proboscis of the imago have been overlooked. The book, however, provides up to date work and illustrates the recent advances made in biochemical and chemosensory systems.

Much physiological information has been drawn from papers dealing with the effects of insecticides. It has been necessary for the biochemist to establish variables in the normal functioning of the insect such as respiration, haemolymph contents, hydrogen ion concentration, muscle and nerve action, before the effects of novel compounds could be ascertained. We owe therefore, much to the biochemist for our understanding of the workings of *P. brassicae*.

Since the 1960s, and through the work of Dr. W.A.L. David and Mr. B.O.C. Gardiner at Cambridge, England, *P. brassicae* has been shown to be the ideal insect for use on synthetic diets. The development of cabbage-free diets, on which the insect appears to grow normally, obviates the necessity for winter cabbage. In the past it had always been possible to restock laboratory cultures from wild collected specimens, but in recent years the insect has not been abundant enough in Europe. Thus the use of artificial diets is of inestimable value in continuous breeding cultures. Its decline in numbers may be traced back to 1955 when the population of *P. brassicae* was reported to have been severely depleted by a natural attack of granulosis virus. The use of organic insecticides since 1942 may have adversely affected butterfly populations as a whole but this has never been quantified.

The text has been prepared with the assistance of many experts including three co-authors: Mr. R.I. Vane-Wright (British Museum (Natural History)) has written on nomenclature in Chapter 1; Dr. M.R. Shaw (Royal Scottish Museum) the section on hymenopterous parasites in Chapter 14, and Dr. H.D. Burges (Glasshouse Crops Research Institute) the section on bacterial control in Chapter 15.

It is hoped that the need for such a laboratory source book on *P.*

brassicae, as a typical butterfly, will have been partially fulfilled and that research scientists, postgraduates, undergraduates, pest control officers, laboratory staff and entomologists generally will find useful some of the facts and figures provided in the text and literature cited.

John Feltwell
Henley Down, Battle

Acknowledgements

Many people have materially assisted in the production of this book and, indeed, without their help the book could not have been written. The Hon. Miriam Rothschild has been exceptionally helpful and has encouraged the project throughout its preparation from idealistic aim to reality. Brian Gardiner of the Agricultural Research Council Unit at Cambridge kindly read several chapters and gave helpful advice.

I am also particularly indebted to Dr. H.D. Burges of the Glasshouse Crops Research Institute, Dr. Mark Shaw of the Royal Scottish Museum and Mr. R.I. Vane-Wright of the British Museum (Natural History) who have kindly acted as co-authors for bacterial control, hymenopterous parasites and nomenclature sections respectively. Dr. Kenneth Spencer, the scientific editor of this series, has been extremely helpful in co-ordinating the structure of the book and has also advised on most chapters.

The specialised nature of each chapter has made it necessary to consult various experts in those fields, and indeed, many of these experts have themselves published important work on *P. brassicae*. Much of their time has been spent checking drafts of the chapters and I am extremely grateful to them all. These include Mr. John Badmin (Shell, Sittingbourne) on development and biochemistry; Dr. Robin Baker (University of Manchester) on migration; Dr. Gérard Chovet (Université Pierre et Marie Curie, Paris) on morphology and anatomy; Dr. John Cranham (East Malling Research Station, Kent) on chemical control; Dr. J. Fourche (Université de Lyon) on physiology and hormones; Mr. R.A. French (Rothamsted Experimental Station, Harpenden) on migration; Miss Marion Gratwick (Ministry of Agriculture, Fisheries and Food, Harpenden) on life history; Mr John Heath (Monks Wood Experimental Station) on distribution; Dr. N.P. Kristensen (Zoological Museum, København, Denmark) on morphology and anatomy; Dr. René Lafont (École Normale Supériéure, Paris) on hormones; Dr. Jacques Lhonoré (Université Pierre et Marie Curie, Paris) on morphology, anatomy and physiology; Dr. C. Payne (Glasshouse Crops Research Institute) on virus control; The Hon. Miriam Rothschild (Ashton, Peterborough) on foodplants; Dr. Art Shapiro (Davis, California) on physiology (diapause), Dr. Kenneth Spencer (University of Exeter) on foodplants and economic importance; Dr. D.C. Twinn (May & Baker, Essex) on chemical and integrated control and Dr. Ian Watkinson (Shell, Sittingbourne) on hormones.

I am particularly grateful to Geoffrey Burton (Minster, Isle of Sheppey) who so patiently read through the manuscript highlighting those inadvertent and repetitious slips; and to Carolyn Boulton for proofreading the manuscript in its latter stages. The assistance of library staff has been quite exceptional and I am particularly grateful to Mrs. B.G. Leonard of the Royal Entomological Society of London; Miss Pamela Gilbert and Miss Ann Lum of the British Museum (Natural History), Entomological Library and Mrs. M. Wanstall and Mrs. C. Fagg of the Shell Research Laboratory Library at Sittingbourne, Kent. Thanks are due to Cassell's Publishing house for permission to reproduce Robert Graves' poem.

A work of this nature could not have been undertaken without the assistance of many other people whose names are too numerous to mention and who have been extremely helpful in making suggestions and seeking out information on my behalf. To all of these people who have endured my repeated requests for literature I express my thanks. I would also like to thank my wife, Carol for tolerating my *P. brassicae* obsession over the years, and for typing most of the manuscript, with which she is now very familiar.

Finally, I would like to thank the Royal Society of London for a European Exchange Grant and the Centre for European Agricultural Studies, Wye, Kent, for a grant which supported a visit to 30 research stations in five countries in 1975 in pursuance of information for this book, and to the Linnean Society of London for a grant in 1978 towards cataloguing the references.

Although many people have guided or influenced my thoughts on the subjects of this book, the author holds total responsibility for any inaccuracies and opinions expressed.

Contents

List of plates

Grateful thanks are due to the late Dr. T.R.L. Bigger of the Medical Research Council, Harwell for Plate IV, Dr. H.D. Burges of the Glasshouse Crops Research Institute for Plate X which is by courtesy of Professor J.R. Norris of the Meat Research Institute, Mr. John Heath of the Institute of Terrestrial Ecology, Monks Wood Experimental Station for Plate II, Dr. Jacques and Denise Lhonoré of the Université Pierre et Marie Curie, in Paris for Plate V, VI and VII, and to Professor L.M. Schoonhoven of the Agricultural University Wageningen for Plates VIII and IX.

List of tables and figures

Tables

Figures

1. Nomenclature

SECTION A. CLASSIFICATION OF PIERIS BRASSICAE*
J.S.E. Feltwell and R.I. Vane-Wright of the British Museum (Natural History), London

Classification and derivation of Pieris brassicae

The Large White butterfly, the subject of this monograph, is an insect of the order Lepidoptera, within which it belongs to the suborder Ditrysia. Placed with the true butterflies in the superfamily Papilionoidea, the Large White is the "typical" species of the family Pieridae, a group mostly comprising the familiar whites (Pierinae) and sulphurs (Coliadinae). The technical or "scientific" name of the Large White is *Pieris brassicae* (Linnaeus, 1758).

The modified Latin word *Pieris* (plural Pierides) comes from the Greek Pieria. The Pierides were a group of nine godesses, "the nine Muses" or "the Nine", daughters of Zeus and Mnemosyne, inspirers of poetry and music, who worshipped at their birthplace on Mount Pieris in the district of Thessaly (Oxford University Entomological Society, 1858; Oxford English Dictionary, 1933); *brassicae* comes from the Latin brassica, meaning cabbage.

Generic names and classification

Names for all insects, including *Pieris brassicae*, moved from early descriptive accounts (Table 1.1) to a binominal system of nomenclature

---

*For James Francis Stephens F.L.S., who described the spring and summer broods of *Pieris brassicae* as two distinct species in his work *Illustrations of British Entomology* ... (1827), and to "Bombyx Atlas" (1852), who, 25 years later supported Stephens' views by enumerating the differences between them.

1

Table 1.1 Early descriptions of the Large White.

Christopher Merrett (1667)	*Capite, alisq, lacteis quibus maculae fuscae & nigricantes*
James Petiver (1717)	*Pieris alba major apicibus nigris*
John Ray (1710)	*Pieris brassicaria alba major vulgatissimus*
Eleazar Albin (1720)	*Papilio albus vulgaris major*
Carl Linnaeus (1746)	*Pieris hexapus, alis rotundatis albis: primariis bimaculatis: apicibus nigris: major*
William Turton (1802)	Wings rounded entire white; tip of the upper-pair brown and (in the male) 2 brown spots. Lower wings with a brown spot on the outer margin, beneath pale sulphur. Larva cinereous dotted with black, with 3 sulphur lines, tail black: pupa pale green with 3 yellow lines and 3 globular segments
James Stephens (1827)	*Alis albis, anticus supra cinerascente nebulis nigris ciliisque flavo-albidis subtus maculis duabus nigris, posticis subtus lutescentibus, nigro valde irroratis*

devised by Linnaeus in 1758. At least ten generic names have been applied to the species (Table 1.2).

Linnaeus used the generic name *Papilio* for all the butterflies that he recognised, calling the Large White *Papilio brassicae*. By the early 1800s the need to divide Linnaeus' huge genus *Papilio* was acute, and many new genera were soon proposed, notably by Jacob Hübner. However, it was Franz von Paula Schrank who, in 1801, proposed the genus ·*Pieris* to include the Large White. A little later, Latreille (1810) fixed *Papilio brassicae* Linnaeus, 1758, as the type-species of *Pieris* Schrank. The names *Pieris* Schrank, and *Pieris brassicae* (Linnaeus) have now been placed on the official indexes of generic and specific names in zoology, by the International Commission for Zoological Nomenclature.

The majority of other changes in generic combination for *brassicae* were introduced with little discussion. As can be seen from Table 1.2, all are invalid, either because they are objective synonyms of *Pieris* (i.e. they have the same type-species), or they involve the use of generic names junior to *Pieris*. For a fuller description of the generic names referred to in Table 1.2, the reader is referred to Hemming (1967).

The genus *Pieris*, in the strict sense, is now considered to consist of a single superspecies, comprising three semispecies: *brassicae*, *deota* de Nicéville and *brassicoides* Guérin Méneville (Bernardi, 1947; Lagnel, 1966). *P. brassicoides* is Afrotropical, being restricted to highland regions in Ethiopia and Tanzania. *P. brassicae* and *P. deota* are Palaearctic

Table 1.2 Generic names applied to *P. brassicae.*

Papilio brassicae Linnaeus, 1758. (Placed on Official List of Specific Names in Zoology, name no. 108, *Opinion* 278; type-species of *Papilio* Linnaeus, 1758, *Papilio machaon* Linnaeus by selection of Latreille, 1810).

Pieris brassicae (Linnaeus); Schrank, 1801. (Type-species of *Pieris* Schrank, 1801, *Papilio brassicae* Linnaeus, by selection of Latreille, 1810; *Pieris* Schrank placed on Official List of Generic Names in Zoology, name no. 704, *Opinion* 278.)

Mancipium brassicae (Linnaeus); Hübner, (1806). (Type-species of *Mancipium* Hübner, (1806), *Papilio brassicae* Linnaeus by monotypy; *Mancipium* Hübner, (1806), placed on Official Index of Rejected and Invalid Names in Zoology, name no. 84, *Opinion* 278.)

Pontia brassicae (Linnaeus); Fabricius, 1807. (Type-species of *Pontia* Fabricius, 1807, *Papilio daplidice* Linnaeus, by selection of Curtis, 1824.)

Danaus brassicae (Linnaeus); Oken, 1815. (Type-species of *Danaus* Oken, 1815 (not *Danaus* Kluk), *Papilio brassicae* Linnaeus, by selection of Crotch, 1872; *Danaus* Oken is an invalid homonym of *Danaus* Kluk, 1802, and an objective synonym of *Pieris* Schrank.)

Ganoris brassicae (Linnaeus); Dalman, 1816. (Type-species of *Ganoris* Dalman, 1816, *Papilio brassicae* Linnaeus, by original designation; *Ganoris* Dalman is an objective synonym of *Pieris* Schrank.)

Catophaga brassicae (Linnaeus); Hübner, (1819). (Type-species of *Catophaga* Hübner, (1819), *Papilio paulina* Cramer, by selection of Scudder, 1875.)

Andropodum brassicae (Linnaeus); Hübner, 1822. (Type-species of *Andropodum* Hübner, 1822, *Papilio brassicae* Linnaeus, by selection of Hemming, 1933; *Andropodum* Hübner is an objective synonym of *Pieris* Schrank.)

Tachyptera brassicae (Linnaeus); Berge, 1842. (Type-species of *Tachyptera* Berge, 1842, *Papilio brassicae* Linnaeus, by selection of Hemming, 1934a, b; *Tachyptera* Berge is an objective synonym of *Pieris* Schrank.)

Synchloe brassicae (Linnaeus); Wood, 1867. (Type-species of *Synchloe* Hübner, 1818, *Papilio callidice* Hübner, by selection of Butler, 1870.)

---

species, *brassicae* extending from the Azores, Ireland and the Canaries east to China. *P. deota* (the Kashmir White) is allopatric to *brassicae*, and restricted to high regions in the area from Kashmir to Kirgizia. The dentate male genital valvae of *brassicae* and *deota* (Lagnel, 1966) are probably synapomorphous, suggesting that the two form a sister-species pair. If so, this gives a simple vicariance pattern, *brassicoides* being the Afrotropical sister-group of the Palaearctic (*deota + brassicae*).

For over 50 years some 20 other species have been included additionally in the genus *Pieris*. Bernardi (1947) divided this assemblage into five groups, *Pieris* sensu stricto to contain the three species already discussed, and four other groups to receive the remaining 20 or so species. Since then there has been a general trend to regard these groups as full genera. In particular, Kudrna (1974) elevated the name *Artogeia* Verity, 1947 (originally proposed by Verity as a subgenus of *Pieris*; type-species *Papilio napi* L.) to full generic rank for Bernardi's group 2, the "*rapae* group".

3

This action thus removes the Green-veined White and the Small White (and a number of similar species) from *Pieris* sensu lato, to become *Artogeia napi* (L.) and *A. rapae* (L.) respectively. This has been followed by Higgins (1975) in dealing with the classification of the European fauna. The generic names *Talbotia* Bernardi, 1958, *Pontieuchloia* Verity, 1908, and *Synchloe* Hübner, 1818, are available, respectively, for Bernardi's remaining groups 3, 4 and 5.

Although the characters used by Bernardi (1947) and Lagnel (1966) to define *Pieris* sensu stricto are probably apomorphous (i.e. unique to this group), giving confidence that *brassicae*, *deota* and *brassicoides* form a true clade or "natural group", it is not clear that this is the case with respect to the groupings *Artogeia*, *Talbotia*, *Pontieuchloia* and *Synchloe* – a group of species which Bernardi himself (1947: 71) described as "*très homogène*". The characters used to define *Artogeia* (typical pierine aedeagus, simple valvae, short androconia) might well be symplesiomorphous ("primitive") to the whole complex; if so, these characters provide no evidence that *Artogeia* is a monophyletic assemblage or natural group. A thorough re-investigation of these groups, and such "related" genera as *Pontia* and even *Reliquia* (Ackery, 1975; Shàpiro, 1978), involving an attempt to recognise apomorphous characters, is required.

Finally, it may be noted that Kawazoé & Wakabayashi (1976), in their excellent treatment of the Japanese fauna, revert to treating both *Synchloe* and *Artogeia* as subgenera of *Pieris*. This is a practical position, with much to recommend it in our present ignorance of the phylogenetic relationships within this group.

Common names

The common names applied to *P. brassicae* in various countries have been collected together in Table 1.3.

The earliest vernacular name given to this species in Britain appears to have been the Greater White Butterfly, as used by Petiver (1695). Many of the later common names are permutations of this "theme". Thus Bradley (1718) called it the Common White Cabbage Butterfly, Rees (1819) the Common Large White Butterfly, Kirby (1856) the Cabbage Butterfly, Ormerod (1882) the Large Cabbage Butterfly, the Large White Cabbage Butterfly, and the Large Cabbage White Butterfly, and Soenen (1947) the Large Garden White. Today it is "officially" called the Large White (Thomas *et al.*, 1968). In the literature the Large White has occasionally been called or mistaken for the Cabbage Worm or Imported Cabbage worm, which is *Pieris* (*Artogeia*) *rapae*, the Small White.

Table 1.3 Common names of *P. brassicae*.

Country	Common name
Czechoslovakia	Bĕlásek selný
Denmark	Kålsommerfugle
France	Le grand papillon blanc du chou
	Piéride du chou
Germany	Grosser Kohlweissling
Great Britain	Large White
Hungary	Káposztalepke
	Káposztopillangó
Italy	La Cavolaia
Malta	Farfett tal Cromb
Netherlands	De groote Witjes-Vlinder
	Groot koolwitje
Poland	Bielinek kapustnik
Spain	La gran mariposa blanco de la col
Sweden	Kalfjaril
USSR	Kåpustnaya belyanka

Pieris brassicae and its subspecies

Introduction

According to Talbot (1939: 427–9), the distribution of *P. brassicae* is "all Europe to Turkestan, Baluchistan, Himalayas ... into Tibet and Yunnan". The subdivision of *brassicae* into subspecies or races is in an unsatisfactory state. Up to 10 or more races have been recognised by some workers (Graham-Smith & Graham-Smith, 1930), Lagnel (1966) recognised four subspecies and Higgins (1975) recognised just two for the whole of Western Palaearctic. In contrast, Kudrna (1973) elevated one of the races recognised by Graham-Smith & Graham-Smith to full specific status. Given the peripheral distribution of some of the populations, the seasonal variation of *brassicae*, with up to seven or even eight generations possible (Gardiner, 1979: 26), and the migratory habits of the butterfly, it is hardly surprising that differences of opinion exist.

In 1974 *P. brassicae* reached Chile, probably by importation (Gardiner, 1974b). It will be interesting to see if the Chilean population diverges. Gardiner noted some Chilean specimens with a dark green underside to the hind-wing, a rather unusual variation found in some European populations.

Fischer (1925) tried to recognise three groups of *brassicae* "subspecies" (including *brassicoides*), according to similarity of colour and markings, but confused the matter by equating aberrational names with formally proposed species or subspecies names. Graham-Smith & Graham-Smith (1930) provide a useful summary of the races, placing them in four groups: European, Mediterranean, Atlantic Island and Eastern. Based on this, and taking into account the views of Higgins (1975) and Gardiner (1979), we suggest the following division into three major subspecies; *brassicae brassicae, brassicae cheiranthi* and *brassicae nepalensis*. Each of these might be divisible further into "subraces", but the characters of these are very weak, generally quite insufficient to identify individual specimens. No attempt is made here to characterize these different groupings; this survey is merely an attempt to reflect some consensus of opinion and should be read in conjunction with Gardiner's (1979) general review of *P. brassicae* variations. Brief notes are appended on those nomenclaturally available names treated as synonyms, so that they can be related to the scheme outlined. Until the geographical variation (both phenotypic and genetic) has been studied in far greater detail than at present, and some attempt made to assess effective gene flow between populations, it is doubtful if any more satisfactory, biologically meaningful division can be made.

With these strong reservations in mind, the three major subspecies are listed below together with their included "subraces".

Pieris brassicae (Linnaeus)
Papilio brassicae Linnaeus, 1758. (type-locality: Sweden).
Pontia chariclea Stephens, 1827. (type-locality: Great Britain).

As already noted above, *chariclea* of Stephens represents no more than the spring generation of *brassicae* in north-western Europe.

Pieris brassicae brassicae (Linnaeus)

Following Higgins (1975), we include under subspecies *brassicae* all Large Whites from mainland Europe, the Mediterranean, northern Africa, Turkey and Asia Minor, eastwards to the Urals, Caucasus and Zagros mountains. Within this, adapting the scheme of Graham-Smith & Graham-Smith (1930), it might be possible to recognise three subraces north European, Mediterranean and Syrian, but this is very doubtful.

North European subrace
According to Graham-Smith & Graham-Smith (1930), *Pieris brassicae brassicae* sensu stricto occurs "northwards from central Italy". The

6

boundary between it and the following "subrace" must be very imprecise, and may be chimerical.

Mediterranean subrace

The Mediterranean subrace covers *brassicae brassicae* from the Mediterranean region south of central Italy, including north Africa, and eastwards to Turkey and the Zagros mountains. It includes four taxa originally proposed as subspecies, of which *italorum* Stauder is the oldest available name.

Pieris brassicae italorum Stauder, 1921 (type-locality: Italy, Appenines of Calabria) was described from four male and six female specimens captured in mid June on Mt. Martinello, and between the end of June and July 20th on Aspromonte, between 400–1600 m. Stauder indicates that *italorum* was a mixture of the *chariclea*, *catoleuca* and *lutea* phenotypes. Some of the females had the wings similar to *wollastoni*. Gardiner (1979: 34) treats *italorum* as an inconstant, environmentally-induced local form.

Mancipium brassicae verna Verity, 1923 (type-locality: Sicily). Verity (1923: (20)) refers back to "verna Zeller, 1847". However, it is clear that Zeller (1847), in describing spring forms of *brassicae* from Sicily, uses the word "verna" only to denote spring individuals. He is caustic about Stephens' *chariclea*, and it is evident that Zeller had no intention of introducing a new name. He uses the word "verna" and "aestiva" as terms for the spring and summer forms of several butterflies, including other species of *Pieris* s.l. The use of the name *verna* as an available name dates from Verity (1923); it is junior to *italorum* Stauder. Unfortunately Graham-Smith & Graham-Smith (1930) followed Verity in attributing *verna* to Zeller, which has been repeated by a number of authors, including Gardiner (1979: 42) who, following Graves and Verity, considers it a synonym of "aestiva Zeller", thus compounding the misunderstanding even further.

Mancipium brassicae cyniphia Turati, 1924 (type-locality: Libya, Benghazi) was described from a single (very large) female specimen captured at Benghazi on 20 May 1922. Turati believed the race to be present throughout Libya. The insect described is similar to the *lepidii* Röber phenotype. Gardiner (1979: 31) considers this no more than an environmental form of *brassicae brassicae*.

Mancipium brassicae subtaeniata Turati, 1929 (type-locality: Greece, island of Rhodes) was based on one male and four females captured by Ghigi on Rhodes, and two other specimens from Mt. Santo Stefano. All were described as very similar, Turati's photograph of the underside of a female showing a black bar joining the main post-discal and posterior black spots of the fore-wing. Gepp & Stark (1974) record this phenotype from mainland Greece. Gardiner (1979) does not list this taxon.

Syrian subrace

This subrace extends from the Taurus Mountains south into Asia Minor, including Cyprus. It is doubtful whether it is sharply separable from the Mediterranean subrace. Graham-Smith & Graham-Smith (1930) did not distinguish it from the Mediterranean group (including *catoleuca* in the European group, race *italorum* (as *verna*); and *cypria* in the Mediterranean group, race *cypria*). However, Larsen (1974) separates it as a subspecies (for which he uses the available name *catoleuca* Röber), although at one and the same time he admits the weakness of the separation.

Pieris brassicae var. *catoleuca* Röber, 1896 (type-locality: Turkey, Taurus Mts.) was described from four male and two female specimens. Bodenheimer (1935) indicates that the Syrian subrace is common between September and December, and again in April, noting that it is "absolutely restricted to the neighbourhood of human settlements and feeds on *Capparis*, *Beta* and *Tropaeolum*". Gardiner (1979: 30) considers *catoleuca* to be typical of *brassicae* from southern Europe.

Pieris brassicae form *cypria* Verity, 1908 (type-locality: Cyprus) was described as a dwarf race (wingspan about 45 mm), but both Turner (1920) and Bretherton (1954) report normal sized or even very large *brassicae* specimens from the island. Gardiner (1979) does not list this taxon.

Pieris brassicae cheiranthi (Hübner)
Catophaga cheiranthi Hübner, 1806 (type-locality: Canary Islands).

Higgins (1975) includes all the Atlantic Island races of Graham-Smith & Graham-Smith (1930) under this name. So constituted, the subspecies is divisable into three subraces, all of which are frequently regarded as separate subspecies or even species. But if Gardiner (1979) is correct concerning the recent introduction of *brassicae* into Madeira and the Canaries, then there is little justification for using this grouping of Atlantic populations, other than a convenient phenetic similarity – not even well shown by *brassicae* from the Azores.

Azores subrace
Pieris brassicae azorensis Rebel, 1917 (type-locality: Azores) is often still regarded as a full subspecies (e.g. Kudrna, 1973). Large Whites from the Azores have been reported to be like the *chariclea* form with the underside of the hind-wings dark. The butterflies are possibly larger than mainland *brassicae brassicae*, and apparently have the same form of male genital valvae as the Madeiran subrace (see figs. in Kudrna, 1973). Gardiner (1979: 29) appears to regard *azorensis* as an unstable island form, not regularly distinguishable from *brassicae brassicae*.

Madeira subrace

Pieris brassicae wollastoni Butler, 1886 (type-locality: Madeira) is also often regarded as a full subspecies (Altena, 1949; Lagnel, 1966; Kudrna, 1973). The Large White population restricted to Madeira (Cockerell, 1923) was described by Butler (1886) as differing in many aspects of colour and pattern both from *cheiranthi* s.s. and *nepalensis*, to which he considered it to be intermediate in size. Specimens in the British Museum (Natural History) (BM(NH)) tend, in the opinion of the authors, to look much like *brassicae brassicae*. Gardiner (1979: 42, text under *vasquezi* due to *lapsus*) considers *wollastoni* to be a "race" characterized by a high frequency of ab. *fasciata* also having the underside darker than *brassicae brassicae*; he speculates that it has evolved on Madeira during the last few hundred years.

Canaries subrace

Pieris brassicae cheiranthi s.s. is found on the Canary Islands, including Tenerife and Gran Canaria. Altena (1949) regarded it as a separate insular subspecies. Kudrna (1973) regarded this isolate as a full species and was followed in this by Schurian (1975). However, the features of the male genital valvae described by Kudrna as diagnostic for *cheiranthi* can readily be found in N. European *brassicae brassicae* specimens (Drosihn, 1933; Altena, 1949; Niels Wolff & E. Schmidt Nielsen, pers. comm.). The colouration of *cheiranthi* is distinctive only in that most Canary Island specimens belong to a particular extreme (large size, yellowish hind-wings, very pronounced black markings) of the normal variation of *brassicae* (Gardiner, 1979: 44). This could be explained by a founder effect, or a genetic "bottleneck" in the origin of the present population. The Canary Island Large White is not known to possess, nor would it appear to have, any specific mate recognition system separating it from *brassicae*, or any post-mating isolating mechanism either. Gardiner (1965) obtained viable F_1 hybrids from a male *cheiranthi* x female *brassicae brassicae* (Cambridge, England) cross, which he was able to propagate to an F_8 generation by backcrossing to *brassicae brassicae* – and although by the F_8 mortality had increased to the point where few butterflies survived, he reports high fertility to the F_5 stage. Kudrna (1973) notes that *cheiranthi* will only feed on *Tropaeolum*, not cabbage (Fernández, 1970). No details of the experiments are given by Fernández, and cabbage is evidently rare on the Canaries (Manley, pers. comm.). Kudrna also claims that the colour of *cheiranthi* larvae is very different from *brassicae*; if this statement is based on the colour photograph in Fernández (1970), it should be noted that the colour-balance of the reproduction is heavily blue-biased. However, Gardiner (1979: 30) who has performed extensive breeding experiments with *cheiranthi*, confirms that this "race" does

9

prefer *Tropaeolum*, and that if fed on cabbage it suffers high mortality. He further confirms that the larvae and pupae show a marked difference from the nymotypical form. Gardiner is of the opinion that *cheiranthi* has evolved rapidly (partly through genetic drift) from *brassicae* stock introduced to the Canaries during the sixteenth century, along with Nasturtium from the Americas.

Clearly, there is some genetic differentiation of the Canary Large White population, but not sufficient to warrant total separation from *brassicae* s.l., of which it is still best regarded as an insular replacement. For a good review of literature on *cheiranthi*, the reader is directed to Schmidt-Koehl (1971).

Pieris brassicae nepalensis Gray

Pieris brassicae var. *nepalensis* Doubleday, 1844 [*nomen nudum*].
Pieris brassicae var. *nepalensis* Gray, 1846. (type-locality: Nepal).

This subspecies corresponds to the "Eastern Races" of Graham-Smith & Graham-Smith (1930). Talbot (1939: 429) describes the distribution of *brassicae nepalensis* as "Baluchistan and Chitral to Assam, and the plain adjoining the Himalayas, extending into Tibet and Yunnan". Gardiner (1979: 37) accepts *nepalensis* as the "Himalayan race", but indicates that little is known about it, and that the *nepalensis* facies is neither stable nor unique; however, elsewhere in the same paper (Gardiner, 1979: 26) he claims that *nepalensis* shows "a consistent major variation in the extent of black markings", but does not specify precisely what this difference is.

Himalayan subrace

This covers the Large White as it extends from Baluchistan, through southern Afghanistan, northern India, Nepal and Tibet to Yunnan. Gray originally described *nepalensis* from specimens captured in Nepal by Hardwick. Moore (1882) states that it is common in the north west Himalaya, where it has 2–3 broods each year. Li (1962) has recorded it from 13 areas of Yunnan. The Himalayan subrace is doubtfully divisible from the following subrace.

Turkestan subrace

Pieris brassicae ottonis Röber, 1907 (type-locality: U.S.S.R., Fergana, in Kirgizia) was introduced as a replacement name for the preoccupied name *Pieris brassicae* var. *brassicoides* Staudinger, 1901 (not *brassicoides* Guérin Méneville). The Large White populations from the Tashkent area, through much of Afghanistan, and probably into Iran, can be and often are separated under the subspecies name *ottonis*. However, according to Graham-Smith & Graham-Smith (1930), specimens from Turkestan "do not seem to show special features" when compared with *nepalensis* s.s. Presumably it is this Turkestan subrace that is replaced at higher altitudes in the Pamirs and eastern Kirgizia by *Pieris deota*. Gardiner (1979: 39) considers *ottonis* merely a name for the spring form, and, as such, a synonym of *chariclea*; it is not clear if Gardiner places the Fergana population with *brassicae brassicae* or *brassicae nepalensis*.

SECTION B. INFRASUBSPECIFIC VARIATION AND NOMENCLATURE
J.S.E. Feltwell

Aberrations

Attention has been drawn to the uniformity of colour and lack of variation in *P. brassicae* brought about by its highly migratory habit in Europe (Verity, 1916; Petersen, 1947), which has resulted in a truly panmitic population. Adkin (1918) noted that he could not find any significant difference in the colour of several hundred specimens observed feeding at flowers.

Considerable attention has been given to the aberrations of *P. brassicae* by a number of workers. Rocci (1919) in Italy described six aberrations, Graham-Smith & Graham-Smith (1930) working in Scotland published details on 25 aberrations including eight new ones; Lempke (1936, 1953) in the Netherlands published details on seasonal variation and cited references to 26 aberrations and described five others. For each of these aberrations Lempke gave the location of the type specimens. The following year Caruel (1954) in France classified 27 aberrations of *P. brassicae* according to their colouration and divided them into three groups, a) aberrations of the ground colour (5), b) aberrations of the black pattern (14), and c) underside aberrations (8).

There are, however, 47 aberrations of *P. brassicae* listed in the BM(NH) working catalogue (unpublished). This includes almost as many synonyms as there are aberrations. The aberrations can be broken down to 36 upperside and 11 underside aberrations.

Table 1.4 lists 61 aberrations, comprising all those cited by the BM(NH) together with other non-British aberrations. It must be stressed that a more complete study of the aberrations of *P. brassicae* would provide ample scope for a research project.

There are very few aberrations of *P. brassicae* which are striking in appearance although there are many described. The ones which stand out, when a direct comparison of several aberrations can be made as in the BM(NH), are *albinensis*, *coerulea*, *flava*, *lacticolor* and *nigroviridescens*.

Table 1.4 Aberrations. This list is based on that of the British Museum (Natural History) with additions. Those indented indicate synonyms. See also list of Gardiner (1979) which includes seasonal forms and subspecies.

albinensis Gardiner 1962
alligata, see *fasciata*
alpina Rocci 1919
anthrax, see *nigroviridescens*
autumnalis Rocci 1919
aurea, see *flava*
aversomaculata Lempke 1953
basi-nigrescens Graham-Smith &
   Graham-Smith 1930
biligati, see *striati*
bi-nigronotata Graham-Smith &
   Graham-Smith 1930
brassicae Linnaeus 1758
   =*lepidii* Röber 1907
brassicae-flava, see *flava*
carnea Graham-Smith &
   Graham-Smith 1930
chariclea Stephens 1828
cheiranthi Hübner 1806
coerulea Gardiner 1963
colliurensis Gélin 1914
   =*fischeri* John 1922
cyniphioides Rocci 1930
elongata Gélin 1914
emigrisea, see *vazquezi*
fasciata Kiefer 1918
   =*alligata* Cabeau 1924
fischeri, see *colliurensis*
flava Kane 1893
   =*aurea* Mosley 1896
   =*brassicae-flava* Fischer 1925
   =*flavus* Frohawk 1938
   =*lutea* Röber 1907
flava Kroulikovsky 1902
flavopicta Rocci 1919
   =*jauni* Gardiner 1963
flavus, see *flava*
fuscosignata Lempke 1953
glaseri Müller 1925
griseopicta Rocci 1919
gynandra Rocci 1930
henriettae, see *vazquezi*
hiemalis Turati 1936
infra-fasciata Graham-Smith &
   Graham-Smith 1930
infratrinotota Caruel 1954
insupermaculata Cretschmar 1935
jauni, see *flavopicta*
lacticolor Lempke 1953
lepidii, see *flava*
longomaculata Lempke 1953

lutea, see *flava*
major Verity 1947
marginata Graham-Smith &
   Graham-Smith 1930
marginavenata McLeod 1968
maria Van Mellaerts 1926
   =*supra-fasciata* Graham-Smith &
   Graham-Smith 1930
meridionalis Rocci 1919
minor Ksenžopoliskij 1912
   =*nana* Rocci 1919
nana, see *minor*
nigrescens Cockerell 1889
   =*obscurata* Oberthur 1896
nigronotata Jachontov 1903
nigropunctata, see *posteromaculata*
nigroviridescens Rocci 1919
   =*anthrax* Graham-Smith &
   Graham-Smith 1930
obenbergi Tykač 1947
obscurata, see *nigrescens*
ocellata-loberi Kraut 1938
pallida Graham-Smith &
   Graham-Smith 1930
parvomaculata Rocci 1919
perflava Lempke 1953
Plasschaerti Dufrane 1912a,b
posteromaculata Verity 1911
   =*nigropunctata* Walcourt 1920
postice-ochreata Verity 1919
pseudocatoleuca Stauder 1921
punctigera Graham-Smith &
   Graham-Smith 1930
ramnei Knop 1923
reducta Fritsch 1913
semi-nigrescens Graham-Smith &
   Graham-Smith 1930
separata Pionneau 1928
striata Rocci 1919
   =*biligata* Cabeau 1925
sublutea Turati 1925
supra-fasciata, see *maria*
tertia Verity 1919
trimacula Rocci 1919
ultimogenita Verity 1947
vazquezi Oberthur 1914
   =*emigrisea* Rocci 1919
   =*henriettae* Pionneau 1924
venata Verity 1908
vernalis Turati 1925

Most of the aberrations have been described only on the basis of subtle differences in intensity and distribution of the black and white scales.

Occasionally, local populations are accidentally found to give rise to a high proportion of aberrations and such was the case experienced by Graham-Smith & Graham-Smith (1930) in Aberdeenshire. Out of 907 specimens which they bred, 63 were divisible amongst 11 different aberrations. It is evident that the genetic control of these aberrational forms may be fairly complex. These authors described in detail aberrations then known from Europe, Mediterranean, Atlantic and eastern regions. Chalmers-Hunt (1960) recorded seven aberrations of *P. brassicae* from Kent.

Gardiner (unpublished) also found that a surprising range of aberrations could be obtained from individuals of ab. *fasciata*; when they were selfed they produced all *fasciata* in the F_1 and an apparently infinite variation of forms in the F_2. It is interesting to note in considering control of aberrations, that Pol & Ponsen (1963) showed that certain aberrations in pigmentation may be related to granulosis infection. They experimentally infected larvae of *P. brassicae* and fed them on the white leaves of white cabbage. This resulted in white blotches overlying the black at the apices of the fore-wings of the imago.

Brian Gardiner recently reviewed the variations of *P. brassicae* and pointed out that of the "seventy races, forms, varieties or aberrations known fifty are said to be found in the British Isles" (Gardiner, 1979). Eighty-one synonymous names are mentioned in his alphabetical list of aberrations, including homeotics, gynandromorphs, teratological specimens; there are also a few subspecific names included, such as *azorensis*, *catoleuca* and *nepalensis*, and the names of seasonal forms such as *ottonis* and *aestiva*. Thirteen aberrations were illustrated.

Gardiner (1979) recognises four groups of *P. brassicae* variation, a) those in which there is a major colour change and which is controlled by a recessive allele, b) those in which the amount and intensity of the black markings varies, c) those in which the underside colour varies, and d) those due to various other factors (miscellaneous) e.g. gynandromorphs. He argues that *P. brassicae* exhibits some degree of plasticity in that environmental conditions can influence the colouration of the butterflies. Through his own breeding experiments Gardiner has been able to select out different forms. Extreme environmental conditions may produce a locally-different aberration, which may cause some entomologists to designate a new aberration. This Gardiner argues is how some aberrations, for instance *alpina*, have been named.

It is not the intention of the author here to write short notes on all the aberrations attributed to *P. brassicae*. However, short notes follow on three interesting aberrations, namely *coerulea*, *major*, *minor* and details on gynandromorphs are included.

Some major aberrations

ab. coerulea Gardiner, 1963

Imagines have been illustrated by Gardiner (1963) and Russwurm (1978) and wing scales by Feltwell & Rothschild (1974, Plate 1a, p. 456). A blue form of the butterfly was first figured by Frohawk (1914, Plate 3, Fig. 20) and was probably similar to the specimens exhibited at the South London Entomological and Natural History Society in 1909 and 1915. Newman & Leeds (1933) also mention a specimen with pale blue undersides caught in August 1908 which may refer to the same specimen.

Since all aberrations of *coerulea* breed true one may assume that it is controlled by an autosomal recessive gene. The blueness is thought to be caused by a number of factors: a) the wings lack yellow pigments (Gardiner, 1963), thought to be pterins (Harmsen, 1964); b) lack of carotenoids (Feltwell & Rothschild, 1974); c) defective scales which permit biliverdin pigment in the wing membrane to be visible; and d) semi-transparent scales which produce modified iridescence.

Further characteristics of ab. *coerulea* have been described by Feltwell & Rothschild (*loc. cit.*) and others, a) lack of yellow scales or setae on the wings and body, b) great transparency of the white scales, c) compound eyes much paler in hue than normal, d) wings more blue in the female, e) scale sockets containing biliverdin around their bases are more visible (because overlying scales detach), f) pupae more toxic than normal (injection into mice test; Marsh & Rothschild, 1974), g) haemolymph pale bluish green, h) melanin at apical tips and black markings of the female much less intense than normal, and i) *coerulea* survives better at lower temperatures than normal *P. brassicae*. At 21°C (70°F) normal *P. brassicae* survives extremely well, but *coerulea* has low viability above 15.5°C (60°F) (Rothschild, 1976, pers. comm.).

Tubbs (1978) found that a serious scale defect accompanied the double aberration of *coerulea + albinensis*, where males were unable to free themselves from their pupal cases.

ab. major Verity, 1947

Frohawk (1934) quoted a figure of 76 mm for the wingspan of a female *P. brassicae* but this is regarded by Gardiner (1963) as a possible error. Both sexes have wingspans which vary up to about 70 mm and an insect with 76 mm would certainly be a giant. However, numerous references occur to specimens being seen which are "very large forms" and indeed, Bretherton (1954) recorded a very large form of *P. b. catoleuca* from Cyprus. The figures quoted by different authors for males and females of *P. brassicae* have been included in Table 7. and comment made on them in Chapter 7. Experimentally, ab. *major* can be produced by thinning out larvae to 100

14

per m^2 and keeping the temperature at 10–15°C (Gardiner, 1978, pers. comm.).

ab. minor Ksienchopolsky, 1911

Specimens of this aberration have a wingspan in the order of 37 mm in males and 38 mm in females, although anything between 40–50 mm can be considered as *minor* (Gardiner, 1963). There are three ways in which this aberration may be produced. First, Gardiner has observed that "backward" larvae mature to produce small imagines. Second, crowding the larvae to 500 per m^2 (Gardiner, 1978, pers. comm.), and third, starving fifth instar larvae produces small imagines (Gardiner, 1963). In the wild, dwarfs have been recorded from Ireland (Barrett, 1893) and the Netherlands (Wageningen Museum, Agricultural University). South (1891) stated that specimens caught at 3354–5183 m (!!) in India were smaller than British examples.

Gynandromorphs

The gynandromorphs described by Gardiner (1957, 1973) were of two types: partial and complete bilateral, and mosaic. The gynandromorphs were thought to be produced by genetic damage caused by virus, presumably in the same way that *coerulea* may arise. It is interesting to recall that Frohawk (1921) found the incidence of gynandromorphs in the three species *P. napi*, *P. rapae* and *P. brassicae* to be very rare and that he only ever knew of one specimen.

Un-named aberrations

The literature is rich with many descriptions of *P. brassicae* found with particular atypical colourations which deviate slightly from the norm. Wings suffused with pink have been found fairly frequently (Weizmantel, 1900; SLENHS, 1916; Newman, 1917), all which could be similar to ab. *carnea* found by Rait-Smith (1938). Wings suffused with yellow from pale to sulphur and pale buff have been described also by Rait-Smith (1930), Greer (1922) and Eastwood (1940). Green veins have been recorded by Webb (1889), Mosley (1889) and Blandford (1897). Perhaps most descriptions of aberrations are those which indicate various degrees of greyness over the wings (Williams, 1893; SLENHS, 1932), melanic form (Berliner Entomologische Gesellschaft, 1919), "blackness" (Bramwell, 1869; Mosley, 1903; Carter, 1913; Turner, 1925; Neave, 1927; Bell, 1907–8; SLENHS, 1902; Mathew, 1892; Eckstein, 1921; Sheldon, 1922; Gardiner, 1974a). Seitz (1901) published a useful list of forms and aberrations of *P. brassicae*.

15

Other aberrations of note

A number of distinctive aberrations of *P. brassicae* have been sold at various sales (Table 1.5). A specimen of ab. *flavescens* fetched £1.40 (£1.8s) in 1948 and two other specimens which showed *flavescens* features on the undersides of the wings went for £1.20 during the sale of J.N. Marcon's Collection of British Butterflies (Cockayne, 1948). At the sale of the Bright Collection, a specimen of ab. *carnea* of a deep pink shade was sold for £9 (Rait-Smith, 1938; Bright, 1942). According to Wood (1972) a black *P. brassicae* was sold in London for £44 but neither Wood nor Feltwell have been able to trace the record.

Seasonal variation

There are a number of aberrations of *P. brassicae* which have been attributed to seasonal variation. The difference between the markings on spring and summer individuals may be quite pronounced and it was probably this factor which caused some workers in the nineteenth century to state that the spring form was a different species. Indeed, the names

Table 1.5 Sale of aberrations and varieties.

Collection	Aberration or description	Price	Authority
Bright	ab. *carnea*	£9	Rait-Smith, 1938
Bright	ab. *fasciata*	£1.4s (£1.20)	Rait-Smith, 1938
Bright	pale blue undersides	£5	Rait-Smith, 1938
Bright	male suffused with pinkish sepia	£3.3s (£3.15)	Rait-Smith, 1938
Crabtree	Fawn coloured examples	£14.10s (£14.50)	Crabtree, 1947
Farn	series with minor abs.	16/- (80p)	Farn, 1922
C.F. Johnson	series containing a pale buff example	35/- (£1.75)	Rait-Smith, 1930
J.N. Marcon	ab. *flavescens*; 2 others with same features on undersides	£1.8s (£1.40) £1.4s (£1.20)	Cockayne, 1948
B.H. Smith	lemon tinted	£1.14s (£1.70)	Rait-Smith, 1947
Stiff, Joy, Leeds & others	pale yellow grey	£2.4s (£2.20) £2.4s (£2.20)	Rait-Smith, 1943
Whitehouse	pale lemon	£1.12s (£1.60)	Rait-Smith, 1944a
Whitehouse	3 specimens, two of which were banded	£3	Rait-Smith, 1944b

which should be given to the seasonal forms of *P. brassicae* seem to have been a point for discussion over at least a hundred years. James Stephens (1827), in his book *Illustrations of British Entomology*, originally described the first (spring) generation of *P. brassicae* as a distinct species and named it *Pontia chariclea* (cf. Verity, 1913; South, 1936; Riley, 1925). Twenty-five years later, another entomologist writing under the pseudonym of Bombyx Atlas (1852) supported Stephens' view and wrote, "I have not the slightest shade of hesitation in pronouncing *Pieris chariclea* to be a perfectly distinct species from *Pieris Brassicae*" (sic). This anonymous entomologist then set out a table giving ten points of difference between the two species, mostly on colouration (Verity, 1913).

Further remarks were made on the nomenclature which should be applied to seasonal forms of *P. brassicae* by Röber (1907), Seitz (1901), Verity (1913) and Rocci (1933).

This spring generation called *chariclea* Stephens (1828) was also referred to as *vernalis* or *chariclea* Stephens (Seitz, 1901), and was noted as the typical first generation throughout Italy by Rocci (1933). It is perhaps worth noting that the male specimen of *P. brassicae* in Linnaeus' collection is a spring form.

The summer generation, *lepidii* Röber 1907 was also referred to as "*aestralis lepidii*" Röber or "*aestivalis*" by Seitz (1901), but now it is regarded to be the synonym of *P. b. brassicae* Linnaeus (1758) (MN(NH) Unpublished Working Catalogue).

Rocci (1933) recorded *lepidii* in northern Italy in the second generation during July to August and in the third generation during September. He also referred to aestiva Zeller (sic) as the second generation in the southern part of Italy in March and April.

Ubaldo Rocci (1933) gave names to each of the three generations of *P. brassicae* in northern Italy and the four occurring in both the central and southern regions. As an example, the central region has two other seasonal forms attributed to it which have not been mentioned here; ab. *tertia* Verity (1919) in the third generation (Sept–Oct.) and ab. *autumnalis* Rocci (1919) in the fourth generation during November. Turati (1929) furthermore referred to ab. *hiemalis* as the third generation in Italy. Today, some of these aberrational names have been retained, for instance "*vernalis*" and "*aestivalis*" have been recorded in Poland (Starega, 1975, pers. comm.).

First generation imagines of *P. brassicae* usually have much paler apical wing markings than second generation imagines (Frohawk, 1934; Ford, 1945, 1975), indeed Lister (1917) working on seasonal variation in *P. rapae* coined the phrase "faint, frail and first" to remember which phenotype came first in the year. First generation imagines have also been recorded as being smaller than second generation imagines in both Italy

(Messina) by Zeller (1847), Malta (Mathew, 1898), and in Germany Blunck (1950) thought that second generation imagines were stronger migrants than those of the first generation.

On the other hand the markings on French spring imagines were found to be more intense than on English ones (Sevastopulo, 1924), and second generation imagines in Ireland were considered to be larger than English summer imagines (Hodgson, 1933). However, without any quantitative evidence the truth of this apparent size difference remains in the eye of the beholder. In the south of France there is a partial reversion to spring forms in the late autumn amongst many pierid species including *P. brassicae* (Morris, 1922).

Pictet (1922) made some observational comments on first generation imagines which had hatched from overwintering pupae and which he termed "pseudo hivernale". They had: a) pale markings on the fore-wings, b) were no bigger than male *P. rapae*, c) the hind-wings had black scales which reflected blue and yellow giving the wing a greenish hue, and d) some specimens had dark grey hind-wings and one specimen had almost transparent wings.

Larvae of *P. brassicae* in Great Britain, as in western Europe, have the ability to withstand great fluctuations in temperature, which often include frost conditions during the autumn and winter, aided by special super-cooling substances in their haemolymph (see chapter 8). In many cases the development time of the larvae is considerably increased. In Great Britain late larvae have been seen at the end of October (Bull, 1940), on 8th December (Day, 1915), 17th December (Nicholson, 1939), at Christmas (Youden, 1975) and early January (Newman, 1942).

It is most probable that some extrinsic factor such as temperature governs the subtle colour changes found in the seasonal variants, by setting off or suppressing certain physiological mechanisms. Lister (1917) maintained that the low temperatures and frosts which the pupae of the first generation experience are responsible for the lack of markings in *P. rapae* and conversely, the high temperatures and increase in the photoperiod favours the development of the pigments in the wing; the critical time for colour differentiation occurs in the pupae during the latter half of the development. Bertolini (1927) also investigated external factors on seasonal variation in relation to the histological structures of the integument.

References cited

Ackery, P.R., 1975. A new pierine genus and species with notes on the genus *Tatochila* (Lepidoptera: Pieridae). *Bull. Allyn Mus.* (30): 9 pp., 1 pl.
Adkin, R., 1918. The abundance of white butterflies in 1917. *Entomologist* 51: 36–39, 56–59.

Albin, E., 1720. *A natural history of British insects.* Vol. 1. Printed for the author, London.

Altena, C.O. van R., 1949. Macrolepidoptera collected in Tenerife and La Palma in the spring of 1947. *Tijdschr. Ent.* 91: 12–22.

Barrett, C.G., 1893. *The Lepidoptera of the British Isles.* I. Rhopalocera. Reeve & Co., London.

Bell, S.J., 1907–8. Minutes of December 18th and January 17th, 1908, *Entomologist* 40: 43; 41: 70.

Berge, F., 1842. *Schmetterlingsbuch. Naturgeschichte der Schmetterlinge.* Hoffmann, Stuttgart.

Berliner Entomologische Gesellschaft, 1919. Sitzungberichte vom 30. Januar 1917. *Int. Ent. Z.,* 13: 17–18.

Bernardi, G., 1947. Révision de la classification des espèces holoarctique des genres *Pieris* Schr. et *Pontia* Fabr. *Miscnea ent.* 44: 65–79.

—— 1958. Taxonomie et zoogéographie de *Talbotia naganum* Moore (Lepidoptera, Pieridae). *Revue fr. Ent.* 11: 420–431.

Bertolini, F., 1927. Biology and histology investigation on *Pieris brassicae.* (In Italian.) *Redia* 16: 29–39.

Blandford, W.F.H., 1897. Report of meeting, November 17th, 1897. *Entomologist* 30: 324.

Blunck, H., 1950. Contribution to knowledge of fluctuations in numbers of *P. brassicae* with special reference to drought year 1947. *Z. angew. Ent.* 32: 141–171.

Bodenheimer, F.S., 1935. *Animal Life in Palestine.* L. Mayer, Jerusalem.

Bombyx Atlas, 1852. On the difference between *Pieris* (*Pontia*) *brassicae* and *Pieris chariclea,* the Large Cabbage White Butterfly by Bombyx Atlas (B.R.M.). *Naturalist* 2: 178–180.

Bradley, R., 1718. *New improvements of planting and gardening both philosophical and practical, explaining the motion of the sapp and generation of plants.* Part 3. Mears, London.

Bramwell, J.M., 1866. Black specimens of *Pieris brassicae* and *P. rapae. Entomologist* 4: 258–259.

Bretherton, R.F., 1954. A week's butterfly collecting on Cyprus. *Entomologist* 87: 207–211.

Bright, P.M., 1942. Sale of Bright Collection. *Entomologist's mon. Mag.* 78: 44–45.

Bull, G.V., 1940. Late emergence or extra broods. *Entomologist* 73: 277.

Butler, A.G., 1870. A review of the genera of the sub family Pierinae. *Cistula Ent.* 1: 33–58.

—— 1886. Description of a hitherto unnamed butterfly from Madeira. *Ann. Mag. nat. Hist.* (5) 17: 430.

Carter, J.W., 1913. Exhibit. *Entomologist's mon. Mag.* 49: 40.

Cabeau, B.M., 1924–1925. Aberrations des Lépidoptères. *Revue mens. Soc. ent. Namur.* 24: 17–18, 25, 34–35, 50, 66–67; 25: 38.

Caruel, M., 1954. *Pieris brassicae* with ab. n. *infratrinotata. Revue fr. Lep.* 14: 147–153, 261–264.

Chalmers-Hunt, J.H., 1960. Lepidoptera of Kent. *Entomologist's Rec. J. Var.* 72: 18–19.

Cockayne, E.A., 1948. Sale of Rev. J.N. Marcon's collection of British Butterflies. *Entomologist's mon. Mag.* 84: 93.

Cockerell, T.D.A., 1889. On the variation of insects. *Entomologist* 22: 54–56.

—— 1923. The lepidoptera of the Madeira Islands. *Entomologist* 56: 243–247.

Crabtree, B.H., 1947. Sale of the Crabtree collection. *Entomologist* 80: 151–152. (Article written by W. Rait-Smith.)

Cretschmar, M., 1920. Besprechung über die Farbe der Puppen der *Pieris brassicae* und die Schutzfarbung. *Int. ent. Z.* 14: 125–127.

Crotch, G.R., 1872. On the generic nomenclature of lepidoptera. *Cistula ent.* 1: 59–71, 91–92.

Curtis, J., 1824. *British Entomology* 5: pl. 48. The author, London.

Dalman, J.W., 1816. Forsök till systematisk Uppstallning af Sveriges Fjarilar. *K. svenska Vetenskad.* 33: 48–100, 199–225.

Day, F.M., 1915. Late larvae of *Pieris brassicae*. *Entomologist's mon. Mag.* 608: 19.

Doubleday, E., 1844. *List of specimens of Lepidoperous insects in the collection of the British Museum.* Part 1. p. 32. British Museum, London.

Drosihn, K.J., 1933. Über Art-und Rassenunterschlede der männlichen Kopulationsapparate von Pieriden (Lep.). *Ent. Rdsch.* (suppl.) 50: 135 pp.

Dufrane, A., 1912a. A propos de l'ab. *Plasschaerti* Dufrane de *Pieris brassicae* L. *Lambillionea* 3040–3042.

—— 1912b. Piérides de Belgique. *Revue mens. Soc. ent. Namur.* 12: 23–25.

Eastwood, J.E., 1940. A curious form of *Pieris brassicae*. *Entomologist* 73: 115.

Eckstein, K., 1921. *Die Schmetterlinge.* Pestalozzi, Wiesbaden.

Fabricius, E., 1807. Die neueste Gattungs-Eintheilung der Schmetterlinge aus den Linneischen Gattungen *Papilio* und *Sphinx. Magazin für Insektenkunde Mag.* 6: 277–295.

Farn, A., 1922. The Farn collection of British Lepidoptera. *Entomologist* 45: 91–93.

Fernández, J.M., 1970. *Les Lépidoptères diurnos de las Islas Canarias.* Enciclopedia Canaria, Anlade Cultura Tenerife.

Feltwell, J.S.E. & Rothschild, M., 1974. The carotenoids of 38 species of lepidoptera. *J. Zool. Lond.* 174: 441–465.

Fischer, E., 1925. Alte und neue Formen der *Pieris brassicae*. *Int. ent. Z.* 19: 66–68.

Ford, E.B., 1945. *Butterflies.* Collins, London.

—— 1975. *Butterflies.* Collins, The Fontana New Naturalist Series.

Fritsch, W., 1913. Lepidopterologische Beobachtungen. Wärmebedürfnis der Tagfalter. *Ent. Rdsch.* 30: 46–47.

Frohawk, F.W., 1914. *A natural history of British Butterflies.* Hutchinson & Co., London.

—— 1921. *Pieris rapae* gynandrous abberation. *Entomologist* 54: 201.

—— 1934. *British Butterflies.* Ward, Lock & Co., London & Melbourne.

—— 1938. *Varieties of British Butterflies.* Ward, Lock & Co., London.

Gardiner, B.O.C., 1957. The annual exhibition, 29th Oct. 1955. *Proc. & Trans. S. Lond. ent. nat. hist. Soc.* 1955. p. 40.

—— 1962. An albino form of *Pieris brassicae* L. (Lep: Pieridae) *Ent. Gaz.* 13: 97–100.

—— 1963. Genetic and environmental variation of *Pieris brassicae*. *J. Res. Lepid.* 2: 127–136.

—— 1965. Hybrids between typical *Pieris brassicae* L. and race *cheiranthi* Hueb. *Proc. 12th Int. Congr. Ent. 1964,* 261.

—— 1973. Gynadromorphs in *Pieris brassicae* L. *J. Res. Lepid.* 11: 129–140.

—— 1974a. An unusual aberration of *Pieris brassicae* (L.) *Ent. Gaz.* 25: 186.

—— 1974b. *Pieris brassicae* L. established in Chile; another palaearctic pest crosses the Atlantic (Pieridae). *J. Lepid. Soc.* 28: 269–277.

—— 1979. A revision of variation in *Pieris brassicae* (L.) (Lep: Pieridae). *Proc. Brit. ent. nat. hist. Soc.* 12: 24–26.

Gélin, H., 1914. Note sur quelques formes de Lycènes et de Piérides de la faune française. *Bull. Soc. ent. Fr.* 6: 183–187.

Gepp, J. & Stark, W., 1974. Ökologie und Zoogeographie der bekannten Pieridae der Griechenland. *Ber. Arb. Ökol. ent. Graz* 4: 1–5.

Graham-Smith, G.S. & Graham-Smith, W., 1930. *Pieris brassicae* L. with special reference to aberrations from Aberdeenshire. *Entomologist's Rec. J. Var.* 41: 157–161, 173–180; 42: 1–7, 17–22.

Gray, G.R., 1846. *Lepidopterous insects chiefly fro... Nepal.* Longman, Brown, Green and Longman, London.

Greer, T., 1922. Lepidoptera in E. Tyrone, 1922. *Entomologist* 55: 258–259.

Harmsen, R., 1964. Genetically controlled variation in pteridine content of *P. brassicae* L. *Nature, Lond.* 204: 1111.

Hemming, F., 1933. On the types of certain butterfly genera. *Entomologist* 66: 196–200.

—— 1934a. Notes on nine genera of butterflies. *Entomologist* 67: 37–38.

—— 1934b. *Generic names of Holarctic butterflies.* I. 1758–1863. British Museum (Natural History), London.

—— 1967. *Generic names of the Butterflies and their type species.* Bull. Brit. Mus. Nat. Hist., Ento., Supplement 9. pp. 362–363.

Higgins, L.G., 1975. *The classification of European butterflies.* I. 1758–1863. British Museum (Natural History), London.

Hodgson, S.B., 1933. Rhopalocera of the Mullet Peninsula, Co. Mayo. *Entomologist* 66: 134–135.

Hübner, J., 1806–1826. *Sammlung exotischer Schmetterlinge.* I: vi. 213 pls. Augsburg.

—— 1820. *Verzeichnis bekannter Schmetterlinge.* p. 93 Augsburg.

—— 1924. *Nachträge zur Sammlung europäischer Schmetterlinge.* Augsburg.

Jachontov, A.A., 1903. An aberration of *Pieris brassicae* L. (Lep: Pieridae) (in Russian). *Russk. ent. Obozr.* 3: 38.

John, K., 1922. Neue Lepidopteren-Formen. *Ent. Z.* 36: 33–34.

Kane, W.J. de V., 1893. A catalogue of the lepidoptera of Ireland. *Entomologist* 26: 117–121.

Kawazoé, A. & Wakabayashi, M., 1976. *Coloured illustrations of the butterflies of Japan* (in Japanese). viii, 422 pp. Hoikusha Publishing Co., Osaka.

Kiefer, H., 1918. Eine neue Form von *Pieris brassicae* L. *Z. Österr. ent. Ver.* 3: 122–123.

Kirby, W.F., 1856. *Introductions to entomology of elements of the natural history of insects.* Longmans, London.

Kluk, K., 1802. Zwierzat donowych i dzikich osobliwie krajowych. Rozdzial 4. O. Owadzie Motylowym (Lepidoptera). *Zwierz. Hist. nat. pocz. gospod.* 79–111.

Knop, T., 1923. *Pieris brassicae* L. ♀ ab. *Rammei* nov. ab. (Note). *Ent. Z.* 36: 68.

Kraut, W., 1938. *Pieris brassicae* L. at. ocellata loberi: Kraut ab. nova. *Ent. Z.* 51: 353.

Kroulikovskij, L., 1902. Small notes on lepidoptera IV (in Russian). *Russk. ent. Obozr.* 2: 22.

Ksenžopolïskij, A.V., 1911. *Ges. erf. Wolhyniens* 8.

—— 1912. Die Rhopaloceren von Südwest-Russland. *Trudy Obsc. izsl. Volyni Zitomir* 8: 1–76.

Kudrna, O., 1973. On the status of *Pieris cheiranthi* Hübner (Lep: Pieridae). *Ent. Gaz.* 24: 299–304.

—— 1974. *Artogeia* Verity, 1947. Gen. rev. for *Papilio napi* Linnaeus (Lep: Pieridae). *Ent. Gaz.* 25: 9–12.

Lagnel, M., 1966. Note sur l'armure genitale mâle et femelle du sous genre *Pieris* Schrank. *Bull. Soc. ent. Fr.* 71: 91–94.

Larsen, T.B., 1974. The Butterflies of the Lebanon. C.N.R.S., Lebanon.

Latreille, P.A., 1810. *Considérations générales sur l'ordre natural des animaux.* 444 pp. F. Schoell, Paris. pp. 440, 351.

Lempke, B.J., 1936. Catologus der Nederlandsche macrolepidoptera, *Ent. Ber.* 79: 238–315.

—— 1953. Catologus der Nederlandsche macrolepidoptera. *Ent. Ber.* 14: 239–305.

Li, C-L., 1962. Results of the zoological-botanical expedition to south west China, 1955–1957. (Lepidoptera: Rhopalocera) (in Chinese). *Acta ent. sinica* 11: 172–198.

Linnaeus, C., 1746. *Fauna Svecica.* Edition I. p. 244. Stockholmiae.

—— 1758. *Systema Nature.* 1. Regnum Animale. Edition 10. 824 pp. p. 467. Holmiae.

Lisney, A.A., 1940. New records for Irish lepidoptera. *Entomologist* 73: 123–129.

Lister, J.J., 1917. Note on the influence of temperature on the development of pigmentation in *Pieris rapae* L. *Entomologist* 50: 241–244.

McLeod, L., 1968. A new aberration of *Pieris brassicae* (Linnaeus). Lepidoptera: Pieridae. *Entomologist's Rec. J. Var.* 80: 127–129.

Marsh, N. & Rothschild, M., 1974. Aposematic and cryptic lepidoptera tested on the mouse. *J. Zool. Lond.* 174: 89–122.

Mathew, G.F., 1892. Abundance of the larvae of *Pieris brassicae*. *Entomologist* 25: 287.

—— 1898. Notes on lepidoptera from the Mediterranean. *Entomologist* 31: 77–84.

Merrett, C., 1667. *Pinax rerum naturalium Britannicarum, continens vegetabilia, animalia et Fossilia, in hac Insula reperta Inchoatus.* Pulleyn, London.

Moore, F., 1882. 1. List of the Lepidoptera collected by the Rev. J.H. Hocking, chiefly in the Kangra District, N.W. Himalayas; with descriptions of new genera and species. *Proc. zool. Soc. Lond.* Part 1. 234–262.

Morris, C.E., 1922. *Colias edusa* migration in Cannes district. *Entomologist* 55: 35–36.

Mosley, S.L., 1889. Variation of *Pieris brassicae*. *Entomologist* 22: 112.

—— 1896. An illustrated catalogue of varieties of British Lepidoptera. *Nat. J.* 3rd. suppl. 5: 1–28.

—— 1903. *Pieris brassicae*. *Entomologist's Rec. J. Var.* 15: 167.

Müller, E., 1925. Eine neue Form der *Pieris brassicae* L. Notiz. *Int. ent. Z.* 18: (46) 277.

Neave, S.A., 1927. Proceedings of the Entomological Society. *Entomologist* 60: 22–23.

Newman, L.H., 1917. Note on *Pieris brassicae* L. *Trans. ent. Soc.* Lond. p. 13.

—— 1942. Note. *Entomologist* 75: 76.

Newman, L.W. & Leeds, H.A., 1913. *Text book of British Butterflies & Moths.* Gibbs & Bamforth, St. Albans. pp. 216.

Nicholson, C., 1939. Larvae of *Pieris brassicae* (?) in December. *Entomologist* 72: 111.

Nordman, A.F., 1935. Verzeichnis der von Richard Frey und Ragnar Storå auf den Kanarischen Inseln gesammelten Lepidopteren. *Commental. biol.* 5: 1–20.

Oberthur, C., 1896. Études d'entomologie de la variation chez les Lépidoptères. *Étude d'entomol.* 20: 1–74.

—— 1914. Les Lépidoptères de la Californie. *Étude de lépidopt. Comp.* 9: 73–89.

Oken, L., 1815. *Okens Lehrbuch der Naturgeschichte III Zoologie* (1), xxviii +842 pp.+iv, 40 pls. Leipzig & Jena.

Ormerod, E.A., 1882. *Notes and observations on injurious insects during 1882.* Simpkin Marshall, London.

Oxford English Dictionary, 1933. Clarendon Press, Oxford. Vol. 1.

Oxford University Entomological Society, 1858. *Accentuated list of the British Lepidoptera.* J. van Voorst, London.

Petersen, B., 1947. Die Geographische Variation einiger Fennoskandischer Lepidopteren. *Zool. Bidr. Upps.* 26: 329–531.

Petiver, J., 1695. *Rariora Naturae*, Londini.

—— 1717. *Papilionum Brittaniae.* Petiver, London.

Pictet, A., 1922. Recherches sur l'hibernation de *Pieris brassicae* à l'état de chenille. *Bull. Soc. lépidopt. Genève* 5: 47–57.

Pionneau, P., 1924. Nouveautés lépidoptérologiques pour la faune girondine avec la description de deux formes nouvelles. *Misc. Ent.* 27: 57–58.

—— 1928. Sur quelques Rhopalocères Paléarctique. *L'Exchange* 44: 3–4.

Plant, J., 1847. Note on a singular variety of the large cabbage white. *Zoologist* 1: 471.

Pol, P.H. & Ponsen, M.B., 1963. Pathologische verschijnselen bij vlinders van *Pieris brassicae* L. verband houdende met een granula (virus) infectie van rupsen. *Ent. Ber., Amst.* 23: 106–108.

Rait-Smith, W.R-S., 1930. The C.F. Johnson Collection. *Entomologist* 63: 90–93.

—— 1938. The Bright Collection. *Entomologist* 71: 187–190.

—— 1943. The sale of the Stiff, Leeds, Joy and others collection. *Entomologist* 76: 126–128.

—— 1944a. Sale of the Whitehouse collection; first portion. *Entomologist* 77: 43–46.

—— 1944b. Sale of the Whitehouse collection; second portion. *Entomologist* 77: 93–96.

—— 1947. The sale of the B.H. Smith collection. *Entomologist* 80: 197–199.

Ray, J., 1710. *Historia Insectorum.* p. 114.

Rebel, H., 1917. Siebenter Beitrag zur Lepidopterenfauna der Kanaren. *Annl. natur. Mus. Wien* 31: 1–62.

Rees, A., 1819. *Cyclopaedia or Universal dictionary of arts, sciences and literature.* Vol. 26. Longmans, Hurst, Rees, Orm, Brown, London.

Riley, N.D., 1925. Seasonal variation in butterflies. *Proc. S. Lond. ent. nat. Hist. Soc.* 63–81.

Röber, J., 1896. Neue Schmetterlinge aus dem cilicischen Taurus. *Ent. Nachr.* 22; 81–88.

—— 1907. Pieridae. *In* Seitz, A. *Die Gross-Schmetterlinge der Erde.* Section i, Vol. 1: pp. 39–74.

Rocci, U., 1919. Osservazioni sui Lepidotteri di Liguria. *Atti. Soc. Ligui Sci. nat.* 19: 1–44.

—— 1930. Sulla forma autumnale in Lombardia, di *Mancipium brassicae* L. (Lep.). *Boll. Soc. ent. ital.* 62: 15–17.

—— 1933. Lepidotteri di Liguria. *Mem. Soc. ent. ital.* 8: 90–97.

Russwurm, A.D.A., 1978. Aberrations of *British Butterflies.* E.W. Classey, Faringdon. pp. 151.

Schmidt-Koehl, W., 1971. Zur Rhopalocerenfauna der Kanareninsel Teneriffa (Insecta: Lepidoptera). *Mitt. ent. Ges. Basel* 21: 29–91.

Schrank, F. von P., 1801. *Fauna Boica* 2: viii + 374 pp. J.W. Krull, Ingolstadt p. 152.

Schurian, K., 1975. Notizen an *Pieris cheiranthi* (Lepidoptera: Pieridae). *Ent. Z. Frank. a.M.* 85: 252–256.

Scudder, S.H., 1875. Historical sketch of the generic names proposed for butterflies. *Proc. Am. Acad. Arts Sci.* 10: 91–293.

Seitz, A., 1901. *Macrolepidoptera of the World.* 1: 45.

—— 1929–1932. *The Macrolepidoptera of the World.* The *Palaearctic Butterflies.* Supplement. Alfred Kernen Verlag, Stuttgart.

Sevastopulo, D.C., 1924. Notes on the lepidoptera of St. Germain, Seine and Oise. *Entomologist* 57: 113–116.

Shapiro, A.M., 1978. The life history of *Reliquia santamarta*, a neotropical alpine pierine butterfly (Lepidoptera: Pieridae). *Jl N.Y. ent. Soc.* 86: 45–50.

Sheldon, W.G., 1922. Notes on the lepidoptera of the Assynt District of Sutherland. *Entomologist* 55: 30–35.

Soenen, A., 1947. *Lijst der Voornaamste Vijanden in Land en Tuinbouw.*

South, R., 1891. On the distribution in eastern asia of certain species of lepidoptera occurring in Britain. *Entomologist* 24: 81–86.

—— 1936. *Butterflies of the British Isles.* Warne, London.

SLENHS, 1902. Meeting Nov. 28th 1901. *Entomologist* 35: 49–51.

—— 1909. Meeting July 22nd 1909. *Entomologist* 42: 262.

—— 1915. Meeting May 13th 1915. *Entomologist* 48: 174.

—— 1916. Meeting Nov. 25th 1915. *Entomologist* 49: 42–44.

—— 1921. Meeting Feb. 10th. *Entomologist* 54: 1–152.

—— 1932. Minutes. *Entomologist* 65: 191–192.

Stauder, H., 1921. Neues aus Unteritalien. *Dt. ent. Z. Iris* 35: 26–31.

Staudinger, O., 1901. *In* Staudinger, O. & Rebel, H., *Catalogus der Lepidopteren des palaearctischen Faunengebebietes.* (1), xxxii + 411 pp. Friedländer & Son, Berlin.

Stephens, J.F., 1827–8. *Illustrations of British Entomology.* Haustellata. Vol. 1. Baldwin & Cradock, London. 152 pp. 12 pl.

Talbot, G., 1939. *Butterflies.* 2nd Edition. *In* Blandford, W.T. *Fauna of British India.* Vol. 1. xxxix + 600 pp. 3 pls., 1 map.

Thomas, I., Janson, H.W. & Aitken, A.D., 1968. *Common names of British insects and other pests.* Minist. Agric. Fish. Fd. 6: iv, 72 pp.

Tubbs, R., 1978. The breeding of butterflies with special reference to thé genetics of aberrational forms. *Proc. Trans. Brit. Ent. nat. Hist. Soc.* 11: 77–87.

Turati, E., 1924. Spedizione lepidotterologica in Cirenaica 1921–1922. *Atti. Soc. ital. di Storia Nat.* 63: 21–191, 6 pls.

—— 1925. Missione Zoologica del Dr.E. Festa in Cirenaica: Lepidotteri. *Boll. Mus. di Zool. ed Anat. comp. della R. Univ. Torino* 39: 1–9.

—— 1929. Lepidotteri. *Archo. zool. ital.* 13: 179–186, pl. 27; pl. 1. fig. 3.

—— 1936. Contributi alla fauna Cirenaica. *Mém. Soc. ent. ital.* 15: 55–77.

Turner, H.J., 1920. The butterflies of Cyprus. *Trans. ent. Soc. Lond.* 1920: 170–207.

—— 1925. Societies. *Entomologist* 58: 172–174.

Turton, W., 1802. *A general system of nature*, animal kingdom 3: insects (2), frontispiece + 376 pp. Lackington, Allen & Co., London.

Tykač, J., 1947. *Pieris brassicae* L. gen. *aest. lepidii* Rob. ab. *obenbergeri* n. ab. (Lep: Pieridae). *Čas. čsl. Spol. ent.* 44: 119–120.

Van Mellaerts, I., 1926. Aberrations de Lépidoptères. *Lambillionea* 1: 82–84.

Verity, R., 1908 (1905–1911). *Rhopalocera Palaearctica.* Iconographie et description des papillons diurnes de la region palaéarctic Rhopalocera. 2 vols. Florence.

24

—— 1913. Revision of the Linnaean types of Palaearctic Rhopalocera. *J. Linn. Soc. Lond.* 32: 173–191.

—— 1916. British races of butterflies; relationships and nomenclature. *Entomologist's Rec. J. Var.* 28: 97–102.

—— 1919. Seasonal polymorphism and races of some European Grypocera. *Entomologist's Rec. J. Var.* 31: 87–89.

—— (*in* Verity, R. & Querci, O.) 1923–4. Races and seasonal polymorphism in the Grypocera and of the Rhopalocera of Peninsular Italy. *Entomologist's Rec. J.* 35 (suppl.): 1–20; 36: (suppl.): 21–44.

—— 1947. Revision of Italian Pieridae. *Diurne farfalle ducre d'Italia* 3: 111–307.

Walcourt, de H., de B., 1920. Présence d'un point noir dans l'espace compris e tre les nervures 3 et 4 du dessus des ailes postérieures chez les mâles de nos trois Pierides communes. *Rev. Mens. Soc. ent. Namur.* 20: 15.

Webb, S., 1889. *Pieris brassicae* with veins of wings green. *Entomologist* 22: 138.

Weizmantel, V., 1900. *Rovart. Lap.* 7: 175.

Williams, H., 1893. Meeting July 13th 1893. *Entomologist* 26: 254.

Wood, G.L., 1972. *The Guinness Book of Animal Facts and Feats.* Guinness Superlatives Limited, Enfield.

Wood, T.W., 1867. Remarks on the colouration of chrysalides. *Proc. ent. Soc. Lond.* 99–101.

Youden, G.H., 1975. Larvae of *P. brassicae* L. at Christmas. *Entomologist's Rec. J. Var.* 87: 92.

Zeller, P.C., 1847. *Bemerkungen über die auf einer Reise nach Italien und Sicilien beobachteten Schmetterlingsgarten.* Isis von Okai, Leipzig.

Further references

Ashby, E.B., 1933. *Trans. 5th Int. Cong. Ent.* 2: 65–78 ("var. *chariclea*").

Bell, S.J., 1909. *Entomologist* 42: 21 (ab. with narrow black hind-wing border).

Bender, R., 1963. *Z. Wien Ent. Ver.* 48: 11–20 (*catoleuca, verna* on Rhodes).

Benson, R.B., 1923. *Entomologist* 56: 238 (imago (female) with larval head).

Bernardi, G., Herbulot, C. & Picard, J., 1951. *Rev. Fr. lep.* 11: 420–431 (*brassicae* as separate sub-genus from *Artogeia rapae, napi*).

Bernardi, G., 1958. *Rev. Fr. ent.* 25: 125–128 (*nepalensis*).

Bramwell, J.M., 1869. *Entomologist* 4: 258 (black specimens).

Carpenter, G., 1906–7. *Proc. Trans. S. Lond. ent. nat. Hist. Soc.* 90 (ab. with distal spot connected to apical patch).

Claret, J., 1973. *Archs. Zool. exp. gén.* 114: 271–275 (mutant with melanisation of veins).

Cockerell, T.D.A., 1912. *Entomologist* 45: 322–323 (ab. *obscurata, nigrescens*).

D'Auriol, G.L., 1917. *Archs. Sci. Phys.* 44: 32–47 (vars. in Jura).

ESL (Entomological Society of London), 1902. *Entomologist* 35: 47–49 (greenish underwings).

Fischer, E., 1952. *Bull. Soc. ent. Mulhouse* 1952: 5–8 (pierids of France and Palaearctic).

Goodson, A.L., 1960. *Entomologist* 93: 146–149 (imago with larval head).

Gregson, C.S., 1872–3. *Entomologist* 6: 286–288 (ab. with bright emerald green on underside).

Hemming, F., 1934a. *Generic names of Holarctic Butterflies* I. 1758–1863. British Museum, London.

—— 1934b. *Entomologist* 67: 49–53 (list of British Butterflies revised).

—— 1956. Bull. zool. Nom. 12: 291–306 (proposal to validate "Pieridae" as opposed to "Pierididae").

Heslop, I.R.P., 1964. *Revised Indexed check list of British Butterflies.* Library edition, Burnham on Sea.

Jones, H.P., 1918. *Entomologist* 51: 248–252 (minor aberrations of butterflies).

Kaucki, T.F., 1929. *Polskie Pismo ent.* 7: 180–199 ("ab." *wollastoni* in Madeira).

Kimber, D.J., 1945. *Entomologist* 78: 107 (ab. with sulphur yellow ground colour, wingtips with dark chocolate brown).

Krzywicki, M., 1962. *Klucze do oznacznia owadów Polski* 27: (65–66): 45 pp. (key to pierids).

Lancashire & Cheshire Entomological Society, 1892. *Entomologist's Rec. J. Var.* 3: 43–48 (proposal that *P. brassicae* is in same genus as *P. rapae* and *P. napi*).

—— 1933. *Entomologist* 66: 69–70 (dumb-bell form exhibited).

Mousley, H., 1903. *Entomologist's Rec. J. Var.* 15: 167 (male with black mark).

Newman, L.H., 1917. *Trans. ent. Soc. Lond.* 1916 (aberration).

Oudemans, J.Th., 1905. *Tijdschr. Ent.* 48: 1–21 (aberration).

Pickett, C.P., 1903. *Entomologist's Rec. J. Var.* 15: 190 (ab. with small black spot).

Powall, J., 1872–73. *Entomologist* 6: 315 (ab. with green colouration).

Reuss, F.A.T., 1935. *Int. ent. Z.* 29: 331 (*conspicua-spinosa, inconspicua-spinosa*).

South London Entomological and Natural History Society (SLENHS), 1907. *Entomologist* 40: 18–20 (variation with discal spots united).

—— 1917. *Entomologist* 51: 23, 46, 116–117 (many aberrations exhibited).

—— 1918a. *Entomologist's mon. Mag.* 54: 41, 42, 66, 186–187 (paper read on variation in pierids).

—— 1918b. *Entomologist* 51: 190–191 (black spot on fore-wing).

—— 1919. *Entomologist* 53: 69–71 (green specimen).

—— 1921. *Entomologist* 54: 1–152, 78–79 (several abs. exhibited).

—— 1940. *Entomologist* 53: 140–141 (aberration exhibited).

—— 1949. *Entomologist* 82: 214 (black aberration exhibited).

—— 1952. *Entomologist's mon. Mag.* 88: 25 (dwarf exhibited).

Stauder, H., 1921. *Z. wiss. Insekteneb.* 16: 151 (ab. *pseudocatoleuca*).

—— 1921. *Dt. ent. Z. Iris* 35: 26–27 (ab. *italorum*).

Strand, E., 1909. *Int. ent. Z.* 3: 78–79 (*catoleuca* in Asia Minor).

Weisman, A., 1882. *Studies on the theory of Descent.* Sampson, Low, Marston, Searle & Rivington, London (pierids and seasonal dimorphism).

Wheeler, G., 1916. *Entomologist's mon. Mag.* 52: 119 (abs. with pink colouration).

Wilde, H., 1957. *Ent. Z.* 67: 257–258 (specimen with heavy black scales).

Woodforde, F.C., 1920. *Entomologist* 53: 152–157 (generation of small imagines).

Wynne, A.S.B.F.P., 1944. *Entomologist* 77: 60 (deformed imago).

Yorkshire Naturalists Union and Entomological Society, 1917. *Entomologist's mon. Mag.* 53: 20 (small male and large female).

Ziegler, I., 1961. *Adv. Genet.* 10: 349 (*P. b. cheiranthi* – pteridine pigmentation).

2. Distribution

Introduction

The Large White is often quoted to have a typical Palaearctic distribution from north Africa across Europe and Asia to the Himalaya mountains (Frohawk, 1934; Higgins & Riley, 1970), and indeed the recent published world distribution map of *P. brassicae* (Commonwealth Institute of Entomology (C.I.E.), 1976) corroborates this. However, there is a very patchy distribution of records across this region with perhaps thousands of recorded sightings of *P. brassicae* in Europe and considerably fewer further east. This would tend to suggest that more information is still needed from such countries as Japan and China, where some workers believe that *P. brassicae* occurs, before any hard and fast distribution line is drawn across this part of the world.

Data for this chapter has come from several sources, not least of which has been the mass of information present in notes and communications published in the entomological journals. It cannot be said however, that all the references relating to *P. brassicae* have been taken from all possible journals, but publications such as *The Entomologist* and *Entomologist's Monthly Magazine* have been thoroughly searched and many Ministry of Agriculture, Fisheries and Food publications and records of local natural history societies have been consulted.

Much additional information has come from many of the committee members and their colleagues of the European Invertebrate Survey (E.I.S.) (International Commission for Invertebrate Survey), which has its secretariat at Monks Wood Experimental Station, Institute of Terrestrial Ecology, Abbots Ripton, Huntingdon, Cambridgeshire, England. Advice has been sought from most of the representatives of the 24 participating countries and much useful information is presented as personal communications.

Distribution maps are available for only a limited number of countries throughout the range of *P. brassicae*, all of them in western Europe.

However, the information they present shows more the distribution of entomologists than of *P. brassicae* (with the exception of the Iberian peninsula) and no real value would be gained from preparing a composite map using the data from each of the countries concerned. As an example, *P. brassicae* is clearly present in several of the unrecorded *départements* of France (Le Cerf, 1972).

Much incidental information with regard to distribution in various countries can be found in other chapters, for instance in Chapter 13 (Economic Importance) and Chapter 11 (Migration).

World range

Two maps indicating the world range of *P. brassicae* have been published, the first by the Russian Kurentzov (1929), the second more recently by the C.I.E. in London (C.I.E., 1976). The two are generally in agreement with each other, indicating that *P. brassicae* has a typical Palaearctic distribution throughout Europe, USSR, around the Mediterranean, including the north African coast, the Middle East and through the north of India and perhaps Pakistan and into the Himalayas. Beyond these regions *P. brassicae* occurs as a subspecies, *P. b. nepalensis*, extending into China (cf. Chapter 1).

Pieris brassicae appears to be restricted in its world range by temperature and humidity which present various physiological barriers. Kurentzov's map indicates that the northern range of *P. brassicae* is limited by the $-20°C$ winter (January) isotherm, a fact corroborated by Blunck (1953) who maintained that *P. brassicae* is restricted between the winter isotherm of -16 to $-20°C$ and the summer (July) isotherm of $+28°C$. *P. brassicae* cannot tolerate temperatures below $-13°C$ and above $28°C$ at 60% relative humidity in the wild, and even at lower humidity, oviposition is inhibited (Klein, 1932). Bonnemaison (1965), however, found that the pupae of *P. brassicae* could withstand temperatures down to $-22.5°C$. Haig-Thomas (1938) recorded *P. brassicae* common in Lappland (Maalselven, latitude 69°N) where summer temperatures reach 35°C in the shade.

The C.I.E. (1976) lists 41 countries from which *P. brassicae* has been recorded, including Chile as a new locality in the New World since they published their earlier edition in 1953 (C.I.E., 1953) (Table 2.1). *P. brassicae* was first noticed in Chile in 1971 and has spread considerably since (Feltwell, 1978b). Gardiner (1974) published a distribution map of *P. brassicae* in South America. The C.I.E. (*loc. cit.*) also indicated a population of *P. brassicae* in Ethiopia but there are no recent records to substantiate this.

Table 2.1 List of countries and islands from which *P. brassicae* has been recorded. (based on C.I.E. (1976) with amendments; ? indicates doubtful occurrence).

Europe	Switzerland	Novosibirskaya
Austria	Yugoslavia	Armenia SSR
Azores		Kazakhskaya SSR
Balearic Islands		Kirgiskaya SSR
Belgium	*Asia*	Lithuania SSR
Bulgaria	Afghanistan	Tartarskaya SSR
Channel Islands	Bangladesh	Tadzhikistan SSR
Corsica	Bhutan	Turkmenskaya SSR
Crete	Burma	Uzbekskaya SSR
Czechoslovakia	China	Voronezhskaya SSR
Denmark	Cyprus	
Faroes ?	Hong Kong	
Finland	India	*Africa*
France	Iran	Algeria
Germany	Iraq	Libya
Great Britain	Israel	Morocco, including
Greece	Jammu & Kashmir	Tangier
Hungary	Jordan	Tunisia
Iceland ?	Lebanon	
Ireland	Nepal	
Italy	Pakistan	*Atlantic islands*
Malta	Sikkim	Canary Islands
Netherlands	Syria	Madeira
Norway	Tibet	
Poland	Turkey	
Portugal		*South America*
Rumania		Chile
Sicily	*USSR*	
Spain	European part	
Sweden	Georgia & Caucasus	

The exact distribution of *P. brassicae* in the east of its world range is not at all clear. Recent records exist for *P. brassicae* in western Siberia, and in Yunnan in China but there appears to be no account of *P. brassicae* in Japan.

In Europe much information on the distribution of *P. brassicae* is available and distribution maps have been published for a number of countries (Table 2.2). For those countries not listed in Table 2.2, locality lists may be obtained from the representative committee members of E.I.S.

Only Great Britain, Ireland and France have sufficient information on the distribution of *P. brassicae* to warrant the information being presented according to county, region or *département* respectively (Tables 2.3; 2.4).

29

Table 2.2 Distribution maps of *P. brassicae.*

Published	
Austria	(Kusdas & Reichl, 1973;
	Berger, 1973 – unpublished)
Belgium	(Verstraëten, 1970)
Denmark	(Calov, 1975 – unpublished)
Finland	(Nordström *et al.*, 1955)
France	(Le Cerf, 1972)
Germany BRD	(Schreiber, 1976; cf. Schmidt-Koehl, 1971)
Great Britain & Ireland	(Heath, 1973)
Greece	(Koustaftikis, 1974c)
Iberian peninsula	(Gómez-Bustillo & Rubio, 1974)
Norway	(Aagaard & Gulbrandsen, 1976)
Rumania	(Niculescu, 1963)
Scandinavia	(Nordström *et al.*, 1955)
World	(C.I.E., 1976; Kurentzov, 1929)

Europe

Albania

Recorded from six localities by Rebel & Zerny (1931).

Austria

Many migrations of *P. brassicae* have been recorded in Austria, even high up in the Tyrol (Rowland-Brown, 1904; Williams, 1958) and the Dolomites (Gurney, 1927; Cooke, 1927). Kusdas & Reichl (1973) quoted a period of thirteen years, between 1917 to 1955, when *P. brassicae* was abundant (see Chapter 11) and Berger (1975, pers. comm.) stated that it was particularly abundant in three regions, in the north near Linz, in the east around Eisenstadt, and in the south around Graz (cf. Glaeser, 1974).

Belgium

Old records for the distribution of *P. brassicae* in Belgium are those of Dubois & Dubois (1874) and Manning Stanton (1925) which covers the Belgian coast.

Distribution maps were produced for Belgium by Leclerq (1970) and Verstraëten *in* Leclercq (1970). These showed areas from which records appear to be lacking, mainly the south-east, the north and the east.

Bulgaria

Pieris brassicae was recorded from 17 localities in the Balkans by Rebel (1904). Two reports of the same period stated that *P. brassicae* was scarce everywhere in June and July 1933 (Thomas, 1936) and not common in

Sliven on the southern edge of the Balkans, and Kostenetz Bania (Straubenzee, 1933).

Recently Kwartirnikova (1978, pers. comm.) stated that *P. brassicae* is widespread all over Bulgaria mainly in fields but also in the mountains up to 1600 m. Although there are three generations a year of *P. brassicae* in Bulgaria Kwartirnikova noted that the density of *P. brassicae* populations has been decreasing over the last decade.

Czechoslovakia

Recorded as widespread and common throughout the country (Kudrna, 1974a), and in South Bohemia (Kudrna, 1959, 1970, 1971; cf. Bretherton, 1949). Spitzer (1975, pers. comm.) stated that *P. brassicae* was a menace to *Brassica* crops.

Fennoscandia

The map of Nordström *et al.* (1955) for the whole of Scandinavia showed that *P. brassicae* was mostly restricted to the south of Sweden and Finland. Haugen (1976, pers. comm., unpublished MS), recorded it as common in southern Scandinavia but "becoming rarer to the north, sporadic at 65°N". Occasional records from the central montane area may represent specimens which have flown there on migration (cf. Chapter 11). He also found that *P. brassicae* was largely absent or uncommon in Norway, particularly in the south-east. However, the distribution map published by Aagaard & Gulbrandsen (1976) shows *P. brassicae* around the south-eastern coastal region of Norway, and around Trondheim and Oslo. Early reports mentioned that *P. brassicae* was scarce in Norway (Cooper, 1894) and that it was conspicuously absent in south and northern Norway (Chapman, 1899).

According to Utrio (1975, pers. comm.) the northern limit of *P. brassicae* in Norway is the Arctic Circle. Aagaard & Gulbrandsen's map shows old localities (pre-1960) near Vadso. There are several accounts of *P. brassicae* being present in Lappland, which covers the northern regions of Norway, Sweden, Finland and the USSR, but all of these are pre-1940 records (Rowland-Brown, 1906; Sheldon, 1912; Rosa, 1920; Thomas, 1938a,b).

In Finland *P. brassicae* appears to have a fairly complete coverage of the south-east of the country but it has patchy distribution elsewhere (Calov, 1975, unpublished map); however Svendsen (1976) reported it common all over the country.

France

Accounts of *P. brassicae* in France are plentiful, particularly in areas popular with tourists. Le Cerf (1972) published a distribution map for

Table 2.3 Distribution of *P. brassicae* in France.

Département (Regions marked as*)	Authority
Alpes-Maritimes	Bromilow, 1893a,b; Bretherton, 1962; Morris, 1919, 1922
Aude	Lack & Lack, 1951
Aveyron	Rowland-Brown, 1905; Snow & Ross, 1952
*Brittany	Blackie, 1920; Cooke, 1924b; Hitier, 1918
Calvados	Lane, 1955, 1957
Cantal	Burton (1981, in preparation)
Creuse	Chazaud, 1977
Charente	Simes, 1932
Dordogne	Bretherton, 1975
Gard	Aldin, 1931; Feltwell, 1977a,c,d; Oldaker, 1922
Gironde	Burton & Owen, 1954
Haute-Savoie	Wiltshire, 1973
Hautes-Pyrénées	Brown *et al.*, 1958; Cooke, 1925; Molesworth-Muspratt, 1959
Hérault	Feltwell, 1977a
Landes	Burton & Owen, 1954
Lot	Rowland-Brown, 1907
Lozère	Rowland-Brown, 1909
Nièvre	Burras, 1923
*Provence	Cooke, 1926; Hanson, 1941
Pyrénées-Atlantiques	Sheldon, 1908a
Pyrénées-Orientales	McClymont, 1917, Nabokoff, 1931; Bretherton *et al.* 1952; Dufray, 1968
Saône-et-Loire	Constant, 1866
Somme	Graham, 1919
Vaucluse	Oldaker, 1922; McLeod, 1972
Vosges	Gibbs, 1909

France covering many *départements*, a few not being represented on account of paucity of records. Table 2.3 lists a selection of records of *P. brassicae* in France by *département*, but by no means represents a thorough search of the literature.

Some records are from regions of France which comprise several *départements*, i.e. Riviera and Provence (4 *départements* each), Cévennes (3), Pyrénées (4), Brittany (5) and so on. Further details of the distribution of *P. brassicae* are included in Chapter 11 (Migration).

Germany
Professor Blunck at Kiel (N. Germany) made a detailed study of migration in *P. brassicae* and recorded the insect as common in that part

of Germany. He recorded a 3–10 million strong migration of *P. brassicae* (Blunck, 1953, 1954). Migrations of this magnitude or less, which are often cited for this region, were not witnessed by the author during the same months 23 years later (Feltwell, 1977b). A more recent distribution map for *P. brassicae* in Saarland was published by Schmidt-Koehl (1971). *P. brassicae* was also recorded in southern Hanover (Finke, 1938). Several other details of distribution of *P. brassicae* in Germany may be found in Chapter 11 (Migration).

Great Britain & Ireland

A distribution map of *P. brassicae* combining information from England, Scotland, Wales and Ireland was published in Heath (1970, 1973) but the most recent edition of this map (January, 1978) is represented here in Plate II. The distribution of *P. brassicae* is dealt with by county and then by county of both England and Scotland. It must be reiterated however, that *P. brassicae* probably occurs all over Great Britain but has yet to be recorded from each ten kilometer square. Of particular importance is the bibliography of county distribution lists in the first volume of *The Moths and Butterflies of Great Britain and Europe* (Heath, 1976).

England

From Plate II it is clear that *P. brassicae* has a fairly complete coverage in the south of England, but above a line from the River Wash to the River Severn, the records are fairly patchy.

County distribution records are in the process of being shown as separate maps, as already has been done for Kent (Philp, 1978, unpublished, Maidstone Museum) and for Staffordshire where it has been recorded for all but one of the ten kilometer squares (Warren, 1975). In Cheshire *P. brassicae* has been recorded for all of the thirty 10 kilometer squares (Rutherford, 1978, pers. comm.). In Warwickshire, *P. brassicae* has been recorded from all but two of the twenty 10 kilometer squares but here this lack of records is probably due to lack of recorders from these regions (Steeden & Steeden, 1977 and 1978, pers. comm.). A distribution map for *P. brassicae* on Porton Down in Wiltshire was published by Morris (1977).

There are so many notes and communications on the presence or absence of *P. brassicae* from different regions of England that it is not the intention of the author to quote all of these single observations here (see Further references). However, as a selection, a few of these references, which include papers on the lepidoptera or rhopalocera of the respective county, are included in Table 2.4 by county.

P. brassicae has been cited on many islands off the coast of England, for instance on the Isle of Wight (Cornell, 1919; Fearnclough, 1972), Isle of

Table 2.4 Distribution of *P. brassicae* in England. NB. Middlesex no longer exists as a county and is now part of Surrey and London, while parts of Staffordshire, Shropshire and Warwickshire are now West Midlands.

County	Notes and quotations	Authority
Bedfordshire	more or less common	Verdcourt, 1945
Berkshire	abundant – in 1945	Kettlewell, 1955
Buckinghamshire	south east	Worms, 1949
	common in all parts	Ansorge, 1969
Cambridgeshire	present	Gardiner, 1963
Cheshire	distribution map in preparation	
	"generally common"	Rutherford in prep. 1980
Cornwall	present	
	common in Launceston area	Wild & Marshall, 1973 Spencer, 1979 pers. comm.
Cumberland	common in 1925 and 1926	Day, 1921; Ford, 1924, 1927
Derbyshire	very common everywhere	Haywards, 1926
Devonshire	10 miles (16 km) radius of Taunton	Butler, 1915; Blair, 1925; Worms, 1968
Devonshire	common except in central Dartmoor	Stidson, 1952
Dorset	1929	Andrews, 1930
Durham	becoming scarce	Dunn, 1975 pers. comm.
Essex	present	Johnson, 1908
	London area	Worms, 1949
Gloucestershire	present	Hodgson, 1957
Hampshire	common everywhere	Goater, 1974
Herefordshire	Ross-on-Wye	Leeds, 1938
Hertfordshire	1941	Byles, 1942; Worms, 1949
Huntingdonshire	common	Blaikie, 1949
Kent	London area	Worms, 1949
Kent	distribution map (unpublished)	Philp, 1978, pers. comm.
Kent	many localities listed	Chalmers-Hunt, 1962
Lancashire	abundant throughout county but more in Cheshire and S. Lancashire than N. Lancashire	Ellis, 1940
Lancashire	distribution map	Steeden & Steeden, 1977
Leicester	1933	Buckler, 1934
Middlesex	London area	Worms, 1949
Norfolk	Wicken Fen	Tutt, 1892
Northamptonshire	1950	Payne, 1951
Northumberland	present	Church, 1959
Nottinghamshire	very plentiful in all parts	Carr, 1916

Table 2.4 continued

County	Notes and quotations	Authority
Oxfordshire	present	Bussey-Bell, 1915; Prior, 1968
Shropshire (central)	1948–1951	Smith, 1952
Somerset	present	Bretherton, 1944
Staffordshire	distribution map	Warren, 1975
Surrey	1946–1948	Stallwood, 1949
Surrey	London area	Worms, 1949
Surrey	20 miles (32 km) radius of London	Stafford, 1922
Surrey	usually common	Evans & Evans, 1973
Sussex	common to abundant	Haes, 1977
Sutherland	east	Rollason, 1907
Warwickshire	distribution map in preparation	Smith, 1978
Westmorland	1925	Birkett, 1954
Wiltshire	generally distributed	Worms, 1962
Wiltshire	distribution map of Porton Down	Morris, 1977
Yorkshire	present	Owen, 1951

Sheppey (Betts, 1920), Isle of Man (Clarke, 1893; Cowin, 1938; Hedges, 1947 "all over the island"; Chalmers-Hunt, 1970), Farne Islands (Wooff, 1958) and the Scilly Isles (Adkin, 1911; Blair, 1925; Granville-Clutterbuck, 1940; Summers, 1975; Beavis, 1976).

P. brassicae has also been recorded as common on the Channel Islands (Long, 1970, cf. Sich, 1883). Other records are given for Jersey (Piquet, 1873; Riley, 1925; Le Quesne, 1946) and Sark (Newman, 1931).

Ireland
Heath's (1978, unpublished) distribution map of Great Britain & Ireland shows that *P. brassicae* is well recorded around Dublin and Cork, and around the coast on the south and east. The rather patchy distribution of *P. brassicae* throughout the rest of Ireland probably represents observer deficiency (Hillis, 1978, pers. comm.). He regards *P. brassicae* as "a near ubiquitous opportunist" which detects foodplants readily and colonizes these cultivated areas accordingly. An earlier distribution map for Ireland based on the same grid system appeared in Crichton & Maloney's (1975) *Provisional Atlas of Butterflies in Ireland*. Donovan (1901, 1936) gives several localities for *P. brassicae*.

Old records for some of the counties of Ireland include Co. Sligo (Kone, 1901), West Meath, where it was abundant in 1905 but "had been scarce for the last few years" (Battersby, 1905), Co. Mayo (Hodgson, 1933),

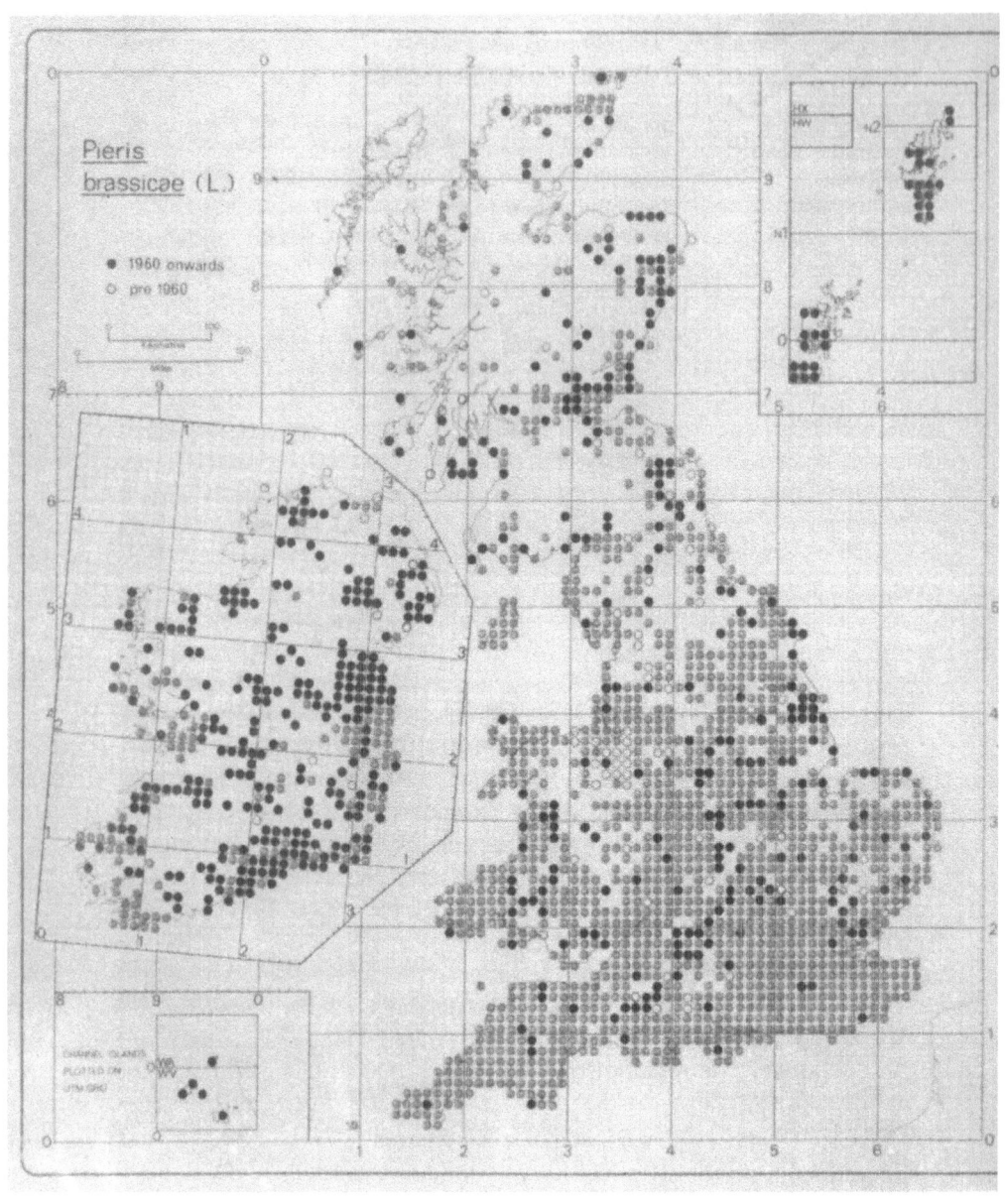

Plate II Distribution of *Pieris brassicae* (L.) in Great Britain and Ireland, 19
(Courtesy of John Heath, Monks Wood Experimental Station)

Co. Down (Johnson, 1927) Co. Cork (Huggins, 1953) and Co. Kildare (Lisney, 1940).

More recently Baynes (1973 cf. 1957, 1964) stated that *P. brassicae* occurred as far north as Tory Island, 14 km off Co. Donegal, and that in Ireland, *P. brassicae* was plentiful in 1959, scarce in 1960 and built up again, augmented by immigrants in 1962. Baynes (1969) also recorded *P. brassicae* on the Dingle peninsula, Co. Kerry in 1968 and very plentiful at Cloyne, Co. Cork in August 1970 (Baynes, 1971).

Scotland

P. brassicae may be encountered all over Scotland, but the most recent distribution map for Great Britain shows a patchy distribution in Scotland, with records lacking from the Southern Uplands, the Grampians and the North West Uplands.

Recent records exist for *P. brassicae* in the Shetland Islands (61°N), although Beirne (1945) stated that like South, he could find no original records. Ralston (1959) recorded *P. brassicae* in the Shetlands and Williams (Index cards:- letter received from C.B. Williams) stated that it was seen on the mainland in 1970.

There are several accounts of *P. brassicae* in the Orkneys where it is regarded as being common most years (see Lorimer, 1970 for a review of butterflies covering the period 1868 – 1968; Traill 1869; Cheesman, 1898). The insect has been recorded from the island of Hoy several times (Weir, 1882; Gregson, 1885; South, 1895; Howard, 1975; cf. Shapiro, 1970).

Many records exist for the presence of *P. brassicae* in the Hebrides, principally through the work of Heslop-Harrison and Campbell. Campbell (1936, 1945, 1946, 1949a) reported *P. brassicae* in the Hebrides on several occasions as a "resident migrant". In the Outer Hebrides, it was recorded from the islands of Lewis, Harris and Scarp, where it was found breeding (Heslop-Harrison, 1938, 1942). Williams (*loc. cit.*) noted that *P. brassicae* was rarely found on Lewis in 1948.

In the Inner Hebrides, it has been recorded from the islands of Canna, Eigg, Gunna, Raasay, Rhum, Scalpay and Tiree (Heslop-Harrison, 1937, 1939, 1941, 1945, 1946, 1947, 1949a,b,c, 1950a,b, 1952–56, 1958) and on Arran (Stewart, 1925). Steele & Woodruffe (1969) stated that *P. brassicae* occurred regularly on Rhum, and that it sometimes breeds at Kinloch. However, in 1962–64 no breeding accounts were noted. Kevan (1941) also recorded *P. brassicae* on Eigg. Campbell (1951, 1970, 1978) reported that it was a "commensal resident" on Canna, and that the cold spring of 1949 wiped out the colony there. More recently, *P. brassicae* has been recorded on Canna in late April 1971, and was abundant in April 1974 (Campbell, 1972, 1975). *P. brassicae* was noted on Iona during a heatwave in June 1975 (Luckens, 1976).

On the mainland of Scotland accounts of *P. brassicae* have been recorded from Aberdeenshire (Palmer, 1974), Argyllshire during 1918–1922 (Campbell, 1949b), Berwickshire (Long, 1956, where it was described as indigenous throughout), Dumfrieshire (Murray, 1934), Invernesshire (Harper, 1968; Howard, 1978), Kincardineshire (Palmer, 1974), Stirlingshire (Coates, 1968, not seen every year) and Sutherland (Heslop-Harrison, 1940, on the island of Handon off the west coast).

Wales

Early records are for Porthcawl (Crowther, 1932), Llandudno (Ellis, 1948) and Anglesey (Baynes, 1909). It was also recorded from the Bangor region of Caernarvonshire in August 1963 (Morgan, 1965) and on the islands of Skokholm (Conder, 1949) and Bardsey (Askew, 1974). Smith (1948) made a study of the lepidoptera of the counties of Flintshire, Denbighshire, Caernarvonshire, Anglesey and Merionethshire and recorded *P. brassicae* unusually common throughout (1947 was an exceptionally good year for butterflies throughout Great Britain).

A marked migration was noted in the Moorlands near Llyn Cwmynach, Merioneth in 1970 (Young, 1974) and *P. brassicae* was noted in Radnorshire in August 1973 (Evans, 1973).

Greece

The distribution map of *P. brassicae* in Greece indicating 21 localities is by no means complete (Koutsaftikis, 1974c) and can be supplemented by the following information. An early reference quoted *P. brassicae* as being common in Macedonia in 1916–17 (Barraud, 1918). Van Straubenzee (1932) spent three months collecting in Greece and reported *P. brassicae* common everywhere, and Dacie *et al.* (1972) found it widespread and quite common in northern and central Greece. Bretherton (1968) also recorded *P. brassicae* with general distribution. Coustis (1969) listed four mountain areas where *P. brassicae* was encountered: Olympus, Parnassius, Pindus and Taygetus. According to Gepp & Stark (1974) and Koutsaftikis (1974b) the distribution of *P. brassicae* in Greece is governed by temperature and humidity; it is restricted to areas with temperatures between 18–42°C and at a relative humidity of 35–68%.

Distribution records for *P. brassicae* on the Greek Islands are fairly extensive; Corfu (Norris, 1891) where larger than normal forms were found (Lipscomb, 1977), Rhodes (Betts, 1922), Sifnos (Coutsis, 1978) where small forms were found, Skyros (Coutsis, 1976; cf. Gepp & Stark, 1974), Simi & Kastelloriso (Koutsaftikis, 1974a, 1975).

Hungary

During a visit to Hungary, Rothschild & Wertheimstein (1913) noted that *P. brassicae* was "generally abundant" in the Cséhtelek region in the centre of the country. Very few specimens of *P. brassicae* were seen at Budapest by Cooke (1924a).

Iberian Peninsula

P. brassicae has been recorded from the entire Iberian peninsula (Agenjo, 1947; Gómez-Bustillo, 1974, see distribution map).

Portugal

Two early accounts mention that *P. brassicae* could be found in and around Lisbon and at Estoril (Jackson, 1937; Browne, 1924), while Zerkowitz (1946) in *Lepidoptera of Portugal* stated that *P. brassicae* is on the wing in very many places throughout the year. Steiniger (1972) recorded *P. brassicae* from several places in 1971.

Spain

There are many early accounts of *P. brassicae* from many different provinces, around Jerez (Badajoz) and Ronda (Málaga) in southern Andalusia (Lang, 1902), near Gibraltar (off Cádiz) (Sowerby, 1907), in and near the Pyrenees (Phillips, 1960) and at Vigo (Galicia) (Mathews, 1898). Gurney (1924) reported *P. brassicae* very scarce in 1922 and Cooke (1928) reported it as abundant. It has also been reported from southern and central Spain in the province of Granada (Walker, 1891; Rosa, 1923) and in south west Spain (Owen, 1951). More recently *P. brassicae* has been reported from the Cantabrian mountains (Gómez-Bustillo, 1971) and around Barcelona (Kudrna, 1973a,b), and from The Provinces of Catalonia (Manley *et al.*, 1977, cf. Manley & Allcard, 1970) and Valencia (Dacie, 1974).

Iceland & Faroes

P. brassicae has been recorded as far north as Iceland (65°N) where one specimen (dated 31 March 1956) is preserved in the Museum of Natural History at Reykjavik and was thought to have been brought there by ship; and another from northern Iceland, but its reliability has not been substantiated (Olafsson, 1978, pers. comm.). Dannreuther (1945) noted a female *P. brassicae* in the Faroes (63°N) in 1944, the first seen there for three years, possibly imported on fresh cabbage.

Italy

Early accounts of *P. brassicae* from Italy are those for Spezia (de la Garde, 1892), on the island of Capri (Seymour-Brown, 1903) and Venice (Gurney,

1913; Hicks, 1914) when *P. brassicae* was regarded as a pest on cabbage.

More recent information from Italy confirms that exact locality citations for Italy are scarce, and that *P. brassicae* is widespread, but not very common, occurring from sea level up to about 2200 m in both the Alps and the Appenines (Balletto, 1978, pers. comm.). Balletto *et al.* (1978) studied the lepidopterous fauna of 108 biotopes along the Appenines and seven localities of the neighbouring Ligurian Alps and found that *P. brassicae* occurred in nearly all of these biotopes, including all the seven vegetational types encountered; *Mesobrometi, Xerobrometi, Nardus stricta* Linnaeus and *Formazioni nitrofile* associations, *Festuca violacea* (Clairville) grasslands, *Macereti, Sesleria appenina* Ujhelyi -*Carex kitaibeliana* Degener *ex* Becherer grasslands and beech and juniper. Worms (1974) recorded *P. brassicae* from Monte Argentario in the Tuscany mountains.

Mediterranean Islands

P. brassicae has been recorded often from many of the larger islands of the Mediterranean; however, information on its presence on the smaller islands is more difficult to locate in the literature. Only the larger Mediterranean islands are dealt with here, the smaller islands are dealt with according to the country to which they belong.

P. brassicae is recorded from all the Balearic islands by Gómez-Bustillo (1974 cf. Rebel, 1926, 1934) and was found recently on Mallorca (Perceval, 1974).

P. brassicae has also been recorded from Corsica (Gurney, 1914; Cooke, 1929; Panchen & Panchen, 1973) where it had been seen since 1900 (Bretherton *et al.*, 1962). However, on Sardinia *P. brassicae* is not included in a list of lepidoptera of the island by Hartig & Amsel (1951) (only *Pieris rapae*) but is included by Bytinski-Salz (1934) as an aberration, *P. b. verna*, previously not recorded from the island.

On Sicily, *P. brassicae* has been noted several times (Fountaine, 1897; Barrett, 1914; Bretherton, 1956; Valletta, 1976; Bigger, 1977), and also on Malta (de la Garde, 1892; Mathew, 1898; Fletcher, 1904; Borg 1932) where it is common all the year, is reinforced by immigrants and where its numbers decrease in the summer due to lack of vegetation (Valletta, 1966). Röber (1907) noted the subspecies *P. b. wollastoni* from Malta.

The Netherlands

An early record exists for Alkmaar by Caland (1901). No distribution map for *P. brassicae* exists for the Netherlands but the insect has been recorded as an occasional pest by Van Driest (1974, pers. comm.). de Jong (1978, pers. comm.) stated that *P. brassicae* was "quite common" and Lempke (1978, pers. comm.) stated that it appears erratically in the

Netherlands, that reinforcement is of little importance and that there are no large swarms as once was the case.

Poland

Apart from Chlodny's (1967) work on the energetics of *P. brassicae* at Poznan, there is little information from Poland. However, Kagan (1977) stated that *P. brassicae* was a pest on brassicas in that country, and both Przybylski (1968) and Starega (1976, pers. comm.) found *P. brassicae* a pest in Poland. Mikolajozyk (1978, pers. comm.) stated that *P. brassicae* was widespread in Poland.

Rumania

The map published by Niculescu (1963) shows that *P. brassicae* is generally distributed throughout most of the country, but is slightly less widespread in the west. In Rumania *P. brassicae* is certainly less common than either *P. napi* or *P. rapae*. Mustaţa & Andriescu (1973) mentioned four periods between 1948 and 1972 when *P. brassicae* was particularly abundant.

Switzerland

An early account mentioned *P. brassicae* on migration in Switzerland (Pictet, 1918). Most recent information available confirmed that *P. brassicae* is widespread throughout the country but is scarcer than it was twenty years ago (Sauter, 1978, pers. comm.).

Turkey

The three pierids *P. brassicae*, *P. rapae* and *P. napi* are found generally throughout the country but *P. brassicae* is the predominant species in western Anatolia including the Marmara region (Lodos, 1978, pers. comm.).

Yugoslavia

Pieris brassicae has been encountered in the province of Montenegro (Crna Gora) (Gibbs, 1913; Bretherton, 1973) and a migration of *P. brassicae* was described by Groves (1956) in this country.

Middle East & North Africa

The southern limit of *P. brassicae* extends into four north African countries according to Kurentzov (1929) and C.I.E. (1976): Algeria, Libya, Morocco and Tunisia. In Libya *P. brassicae* is found rarely south of 32°N latitude (i.e. a typical Palaearctic range).

However, earlier literature indicated that *P. brassicae* may possibly have had a much wider distribution in north Africa, as it was stated by Zeller (1847) that it was found in Egypt and the Barbary States. Westwood (1854) also recorded *P. brassicae* from Egypt and the Barbary States but he may have based his information on that of Zeller. In any case, the Barbary States as they were known then (Sudan today) extend southwards from Egypt to the west of the Red Sea. If *P. brassicae* had been distributed in Egypt and Sudan, this would have given it an almost complete range around the Mediterranean.

Pieris brassicae has been recorded many times from Algeria (Fountaine, 1906; Gibbs, 1911; Longstaff, 1912; George, 1927, 1928) and Morocco (Lagnel, 1966; Kudrna, 1972). Meade-Waldo (1901, 1902, 1905) did much of his collecting around Tangier, but noted that *P. brassicae* was absent from the central plains in summer. He thought that *P. brassicae* resembled *P. b. wollastoni* Butler in Tangier but it did not exhibit much seasonal differences. *P. brassicae* is replaced by the subspecies *P. b. cheiranthi* (Hübner) on the Canary Islands (Rebel, 1917).

Cyprus

Two subspecies of *P. brassicae* are recorded from Cyprus, *P. b. cypria* (Verity, 1905–11), and *P. b. catoleuca* (Rebel, 1939; Bretherton, 1954). Several localities are quoted by these workers for *catoleuca*. Other records for Cyprus are Morris (1928a,b) who found *P. brassicae* a pest on cauliflowers, and Bretherton (1974) who revisited the same area, Kyrenia, where *P. brassicae* was again flying.

Egypt

Larsen (1975) noted that occasional specimens of *P. brassicae* make their way to Egypt "where it is not a permanent resident".

Iran

Pieris brassicae was recorded as a pest in the north of the country in 1927–1928 (Moritz, 1928) and noted at Sersang in Kurdistan by Higgins (1958).

Iraq

In 1940, Wiltshire noted that *P. brassicae* was confined to the mountains and did not occur on the plains and he confirmed this point in his book on the lepidoptera of Iraq (Wiltshire, 1957). He regarded *P. brassicae* in Iraq as "Euroriental" probably of Anatolian-Iranian origin. Talhock (1969) also confirmed that *P. brassicae* was confined to the mountains and said that it was excluded from Arabia because of the high summer temperatures. It was recorded from Mesopotamia, which today is both in Syria and Iraq, by Stoneham (1920).

Israel

The southern limit of *P. brassicae* in the Middle East is thought to occur in Israel (Messenger, 1959). *P. brassicae* has been recorded in the north of the country as a migrant (Bytinski-Salz, 1966), the Holy Land (Barraud, 1920) and around Jerusalem (Buxton, 1923; cf. Bodenheimer, 1935). *Pieris* sp. were seen at Lake Tiberius (Lane, 1961; cf. Gardiner, 1978).

Jordan

The subspecies *P. b. verna* was thought to occur in all months of the year except June, August and September in "Transjordan" (Hemming, 1934).

Lebanon

Pieris brassicae is considered to be present throughout Lebanon (Larsen, 1974).

Syria

Pieris brassicae was not recorded in a list of Syrian insects published by Hariri (1971).

Indian subcontinent

Moore (1904) cited a page of references to various localities of *P. brassicae* in India, including Afghanistan (cf. Ahmad, 1940; Naumann, 1974), Kashmir and Sikkim (Williams, 1958), while Maxwell-Lefroy (1909) stated that *P. brassicae* was found within 16 km of the Himalayas in eastern Bengal (=Bangladesh), Bihar, the United Provinces and eastern Punjab. He mentioned that it was "extremely" abundant during cold weather, and that its sporadic appearance was due to high parasitism and its irregular migrations from the hills to the plains. Chuang-Lung (1962) recorded *P. b. nepalensis* in the southern Himalayas and Tibet where large breeding grounds occur. Other localities mentioned are western Karakorum in northwest India (Evans, 1927), Punjab, Lyallpur and in the plains (Husain, 1924; Rataul, 1959), while more recently *P. brassicae* has been found at Katrain in the Kulu Valley (Chandra & Lal, 1977).

USSR

The world distribution map produced by the C.I.E. (1976) shows the distribution of *P. brassicae* in USSR as west of a line drawn from Finland passing along the Volga river, through the northern half of the Caspian Sea and well south of the Aral Sea. However, *P. brassicae* has been reported as a pest in the Chu Valley, in Kirgizia (Moiseeva *et al.*, 1975) and the distribution line should therefore be moved north about 240 km to the east of the Aral Sea. Nabokoff (1920) noted that *P. brassicae* was absent in the Crimea, and Klein (1932) gave six localities for *P. brassicae* in the USSR – Kasan, Leningrad, Lugansk, Astrachan, Taschkent and Baku.

Both Shapiro (1976) and Effremova (1976) stated that *P. brassicae* is found in central Asia and Shapiro also quoted the Far East and Siberia. In 1963 *P. brassicae* was reported on cabbage in the Novosibirsk region of western Siberia, an area from which it had not formerly been reported (Golubkin, 1964). Kadamshoev (1971) also stated that *P. brassicae* occurred in western Pamir, a high altitude area in central Asia, adjoining Afghanistan.

There are several localities mentioned, between 40–60°N, where *P. brassicae* has been or is still a pest (Standel, 1958; Sedykh, 1974; (10 localities); Sedykh & Sedykh, 1959; Maslennikova, 1958, 1959; Maslennikova & Mustafayeva, 1971; Voltukhov, 1975; Mepzheevskya *et al.*, 1976).

Far East

China

The C.I.E. (1976) world distribution map indicates that *P. brassicae* is present in China and Burma (Table 2.1). It is intriguing to note that Carl Linnaeus (1761) mentioned in his own handwriting in the margin of one of his books that *P. brassicae* was also found in China, but he gave no specific region, nor source of his information. Chuan Li (1958) also listed *P. brassicae* in a book on Chinese butterflies, but did not make any comment about its distribution.

It is very likely that the north-eastern range of *P. brassicae* is greater than is indicated by either world maps and that it extends further into China and the eastern territories of the USSR. In a book on the butterflies of the Far Eastern USSR, Kurentzov (1970) did not mention *P. brassicae*; the only pierids present being *P. canidia* Sparrman, *P. rapae*, *P. napi* and *P. melete* Ménétriés.

Westwood (1854) stated that *P. brassicae* occurred in Siberia, Japan and Nepal; while on the contrary, Smith (1891) stated that it did not occur in China, Korea or Japan.

Java, Korea, Japan

Heath (1978, pers. comm.) confirmed the absence of *P. brassicae* from Korea, and Kudrna (1974b) did not record *P. brassicae* in his annotated list of the butterflies of Japan. It is worth pointing out that there are three specimens of what looks similar to *P. brassicae*, caught in Japan, in the Wageningen Agricultural Museum (personal observation). *P. brassicae* is however, not present in a collection of butterflies from Java, in the same museum.

Australia & New Zealand

Pieris brassicae is absent from Australia and New Zealand (Bartholemew, 1911), its presence in New Zealand reported by Ford (1976) was presumably in error for *P. rapae*.

South America: Chile

The earliest account of *P. brassicae* in Chile is February 1971 when it was reported from Valparaiso (Campos, 1978, pers. comm.). Larvae were noted in late summer 1971, feeding on crucifers in the grounds of the Botanic garden in Viña del Mar, near Valparaiso (González, 1972). By early spring 1972 imagines were seen flying in adjacent coastal regions near Valparaiso and in the same year living specimens were sent to Gardiner in Cambridge from Santiago (about 70 km south of Valparaiso) where they were flourishing in the gardens of that city (also confirmed in Ripa, 1978). In June 1973 *P. brassicae* was reported destroying crucifers in other provinces. The specimens were essentially the same as those from Europe except for a dark green colour on the undersides of the hind-wing which may have been a local adaptation (Gardiner, 1974).

The Large White is thought to have been introduced to Chile accidentally as diapausing pupae in the holds of ships (González, *loc. cit.*), which may have been a result of trade with eastern Europe (Gardiner, *loc. cit.*). However, it is worth pointing out that Anonymous (1920), writing in a Uruguayan Ministry of Agriculture publication described methods for eliminating the Large White (named as *P. brassicae* in the text, together with a photograph of *P. brassicae*). It appears on investigation that native Uruguayan species were often ascribed names of similar looking European species during this period (Feltwell, 1978b).

References cited

Aagaard, K. & Gulbrandsen, J., 1976. *Study of Norwegian butterflies* (in Norwegian). University of Trondheim.

Adkin, B.W., 1911. *Colias edusa* and *Sphinx convolvuli* in Sicily. *Entomologist* 44: 324.

Agenjo, R., 1947. *La Catálogo ordenador de los Lepidópteros de España*. Madrid. Graellsia.

Ahmad, T., 1940. A survey of the insect fauna of Afghanistan. *Indian J. Ent.* 2: 159–176.

Aldin, A.D., 1931. Euzet-les-Bains. *Revue fr. Lep.* 5: 298–304.

Andrews, H.L., 1930. Lepidopterological and other notes from Dorset. *Entomologist* 63: 66–69.

Anonymous, 1920. La Oruga de los Coles. *Pieris brassicae*. Uruguay: Minist. Indust., Defensa Agricola, Bol. Mens., Montevideo, i, no. 9, Sept. 1920, 223–224.

Ansorge, E., 1969. *Macrolepidoptera of Buckinghamshire. Bucks. arch. Soc.*

Askew, R.R., 1974. Insects from Bardsey Island. *Ent. Gaz.* 25: 45–51.

Balletto, E., Tusco, G., Barberis, G. & Russaro, B., 1978. Aspetti dell' ecologia dei Lepidotteri Ropaloceri nei consorzi erbacei alto appenninici. *Animalia (Catania)* 4 (3): 277–343.

Barraud, P.J., 1918. Notes on lepidoptera observed in Macedonia 1916, 1917. *Entomologist* 51: 59–63.

—— 1920. Entomology in the holy land. *Entomologist* 53: 225–227.

Barrett, J.P., 1914. Butterfly collecting in Sicily and Calabria in 1912–3. *Entomologist* 47: 152–156.

Bartholemew, 1911. *Physical Atlas.* V. Zoogeography.

Battersby, F.J., 1905. Abundance of *Pieris brassicae* in west Meath. *Entomologist* 38: 238.

Baynes, E.S.A., 1909. Lepidoptera from the Isle of Anglesey. *Entomologist* 42: 285–286.

—— 1957. The lepidoptera of Tory Island, Co. Donegal. *Entomologist* 90: 310–313.

—— 1964. *A Revised Catalogue of Irish Lepidoptera.* Hampton, Middlesex.

—— 1969. Report on migrant insects in Ireland for 1968. *Ir. Nat. J.* 16: 145–147.

—— 1971. Report on migrant insects in Ireland for 1970. *Ir. Nat. J.* 17 (2): 32–33.

—— 1973. *A Revised Catalogue of the Irish Macrolepidoptera and Supplement.* 1970. E.W. Classey, London.

Beavis, I.C., 1976. Butterflies on the Isles of Scilly during August 1975. *Entomologist's Rec. J. Var.* 88: 194–195.

Beirne, B.P., 1945. The Lepidoptera of Shetland. *Entomologist's Rec. J. Var.* 57: 37–40.

Betts, E.B., 1920. Butterflies observed in the Isles of Sheppey, Kent. 1919. *Entomologist* 53: 67.

Betts, E.B.C., 1922. Rhopalocera from Rhodes, Samos, Gallipoli etc. *Entomologist* 55: 137–138.

Bigger, T.R.L., 1977. Some notes on Sicilian butterflies (Lepidoptera: Pieridae). *Ent. Gaz.* 28: 211–213.

Birkett, N.L., 1954. The present status of the butterfly population of the Kendal District. *Entomologist's mon. Mag.* 90: 293.

Blackie, J.E.H., 1949. The butterflies of the Monks Wood District. *Entomologist* 82: 54–58.

Blair, K.G., 1925. The lepidoptera of the Scilly Isles. *Entomologist* 58: 3–10.

Blunck, H., 1953. *Tierische Schädlinge an Nutzpflanzen* I. Teil; 5. Aufl; 2. Lief. Lepidopteren und Trichopteren. Berlin.

—— 1954. Beobachtungen über Wanderflüge von *Pieris brassicae* L. *Beitr. Ent.* 4: 485–528.

Bodenheimer, F.S., 1935. *Animal Life in Palestine.* L. Mayer, Jerusalem.

Bonnemaison, L., 1965. Insect pests of crucifers and their control. *A. Rev. Ent.* 10: 233–256.

Borg, P., 1932. Lepidoptera of Maltese Islands. Government Printing Office, Malta.

Bretherton, R.F., 1944. April Butterflies. *Entomologist* 77: 84.

—— 1949. Spring butterflies in Bohemia. *Entomologist* 82: 254–255.

—— 1954. A week's butterfly collecting in Cyprus. *Entomologist* 87: 207–211.

—— 1956. Sampling butterflies in Sicily. *Entomologist's mon. Mag.* 77: 168–176.

—— 1962. April butterflies in Provence 1962. *Entomologist's Rec. J. Var.* 74: 144–147.

—— 1968. More about Greek butterflies, June 1968. *Entomologist's Rec. J. Var.* 80: 273–281.

—— 1973. Montenegro, July 1972. *Entomologist's Rec. J. Var.* 85: 1–12.

—— 1974. Butterflies in Cyprus, June 1973. *Entomologist's Rec. J. Var.* 86: 1–5

—— 1975. Lepidoptera in Dordogne, South west France in May 1974. *Entomologist's Rec. J. Var.* 87: 33–37.

Bretherton, R.F., De Worms, C.G. & Johnson, G., 1962. Butterflies in Corsica 1962. *Entomologist's Rec. J. Var.* 75: 93–104.

Bretherton, R.F., Ellison, R.E., Manley, W.B.L., 1952. Lepidoptera in the east Pyrenees, 1951 (Porté Puymorens & Vernet-Les-Bains. June 30–July 19) and the Forêt de Rambouillet, near Paris (July 16–20). *Entomologist* 85: 222–229.

Bromilow, F., 1893a. *Butterflies of Riviera.* Galignani, Nice.

—— 1893b. Early spring Rhopalocera on the Riviera. *Entomologist's mon. Mag.* 29: 86–87.

Brown, R.G.B., Ashmole, N.P., Campbell, R.P., 1958. Insect migration in the Pyrenees in the autumn of 1955. *Entomologist's mon. Mag.* 94: 217–266.

Browne, C.A.R., 1924. Rhopalocera near Lisbon: November-March. *Entomologist* 57: 91–92.

Buckler, H.A., 1934. Butterflies in Leicester, 1933. *Entomologist* 67: 103.

Burras, A.E., 1923. An entomological visit to the Morvan (Central France). *Entomologist* 56: 134–139.

Burton, G.N., 1981. Les Rhopalocères de Cantal aux mois de juillet et août (in preparation).

Burton, J.F. & Owen, D.F., 1954. Insect migration in S.W. France. *Entomologist's mon. Mag.* 90: 66–69.

Bussey-Bell, J.W., 1915. Butterflies of the Oxfordshire chilterns. *Entomologist* 48: 157–159.

Butler, W.B., 1915. Butterflies of the Taunton district. *Entomologist* 48: 123–124.

Buxton, P.A., 1923. Applied entomology of Palestine, being a report to the Palestine Government. *Bull. ent. Res.* 14: 289–339.

Byles, R.S., 1942. Radlett butterflies of 1941. *Entomologist* 75: 23.

Bytinsky-Salz, H., 1934. Ein Beitrag zur Kenntnis der Lepidopteren Sardiniens. *Int. ent. Z.* 28: 89–106.

—— 1966. Observations on migrating moths. *Isr. J. Ent.* 1: 193.

Caland, M., 1901. Macrolepidoptera waargenomen in de anstreken van's Hertogenbosch en Alkmaar. *Tijdschr. Ent.* 44: 46–53.

Campbell, J.L., 1936. Immigrant lepidoptera in Outer Hebrides, 1936. *Entomologist* 69: 265–266.

—— 1945. Hebridean notes: Lepidoptera observed in May, 1945. *Entomologist* 78: 109.

—— 1946. Catalogue of a collection of Macro-lepidoptera made in the Hebrides between 1936 and the present date. *Entomologist* 74: 49–53.

—— 1949a. Migrants in the Hebrides. *Entomologist* 82: 189.

—— 1949b. Macrolepidoptera from Knapdale (Argyllshire). *Entomologist* 82: 234–235.

—— 1951. An experiment in marking migratory butterflies. *Entomologist* 84: 1–6.

—— 1972. Isle of Canna Notes for 1970 and 1971. *Entomologist's Rec. J. Var.* 84: 196–198.

—— 1970. Macrolepidoptera Cannae, Butterflies and moths of the Isle of Canna, Inner Hebrides. *Entomologist's Rec. J. Var.* 82: 1–27.

—— 1975. Isle of Canna report, 1972–1974. *Entomologist's Rec. J. Var.* 87: 10–12.

—— 1978. *Pieris rapae* L. on the Isle of Canna (note). *Entomologist's Rec. J. Var.* 90: 279.

Carr, J.W., 1916. *The Invertebrate Fauna of Nottinghamshire*, J. & H. Bell, Nottingham 138–258.

Chalmers-Hunt, J.M., 1962. The Butterflies and moths of Kent. I. Rhopalocera. *Entomologist's Rec. J. Var.* Supplements 1–144.

—— 1970. The butterflies and moths of the Isle of Man. *Proc. Trans. Br. ent. nat. Hist. Soc.* 18: 1–171.

Chandra, J. & Lal, O.P. 1977. Mating behaviour of cabbage butterfly, *Pieris brassicae* (Linnaeus). *Indian J. Ent.* 38 (2): 197–198.

Chapman, T.A., 1899. Butterflies in south and north Norway. *Entomologist's mon. Mag.* 35: 20–28.

Chazaud, P., 1977. Contribution à l'étude des Macrolépidoptères de la Creuse, la Montagne de Saint-Goussaud. *Bull. Soc. lépidopt. Fr.* 1: 145–148.

Cheesman, W., 1898. Lepidoptera captured in the Orkney Islands. *Entomologist's Rec. J. Var.* 10: 204–206.

Chlodny, J., 1967. The energetics of the development of cabbage white *Pieris brassicae* L. (Lepidoptera). *Ekol. Pol.* Ser. A. 15: 553–561.

Chuan Li, 1958. *Butterflies*. (In Chinese.)

Chuang-Lung, L., 1962. Results of the Zoologico-Botanical expedition to south-west China, 1955–1957. (Lepidoptera, Rhopalocera). *Acta ent. Sinica* 11: 172–198.

Church, H.F., 1959. Some further observations on sex ratios in butterflies. *Entomologist* 92: 33.

Clarke, H.S., 1893. Rhopalocera of the Isle of Man. *Yn. Lioar Manninagh* 2: 100–103.

Coates, D.L. 1968. Lepidoptera for the Stirling area. *Entomologist's Rec. J. Var.* 80: 7–12.

Commonwealth Institute of Entomology, 1953, 1976. Distribution maps of insect pests. Series A. 25–30. *Pieris brassicae.*

Conder, P.J., 1949. Observations on a migration of *Pieris brassicae* L. at Skokholm Island, Pembrokeshire, in August 1947. *Proc. R. ent. Soc. Lond.* Ser. A. 24: 35–38.

Constant, A., 1866. *Catalogue des Lépidoptéres de Saône et Loire.* Dejussieu, Autun.

Cooke, B.H., 1924a. Butterflies of Austria and Hungary. *Entomologist* 57: 74–78.

—— 1924b. Lepidoptera on the coast of Brittany. *Entomologist* 57: 203–205.

—— 1925. A summer in the Hautes Pyrénées. *Entomologist* 58: 87–96.

—— 1926. Rhopalocera in Provence and the Basse Alpes, 1925. *Entomologist* 59: 210–219.

—— 1927. Rhopalocera in the Dolomites and South Tyrol, 1926. *Entomologist* 60: 127–132.

—— 1928. An entomological motor tour in Spain in 1927. *Entomologist* 61: 176–182.

—— 1929. Spring collecting in Corsica (Lepidoptera). *Entomologist* 62: 79–83.

Cooper, C.F., 1894. Scarcity of *Pieris brassicae* and *Pieris rapae*. *Entomologist* 27: 348.

Cornell, E., 1919. Some notes on the butterflies of the south coast of the Isle of Wight, 1919. *Entomologist* 52.

Coutsis, J.G., 1969. List of Grecian butterflies. *Entomologist* 102: 264–268.

—— 1976. Spring butterflies on the island of Skyros, Greece. *Entomologist's Rec. J. Var.* 88: 33–37.

—— 1978. Spring butterflies on the Greek Island of Siknow. *Entomologist's Rec. J. Var.* 90: 300–301.

Cowin, W.S., 1938. Our Manx butterflies. *J. Manx Mus.* 4: 34.

Crichton, M. & Maloney, E., 1975. *Provisional Atlas of Butterflies in Ireland.* St. Martin's House, Dublin.

Crowther, G.F., 1932. Relative abundance of lepidoptera. *Entomologist* 65: 246.

Dacie, J. & M., 1974. Albarracin and vicinity, Spain, in July 1973. *Entomologist's Rec. J. Var.* 86: 208–213.

Dacie, J.V., Dacie, M.K.V. & Grammaticos, M.D., 1972. Butterflies in northern and central Greece. *Entomologist's Rec. J. Var.* 84: 257–266.

Dannreuther, T., 1945. Migration records, 1944. *Entomologist* 78: 49–56.

Day, F.H., 1921. Butterfly notes from Cumberland. *Entomologist* 54: 99–100.

De La Garde, P., 1892. Mediterranean Lepidoptera. *Med. Nat.* February, 123–134.

Donovan, C., 1901. A list of the lepidoptera of county Cork. *Entomologist* 34: 333–336.

—— 1936. *A Catalogue of the Macrolepidoptera of Ireland.* Burrow & Co. Ltd., Cheltenham & London.

Dubois, C.F. & Dubois, A., 1874. *Les Lépidoptères de la Belgique.* Libraire Muquardt, Bruxelles, Leipzig, Gand.

Dufray, C., 1968. *Faune terrestre d'eau douce des Pyrénées-Orientales I. Macrolépidoptères.* Univ. Paris.

Effremova, T.G., 1976. Pests of cabbage. (In Russian.) *Zashch. Rast.* (8): 58–59.

Ellis, J.W., 1940. *The Lepidopterous Fauna of Lancashire and Cheshire.* Liverpool.

Ellis, G., 1948. Early butterflies at Llandudno. *Entomologist* 81: 180.

Evans, W.H., 1927. Lepidoptera: Rhopalocera obtained by Mme L. Visser-Hooft of the Hague (Holland) during an expedition of previously unknown country in the Western Karakorum, N.W. India. *Tijdschr. Ent.* 70: 158–162.

Evans, L.K., 1973. An entomologist in Radnorshire, 1971. *Entomologist's Rec. J. Var.* 85: 33–40.

Evans, L.K. & Evans, K.G.W., 1973. A survey of the Macro-Lepidoptera of Croydon and N.E. Surrey. *Proc. Croydon nat. Hist. Sci. Soc.* 14: 273–408.

Fearnclough, T.D., 1972. The butterflies of the Isle of Wight. *Entomologist's Rec. J. Var.* 84: 102–109.

Feltwell, J.S.E., 1977a. Check list of the Rhopalocera of the Parc National des Cévennes. *Ent. Gaz.* 28: 85–97.

—— 1977b. Migration of whites. *Entomologist's mon. Mag.* 112: 88.

—— 1977c. Skeletonisation by *Pieris brassicae. Entomologist's mon. Mag.* 112: 104.

—— 1977d. Migration of *Hipparchia semele.* L. *J. Res. Lepid.* 15: 83–91.

—— 1978a. Butterflies of the cols of the Cévennes. *Entomologist's Rec. J. Var.* 90: 33–36.

—— 1978b. *Pieris brassicae* in South America. *Entomologist's Rec. J. Var.* 90: 330.

Finke, K., 1938. *Die Grosschmetterlinge Südhanovers.* Häntzschel, Göttingen.

Fletcher, T.B., 1904. A preliminary list of the lepidoptera of Malta. *Entomologist* 37: 273–276.

Ford, H.D., 1924. Notes from the log-book of a Cumberland garden. *Entomologist* 57: 34–37.

—— 1927. Notes on Cumberland Lepidoptera. *Entomologist* 60: 17–18.

Ford, R.L.E. 1976. The influence of the Microgasterini on the populations of British Rhopalocera (Hymenoptera: Braconidae). *Ent. Gaz.* 27: 205–210.

Fountaine, M.E., 1897. Notes on the butterflies of Sicily. *Entomologist* 30: 4–11.

—— 1906. Algerian butterflies in the spring and summer of 1904. *Entomologist* 39: 84–89.

Frohawk, F.W., 1934. *British Butterflies*. Ward, Lock & Co. Ltd., London & Melbourne.

Gardiner, B.O.C., 1963. The Butterflies of Cambridgeshire. *Nature, Cambs.* 6: 31–36.

—— 1974. *Pieris brassicae* L. established in Chile: another Palaearctic pest crosses the Atlantic (Pieridae). *J. Lepid. Soc.* 28: 269–277.

—— 1978. Instar number and pupal colouration in Palestinian *Pieris brassicae*. *Proc. & Trans. Brit. ent. nat. Hist. Soc.* 11: 21–23.

George, L., 1927. Observations sur la biologie de deux Hyménoptères entomophages. *Bull. Soc. Hist. nat. afr. N.* 18: 55–71.

—— 1928. Sur la biologie de l'*Apanteles glomeratus* L. *Bull. Soc. Hist. nat. Afr. N. Algiers* 19: 104–112.

Gepp, J. & Stark, W., 1974. Ökologie und Zoogeographie der bekannten Pieridae der Griechland. *Ber. Aro. Okol. entomol. Graz.* 4: 1–5.

Gibbs, A.E., 1909. Five weeks in the Vosges. *Entomologist* 42: 153–156.

—— 1911. An Algerian holiday. *Entomologist* 44: 170–174.

—— 1913. Butterfly hunting in the Balkans. *Entomologist* 46: 122–130.

Glaeser, G., 1974. The occurrence of significant harmful factors in plant crops in Austria in 1973. *Pflanzenschutzberichte* 44: 113–126.

Goater, B., 1974. *The Butterflies and Moths of Hampshire and the Isle of Wight*. E.W. Classey, Berks.

Golubkin, V.G., 1964. The cabbage butterfly in the Novosibirsk region (in Russian). *Zashch. Rast.* 1964: 56.

Gómez-Bustillo, G., 1971. Por un mejor conocimiento de los ropalóceros españoles. IV. Los Picos de Europa, centro de la Cordillera Cantabrica. *Soc. Cienc. nat. Arazadi. R.S.V.A.P.* No. 19. 33–39.

Gómez-Bustillo, M.R.G. & Rubio, F.F., 1974. *Mariposas de la Península Ibérica. Ropalóceros.* 2 volumes. Ministerio de Agricultura, Instituto naacional Para la Conservación de la Naturaleza. Estación central de Ecología.

Gonzalez, R.H., 1972. Chile. Oriental fruit moth on peach. Large cabbage butterfly. *FAO Plant Prot. Bull.* 20: 89–91.

Graham, J.A., 1919. Rhopalocera of the Doullens District, Somme, France, May– August, 1918. *Entomologist* 52: 131–134.

Granville-Clutterbuck, C.G., 1940. Lepidoptera in the Scilly Isles in August 1939. *Entomologist* 73: 129–131.

Gregson, C.S., 1885. Notes of lepidoptera taken by E.R. Curzon, Esq., at the Island of Hoy, one of the Orkney Islands, during the summer of 1885. *Young Nat.* 6: 272–278.

Groves, E.W., 1956. A migration of *Pieris brassicae* L. in Yugoslavia. *Entomologist's Rec. J. Var.* 68: 274.

Gurney, G.H., 1913. Butterflies near Venice. *Entomologist* 46: 232–234.

—— 1914. An account of an entomological trip to Corsica. *Entomologist* 47: 147–151.

—— 1924. An entomological journey to Spain in 1922. *Entomologist* 57: 124–129.

—— 1927. Three week's butterfly collecting in the Tyrol. *Entomologist* 60: 3–6.

Haes, E.C.M., 1977. *Natural History of Sussex*. Flare Books: The Harvester Press.

Haig-Thomas, P., 1938. Arctic butterflies, and especially those of Maasselven, Lapland, Lat. 69°N. *Entomologist* 71: 1–5.

Hanson, S.M., 1941. Notes on the lepidoptera of Provence, S. France, with special regard to dates of first appearance. *Entomologist* 79: 275–279.

Hariri, G., 1971. A list of recorded insect fauna of Syria. Part 2. *Fac. Agric. Univ. Aleppo.*

Harper, G.W., 1968. The macrolepidoptera of Invernesshire. *Entomologist's Rec. J. Var.* 80: 36–40.

Hartig, F. & Amsel, G., 1951. Lepidoptera Sardinica. *Fragm. Ent. Rom.* 1: 1–152.

Haywood, H.C., 1926. *The Butterflies of Derbyshire.* Derby Entomological Society.

Heath, J., 1970. *Provisional Atlas of Insects of the British Isles.* Monks Wood, Huntingdonshire.

—— 1973. In *South's British Butterflies.* Warne, London. Ed. by T.C. Howarth.

—— 1976. Bibliography. In *The Moths and Butterflies of Great Britain and Ireland.* Ed. by J. Heath. Blackwell Scientific Publications. Oxford. pp. 135–144.

Hedges, A.V., 1947. List of Manx Lepidoptera. *Entomologist* 80: 40–46, 62–66, 89–94.

Hemming, F., 1934. Notes on two collections of butterflies made in Palestine, with a note on the occurrence in Transjordan of an unrecorded species. *Entomologist* 67: 135–138.

Heslop-Harrison, J.W., 1937. Rhopalocera on the island of Scalpay, with an account of the occurrence of *Nyphalis io* on Raasay. *Entomologist* 70: 1–4.

—— 1938. The fate of pupae resulting from the Outer Hebridean *Pieris brassicae* immigration of 1937. *Entomologist* 72: 93.

—— 1939. Immigrant lepidoptera in the Inner and Outer Hebrides during 1939. *Entomologist's mon. Mag.* 75: 252.

—— 1940. Lepidoptera on the Isle of Handa. *Entomologist* 73: 44–45.

—— 1941. More Hebridean days III. The Isles of Tiree and Gunna. *Entomologist* 74: 97–100.

—— 1942. The irregularity of the broods of *Pieris brassicae* L. on the Isles of Lewis, Harris and Scarp in 1941. *Entomologist* 75: 32.

—— 1945. Bedstraw moth (*Celerio galii* Rott.) in Isle of Rhum. *Entomologist's mon. Mag.* 81: 12.

—— 1946. The abundance of *Pieris brassicae* larvae on Rhum in 1945. *Entomologist* 74: 70.

—— 1947. Early spring insects in the Isle of Rhum, with some remarks on the woodland fauna of the island. *Entomologist* 80: 1–4.

—— 1949a. More valuable records. *Entomologist's Rec. J. Var.* 61: 112–113.

—— 1949b. A contribution to our knowledge of the lepidoptera of the isles of Lewis and Harris. *Entomologist* 82: 16–19.

—— 1949c. Rhopalocera in the scottish western isles in 1948, with an account of two new forms of *Pararge aegeria* (Lep. Satyridae). *Entomologist's mon. Mag.* 85: 25–28.

—— 1949d. Lepidoptera in the Outer Hebrides in 1950. *Entomologist* 83: 241–245.

—— 1950a. Observations on the Ranges, Habitats and variations of the Rhopalocera of the Outer Hebrides. *Entomologist's mon. Mag.* 86: 65–70.

—— 1950b. *Pieris brassicae* eaten by a snail. *Entomologist's mon. Mag.* 86: 78.

—— 1952. Lepidoptera in the Isles of Raasay, Rhum (v-c. 104) Lewis, and Harris (v-c. 110) in 1951. *Entomologist* 85: 6–13.

—— 1953. Lepidoptera in the Isles of Lewis and Harris in 1952. *Entomologist* 86: 53–55.

—— 1954. Lepidoptera from the Outer Hebrides in 1953. *Entomologist* 87: 83–86.

—— 1955. Lepidoptera noted in the Outer Hebrides in 1954. *Entomologist* 88: 51–53.

—— 1956. Immigrant lepidoptera in the Outer Hebrides in 1955. *Entomologist* 89: 87.

—— 1958. Lepidoptera in the Scottish Western Isles. *Entomologist* 91: 79–85.

Hicks, J.B., 1914. Butterflies of Venice and neighbourhood. *Entomologist* 47: 206–207.

Higgins, L.G., 1958. Butterflies of Kurdistan. *Entomologist* 91: 38–45.

Higgins, L.R. & Riley, N.D., 1970. *Field Guide to the Butterflies of Britain and Europe.* Collins, London.

Hitier, H., 1918. La Piéride du chou. *J. Agri. Prat. Paris* 31: 37.

Hodgson, S.B., 1933. Rhopalocera of the Mullet Peninsula, Co. Mayo. *Entomologist* 66: 134–135.

—— 1957. Early appearance of some spring insects, 1957. *Entomologist* 90: 189.

Howard, G., 1975. Lepidoptera on Hoy, Orkney. *Entomologist's Rec. J. Var.* 87: 107–109.

—— 1978. Macrolepidoptera of Glengarry and District (West Invernesshire) 1977–78. *Entomologist's Rec. J. Var.* 90: 255–261.

Huggins, H.C., 1953. The lepidoptera of Glengarriff, Co. Cork. *Entomologist* 86: 12–17.

Husain, M.A., 1924. Annual report of entomologist to Government of Punjab-Lyallpur, for the year ending 1924. *Rept. Dept. Agri. Punjab.* 1923 1 (2): 55–90.

Jackson, J., 1937. Butterfly notes from Portugal. *Entomologist* 70: 258–259.

Johnson, T., 1908. *The Butterflies of Essex.* Privately published.

Johnson, W.F., 1927. Lepidoptera in Co. Down, Ireland, 1926. *Entomologist* 60: 114–115.

Kadamshoev, M., 1971. Biology of the cabbage butterfly *Pieris brassicae* L. in western Pamir. *Izv. akad. nauk. Tadk. biol. nauk.* 89–91.

Kagan, K., 1977. Characteristics of the development, intensity of appearance and injuriousness of some main pests of vegetable plants in Poland in 1974 (in Polish). *Biul. Inst. ochr. Roslin.* 60: 287–323.

Kettlewell, H.B.D., 1955. Autumn observations of lepidoptera in north Berkshire, 1954. *Entomologist's mon. Mag.* 91: 85.

Kevan, D.K.M., 1941. The insect fauna of the Isle of Eigg. *Entomologist* 74: 247–254.

Klein, H.Z., 1932. Studien zur Oekologie und Epidemiologie der Kohlweisslinge – I Der Einfluss der Temperatur und Luftfeuchtigkeit auf Entwicklung und Mortalität von *Pieris brassicae* L. *Z. angew. Ent.* 19: 395–448.

Kone, W.J. de V., 1901. *A Catalogue of the Lepidoptera of Ireland.* London.

Koutsaftikis, A., 1974a. Die Lepidopterenfauna der Ostägäischen Insel Simi (Griechenland). *Ann. Mus. Goulandris* 2: 93–98.

—— 1974b. Ökologisch-zoogeographische Übersichtstabelle der Tagfalter (Rhopalocera: Lepidoptera) des Griechischen Festlandes. *Ann. Mus. Goulandris* 2: 99–103.

—— 1974c. *Sistimatiki ecologiki ereyna ton Ropalokeron (Lepidoptera) tis Ipirotikis Ellados.* 155 pp. Athens.

—— 1975. Über die Lepidopterenfauna der Insel Kastelloriso. *Verh. des Sechsten Int. Symp. über Entomofaunistik in Mitteleuropa.* 1975. Junk, The Hague. 313–315.

Kudrna, O., 1959. Fauna Rhopalocer okolí Netolic. S. *Sbornik krajskeho vlativedneko Mus. C. Budejovice* 2: 133–138.

—— 1970. Butterflies of S. Bohemia. *Entomologist's Rec. J. Var.* 82: 323–330.

—— 1971. Butterflies of S. Bohemia. *Entomologist's Rec. J. Var.* 83: 53–67.

—— 1972. On some Moroccan butterflies. *Entomologist's Rec. J. Var.* 84: 267–268.

—— 1973a. On taxonomy and distribution of some Spanish Rhopalocera. *Ent. Gaz.* 25: 15–28.

—— 1973b. Butterflies collected in Catalonia in June 1971. *Entomologist's Rec. J. Var.* 85: 81–84.

—— 1974a. Distribution list of butterflies of Czechoslovakia. *Ent. Gaz.* 25: 161–177.

—— 1974b. An annotated list of Japanese butterflies. *Atalanta B.* 5: 92–120.

Kurentzov, N.J., 1929. *Fauna de l'URSS. I. Lepidoptera.* Leningrad.

Kurentzov, A.I., 1970. *Butterflies of the Far East USSR.* Academy of Sciences of USSR, Siberian Branch.

Kusdas, K. & Reichl, E.R., 1973. Die Schmetterlinge von Ober Österreich. *Auftrag. ent. arb. Landesmus. Linz.*

Lack, D. & Lack, E., 1951. Migration of insects and birds through a Pyrenean Pass. *J. Anim. Ecol.* 20: 63–67.

Lagnel, M., 1966. Note sur l'armure genitale mâle et femelle du sous-genre *Pieris* Schrank. *Bull. Soc. ent.* 71: 91–94.

Lane, C., 1955. Insect migration on the north coast of France. *Entomologist's mon. Mag.* 301–306.

—— 1957. Insect migration at Deauville, France. *Entomologist* 90: 305.

—— 1961. Notes on migrant lepidoptera in Israel. *Entomologist* 94: 286.

Lang, H.C., 1902. Butterfly collecting in southern Andalusia in the spring of 1902. *Entomologist* 35: 228–231.

Larsen, T.B., 1974. *The Butterflies of the Lebanon.* C.N.R.S., Lebanon.

—— 1975. Provisional notes on migrants in Lebanon. *Atalanta* 6: 62–74.

Le Cerf, F., 1972. *Atlas des Lépidoptères de France. Rhopalocères.* Boubée, Paris.

Leclercq, J., 1970. *Atlas provisoire des Insectes de Belgique.* 8 parts. Cartographie des invertèbres européens.

Leeds, F.A., 1938. Notes and observations. *Entomologist* 71: 140.

Le Quesne, J., 1946. The butterflies of Jersey. *Entomologist's mon. Mag.* 82: 22–23.

Linnaeus, C., 1761. *Fauna Svecica, Stockholmiae.* Edition 2. 259–270.

Lipscomb, C.G., 1977. Corfu butterflies in spring 1977. *Entomologist's Rec. J. Var.* 89: 326–328.

Lisney, A.A., 1940. New records for Irish lepidoptera. *Entomologist* 73: 123–129.

Long, A.G., 1956. The macrolepidoptera of Berwickshire. Part 1. *Hist. Berwickshire nat. Club.* 34: 128–151.

Long, R., 1970. Rhopalocera (Lepidoptera) of the Channel Islands. *Ent. Gaz.* 21: 241–251.

Longstaff, G.B., 1912. *Butterfly Hunting in many lands.* Longmans, Green & Co.

Lorimer, R.I., 1970. Orkney lepidoptera 1868–1968. *Ent. Gaz.* 21: 73–101.

Luckens, C.J., 1976. Observations on British butterflies in 1975. *Entomologist's Rec. J. Var.* 88: 145–147.

Manley, W.B.L., Allcard, H.G. & Manley, M.E., 1977. Further notes on some lepidoptera of Catalonia found during June and July of 1975 and 1976. *Ent. Gaz.* 28: 101–107.

Manley, W.B.L. & Allcard, H.G., 1970. *A Field Guide to the Butterflies and Burnets of Spain.* E.W. Classey, Hampton. pp. 192.

Manning Stanton, G., 1925. Rhopalocera of the Belgian coast. *Entomologist* 58: 252–253.

Maslennikova, V.A., 1958. The conditions determining diapause on the parasitic hymenoptera *Apanteles glomeratus* L. (Bracondidae) and *Pteromalus puparum* (Chalcididae) (in Russian). *Ent. Obozr.* 37: 538–545.

―― 1959. The relation between the seasonal cycles of geographical populations of *Apanteles glomeratus* L. and its host *Pieris brassicae* L. (In Russian.) *Ent. Obozr.* 38: 517–522.

Maslennikova, V.A. & Mustafayeva, T.M., 1971. An analysis of photoperiodic adaptations in geographic populations of *Apanteles glomeratus* L. (Hymenoptera, Braconidae) and *Pieris brassicae* L. (Lepidoptera, Pieridae). (In Russian.) *Ent. Rev.* 50: 281–284.

Mathew, G.F., 1898. Notes on lepidoptera from the Mediterranean. *Entomologist* 31: 77–84.

Maxwell-Lefroy, H., 1909. *Indian Insect Life*. Thacker, Spink & Co., Calcutta & Simla.

McClymont, J.R., 1917. Butterfly collecting in the Pyrénées Orientales. *Entomologist* 50: 89–91.

Meade-Waldo, G., 1901. Collecting lepidoptera in Tangier. *Entomologist* 34: 206–207.

―― 1902. Collecting near Tangier in August and September, 1901. *Entomologist* 35: 195–196.

Meade-Waldo, E.G.B., 1905. On a collection of butterflies made in Morocco, in 1900–1902. *Proc. S. Lond. ent. Soc.* 1905: 369–393.

McLeod, L., 1972. Mont Ventoux and the Dentelles de Montmirail. *Entomologist's Rec. J. Var.* 84: 156–163.

Mepzheevskya, O.I., Litvinova, A.N. & Molchaiova, P.V., 1976. *Society Lepidoptera*. (In Russian.) Izdatelvsto, Moscow.

Messenger, P.S., 1959. Bioclimatic studies with insects. *A. Rev. Ent.* 4: 183–206.

Moiseeva, N.V., Mashkina, L.G., Kalashinikova, G.I. & Ratomskaya, A.A., 1975. The effectiveness of the use of *Trichogramma* for control of certain cabbage pests in the Chu Valley, Kirgizia. In *Arthropods of Agricultural Importance*. Ed. by A.I. Protsenko, Ent. issled. Kirghizii Vypusk, Kirghiz. SSR.

Molesworth-Muspratt, V., 1959. Autumnal migrations of butterflies in the central Pyrenees. *Entomologist* 92: 189–196.

Moore, F., 1904. *Lepidoptera Indica*. Reeve, London.

Morgan, M.J., 1965. The lepidoptera of Caernarvonshire with special reference to the Bangor area. *Ent. Gaz.* 16: 43–80.

Moritz, L., 1928. Results of a survey of Acrididae in N. Persia 1927 & 1928. *Ashkhabad Narkomz Turkm. SSR Stazra.* 1928: 1–52.

Morris, C.E., 1919. A season's collecting in the Alpes-Maritimes. *Entomologist* 52: 34–37.

―― 1922. *Colias edusa* migration in Cannes district. *Entomologist* 55: 35–36.

Morris, H.M., 1928a. Entomological notes. *Cyprus agric. J.* 23: 32–33.

―― 1928b. Report of entomologists, 1928. *Report of the Department of Agriculture, Cyprus.* 1928: 43–44.

Morris, M.G., 1977. Butterfly studies on Porton Downs. Royal entomological Society of London workshop "Ecology and Evolution of Butterflies". December, 1977.

Murray, J., 1934. Butterflies in Dumfrieshire in 1933. *Entomologist* 67: 92–93.

Muşţata, G. & Andriescu, I., 1973. Recherches sur le complexe de parasites (Insecta) du Papillon du chou (*Pieris brassicae* L.) en Moldavie (R.S. de Roumania). I. Parasites primaires. *Ecologie Terestra Genet.* 1972–1973. 191–230.

Nabokoff, V.V., 1920. A few notes on Crimea lepidoptera. *Entomologist* 53: 29–33.

—— 1931. Notes on the lepidoptera of the Pyrenees orientales and the Ariège. *Entomologist* 64: 268–271.

Naumann, C.L., 1974. Immigration in Afghanistan 1972. *Atalanta B.* 5: 82–88.

Newman, L.H., 1931. Sark lepidoptera. *Entomologist's Rec. J. Var.* 43: 184–186.

Niculescu, E.V., 1963. *Faune Republicii Populare Romine Insecta XI. Lepidoptera.* (Family: Pieridae). 138–149.

Nordström, F., Opheim, M. & Valle, K.J., 1955. *De Fennoskandiska dagfjärilarnas utbredning.* 175 pp. Lund.

Norris, F.B., 1891. Notes on rhopalocera in Corfu. *Entomologist* 24: 179–180.

Oldaker, F.A., 1922. Rambles in S.E. France. *Entomologist* 55: 127–131.

Owen, D.F., 1951. Some autumn observations on butterflies in N.W. Spain. *Entomologist* 88: 224–226.

Palmer, R.M., 1974. Lepidoptera of Aberdeenshire and Kincardineshire. *Entomologist's Rec. J. Var.* 86: 33–44.

Panchen, A.L. & Panchen, M.N., 1973. Note on the butterflies of Corsica, 1972. *Entomologist's Rec. J. Var.* 85: 149–153, 198–202.

Payne, J.H., 1951. Northamptonshire butterflies in 1950. *Entomologist* 84: 165.

Perceval, M.J., 1974. Butterflies of Majorca. *Entomologist's Rec. J. Var.* 86: 225–234.

Phillips, G.C., 1960. The migration of insects in the Spanish central Pyrenees, Autumn, 1958. *Entomologist's mon. Mag.* 96: 145–148.

Pictet, A., 1918. Les migrations de *Pieris brassicae* en Suisse, en 1917. *Verh. Schw. Naturf. Ges. Aarau.* 1917: 277–278.

Piquet, F.G., 1873. Butterflies of Jersey. *Entomologist* 6: 399–401.

Prior, G., 1968. Field meeting. *Proc. Brit. ent. nat. Hist.* 1: 21.

Przybylski, Z., 1968. Development of the second generation of cabbage white butterfly *Pieris brassicae* L. (Lepidoptera: Pieridae) in agricultural-climatic conditions of Rzeszow region. (In Polish.) *Polski Pismo. ent.* 38: 897–906.

Ralston, G.S., 1959. Migratory butterflies in the north isles of Shetland. *Entomologist* 92: 184.

Rataul, H.S., 1959. Studies on the biology of cabbage butterfly (*Pieris brassicae* L.). *India Hort.* 76 (4): 255–256.

Rebel, H., 1904. Studien über die Lepidopteren der Balkanländer II. Bosnien und die Herzegowina. *Ann. K.K. nat. Hofmus., Wien.*

—— 1917. Siebenter Beitrage zur Lepidopterenfauna der Kanaren. *Ann. K.K. nat. Hofmus.* 31: 1–62.

—— 1926. Lepidopteren von den Balearen und Pityusen. *Dt. ent. Z. Iris* 40: 135–140.

—— 1934. Lepidopteren von den Balearen und Pityusen. *Dt. ent. Z. Iris* 48: 122–138.

—— 1939. Zur Lepidoptern Cyperns. *Mitt. münch. ent. Ges.* 29: 487–564.

Rebel, H. & Zerny, H., 1931. *Die Lepidopteren Albaniens* (mit Berucksichtigung der Nachbarebiete), Wien.

Riley, N.D., 1925. Seasonal variation in butterflies. *Proc. S. Lond. ent. nat. Hist. Soc.* 1924–1925: 63–81.

Ripa, R., 1978. *Studies of the susceptibility of Pieris brassicae (L.) to a granulosis virus.* Ph.D. thesis. University of London (Imperial College).

Röber, J. 1907. Pieridae, *in* Seitz, A. *Die Gross-Schmetterlinge der Erde.* Section I. Vol. 1 (Die palaearctischen Tagfalter). pp. 39–74.

Robson, J.E., 1902. A Catalogue of the lepidoptera of Northumberland, Durham and Newcastle-upon-Tyne. *Trans. nat. Hist. Soc. North'land & Durham.* 12: 1–318.

Rollason, M.A., 1907. Lepidoptera of East Sutherland. *Entomologist* 40: 40–41.

Rosa, A., 1920. Collecting in Finmark, Swedish Lapland, Jemtland, etc. *Entomologist* 53: 109–115.

Rosa, A.F., 1923. Spring lepidoptera in south and central Spain. *Entomologist* 56: 52–57.

Rothschild, N. & Wertheimstein, C. de., 1913. The butterflies of the Cséhtelek district of central Hungary. *Entomologist* 46: 87–89.

Rowland-Brown, H., 1904. Butterfly hunting in the Tyrol. *Entomologist* 37: 222–226.

—— 1905. A butterfly hunt in the Pyrenees. *Entomologist* 38: 273–275.

—— 1906. Some notes on Scandinavia and Lapland butterflies. *Entomologist* 39: 220–227, 242–247.

—— 1907. Butterflies observed during a short tour in southern France in May, 1907. *Entomologist* 40: 149–153.

—— 1909. Some August butterflies of Cantal and Lozère. *Entomologist* 42: 297–302.

Schmidt-Koehl,, W., 1971. Zur Rhopalocerenfauna der Kanareninsel Teneriffe (Insecta: Lepidoptera). *Mitt. ent. Ges. Basel* 21: 29–91.

Schreiber, H. *in* Muller, P., 1976. Der Katalog der Örtlichkeiten im Bundes Republik Deutschland. Vol. 2. Lepidoptera, Saarbrücken.

Sedykh, K.F., 1974. *Distribution list from USSR.* (In Russian.) Syktyvkap.

Sedykh, K.F. & Sedykh. E.D., 1959. Rhopalocera of the Ukhta district, N.E. of the European part of the USSR. (In Russian.) *Rev. ent. URSS* 37: 829–823.

Seymour Browne, C., 1903. A list of the lepidoptera of the island of Capri: with a few notes. *Entomologist* 36: 254–256.

Shapiro, A.M., 1970. Notes on spring butterflies in N.E. Scotland and Orkney. *Entomologist's Rec. J. Var.* 81: 85–86.

Shapiro, V.A., 1976. *Apanteles-* a parasite of the cabbage white butterfly (in Russian). *Zashch. Rast.* 17–18.

Sheldon, W.G., 1908a. Notes from S.W. France. *Entomologist* 61: 294–295.

—— 1908b. *Melitaea aurina* etc. at Barcelona. *Entomologist* 61: 301–302.

—— 1912. Lepidoptera of the Swedish Provinces of Jemtland and Lapland. *Entomologist* 45: 23–27.

Sich, A., 1883. Lepidoptera of the Channel Islands. *Entomologist* 16: 42.

Simes, J.A., 1932. Butterflies in Dépt. Charente, France. *Entomologist* 65: 84–86.

Smith, J.B., 1891. *Insect Life* 3: 218.

Smith, K.G.V., 1952. Notes on the macrolepidoptera of Central Shropshire, 1948–51. *Entomologist* 85: 25–32.

Smith, S.G., 1948. The butterflies and moths found in the counties of Chester, Flintshire, Denbighshire, Caernarvonshire, Anglesey and Merionethshire. *Chester Soc. nat. Sci., Lit. Art, Grosvenor Museum*, Chester. pp. 250.

Snow, D.W. & Ross, K.F.A., 1952. Insect migration in the Pyrenees. *Entomologist's mon. Mag.* 88: 1–6.

South, R., 1895. Orkney Lepidoptera. *Entomologist* 28: 298–300.

Sowerby, F.W., 1907. Short list of lepidoptera collected near Gibralta in March and April, 1907. *Entomologist* 40: 214.

Standel, A.E., 1958. The Butterflies (Lepidoptera, Rhopalocera) of Kharkov. (In Russian.) *Rev. ent. URSS* 37: 900–902.

Stafford, A.E., 1922. Spring rhopalocera in Surrey. *Entomologist* 55: 189.

Stallwood, B.R., 1949. The relative abundance of butterflies in south Middlesex and north Surrey in 1946–1948. *Entomologist* 83: 140–141.

Steeden, C.F. & Steeden, N.J., 1977. West Lancashire Butterflies, a survey of their past and present distribution and status. *The Flyde nat. Soc.* (3) 3–13.

Steele, W.O. & Woodroffe, G.E., 1969. The entomology of the Isle of Rhum Nature Reserve, Lepidoptera. *Trans. Br. ent. nat. Hist. Soc.* 18: 108–127.

Steiniger, H., 1972. Wanderfalterbeobachtungen im Herbst 1970 und 1971 in Portugal. *Atalanta* 4: 43–56.

Stewart, A.M., 1925. Notes from Arran. *Entomologist* 58: 272–273.

Stidson, S.T., 1952. *A list of the lepidoptera of Devon.* Introduction and Part I. The Torquay Times and Devonshire Press Ltd. Torquay.

Stoneham, H.F., 1920. Lepidoptera in Mesopotamia. *Entomologist* 53: 22–23.

Straubenzee, C.H.C. van., 1933. A summer in Bulgaria after butterflies. *Entomologist* 66: 31–35.

Summers, S., 1975. Butterflies in the Isles of Scilly. *Entomologist's Rec. J. Var.* 87: 94–95.

Svendsen, P., 1976. Collecting butterflies in Denmark. *Entomologist's Rec. J. Var.* 88: 47–50.

Talhouk, A.M., 1969. *Insects and Mites Injurious to Crops in Middle Eastern Countries.* Monographien zur angew. Entomologie. No. 21. Verlag Paul Parey, Hamburg & Berlin.

Thomas, P.H., 1936. Bulgarian rhopalocera. June & July, 1933. *Entomologist* 49: 101–103.

—— 1938a. Arctic butterflies, especially those of Maalselven, Lapland, Lat. 69° N. *Entomologist* 71: 1–5.

—— 1938b. The butterflies of Lapland, 1938. *Entomologist* 72: 129–132.

Traill, J., 1869. Notes on the lepidoptera of Orkney. *Entomologist* 4: 197–200.

Tutt, J.W., 1892. Notes on collecting at Wicken Fen. *Entomologist's Rec. J. Var.* 3: 196–202.

Valletta, A., 1966. The Butterflies of the Maltese Islands. *Entomologist's Rec. J. Var.* 78: 38–42.

—— 1976. Collecting lepidoptera and other insects in Sicily in 1975. *Entomologist's Rec. J. Var.* 88: 113–118.

Van Straubenzee, C.H.C., 1932. Three months butterfly collecting in Greece. *Entomologist* 65: 154–159.

Verdcourt, B., 1945. Butterflies in Bedfordshire. *Entomologist's mon. Mag.* 81: 75.

Verity, R., 1905–1911. *Rhopalocera Palaeartica.* Iconographie et déscription des papillons diurnes de la région palaearctique. Florence. 2 volumes.

Verstraëten, C., 1970. *In.* Ledercq. J. *Atlas Provisoire des Insectes de Belgium.* Carte 190. Gembloux.

Voltukhov, V.A., 1975. Economic evaluation of the chemical and biological methods. (In Russian.) *Zashch. Rast.* (5) 29.

Walker, F.A., 1891. Entomology of Granada and neighbourhood. *Entomologist* 24: 160–163.

Warren, R.G., 1975. *Atlas of the Lepidoptera of Staffordshire. Part 1. Butterflies.* Staff. Biol. Recording Scheme.

Weir, J.J., 1882. Notes on the lepidoptera of the Orkney Islands. *Entomologist* 15: 1–5.

Westwood, J.O., 1854. *The Butterflies of Great Britain.* W.S. Orr & Co., Paternoster Row, London.

Wild, E.H. & Marshall, J.E., 1973. Lepidoptera of south and central Cornwall, 1973. *Entomologist's Rec. J. Var.* 85: 273–276.

Williams, C.B., 1958. *Insect Migration*. Collins, New Naturalist Series, London.

Wiltshire, E.P., 1940. Some notes on migrant lepidoptera in Syria, Iraq and Iran. *Entomologist* 73: 231–234.

—— 1957. *The Lepidoptera of Iraq*. N. Kaye Ltd., Ministry of Agriculture, Iraq.

—— 1973. A holiday in or near the western Alps in June–July 1971, with notes on the lepidoptera. *Entomologist's Rec. J. Var.* 85: 41–47.

Wooff, W.R., 1958. *An Ecological Survey of the Insects of the Farne Islands*. Ph.D. Durham.

Worms, C.G.M. de, 1949. The Butterflies of London and its surroundings. *Lond. Nat.* 29: 46–80.

—— 1962. *The Macrolepidoptera of Wiltshire. Wilts. arch. nat. Hist. Soc.* pp. 177.

—— 1968. Mid October butterflies in South Devon. *Entomologist's Rec. J. Var.* 80: 326–327.

—— 1974. Butterflies in Tuscany, May–June 1973. *Entomologist's Rec. J. Var.* 86: 45–48.

Young, M.R., 1974. An account of some of the lepidoptera of the Moorlands near Llyn Cwmynach, Merioneth. *Entomologist's Rec. J. Var.* 86: 10–14.

Zeller, P.C., 1847. Bemerkungen über die auf einer Reise nach Italien und Sicilien beobachteten Schmetterlingsarten. *Isis budiss* 1: 214–243.

Zerkowitz, A., 1946. The Lepidoptera of Portugal. *Jl N.Y. ent. Soc.* 54: 211–261.

Further references

Adkin, R., 1922. *Entomologist* 55: 17–18 (Sussex).

—— 1924. *Entomologist* 57: 165, 205 (Sussex).

—— 1926. *Entomologist* 59: 253–254 (Sussex).

—— 1932. *Entomologist* 65: 52 (Sussex).

—— 1933a. *Entomologist* 66: 93–94 (Sussex).

—— 1933b. *Entomologist* 66: 162 (Cornwall).

Adkin, B.W., 1947. *Entomologist* 80: 147–148 (Kent).

Allan, P.M.B., 1950. *Trans. B. Stortford nat. hist. Soc.* 3–45 (Hertfordshire).

Anderson, J., 1920. *Entomologist* 53: 283 (Sussex).

Anonymous, 1958. *Bull. Kent. Fld Club* 3: 7–8 (Kent).

Arkle, J., 1898. *Entomologist* 32: 298–300 (Cheshire).

—— 1899. *Entomologist* 32: 307–308 (Cheshire).

—— 1901. *Entomologist* 34: 257 (Radnorshire).

Ashby, E.B., 1910. *Entomologist* 43: 317–318 (France).

Bainbridge Fletcher, T., 1925. *Bull. ent. Res.* 16: 177–181 (Assam).

—— 1937. *Amat. ent. Soc.* 2: 6 (London).

Bankes, E.R., 1899. *Entomologist's mon. Mag.* 35: 12 (Dorset).

Barraud, P.J., 1901. *Entomologist* 34: 28 (Kent).

Barraud, P.J. & Blair, K.G., 1903. *Entomologist* 36: 219–220 (Hampshire).

Becker, T., 1925. *Int. ent. Z.* 19: 149–150 (Germany).

Benson, R.B., 1922. *Entomologist* 55: 188 (Hertfordshire).

Bergmann, A., 1952. *Die Grossschmetterlinge Mitteldeutschlands*. II. Urania Verlag, Jena (Germany).

Blackie, J.E.H., 1920. *Entomologist* 53: 277–279 (France).

Blackman, T.M., 1923. *Entomologist* 56: 213–214 (Westmorland).
Blakeborough, T.B., 1901. *Entomologist* 34: 23–24 (Yorkshire).
Bond, L.H., 1926. *Entomologist* 57: 142–143 (Yorkshire).
Bree, W.T., 1832. *Lond. Mag. nat. hist.* 5: 105 (London).
Bretherton, R.F., 1949. *Entomologist* 82: 191 (France).
—— 1966. *Trans. Soc. brit. Ent.* 17: 1–94 (Europe).
Brown, J.W., 1916. *Entomologist* 49: 212–213 (France).
—— 1917. *Entomologist* 50: 189 (France).
Browne, G.B., 1902. *Entomologist* 35: 269–270 (Kent).
Bryantev, B.A., 1925. *Defense des Plantes* 2: 237–241 (USSR).
Bull, G.V., 1924. *Entomologist* 57: 285–287 (Surrey).
—— 1943. *Entomologist* 76: 144 (Kent).
Busbridge, W.E., 1938. *Entomologist* 71: 102 (Kent).
—— 1940. *Entomologist* 73: 149 (Kent).
Bussey Bell, J.W., 1915. *Entomologist* 48: 157–159 (Oxfordshire).
Butler, W.E., 1908. *Entomologist* 41: 62 (Berkshire).
Cameron, A.E., 1941. *Trans. Highland agric. Soc. Scotland* 53: 77–97 (Scotland).
Cardew, P.A., 1927. *Entomologist* 55: 89–90 (N. Ireland).
Carr, F.M.B., 1901a. *Entomologist* 34: 103–112 (Surrey).
—— 1901b. *Entomologist* 34: 319–320 (Devon, Somerset).
—— 1902. *Entomologist* 35: 246–247 (Kent).
—— 1903. *Entomologist* 36: 51–52, 243–246 (Wiltshire).
Chartres, S.A., 1932. *Entomologist* 65: 211 (Sussex).
Christy, W.M., 1918. *Entomologist* 51: 138–139 (Hampshire).
Classey, E.W., 1941. *Entomologist* 74: 215–216 (Surrey).
—— 1944. *Entomologist* 77: 31 (Surrey).
Cleu, H., 1947. *Mem. Mus. natn. Hist. nat., Paris* 20: 141–188 (Durance, France).
Colman, K.E.S., 1940. *Entomologist* 73: 235 (London).
Cornell, E., 1921. *Entomologist* 54: 51 (Isle of Wight).
Croft, E.O., 1912. *Entomologist* 45: 34–36 (Germany).
Crop, C.F.L., 1913. *De Levende Natuur* 17: 24 (s'Gravenhage, Netherlands).
Dick, C.K.G., 1923. *Entomologist* 56: 230–233 (Istanbul, Turkey).
Easton, N.T., 1946. *Entomologist* 74: 37–41 (Sussex).
Eliot, N., 1938. *Entomologist* 71: 77–81 (Cavalaire, France).
Evans, L.J., 1955. *Entomologist's Rec. J. Var.* 67: 100 (Birmingham).
Felton, J.C., 1975. *Trans. Kent Fld Club* 5: 150–174 (Kent).
Fletcher, T.B., 1901a. *Entomologist* 34: 71–73 (Kent).
—— 1901b. *Entomologist* 34: 220–223 (Greece, Malta, Cyprus).
Fletcher, T., 1915. *Rpt agric. Res. Inst. coll. Pusa Calcutta* 62–75 (India).
Ford, H.D. & Ford, E.B., 1932. *Entomologist* 65: 284–285 (Berkshire).
Forsythe, C.H., 1902. *Entomologist* 35: 245–246 (Westmorland).
Foster, A.H., 1907. *Entomologist* 40: 153–155 (Lake District).
Fowler, J.H., 1899. *Entomologist* 32: 267–269 (Hampshire).
Frohawk, F.W., 1899. *Entomologist* 32: 257 (Essex, Sussex).
—— 1908. *Entomologist* 41: 39–40 (Essex).
—— 1909. *Entomologist* 42: 213 (Essex).
—— 1913. *Entomologist* 46: 62 (Berkshire).
—— 1920. *Entomologist* 53: 160 (S.E. & S. counties).
—— 1921. *Entomologist* 54: 147–148 (not given).
—— 1926. *Entomologist* 59: 288 (Dorset, Hampshire, Northumberland).
—— 1936. *Entomologist* 69: 263 (Cork, Surrey).

—— 1937. *Entomologist* 70: 227 (Kent).
—— 1938. *Entomologist* 72: 148–149 (Surrey).
—— 1943. *Entomologist* 76: 145 (Surrey).
Gainsford, A.P., 1973. *Entomologist's Rec. J. Var.* 85: 128–133 (Cornwall).
Gardiner, B.O.C., 1943. *Entomologist* 76: 209 (Kent).
Gibbs, A.E., 1906. *Entomologist* 39: 4–7 (Cornwall).
Golding, A.J., 1903. *Entomologist* 36: 72 (Kent).
Grant, J.H., 1915. *Entomologist* 48: 214–216 (Gloucestershire).
Granville White, A., 1953. *Entomologist* 86: 217 (Surrey).
Graves, P.P., 1923. *Entomologist* 56: 140 (Cornwall).
Greer, T., 1920. *Entomologist* 53: 76–78 (Tyrone), 217–221.
—— 1924. *Entomologist* 57: 17 (Tyrone).
—— 1927. *Entomologist* 60: 100 (Tyrone).
—— 1947. *Entomologist* 80: 183–186 (Tyrone).
Gurney, G.H., 1912. *Entomologist* 45: 96–97 (Digne, France).
—— 1920. *Entomologist* 53: 37–40 (Norfolk).
—— 1928. *Entomologist* 61: 33 (*P. b. cheiranthi* in Canary Islands).
Haas, T., 1925. *Int. ent. Z.* 19: 250–255 (Rufach, Elsass).
Haggett, G., 1948. *Entomologist* 81: 30–34 (Sussex).
—— 1949. *Entomologist* 82: 25–32 (Sussex).
Haig Thomas, P., 1955. *Entomologist* 88: 169–173 (Norway).
Haines, R., 1974. *The Times*, August 24th. Letter (London).
Hamm, A.H., 1945. *Entomologist's mon. Mag.* 81: 58 (Oxfordshire).
Harbich, H., 1967. *Atalanta* 2: 53–65 (Dortmund, Heidelburg, Germany).
Harcourt-Bath, W., 1896. *Entomologist* 29: 9–12 (French Alps).
—— 1897. *Entomologist* 30: 240–242 (Himalayas).
Heslop, I.R.P., 1946. *Entomologist* 79: 140 (Somerset).
Higgins, L.G., 1948. *Entomologist* 81: 25–29 (Granada, Spain).
Hobson, A.D., 1920. *Entomologist* 53: 158 (Cambridgeshire).
Hodgson, S.B., 1935. *Entomologist* 68: 41–42 (Cornwall).
Holford, H.O., 1915. *Entomologist* 48: 55–57 (Majorca).
Hruby, K., 1964. *Prododromus Leidopter Slovenska.* Bratislava. 962 pp.
Hunt, H.F., 1920. *Entomologist* 53: 263 (Sicily).
Imms, A.D., 1898. *Entomologist* 31: 42–44 (Staffordshire).
Jager, J., 1901. *Entomologist* 34: 303–304 (Hessen–Nassau, Germany).
James, R., 1940. *Entomologist* 73: 213 (London).
Jarvis, W., 1909. *Entomologist* 42: 18 (Sussex).
Jary, S.G., 1934. *J. SE. agic. Coll. Wye* 34: 65–69 (Kent).
Jefferys, T.B., 1899. *Entomologist* 32: 22–23 (Somerset).
Johnson, G.F., 1931. *Entomologist* 64: 53–59 (Down, Ireland).
—— 1944. *Entomologist* 77: 149–154 (Italy).
—— 1952. *Entomologist* 85: 217–219 (Libya).
Jones, A.H., 1900. *Entomologist's mon. Mag.* 36: 271–275 (Tyrol, Austria).
Kemp, S.W., 1899. *Entomologist* 32: 260 (Dorset).
Kozhanchikov, I.V., 1936. *Zashch. Rast.* 11: 40–57 (USSR).
Kudrna, O., 1968. *Sbornik Jihoceskeho Musea* 8: 18–23 (Germany).
Kurentzov, N.J., 1929. *Fauna de l'URSS.* I. Lepidoptera. Leningrad.
Lane, C., 1959. *Entomologist's mon. Mag.* 95: 94 (Swiss Alps).
Lane, C. & Rothschild, M., 1960. *Entomologist* 93: 9–12 (Oxfordshire).
Lang, H., 1901. *Entomologist* 34: 263–267 (Austria, Hungary).
Lipscomb, C.G., 1968. *Entomologist's mon. Mag.* 80: 296–297 (Warwickshire).

London Natural History Society, 1915. *Entomologist* 48: 128 (Yorkshire).
—— 1915. *Entomologist's mon. Mag.* 51: 173 (Yorkshire).
Longfield, C., 1929. *Entomologist* 62: 186–187 (S. Ireland).
Lowe, F.E., 1914. *Entomologist* 47: 14–20 (Var, France).
Lowther, R.C., 1931. *Entomologist* 64: 128–135 (N. Lancashire).
Luckens, C.J., 1973. *Entomologist's Rec. J. Var.* 85: 18–21 (Kent).
Lyon, F.H., 1955. *Entomologist* 88: 136 (S. Ireland).
Mac Gillavry, D., 1914. *Tijdschr. Ent.* 57: 89–106 (Netherlands).
Mace, H., 1919. *Entomologist's mon. Mag.* 55: 255–258 (Rumania).
Mann, E.H., 1919. *Entomologist* 52: 2–4 (Lys, N. France).
Mannering, E., 1910. *Entomologist* 43: 318–319 (Cornwall).
Marsh, J.C.S., 1949. *Entomologist* 82: 223–228 (Germany).
Marshall, H., 1945. *Entomologist's mon. Mag.* 81: 94 (Oxfordshire).
Martin, C., 1918. *Entomologist* 51: 146–148 (Somme, France).
Mathew, G.F., 1912. *Entomologist* 45: 135–136 (Essex).
—— 1912. *Entomologist* 45: 343 (Devon).
—— 1919. *Entomologist* 52: 227–228 (Essex).
—— 1920. *Entomologist* 53: 158–159 (Essex).
—— 1921. *Entomologist* 54: 146–147 (Essex).
—— 1924. *Entomologist* 57: 258 (Essex).
Mera, A.W., 1921. *Entomologist* 54: 84–87 (Essex).
Michael, P., 1951. *Entomologist* 84: 186–187 (Hampshire).
Miller, N.C.E., 1919. *Entomologist* 52: 164–166 (Somme, France).
Molesworth Muspratt, V., 1951. *Entomologist* 84: 226–229 (S. France).
Mook, J.H. & Haeck, J., 1965. *Arch. Ned. zool.* 16: 281–293 (Netherlands).
Morley, C., 1899. *Entomologist* 32: 222–224 (Northamptonshire).
Morley, B., 1924. *Entomologist* 57: 109–112 (Yorkshire).
Morris, S., 1934. *Entomologist* 67: 54–56 (Sussex).
Muirhead, J.W., 1911. *Entomologist* 44: 364 (Lancashire).
Musy, M., 1921. *Bull. Soc. Fribourg Sci. Nat.* 25: 6–8 (France).
Nash, A., 1893. *Entomologist's mon. Mag.* 29: 113 (Surrey).
Nathan, L., 1937. *Entomologist* 70: 72 (Lancashire).
—— 1938. *Entomologist* 72: 127–128 (Lancashire).
Neave, B.W., 1918. *Entomologist* 51: 114 (London).
Neave, S.A., 1929. *Entomologist* 62: 143 (London).
Oldaker, F.A., 1904. *Entomologist* 37: 21–23 (Surrey).
—— 1906. *Entomologist* 39: 157–160 (Surrey).
Ormerod, E.A., *Report of observations of injurious insects and common farm pests during the year 1898, with methods of prevention and remedy.* Simpkin, Marshall, Kent & Co., London.
Oudemans, J.T., 1923. *Tijdschr. Ent.* 66: 41–71, 93–97 (Netherlands).
—— 1928. *Tijdschr. Ent.* 71: 84–85 (Netherlands).
Ovenden, J., 1909. *Entomologist's Rec. J. Var.* 21: 30–33 (Kent).
Page, W.T., 1905. *Entomologist* 38: 25 (Essex).
Payne, J.H., 1973. *Entomologist's Rec. J. Var.* 85: 106 (Kent).
Pearson, E.J.W., 1958. *Bull. amat. ent. Soc.* 17: 23–24 (Hampshire).
Peile, H.D., 1929. *Entomologist* 62: 268–269 (Brittany, France).
—— 1930. *Entomologist* 63: 25–29 (Menton, Aix, France).
Philp, E.P., 1972. *In* Collins, M. *et al., Environmental studies at Hothfield.* Kent Education Committee, Maidstone (Kent).
Platt-Barrett, J., 1909. *Entomologist* 42: 42 (Messina, Italy).

Prest, E.E.B., 1902. *Entomologist* 35: 288–289 (Dorset).

Prideaux, R.M., 1915. *Entomologist* 48: 86–89 (N. Downs).

Querci, O., 1932. *Bibliografia dei Ropaloceri del Portugalle.* Arcuives Mus. Bocage Lisboa. 1–112.

Raynor, A.G.S., 1922. *Entomologist* 55: 22 (London).

Redmayne, M.G., 1919. *Entomologist* 52: 139–141 (Macedonia).

Richmond, D.M. & Bevan, K., 1973. *Entomologist's Rec. J. Var.* 85: 121–125 (Wyre).

Robertson, R.B., 1919. *Entomologist* 52: 59–60 (Kent).

Rollason, W.A., 1901. *Entomologist* 34: 23 (Cornwall).

Rosa, A.F., 1902. *Entomologist* 35: 93–97 (Rhône Valley, France).

—— 1929. *Entomologist* 62: 208–210 (Kurseong, India).

Rowland-Brown, H., 1909. *Entomologist* 42: 186, 236 (Middlesex).

—— 1910a. *Entomologist* 43: 203–204 (London).

—— 1910b. *Entomologist* 43: 299–303 (Samoussy, France).

—— 1910c. *Entomologist* 43: 322–327 (SE France).

—— 1914. *Entomologist* 47: 8–14, 126–129 (Issadrome, France).

—— 1914. *Entomologist* 47: 308–313 (Dauphiny, France).

—— 1915. *Entomologist* 48: 77–80 (Buckinghamshire).

—— 1918a. *Entomologist* 51: 1–2 (Great Britain).

—— 1918b. *Entomologist* 51: 244–248 (Dorset).

—— 1919. *Entomologist* 52: 139 (Devon).

Rowland-Brown, H. & Goss, H., 1901. *Entomologist* 34: 161–162 (London).

Schreiber, O., 1950. *Flugbl, Bundesanst PflSchutz., Wien* 64: 2 (Germany).

Scott, E., 1926. *Entomologist* 59: 309–312 (Switzerland).

—— 1964. *Kent Fld Club* 2: 1–98 (Kent).

Scott, E. & Hayesbank, M.B., 1927. *Entomologist* 60: 91–93 (Menton, France).

Seda, A., 1938. *Ochr. Rost. Prague* 14: 16–23 (Czechoslovakia).

Sharp, W.E.S., 1906. "Lancashire", *The Victoria History of the counties of England series* (Lancashire).

Sheldon, W.G., 1913. *Entomologist* 46: 309–313 (Spain).

—— 1908. *Entomologist* 61: 294–295, 301–302 (S.W. France).

Sich, A., 1905. *Entomologist* 38: 258–259 (Surrey).

Siepi, P., 1931. *Ann. Mus. hist. nat. Marseilles* 25: 25–244 (Bouches du Rhône, France).

South London Entomological & Natural History Society, 1913. *Entomologist* 46: 295 (Scilly).

Stafford, A.E., 1941. *Entomologist* 74: 227 (London).

Stowell, E.A.C., 1921. *Entomologist* 54: 122–123 (Hampshire).

Summers, G., 1975. *Entomologist's Rec. J. Var.* 87: 278 (Staffordshire).

Swanson, S., 1940. *Entomologist* 73: 49–50 (Caithness).

Tatchell, L., 1922. *Entomologist* 55: 189 (Dorset).

Temple, V., 1951. *Entomologist* 84: 214–215 (Dorset).

Tulloch, B., 1940. *Entomologist* 73: 249 (Wales).

—— 1942. *Entomologist* 75: 20 (Monmouthshire).

Tulloch, J.B.G., 1933. *Entomologist* 66: 148 (Monmouthshire).

—— 1941. *Entomologist* 74: 18–19 (Monmouthshire).

Turner, D.P., 1902. *Entomologist* 35: 243 (London).

Turner, H.J., 1931. *Entomologist* 64: 96 (London).

Van Wisselingh, T.H., 1949. *Tijdschr. Ent.* 96: 2–22 (Netherlands).

Wagner, O., 1937–38. *Ent. Z.* 51: 149–152 (Germany).

Walker, F.A., 1901. *Entomologist* 34: 355–356 (Paris, France).

Walker, J.J., 1898. *Entomologist's mon. Mag.* 34: 278 (Kent).

—— 1911. *Entomologist's mon. Mag.* 7: 259 (Oxford).

—— 1914. *Entomologist's mon. Mag.* 50: 275–276 (Oxford).

—— 1931. *Entomologist's mon. Mag.* 67: 254–268 (at sea).

—— 1933. *Entomologist's mon. Mag.* 69: 232–233 (Oxfordshire).

Warburton, C., 1925. *Jl. R. agric. Soc. Eng.* 136: 284–291 (Cambridgeshire).

Warnecke, G., 1960. *Bombus* 2: 80 (Hamburg, Germany).

Wattison, J.T., 1928. *Mem. zool. Univ. Coimbra* I: 29 (Portugal).

Wheeler, R., 1947. *Entomologist's Rec. J. Var.* 59: 137 (Sussex).

Whittaker, O., 1899. *Entomologist* 32: 73 (Wales).

Williams, B.S., 1915. *Entomologist* 48: 64–67 (Pembroke).

Wingstrand, K.G., 1949. *Opusc. Ent.* 14: 184 (Hampshire).

Wolfsberger, J., 1971. *Die Macrolepidopteren Fauna des Monte Baldo in Oberitalien.* Museo Covico di Storia Naturale di Verona (Italy).

Woodford, C.M., 1921a. *Entomologist* 54: 119–121 (Sussex).

—— 1921b. *Entomologist* 54: 197 (Sussex).

Woodforde, F.C., 1922. *Entomologist* 55: 22–23 (Caernarvonshire).

Worms, C.G.M. de., 1946. *Entomologist* 79: 73–79 (Surrey).

—— 1956. *Entomologist* 89: 109–113 (Inverness).

—— 1976. *Entomologist's Rec. J. Var.* 88: 164 (Surrey, Sussex).

Worms, C.G.M. de. & Ellison, R.E., 1955. *Entomologist* 88: 99–106 (Tramore).

—— 1957. *Entomologist* 90: 81–91 (Ireland: Clare, Kerry).

Worsley-Wood, H., 1915. *Entomologist* 48: 62 (Tyne & Wear).

Wright, A.E., 1935. *Entomologist* 68: 138 (Westmorland).

—— 1936. *Entomologist* 69: 94–95 (Lancashire).

Wright, J., 1907. *Entomologist* 40: 189 (Isle of Wight).

3. Life history

Original descriptions

17th Century

Only four descriptions or illustrations of *P. brassicae* are known from this century. Maria Graffinn (1679) illustrated all the stages of *P. brassicae* on cabbage (*Brassica oleracea*), including a pupa attached peculiarly to the distal end of a petal. Johann Goedaert's (1700) woodcuts of the stages of *P. brassicae* include, like those of Graffinn, details of parasitism by a hymenopteron. Blankaart (1688) illustrated the life cycle of *P. brassicae* on *Tropaeolum* sp. in colour. Surprisingly Thomas Moufet's book (1634) does not include a plate of *P. brassicae*, although other common insect species are illustrated.

18th Century

Further descriptions of *P. brassicae* in the eighteenth century by John Ray (1710), Merian (1717), Carl Linnaeus (1746) and James Petiver (1795) are found in Table 1.1. Other works in this century, many of them including coloured plates depicting various stages of the life history, were published by Benjamin Wilkes (1747–1749) on *Tropaeolum* sp., Clerck (1759), Sepp (1762), Esper (1777), Rosel (1781) and John Swammerdam (1758). Swammerdam's plate in his *Book of Nature* (1758) is interesting as it includes 17 drawings of the prepupa and various stages of imaginal eclosion.

19th Century

In the nineteenth century more entomologists described the life history of *P. brassicae*: Schrank (1801), Hübner (1806), Herold (1815), Berge (1851), Westwood (1854), Ormerod (1878), Kirby (1898) and South (1899).

Herold and Westwood both illustrated the life cycle in colour on cabbage and rape (?) respectively. A peculiar larva and pupa with nine segments was illustrated by D.I.E. (1883). Theodore Johnson (1878, 1887, 1906, 1908) painted some beautiful illustrations of all the stages of *P. brassicae*.

20th Century

The beginning of the twentieth century brought an increase in the number of descriptions of *P. brassicae*: Fabre (1911), Meyrick (1895, 1928), South (1936), Frohawk (1926a), Hering (1926), Seitz (1929–1932) and these were followed up to the present by Oldham (1950), Blunck (1953), Ministry of Agriculture, Fisheries and Food (MAFF) (1976), Howarth (1973), Guilbot (1975) and Newman (1977). Descriptions of *P. brassicae* also occur in many of the common identification books of respective countries.

Works which stand out as being most comprehensive and useful are Spuler (1908), Krzywicki (1962) who dealt with the biology of *P. brassicae* in Poland, Portier (1949) and Grassé (1975).

Life cycle

The complete life cycle of *P. brassicae* is now described from oviposition to eclosion, including courtship and copulation. Incorporated into this description are data from a variety of sources.

Oviposition

Recognition of the foodplant by the female is the result of a chemical response involving the sensory appreciation of plant odours, the chemical composition of the leaf and plant texture. Mention is made in the next chapter of how female *P. brassicae* selects her specific foodplant using olfactory cues through her antennae, and tactile and chemical cues through her tarsi, which supplement her visual cues.

The female *P. brassicae* has post-imaginal development of the ova, that is the embryo of the ovum develops after hatching (Kaiser, 1949; Benz, 1970) and this pre-oviposition period may last 3–14 days in the laboratory (Street, 1975). However, Gardiner (1977, pers. comm.) found that the female emerges with 50% of the ova fully developed. When the ova are not mature, oviposition apparently will only occur when the optimum temperature is 15.7°C and the humidity is above 78% RH (Avidov & Harpaz, 1969). Gardiner (1977, pers. comm.) stated that *P. brassicae* lay readily at 20% RH. Antoinette Karlinsky's (1977) thesis work at the Université Pierre et Marie Curie in Paris has recently shown that the ova of *P. brassicae* are not mature at emergence, and that they develop rapidly after 24 hours under the control of the corpora allata secretions.

The female will often spend much time fluttering around apparently looking for a suitable leaf on which to oviposit when confronted with a number of plants of the acceptable foodplant. Average numbers of ova which are laid in batches are 26 (max. 227) (Avidov & Harpaz, 1969), 100 or more (Ford, 1945), 37 with a total potential of 750 (David & Gardiner, 1962a), comprising a total of 600 ova laid (Wigglesworth, 1964) or 45–74 as Mukerji (1961) reported from India. Ford (*loc. cit.*) pointed out that *P. brassicae* shared with eight other British butterflies the characteristic of laying its ova in batches.

The rate at which ova are laid in the wild has been observed at 27 in 9 minutes (Gillmer, *in* South, 1924). The ova are glued to the surface of the underside of the leaf and are often found close to the margin; this is a result of the way in which the female bends her abdomen below the leaf (Ford, 1945; Oxford Scientific Films, 1976; Granada TV Film, London, 1960). However, preliminary investigations in the wild show that up to 50% of ova may be on the upper surface (Feltwell, unpublished).

Evidence that imagines of *P. brassicae* use visual, olfactory and tactile stimuli when ovipositing was presented by Rothschild *et al.* (1975) and Rothschild & Schoonhoven (1977) in a series of experiments. Ovipositing females were offered cabbage leaves which carried batches of their own ova, model ova, squashed ova, leaves with ova removed, feeding larvae, damaged leaves, leaves with fluid of pressed female abdomen and leaves treated with extracts from *Asclepias*, *Brassica*, *Calotropia*, *Cannabis*, *Cheiranthus*, *Digitalis*, *Gomphocarpus*, *Hoya*, *Mimosa*, *Nicotiana*, *Stephanotis* and *Tropaeolum* species.

It was pointed out by Hewitt (1917) that *P. brassicae* selects only leaves of cruciferous plants containing mustard oil glycosides on which to oviposit. They also select Tropaeolaceae and Capparaceae foodplants as well as many others (cf. Chapter 4).

A switch mechanism operates in the female when ovipositing (Rothschild & Schoonhoven, 1977). When offered cabbage and *Reseda*, *P. brassicae* oviposit on cabbage first of all, and then switch over to *Reseda* when the former foodplant is overloaded with ova. Experiments with hidden ova and damaged leaves showed that imagines made fewer visits to these foodplants, which indicated that olfactory cues associated with the ova were appreciated by the imago without her landing. On landing, chemoreception through the tarsi occurs, and there may be sensory receptors also in the tip of the abdomen. It also seems possible that a chemosensory pheromone marker is secreted onto the leaves of the foodplant.

Certain plants when grown adjacent to cabbage are antagonistic to ovipositing imagines (cf. Chapter 17). Lundgren (1975) pointed out that seven plant species (Table 17.7) are said to decrease oviposition in *P. brassicae*, so he designed a dual choice chamber with two inlets to test the supposition. A slow air stream was drawn over potted cabbage plants, onto which extracts of the antagonistic plants were applied. The results showed that significantly fewer ova were laid on cabbage when these extracts were applied. Lundgren (*loc. cit.*) explained that the inhibitory reaction was due to secondary plant substances which were presumably volatile. Further species of plant which are reported to act antagonistically are set out in Table 17.7.

Ova maturation

As the orange/yellow ova mature the colouration changes to a darker shade and eventually the black mandibles of the developing larvae are clearly visible beneath the egg case; this is termed the black head stage. The larvae emerge very soon afterwards. Westwood (1854) depicted the larva as it is positioned in the ovum but with the egg case removed (cf. Oxford Scientific Films, 1976). As the larva emerges, it eats the egg case which contains vital nutrients (Fabre, 1912), such as carotenoid pigments which act as vitamin A precursors required for growth. According to Jones (1960), the weight of ova laid in the laboratory in January, July and October does not vary significantly from 2 mg.

Ecdysis

Pieris brassicae typically go through four larval-larval ecdyses and have five instars. This seems to be the case throughout all of its world range, although it was previously recorded from Palestine as having a variable number between five and three when reared between 15–25°C (Klein, 1932a,b). Gardiner (1978) repeated Klein's experiments but did not observe any change from five instars using ova collected in Palestine.

Wandering

Full grown larvae of *P. brassicae* have the habit of walking great distances in search of a suitable pupation site. This process is called wandering. Such is the "purposeful" nature of their walking, that larval wandering has even been described as analagous to rock climbing in Man! (Eliot, 1944). Apparently *P. brassicae* is more agile than *P. rapae* over obstacles and pupae can be found in high situations.

Distances travelled by wandering larvae may be considerable, up to 350 m for a mass movement in Berlin in 1944 (Godan, 1949), and 88 m by

250 larvae which traversed a 15 m wall (Egerton Collier, 1941), and 7 m (Driest, 1966). Masses of larvae wandering over railway lines near Kiev in 1894 were reputed to have held up a locomotive (Guerne, 1894).

The time, and thus the site of pupation, may be influenced by an intrinsic physiological factor; Eliot (1944) suggested that larvae start to pupate only when they have enough materials left to make the silk girdle and cremaster.

Pupation site

Typical pupation sites are under copings of walls, on the top bar of fences, and on undersills (South, 1924); on the bark of beech and chestnut (Paton, 1944; Godan, 1949), and up an electricity pylon (Feltwell, 1977b). Mukerji (1961) reported *P. brassicae* pupating in the soil in India, but this behaviour does not appear to have been witnessed before or since. A set of three diagrams depicting the way in which the larva spins its silk girdle and base were illustrated originally by Réaumur and borrowed by Portier (1949).

The place chosen usually offers some degree of protection against weather and predators (South, 1924). Bird predation can be high. Moss (1933) tested the effect of bird predation on *P. brassicae* by putting pupae either in full view, on trees and fences, out of "normal sight", under eaves and windowsills, and out of sight indoors. In the following spring he found 5.2%, 11.3% and 85.7% of the original pupae left. Ford (1945) however, found that the figure for those in full view and those in partial view was not significant, but the data did show that the least pupae were taken, not surprisingly, from indoors by birds.

In an analysis of the position of pupation with regard to the relative vertical plane and direction of the light, Merrifield & Poulton (1899) found a tendency to direct the head towards light, when reared in cages; most pupating on the roof. They found 36 attached to the roof, 25 on the front, back and sides, 41 on the roof but with head to the front, 11 on the roof with head away from light, 15 obliquely on the roof and 10 in the same position but with heads away from the light. Gardiner (1979) noted that "diapausing larvae" tend to seek horizontal places at the top of the cage in preference to the vertical sides; he explained that this may confer more protection over the winter period.

Eclosion

Introduction

The sequences of eclosion have been described by Cottrell (1964), Nicolson (1976) and Parsons (1978, unpublished MS), and physiological details by Moreau (1973) and Moreau & Gourdoux (1971).

There is still some doubt as to whether air is used in expansion of wings, but it has been established that carbohydrates and fats are metabolised and that a high consumption of oxygen as well as a high increase in internal pressure accompanies this process.

Air swallowing

Prior to eclosion the pupa makes irregular jerking movements of the abdomen thus splitting the pupal case along the proboscis line and across the back of the head. The fresh insect then takes about one minute to free itself from the case (Nicolson, 1976). According to Cottrell (1964) air swallowing commences at this time and ceases when the wings are fully expanded; although air swallowing is regarded as less of an important part of the procedure than muscle movements (Cottrell, unpublished results in Cottrell, *loc. cit.*). Experiments conducted by Cottrell which involved blocking the proboscis of imagines just prior to eclosion confirmed this. He found that seven out of 10 imagines expanded their wings normally, the other three had slight deformities; the latter were found on examination not to have taken in any air. Even an imago ligatured around the neck managed to expand its wings successfully. When the wings are fully expanded the crop was found to be only quarter full of gas. Nicolson (*loc. cit.*) also noted that there was apparently no air swallowing during pupal-imaginal ecdysis.

As soon as the imago has freed itself from the exuvium it climbs to a suitable position where it can expand and dry its wings. During this time the proboscis curls and uncurls giving the impression that air is being drawn in (Parsons, unpublished MS). The wings begin to expand due to haemolymph being forced into the wings by contractions of abdominal muscles, the whole process taking 10 minutes (Nicolson, *loc. cit.*). The tips of the wings are last to expand and these exude a straw coloured fluid (carotenoid?) from small ducts close to the veins. Unwanted reddish-brown products of metabolism left over from the pupa are excreted in the form of meconium shortly after eclosion. This is followed by a watery fluid which originates from the haemolymph and is passed out regularly over a period of three hours in a process called diuresis (Chapter 8).

External and internal pressure

It has been observed in much larger species of lepidoptera, e.g. *Ornithoptera priamus richmondius* Gray (Richmond Birdwing) that increased atmospheric pressure can cause crippledness or failure to eclose. Indeed Pictet (1918a) stated that *P. brassicae* pupae were very susceptible to change in pressure; even a drop of 1 mm mercury would be sufficient to induce imagines to eclose. He also made other statements which today would need to be checked: a) that the rate of eclosion is directly related to the fall in the barometer, i.e. a big depression is followed by many imagines hatching; b) that when pupae are ready to eclose and the pressure rises eclosion is delayed for a day or until the pressure drops; c) a drop in pressure during pigmentation of the pupa speeds up time until eclosion; d) when the pressure remains constant for a long time the pupae speeds up development until eclosion and e) when the pressure remains constant for a long time the pupae are not able to mature.

Parsons (*loc. cit.*) repeated Pictet's (1918a) experiments and observed in the laboratory the eclosion of 20 male and female pupae of *P. brassicae* at an increased pressure or 1.644 lb in^{-2} (equivalent to the effect of bringing pupae down from 1000 m to sea level) when compared to a control group of 20 pupae at normal pressure. Parsons found that there was no detrimental effect of crippling at increased pressure and there was 100% eclosion in both groups. At increased pressure the mean wing expansion time was about a minute faster than at normal pressure (7–9 min).

However, if pupae are subjected to constant high pressure the imagines may be crippled or fail to emerge. At pressures above 845 mm mercury the imagines are again either crippled or fail to emerge and the wing expansion time may be decreased. Thus Parsons concluded that as *P. brassicae* imagines have a comparatively small average wing span of 4 cm, compared to a Richmond Birdwing, and that the body is relatively smaller, the effect of pressure on the surface area available is much smaller and probably negligible for a short time.

The internal pressure of the haemolymph of fifth instar larvae and pupae stays constant throughout the development until eclosion when it increases 18 times (Moreau, 1973, 1974). The maximum pressure reached at any time was 30–5 mm Hg in the abdomen which coincided with muscular contractions of eclosion (Figure 3.1).

Moreau's (1973) thesis was entirely on the physiological aspects of eclosion, such as reactions to different temperatures and humidities,

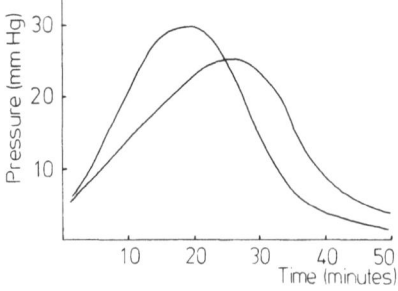

Fig. 3.1 Thoracic and abdominal pressure of *P. brassicae* at imaginal eclosion (after Moreau, 1974)

gaseous exchange, metabolism of glycogen, trehalose and fat, reabsorption of meconium, variations in internal pressure and co-ordination with musculature and the central nervous system. The salient features from his work are that complete expansion of wings occurred most quickly in dry warm conditions (8 min), compared to wet and humid conditions (120 min). High rates of oxygen consumption occurred at eclosion but evolution of carbon dioxide did not vary greatly. High levels of trehalose are metabolised during diapause in *P. brassicae*. Large stores of fat are metabolised rapidly during imaginal life.

Hardening of cuticle

Cottrell (1964) made a study of the hardening and darkening of the cuticle after ecdysis in *P. brassicae* and found that much of the body was sufficiently hardened before ecdysis to prevent any further expansion or change in shape, *viz.* the antennae, proboscis, almost the whole of the thorax, the legs and pteralia, the tergites of the first two abdominal segments and the genitalia. After eclosion each of these become much more hardened and darker. In contrast the remaining parts of the head, the wings, the abdominal sternites 1 and 2 and the abdominal segments 3–7 are all soft at emergence and allow for some growth.

Crippling

Crippling, or the failure of normal wing development, is recorded throughout the lepidoptera. A specific case is in males of *P. brassicae* ab. *coerulea* (Gardiner, 1963). Gardiner found that 2% failed to eclose in the first two generations and subsequently this figure rose to 80%. The affected insects are unable to free themselves properly from the pupal case. During this process successful imagines often loose some of their wing scales. Gardiner also found that the incidence of crippling could be increased by raising the temperature from 12.5°C to 20°C. Feltwell (unpublished results) also found that ab. *coerulea* specimens which had survived a severe virus attack were afflicted with various degrees of crippling. Adding linseed oil to artificial diets has been recommended by David & Gardiner (1966) to alleviate crippling at eclosion.

Reabsorption of meconium

Moreau (1973) showed by using ink markers that some of the meconium is reabsorbed through the gut wall of the imago at eclosion. The function of this previously undocumented phenomenon was thought hypothetically to be either a) reclaiming of essential materials, b) mechanical stimulation of or chemical stimulation of specific receptors in the intestine, or c) conversion of certain elements of the meconium into hormone precursors capable of stimulating the central nervous system in co-ordinating wing expansion.

Courtship

It is a common phenomenon in the lepidoptera that the male seeks the female in the wild by using olfactory and visual cues. The female *P. brassicae* is considered to be more conspicuous (Darwin, 1889) and "more beautiful than the male" (Poulton, 1890). However, Poulton (1890), supported by Prof. Raphael Meldola, stated that the females probably take a more active part in the wooing. In the wild it is common to see the Large White going off into a "courting spiral" or chasing other whites in the vicinity.

In *P. rapae crucivora*, it has been proved that the male uses a visual stimulus to detect the female in the field, and may also use differences in the ultra-violet reflectance of the wings of males and females (the females reflecting more) for intraspecific communication (Obara & Hidaka 1964, 1968).

Mazokhin-Porshniakov (1957) showed that there is the small difference of 3.5% in the reflectivity of the wings of *Pieris*. The greenish-yellow spots on the hind-wing of *P. brassicae* have high surface reflection coefficients. the centre of the fore-wing has 4.5 and the female less than one reflectivity. Precopulation courtship in *P. brassicae* is said to take place in the heat of the day, as Avidov & Harpaz (1969) reported from Israel.

Although androconial secretions have been found in the male of *P. brassicae*, no investigations into any possible aphrodisiac effect have been conducted.

Interspecific courtship may occasionally occur. Adkin (1942) described in Cornwall the courtship between a female *P. brassicae* and a male *P. rapae*: "I noticed a female *brassicae* at rest on a leaf, openly inviting the attention of a male *rapae* which was responding readily, when a male *brassicae* arrived on the scene and challenged the *rapae* fiercely, but this was driven off by the latter, bag and baggage". This challenge was repeated three times with the same result. He also witnessed a *P. brassicae* pursuing a female *G. rhamni*.

Copulation

The sequence of events which lead from courtship to copulation is difficult to observe in the laboratory when virgin imagines are introduced, as they mate so fast (David & Gardiner, 1961). However, Chovet (1974, p. 60) recognised 15 behavioural reactions, which included five major ones, leading up to copulation in *P. brassicae*. These reactions have been drawn up as a flow diagram by Chovet and start from the recognition of the opposite sex. The sequence of events is dependant upon the theoretical emission of a pheromone by the male and its recognition by the female.

Copulation in *P. brassicae* commences with the male landing on or beside the female and curling his abdomen across so that it approaches the female genitalia from below and between her folded wings, and making union. After this he will turn round so that he will be facing away from the female (David & Gardiner, *loc. cit.*; Mukerji, 1961). A marked specimen *in cop.* has been illustrated by Roer (1955).

Rejection by the female is recognised by her opening her wings and raising her abdomen (Benz, 1970), an attitude which may be defensive when adopted by the male when molested by another of the same sex (David & Gardiner. *loc. cit.*). This posture was formerly regarded as female "acceptance" by some authors.

When disturbed, the copulating imagines often take to the air, and there is differing opinion as to which sex carries its partner in *P. brassicae*. Those in favour of the male carrying the female are Donzel (1837), Coltrup (1917), who saw four pairs in the New Forest, Heslop-Harrison (1935), Warren (1938), David & Gardiner (1961), Chandra & Lal (1977) (cf. Wheeler, 1918). Darwin (1889) stated that it was usual in butterflies for the male to carry the female, but in *Pieris* (species not stated) it was the other way round. Poulton (1890) also stated that the female carried the male *in cop.* Donzel (1837) and Warren (1920, 1938) both support the view that all species of the same genus have the same flying habit and this was substantiated by Heslop-Harrison (1935) who found that there were 90 cases of males carrying females in the Pieridae and only one the other way round, which he thought incidentally, was possibly an error. According to Chandra & Lal (1977) mating lasts from two to four hours.

Copulation usually occurs readily between 20–32°C, more so at higher temperatures, but not below 15°C. At lower temperatures the insects tend to mate for longer periods of up to 4 hours than at higher temperatures when copulation may last for an hour (see Figure 12 in Chovet's (1977) thesis). If imagines copulate in the late afternoon they will sometimes remain *in cop.* until the following morning. Mukerji (1961) stated that imagines usually remained *in cop.* for an hour (observations made in India), while Benz (1970, Switzerland) stated that normal copulation lasts about two hours. Poulton (1890) stated that copulation often terminated due to the "coyness" of the male. Artificial pairing is a technique which is usually successful with *P. brassicae*.

Copulation does not normally occur between other species but

Richards (1922) noted a female *P. brassicae* and a *P. rapae in cop.* but no ova were forthcoming. A male *P. brassicae* and a female *Colias croceus* (Geoffroy) were once seen *in cop.* by Valletta (1954), and copulation has also been observed with *Lycaena dispar* (Haworth) (Large Copper), (Portier, 1949; cf. Hering, 1926). Inhibition of interspecific copulation was obtained by Lorkovic (1953) with *P. brassicae* by decapitation experiments. When imagines were decapitated they were able to copulate with different species and produce hybrids.

David & Gardiner (1961) investigated the effects of environmental changes on the mating of *P. brassicae*. The effects tested were changes in temperature, light, atmospheric contamination, low and high humidities, cage size and population density. Various changes which take place in the insect, such as behaviour, age and frequency of copulation were also studied.

Mating is best effected when daylight illumination greater than 200 lumen per square foot is given. David & Gardiner (*loc. cit.*) tested the effect of four different artificial light sources, 500 watt incandescent tungsten filament gas-filled in clear glass; 80 watt "daylight" fluorescent MCF/U; a 400 watt mercury vapour, MA/V; and a 140 watt sodium vapour, SO/H but had poor results in their experiments. The tungsten bulb was the only light source in which only certain individuals would mate. Nothing seems to act better than natural sunlight in stimulating *P. brassicae* imagines to mate. If a cage of butterflies is taken outside into direct sunshine imagines will be seen to mate very readily.

The contamination of the air of a normal laboratory does not alter the mating activity of *P. brassicae*, nor does the size of the cage. They will often mate in a cage of one square foot (David & Gardiner, *loc. cit.*). High humidity does, however, depress copulation.

Age

Males and females do not normally mate from 1–18 hours after eclosion, however, females tend to mate more readily between 46–66 hours after eclosion than males, which tend to mate earlier than females (David & Gardiner, 1961). This disparity may be due to the habit of males eclosing slightly earlier than females from laboratory cultures. Dusaussoy & Delplanque (1964) found that matings tended to occur on the third day, although some would occur earlier on the second day or later on the sixth day. Chovet (1974) believed that the mating response varies according to the individual.

Chandra & Lal (1977) working on *P. brassicae* in India recorded two peaks of mating activity, one from 0800–1000 hours soon after eclosion, the other between 1200–1400 hours. Imagines over four days old did not mate in their laboratory.

Frequency of mating

David & Gardiner (1961) studied the frequency of mating in two groups of 20 males and females over 10 day periods and found that by the 10th day 18 females had mated twice, four mated three times and two four times. Of the males, by 10 days 14 had mated twice or three times. The effective time between matings for females was found to be between six to nine days. Chandra & Lal (1977) believed that only one mating per individual occurred in their laboratory.

Imaginal life

Imaginal females are reported to live in the wild (in Israel) between 5–17 days (Avidov & Harpaz, 1969), but in the laboratory they do appear to live longer, between 5–36 days according to David & Gardiner (1961); however, few reach the end of this period. The length of life is very dependant upon food, temperature and humidity. Freshly emerged imagines can be safely stored for 7–10 days at 12.5°C and 60% relative humidity in dim fluorescent light without any effect on the viability of the ova or mating.

Roosting site

Very little appears to have been written on this topic, although Frohawk stated at a meeting of the South London Entomological and Natural History Society that *P. brassicae* selected pale leaves as roosting sites (SLENHS, 1917). Dannreuther (1933, 1935) recorded that *P. brassicae* slept on thistles, and that some people thought that they selected the same perches or flowering plants as migrant birds. Unlike the vanessids, which are well known to rest head facing downwards, *P. brassicae* always rests head facing upwards. Rothschild (1980, pers. comm.) reports that in a large greenhouse *P. brassicae* roost communally in small groups away from the sun and occasionally on plants.

Voltinism

The only serious study into the number of generations of *P. brassicae* in different areas was carried out by Klein (1932a,b) who published a list of 49 localities (mostly capital cities) of 40 eastern and western European and north African countries, for which he gave the number of generations and the months when *P. brassicae* was likely to be seen on the wing. Table 3.1

Table 3.1 Generations of *P. brassicae*.

Country	Latitude	Number of generations	Authority
Finland (south and central)	63	2	Junnikkala, 1966
USSR (Leningrad)	60	1	Danilevski, 1961
Scotland (Aberdeen)	58	1	Graham-Smith & Graham-Smith, 1930
Scotland (Canna)	57	1	Campbell, 1951
Scandinavia (south)	57	3	Haugum, 1976, pers. comm.
Wales (Abergavenny)	51	2	Tulloch, 1938
Poland	53	2	Starega, 1976, pers. comm.
Netherlands	52	3	Eitschberger, 1968
England (Hampshire)	51	2	Goater, 1974
Luxembourg	51	4	Eitschberger, 1968
Italy (Oberitalien)	48	2	Wolfsberger, 1971
Czechoslovakia	48	2–3	Spitzer, 1975 pers. comm.
Germany (Rhineland)	47	2–3	Blunck, 1950
USSR (Transcaucasia)	43	5	Shapiro, 1976
Palestine (Jaffa)	32	7	Klein, 1932a,b
Palestine	32	5–7	Gardiner, 1978

lists the numbers of generations to be seen in various countries with an indication of the latitude, based on observations made mostly since Klein's time.

The number of generations of *P. brassicae* is dependent on the faculative control of diapause by the daylength. Within these limits the actual number of generations will be dependent upon temperature which can either increase or decrease the development rate. At 20°C in the laboratory, larval and pupal development are complete in 37 days, but at 25°C this time period is reduced to 24 days (David & Gardiner, 1962b). As the centre of the European continent warms up to higher temperatures during the summer than the periphery, it is worth bearing this in mind when looking at the results. It must also be borne in mind that results obtained at the constant diel temperature of a laboratory may not apply to the widely fluctuating diel temperatures of the natural climate.

One generation per year is typical of northern European areas, such as the Shetland Isles, Scotland and the Outer Isles, northern Ireland, USSR (Leningrad) and Germany DDR (Berlin) (Table 3.1), although in these areas and others, when good summer weather conditions prevail there may be two generations. This is true for Belgium, Denmark, Ireland, Netherlands, Norway, Scotland, Sweden and Yugoslavia (Klein,

1932a,b). Occasionally, a third generation may result (Easton, 1946 – in Sussex, England).

Further into the centre of Europe, two generations a year is the rule, but in good years this may rise to three. In Czechoslovakia a rare third generation occurs in the lowland regions of Moravia and Slovakia (Spitzer, 1975, pers. comm.). Three generations a year have been recorded in Italy (Oberitalien), Netherlands and southern Scandinavia (Table 3.1), but in these areas typically the third generation occurs only in very favourable years. For instance, in Scandinavia the first and second generations are in May–June, July–August, and the third generation may be in August–September (Haugum, 1976, pers. comm.). In Poland, the generations start earlier in the year, end of April to second week in June, second week of July to first week of September and the third generation (in a good year) beginning in the middle of September (Starega, 1975, pers. comm.).

In areas where the summer temperatures are even warmer, such as in the south of France, Malta, southern Italy, or when exceptional conditions induce another generation in northern Europe, such as in Luxembourg or Germany DDR, a fourth generation may occur. Four generations may also be found in USSR (Astrakan, Tashkent, Baku), Turkey, Nepal and Tunisia. *P. brassicae* is on the wing throughout the year in Malta (Valletta, 1966) and for 10 months of the year (not December and January) in the south of France.

Five generations are occasionally encountered in Transcaucasia in USSR where *P. brassicae* is a pest (Shapiro, 1976), and in Syria, Libya and north India (Klein, 1932a,b). Data given by Bodenheimer (1930) in Klein, give the precise dates of the five generations in Jerusalem correlated with temperature and humidity.

Seven generations a year appears to be the maximum potential that *P. brassicae* can accomplish in the wild, recorded in Palestine. (Jaffa), although up to six generations a year are found in north Africa (Tangier, Algeria, Tripoli) (Klein, *loc. cit.*). During the hottest months of the year in Palestine, July and August, at 25.3°C the fourth generation *P. brassicae* develop in 25 days; this compares almost exactly with results found in the laboratory by David & Gardiner at this temperature (1962b). It is interesting to note that in the laboratory eight generations of *P. brassicae* can be achieved in continuous culture.

Abundance

Fluctuation in numbers

Pieris brassicae was much more abundant in the past than it is today in western Europe. Early reports mention cabbage fields thick with "whites" and migrations of "whites" as thick as "snowstorms", "blotting out the sun" (Williams, 1930).

Several authors have discussed the fluctuations in abundance of *P. brassicae* in Great Britain and Ireland and have offered explanations (Grover, 1896; Castel Russell, 1925; Gardiner, 1960). Frohawk (1934) astutely observed that "Although the Large White occurs annually throughout Britain in varying numbers and usually abundantly, owing to its migratory habits, it is occasionally scarce all over the greater part of the country, or even over the whole of the British Isles". Castle Russell (*loc. cit.*) wondered whether over-collecting was to blame for the occasional scarcity of *P. brassicae*, and Gardiner (1960) thought that virus disease which arrived in Britain in 1955 was the cause rather than too much parasitism.

In considering the abundance of *P. brassicae* it must be realised that records are biased towards those years when there were many whites and those years when they were reported as being "conspicuous by their absence" (Williams *et al.*, 1942), since those years of great abundance and rarity are usually noted, while years of average numbers are not. Dannreuther & French collated records for Great Britain and Ireland for the period 1933–1961 from which useful information on the abundance of *P. brassicae* can be obtained. Information supplied to both these workers came from many localities throughout the country and the continent, as well as from professional and amateur entomologists.

Table 3.2 indicates the years of comparative abundance and scarcity of *P. brassicae* in Great Britain and Ireland. Many more authorities could be sought from the literature, particularly in the "notes and communications" of many of the smaller entomological journals, but the selection given here is thought to be representative enough to indicate those years of particular note.

There are examples of *P. brassicae* being reported in the same year as abundant in widely spaced localities, as in 1926 in Co. Down, Ireland (Johnson, 1927) and Scarborough, Yorkshire (Walsh, 1952), or scarce in different localities, as in 1894 in Surrey (Grover, 1896) and Devon (Mitchell, 1894). Examples can also be found of *P. brassicae* being reported in the same year as abundant in one place yet scarce in another; for instance, in 1905 *P. brassicae* was plentiful in Ireland yet rare in England (South, 1936), in 1940 abundant in Scotland (Cameron, 1941) but scarce overall (French, 1958), and in 1948 abundant in Kent (Massee, 1949) yet rare in Wiltshire (Kempe, 1949). There are also occasions when the first generation of *P. brassicae* was common and the second generation scarce, as in 1912 (Prideaux, 1913), or vice versa, as in 1942 (Dannreuther, 1943).

Even in the same locality the abundance of *P. brassicae* may fluctuate

78

from year to year; for instance it was extremely abundant in 1939 and 1940 (Goater, 1974) but was virtually absent in Hampshire and the Isle of Wight in 1953. Walsh (1952) noted that in Scarborough *P. brassicae* was abundant in 1909, 1926, scarce in 1927 and 1928 and abundant in 1937, while Baynes (1973) commented on the presence of a large population of *P. brassicae* in Ireland in 1959, followed by a scarcity and then a gradual build up of the population until 1962.

The decline of *P. brassicae* is a phenomenon which has not escaped attention before: Vaughan (1898a,b) noted that *P. brassicae* was not common in Wales and was becoming rarer each year. Abundance of the three pierids in relation to each other each year has also attracted attention: Smith (1950) noted that in 1950 there were more *P. napi* than *P. brassicae* or *P. rapae* at Bicester, Oxfordshire, and in the following year Owen (1952) noted that *P. rapae* was more abundant than *P. brassicae*. He commented that if the pierids were common in August, they were likely to be even more abundant in the autumn. In 1975 and 1976 in Yorkshire both *P rapae* and *P napi* were very common, *P. brassicae* last being common in 1958 (Haxby, 1978, pers. comm.). In the north of England *P. napi* was reported to have replaced *P. brassicae* in the region west of Durham (Dunn, 1975, pers. comm.).

On the European Continent, Feltwell (1977a,b) noted that *P. rapae* was common in many regions from the Baltic to the Mediterranean during the summer of 1973, and that *P. brassicae* was absent from a locality in the north of Germany where it used to be common. Information from Bulgaria is that *P. brassicae* populations have been decreasing during the last decade possibly due to their treatment with insecticides (Kwartirnikova, 1978, pers. comm.), and in Switzerland Sauter (1978, pers. comm.) believed that *P. brassicae* has become scarcer today than 20 years ago. More recently it has been reported that in Norway the population of *P. rapae* has increased so much in the last 30 years that "a hundred could be seen without turning my head" (Lees, 1978, pers. comm.). *P. brassicae* had been scarce in the same locality near Oslo for some years.

Estimating abundance

Information on the abundance of butterflies during the last few years is being assembled by Pollard at Monks Wood Experimental Station, Huntingdonshire. Pollard's (1977a,b) method for estimating the abundance of butterflies uses an index value, which is calculated from the sum of the mean figures for numbers of butterflies seen in a defined area each week between April and September.

Through the years 1973 to 1977 the index values for the two generations

Table 3.2 Abundance of *P. brassicae* in Great Britain & Ireland.

1842	large immigration, Dover (Johnson, 1908)
.	
1864	abundant, Isle of Wight (Hadfield, 1864)
.	
1884	not abundant, Midlands (Hardcourt-Bath, 1886)
.	
1886	very abundant, Sussex (Blaber, 1886, 1888)
.	
1890	scarce, London (Buckel & Prout, 1898)
.	
1892	rather uncommon (Blackmore, 1892)
1893	abundant prior to 1893 (Grover, 1896)
1894	very uncommon, Devon (Mitchell, 1894)
	none seen, Surrey (Grover, 1896)
1895	none seen, Surrey (Grover, 1896)
1896	rare (Pyett, 1898)
1897	numerous, Wales (Bland, 1897)
1898	very common, Staffordshire (Hill, 1899); most abundant, north Wales (Tetley, 1899)
.	
1901	in profusion, Surrey (Oldaker, 1902)
.	
1903	rarely seen, Essex (Mathew, 1903)
.	
1905	plentiful in Ireland, rare in England (South, 1936)
.	

```
1907 ⎫
1908 ⎬ scarce, Hampshire (Corbin, 1912)
1909 ⎬ abundant, Yorkshire (Walsh, 1952)
1910 ⎭
```

1911	very abundant, London, New Forest (Coltrup, 1911–12, Corbin, 1912)
1912	generation I common, generation II scarce (Prideaux, 1913)
.	
1914	very common, Scotland (Macdougall, 1915)
.	
1917	abundant, Oxfordshire (Walker, 1918); plentiful (Porritt, 1917)
.	
1919	exceedingly rare, Sussex (Adkin, 1920); common, Isle of Wight (Cornell, 1919)
.	
1924	very abundant, Oxfordshire (Walker, 1924)
1925	common (Poulton, 1928), very abundant, Westmorland (Birkett, 1954)
1926	very abundant, Yorkshire (Walsh, 1952); scarce (Frohawk, 1926b); very abundant, Co. Down (Johnson, 1927)
1927	scarce, Yorkshire (Walsh, 1952)
1928	scarce, Yorkshire (Walsh, 1952)

Table 3.2 continued

1930	countless thousands (Kettlewell, 1931)
1931	scarce, Essex (MAFF, 1933); conspicuous by its absence (Common, 1931)
1932	1932–34 above normal numbers (MAFF, 1936)
1933	Kent, Sussex (Dannreuther, 1933)
1934	Wales, Pembroke, Leicestershire (Dannreuther, 1934)
1935	below average numbers (Dannreuther, 1935)
1936	many records (Dannreuther, 1937)
1937	abundant, Yorkshire (Walsh, 1952); unusually common (Grant, 1938; Dannreuther, 1937, 1938)
1938	many records (Dannreuther, 1938)
1939	Hebrides (Dannreuther, 1940; 1941)
1940	abundant, Scotland (Cameron, 1941)
1942	generation I scarce, generation II troublesome (Dannreuther, 1943)
1943	above normal (MAFF, 1954); many records (Dannreuther, 1944)
1944	bad year for all butterflies (Dannreuther, 1945)
1945	Rhum (Heslop-Harrison, 1946); many records (Dannreuther, 1946, 1947)
1946	scarce, Chester (CSNHLA, 1947)
1947	Devon (Dannreuther, 1948); Skokholm (Condor, 1949); very common in central London (Worms, 1949)
1948	especially abundant (Massee, 1949); Cheshire (Dannreuther, 1949); almost rare, Wiltshire (Kempe, 1949)
1949	late influx (Dannreuther, 1949)
1950	Firth of Forth (Dannreuther, 1951)
1951	not very abundant (Dannreuther, 1952)
1952	notably scarce, Kent (SLENHS, 1953); excessively common, Essex (Huggins, 1952); poor year (French, 1953)
1953	Dorset, Kent, Yorkshire (French, 1954, 1955)
1955	Essex, (French, 1956)
1957	very bad year, scarce (French, 1958)
1958	very common, Yorkshire (Harby, 1978 pers. comm.)
1959	Shetlands, Ireland (French, 1962a)
1960	scarce (Thimann, 1961); scarce (Alston, 1962); scarce (Pott, 1961); Inverness, Isle of Man (French, 1962a,b)
1961	scarce only in south of England (Alston, 1961) – but plague in Ireland, Surrey, Dorset (French, 1962a)

of *P. brassicae* at Monks Wood gradually increased to a peak in 1975 and declined to zero in 1977 (Figure 3.2; partly from Pollard, 1977). At 24 localities in four regions of England and Wales (E. Midlands & E. Anglia; S. & S. East; S. West & S. Wales; N. Wales & Dyfed) during 1976 and 1977 the index value was consistently higher in the second generation and often, as at Monks Wood, down to zero in the first generation (Pollard, 1978, pers. comm.). The significance of this is not clear but could in some way be due to the migrating habit of *P. brassicae*. During the same period

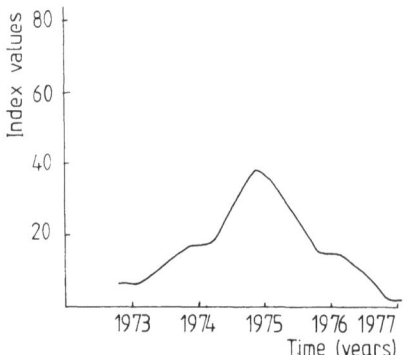

Fig. 3.2 Index values for *P. brassicae*
(after Pollard, 1977a,b and Pollard pers. comm. involving work carried out at the
Institute of Terrestrial Ecology, financed by the Nature Conservancy Council)

the numbers of *P. rapae* fluctuated around zero too. However, Monks
Wood is probably not typical of conditions elsewhere in Great Britain.
Information collected in 1976 and 1977 on *P. brassicae* and *P. rapae* did
not show any compensation in favour of *P. rapae* at the expense of *P.
brassicae* (Pollard, 1978, pers. comm.).

Information from pest reports

Some indication of the size of populations of *P. brassicae* and *P. rapae* can
be assessed from recent accounts of depredations wrought by these species
in the European part of Austria, Czechoslovakia, Hungary, Poland and
USSR (see Chapter 13). Many reports from south west USSR state that
both *P. brassicae* and *P. rapae* are pests in the same region, being
responsible in some years for a 50% loss of the white cabbage crop
(Voltukhov, 1975). In Kazakhstan, the Transcaucasians and the Chu
Valley to the east of the Aral Sea, both *P. brassicae* and *P. rapae* are
important pests of crucifers (Shapiro, 1976; Moiseeva *et al.*, 1975). It has
been calculated that with a 4% loss of a dozen crops throughout its world
range, *P. brassicae* would incur £100m ($224m)* annual loss at current
prices (Feltwell, 1981, in press). Demin (1965) reported that *P. rapae*
caused more damage than *P. brassicae* in the Krasnadar region in USSR.

Increase in population

The population of *P. brassicae* in Great Britain & Ireland is augmented in
years when there are immigrations from the continent. Johnson (1908)
believed that "our home supply" is at times reinforced by vast flights of

*As of 6th Jan. 1980 £1 = 2.24$.

82

these butterflies from the shores of France, and Worms (1962) noted that the population of *P. brassicae* in Wiltshire was supplemented by huge invasions from the continent as occurred in 1955. The years of the major immigrations of butterflies into Great Britain this century have been 1913, 1916–17, 1930, 1937, 1940, 1945, 1947, 1955 and 1958.

Reference to the British Ministry of Agriculture, Fisheries and Food monthly summaries of the incidence of crop pests in England and Wales over the last 30 years indicates that there is a peak in the population of *P. brassicae* every seven to nine years (Rivers, 1978, pers. comm.).

Natural checks of population

Weather

One of the natural checks of population is the vagary of the summer weather; indeed Porritt (1917) has said that a severe winter is the forerunner of a good entomological year. For example, Murray (1934) in Dumfrieshire noted that there had been several bad years leading up to 1933 when *P. brassicae* was abundant; Dannreuther (1945) stated that 1944 was a bad year for all butterflies and Michael (1951) said that 1959 was poor, since 55 species of lepidoptera, including *P. brassicae*, were rarer than in 1958.

The "best" years this century in terms of hottest summers were 1933 followed by 1947 and 1976. In both 1933 and 1947 large immigrations of insects occurred throughout Great Britain. However, the other years of large immigration were not also exceptional summers; 1937 was a "good" summer, 1939 an "average" summer, 1940 an "above average" summer, and 1945 "not exceptional". The worst summers have been 1907, 1909, 1920, 1922, 1954, 1956 and 1972. Between the years 1932 and 1945 inclusive the summer temperatures were all above average (Skelton, 1974, pers. comm.) and may have affected the size of the populations.

Parasites

The hymenopterous endoparasitic ichneumon fly, *Apanteles glomeratus* (L.) is responsible for inflicting heavy losses on larvae of *P. brassicae* and has been known to cause 100% mortality in some countries (Chapter 14). It appears to be density dependent upon its host, that is, as infestations of *P. brassicae* build up, so too does the population of *A. glomeratus*. In India Lal & Chandra (1976) noted that there were greater numbers of *A. glomeratus* in June than in May. *A. glomeratus* is thought to overwinter as a diapausing pupa outside its host corpse. Warren (1975) reported that in

Staffordshire *A. glomeratus* controlled populations of *P. brassicae*, and Goater (1974) noted that it was only after large immigrations in Hampshire that *P. brassicae* "gained temporary ascendency over its parasites".

Apanteles glomeratus is the major hymenopterous parasite of *P. brassicae* but others which inflict some control are the egg parasite *Trichogramma evanecens* Westwood and the pupal parasite *Pteromalus puparum* (L.) There are several other hymenopterous species of parasite which take their toll of ova, larvae and pupae of *P. brassicae*, and make up a very complex system of parasitism and hyperparasitism involving other pierid species. It is interesting to note that Richards (1940) also found that parasites and predators inflicted heavy damage on *P. rapae*.

Pathogens

Naturally occurring pathogenic organisms, such as viruses (granulosis and cytoplasmic polyhedrosis), bacteria (mainly *Bacillus thuringiensis* Berliner) and various species of protozoa and fungi, inflict their own damage on populations of *P. brassicae* in the wild. These organisms may be transmitted from one population to another with immigrants; this was thought to be the mode of entry of the granulosis virus into Britain in 1955 (Smith & Rivers, 1956). The granulosis virus is thought to be density dependent on its host *P. brassicae*, although it appears to be effective at a lower density level than is required for *A. glomeratus* (Evans, 1978, pers. comm.).

Interspecific relationships

World range

Both *P. brassicae* and *P. rapae* have a typical Palaearctic range, although *P. rapae* has successfully established itself in north America and New Zealand, and *P. brassicae* has established itself in Chile during the last six years. The world distribution maps published by the Commonwealth Institute of Entomology (C.I.E., 1952 for *P. rapae* and 1976 for *P. brassicae*) also show that *P. rapae* is widely distributed in China and USSR in the eastern part of their world range.

Life styles / Behaviour

Ova of *P. rapae* are laid singly over the foodplant leaves and therefore need more searching by egg parasites than do those of *P. brassicae*. The larva of *P. rapae* is solitary and moreover spends most of its time between the inner heart leaves of the foodplant and thus has a greater chance of avoiding detection by parasites. By virtue of this habit, *P. rapae* are less susceptible to insecticides such as rotenone, which are applied to the leaves, than *P. brassicae*. *P. brassicae* ova on the other hand are laid generally in clusters on the underside of the leaf and are very susceptible to egg parasites such as *Trichogramma evanescens*, which may parasitise up to 30 host larvae at any one time. The larvae of *P. brassicae* live and feed gregariously on the laminae of outer leaves and are thus very susceptible not only to insecticides but also to hymenopterous parasites. Cabbage plants can readily support both *P. rapae* and *P. brassicae* larvae at the same time and, providing there is enough food, no competition for food appears to exist between the two species. However, if large numbers of larvae are present, as in severe depredations, then competition for food would necessarily take place. In experimental conditions in the laboratory, it has apparently been shown that *P. rapae* larvae will occasionally eat fifth instar larvae of *P. brassicae* in preference to fresh cabbage.

P. napi has a solitary larva which feeds on a variety of cruciferous foodplants including charlock (*Sinapis arvensis* L.) and garlic mustard (*Alliaria petiolata* (Bieberstein) Cavara & Grande) and would only be expected to compete with *P. brassicae* in its first generation in those areas where *P. brassicae* in its first generation in those areas where *P. brassicae* is known to support itself on wild crucifers. It is also a butterfly of shady and damp situations and is rarely if at all found on cabbage.

There is much evidence that *P. rapae* has replaced the native *Pieris* species (*protodice, virginiensis*) in north America.

Effect of chemical control

The chemical control methods recommended for the control of cabbage caterpillars are particularly well documented in Great Britain by the Ministry of Agriculture, Fisheries and Food (MAFF) and are probably typical of the chemical control measures existing on the continent. Before 1942, the MAFF recommended soap and water, salt and water, derris, slaked lime in gardens or lime plus soot, many of these being ineffectual compounds by today's standard. Other recommended plant derivatives were nicotine and pyrethrum (Smith, 1931) and hellebore powder (Curtis, 1877).

Since 1942, the MAFF have persistently recommended one compound, derris, but this was only recommended for garden use with pyrethrum in 1967–69. DDT has been recommended from 1947 until its present banning, while mevinphos and trichlorphon have been recommended from 1962 and 1967 respectively.

These insecticides are just some of the many organochlorines and organophosphorus compounds that farmers have used persistently against *P. brassicae* for the last 30 years. Several companies are involved in the large scale manufacture of many commercial insecticides in the west, and much has been published on the experimental use of these compounds against *P. brassicae*. There is a great lack of information on insecticide use from the USSR. It may be that they have lagged behind the west in development of so many organic insecticides, and this may account for the accounts of depredations in eastern Europe in recent years which have now been largely overcome in western Europe.

It seems likely that two factors have changed in the environment of *P. brassicae* during the last 30 years which have significantly upset its biological equilibrium and caused its numbers to decline. These factors are natural control of granulosis virus and man's chemical war on the cabbage pests which have selectively decreased the *P. brassicae* population of larvae. Other factors such as attack from hymenopterous and dipterous parasites would appear to have played a subsidiary and fluctuating role in regulating population size. Gardiner (1978, pers. comm.) suggested that the present paucity of *P. brassicae* and the lack of large migrations which used to occur are due to its former large breeding areas in the north of Europe having been brought into other use, such as for agriculture, airfields and motorways.

P. rapae is sometimes more common than *P. brassicae*; it has slightly different habits from *P. brassicae*, which may confer upon it certain advantages for survival over *P. brassicae*. There is one other factor which deserves further attention and that is the change of habitat. In Taiwan *P. rapae* has ousted *Pieris canidia sordida* Butler (= *Pieris canidia canidia* Linnaeus) (The Formosan Cabbage Worm) since 1950, preferring open cultivated land to woody areas (Su, 1978, pers. comm.). With the increasing size of monocultures of brassicas in eastern and western Europe, it is possible that this has inadvertently increased the preferred habitat of *P. rapae*.

References cited

Adkin, R., 1920. Random recollections of the season of 1919 at Eastbourne. *Entomologist* 53: 78–80.
Adkin, G.T., 1942. *Pieris brassicae* and *Pieris rapae*. *Entomologist* 75: 91.
Agassiz, L., 1846. *Nomenclatoris Zoologici Index Universalis*. Soloduri.
Alston, H.N.E., 1961. Letters to the editor. *Bull. amat. ent. Soc.* 20: 9.
——— 1962. The Large White Butterfly (*Pieris brassicae* L.). *Bull. amat. ent. Soc.* 21: 17–18.
Avidov, Z. & Harpaz, I., 1969. *Plant Pests of Israel*. Israel University Press.

Baynes, E.S.A., 1973. *A Revised Catalogue of the Irish Macrolepidoptera and Supplement 1970*. E.W. Classey, Faringdon.

Benz, G., 1970. L'influence des stimuli externes sur la gamétogenèse des insectes. *Coll. int. Centre nat. Recherches Sci.* No. 189.

Berge, F., 1851. *Schmetterlingsbuch*. Verlag von Scheitlin, Stuttgart.

Birkett, N.L., 1954. The present status of the butterfly population of the Kendal District, Westmorland. *Entomologist's mon. Mag.* 90: 293.

Blaber, W.H., 1886. Unusual abundance of the larvae of *Pieris brassicae*. *Entomologist* 19: 299–300.

—— 1888. Abundance of Pieridae. *Entomologist* 21: 53.

Blackmore, E.H., 1892. Notes on lepidoptera in Shropshire. *Entomologist* 25: 321.

Bland, F.D., 1897. Notes on the Macro-lepidoptera of the Conway Valley, N. Wales. *Entomologist* 30: 19–20.

Blankaart, S., 1688. *Schouburg der Rupsen Wormen Maden en Vliegende Dierkens*. Hoorn, Amsterdam.

Blunck, H., 1950. Zur Kenntnis des Massenwechsels von *Pieris brassicae* L. mit besonderer Berücksichtigung des Dürrejahres 1947. *Z. angew. Ent.* 32: 141–171.

—— 1953. *Tierische Schädlinge an Nutzpflanzen*. Vol. 1. Lepidopteren und Trichopteren. Berlin.

Buckel, F.J. & Prout, L.B., 1898. The fauna of the London district. *Trans. Cy Lond. ent. nat. Soc.* 51–56.

Cameron, A.E., 1941. Insect and other pests of 1940. *Trans. highl. agric. Soc. Scotland, Edinburgh* 1941: 1–21.

Campbell, J.L., 1951. An experiment in marking migratory butterflies. *Entomologist* 84: 1–6.

Castle Russell, S.G., 1925. The causes of scarcity of *Polyommatus thetis* and *coridon*. *Entomologist* 58: 100–101.

Chandra, J. & Lal, O.P., 1977. Mating behaviour of cabbage butterfly, *Pieris brassicae* Linn. *Indian J. Ent.* 38: 197–198.

Chester Society of Natural History, Literature and Art, 1947. Records of lepidoptera 1947–1953. *Chester Soc. nat. hist. lit. art.* 1947.

Chovet, G., 1974. *Méchanismes d'accouplement, structure et fonctionnement de l'appareil reproducteur mâle de Pieris brassicae (L.) (Lepidoptera: Pieridae)*. Thèse Diplome de Docteur 3 ème cycle, Université de Paris VI.

—— 1977. Les stimulus visuels dans le déclenchment de la parade nuptiale chez *Pieris brassicae* L. *C. r. Acad. Sci. Paris* 284: 2127–2130.

Clerck, C., 1759. *Icones Insectorum Rariorum*. C. Linnai.

Coltrup, C.W., 1911–12. Notes on the season. *Proc. S. Lon. Ent. nat. hist. Soc.* 1912: 63.

—— 1917. Flying habits of butterflies when paired. *Entomologist's Rec. J. Var.* 29: 246.

Common, A.F., 1931. Essex lepidoptera. *Entomologist* 64: 29.

Commonwealth Institute of Entomology, 1952, 1976. Distribution maps of insect pests. Series A.

Condor, P.J., 1949. Observations on a migration of *Pieris brassicae* L. at Skokholm island, Pembrokeshire, in August 1947. *Proc. R. ent. Soc. Lond.* A. 24: 35–38.

Corbin, G.B., 1912. New Forest notes, 1911. *Entomologist* 45: 301–303.

Cornell, E., 1919. Some notes on the butterflies of the south coast of the Isle of Wight, 1919. *Entomologist* 52: 279–280.

Cottrell, C.B., 1964. Insect ecdysis with particular emphasis on cuticular hardening and darkening. *Adv. Ins. Physiol.* 2: 175–218.

Curtis, J., 1877. *Farm Insects.* Blackie & Sons, London.

Danilevski, A.S., 1961. *Photoperiodism and Seasonal Development of Insects.* Oliver & Boyd, Edinburgh & London.

Dannreuther, T., 1933. Migration records. *Entomologist* 66: 186–190.

—— 1934. Migration records. *Entomologist* 67: 10–14.

—— 1935. Migration records. *Entomologist* 68: 5–9, 185–188.

—— 1937. Migration records. *Entomologist* 70: 108–111, 200–202.

—— 1938. Migration records. *Entomologist* 71: 60–66; 72: 9–15.

—— 1940. Migration records. *Entomologist* 73: 29–33, 62–63.

—— 1941. Migration records. *Entomologist* 74: 54–62.

—— 1943. Migration records. *Entomologist* 76: 73–80.

—— 1944. Migration records. *Entomologist* 77: 55–60.

—— 1945. Migration records. *Entomologist* 78: 49–56.

—— 1946. Migration records. *Entomologist* 79: 97–110.

—— 1947. Migration records. *Entomologist* 80: 107–112, 137–144.

—— 1948. Migration records. *Entomologist* 81: 73–83, 110–117.

—— 1949. Migration records. *Entomologist* 82: 73–78, 105–110; 83: 109–114.

—— 1951. Migration records. *Entomologist* 84: 85–90, 102–106.

—— 1952. Migration records. *Entomologist* 85: 125–131.

Darwin, C., 1889. *The Descent of Man and Selection in Relation to Sex.* John Murray, London.

David, W.A.L. & Gardiner, B.O.C., 1961. The mating behaviour of *Pieris brassicae* (L.) in a laboratory culture. *Bull. ent. Res.* 52: 263–280.

—— 1962a. Oviposition and the hatching of the eggs of *Pieris brassicae* L. in a laboratory culture. *Bull. ent. Res.* 53: 91–109.

—— 1962b. Observations on the larvae and pupae of *Pieris brassicae* (L.) in a laboratory culture. *Bull. ent. Res.* 417–436.

—— 1966. Rearing *Pieris brassicae* (L.) on semi-synthetic diets with and without cabbage. *Bull. ent. Res.* 56: 581–593.

Demin, G., 1965. The Cabbage White Butterfly. (In Russian.) *Zashch. Rast. Vredit Bolez* 1965: 39.

D.I.E., 1883. Sur l'accouplement de quelques genres de lépidoptères diurnes, et sur le genre piéridae. *Annls Soc. ent. Fr.* 6: 77–81.

Donzel, M., 1837. Sur l'accouplement de quelques genres de lépidoptères diurnes, et sur le genre pieridae. *Ann. Soc. ent. Fr.* 6: 77–81.

Driest, J. Ph., 1966. Parasiten von Koolrupsen. *Jaarverslag Ann. Ret. Alkmaar* 1966.

Dusaussoy, G. & Delplanque, A., 1964. L'élevage de *Pieris brassicae* L. en toutes saisons: accouplement et ponte en conditions artificielles. *Rev. path. veg. ent. agric. Fr.* 43: 119–134.

Easton, N.T., 1946. Butterfly Jotting in 1945. *Entomologist* 74: 37–41.

Egerton Collier, A., 1941. *Pieris brassicae*, 1939 & 1940. *Entomologist* 74: 70.

Eitschberger, U., 1968. Jahresbericht 1967 der Deutschen Forschungszentrale für Schmetterlingswanderungen. Pieridae. *Atalanta* 2: 168–175.

Eliot, N., 1944. Larval anthropomorphosis. *Entomologist* 77: 22–27.

Esper, E.J.C., 1777. *Schmetterlinge* 1: 52–55.

Fabre, J.H., 1911. *Les Ravageurs.* Delagrave, Paris.

—— 1912. *The Life of the Caterpillar.* Hodder & Stoughton.

Feltwell, J.S.E., 1977a. Migration of whites. *Entomologist's mon. Mag.* 112: 88.

—— 1977b. Skeletonisation by *Pieris brassicae*. *Entomologist's mon. Mag.* 112: 104.

—— 1981. The depredations of the Large White Butterfly (*Pieris brassicae* (L.)). *J. Res. Lepid.* (in press).

Ford, E.B., 1945. *Butterflies.* Collins, London.

French, R.A., 1953. Insect migration records, 1952. *Entomologist* 86: 157–164.

—— 1954. Insect migration records, 1953. *Entomologist* 87: 57–63.

—— 1955. Migration records, 1954. *Entomologist* 88: 123–130.

—— 1956. Migration records, 1955. *Entomologist* 89: 174–180.

—— 1958. Migration records, 1957. *Entomologist* 91: 101–109.

—— 1959. Migration records, 1958. *Entomologist* 92: 164–176.

—— 1962a. Migration records, 1959, 1960, 1961. *Entomologist* 95: 169–177, 204–211; 96: 32–38.

—— 1962b. Migration records, 1960. *Entomologist* 96: 204–216.

Frohawk, F.W., 1926a. *Natural History of British Butterflies.* Hutchinson, London.

—— 1926b. Scarcity of *Pieris rapae* and *Pieris brassicae* in the spring of 1926. *Entomologist* 59: 288.

—— 1934. *British Butterflies.* Ward, Lock & Co. Ltd., London & Melbourne.

Gardiner, B.O.C., 1960. Scarcity of the Large White (*Pieris brassicae* L.) *Bull. amat. ent. Soc.* 19: 99.

—— 1963. Genetical and environmental variation in *Pieris brassicae. J. Res. Lepid.* 2: 127–136.

—— 1978. Instar number and pupal colouration in Palestinian *Pieris brassicae. Proc. & Trans. Brit. ent. nat. Hist. Soc.* 11: 21–23.

—— 1979. Talk given at Pierid workshop held at the Royal Entomological Society, London on 17th October, 1979.

Goater, B., 1974. *The Butterflies and Moths of Hampshire and the Isle of White.* Classey, Berks.

Godan, D., 1949. Massenauftreten von Kohlweisslingsraupen (*Pieris brassicae*) in Berlin. *Biol. Zbl.* 1: 165–166.

Goedaert, J., 1700. *Mitamorphoses Naturelles, Histoire des Insectes.* (3 vols.). The Hague, Netherlands. Vol. 1. 33–38.

Graffinn, M.C., 1679. *Der Raupen Wunderbare Berivandelung und Sonderbare BlumenNahrung.* Graffin, Frankfurt.

Graham-Smith, G.S. & Graham-Smith, S.W., 1930. *Pieris brassicae* L. with special reference to aberrations from Aberdeenshire. *Entomologist's Rec. J. Var.* 41: 157–161, 173–180.

Granada T.V. Film, 1960.

Grant, K., 1938. A migration of cabbage white butterflies in Hertfordshire in May, 1937. *Entomologist* 71: 103–107.

Grassé, P.P., 1975. *Traité de Zoologie.* Masson, Paris. Vol. 8.

Grover, W., 1896. Scarcity of *Pieris brassicae. Entomologist* 29: 126–127.

Guerne, J., 1894. Invasion de chenilles de *Pieris brassicae. Annls Soc. ent. Fr.* 63: 241.

Guilbot, R., 1975. Les Fiches Techniques d'élevage. *OPIE Cahiers de Liaison* No. 8. 19–25.

Hadfield, H., 1864–1865. Abundance of larvae of *Pieris brassicae. Entomologist* 2: 147.

Haes, E.C.M., 1977. *Natural History of Sussex.* Flare Books: The Harvester Press.

Hardcourt-Bath, W., 1886. *Pieris brassicae* in the Midlands. *Entomologist* 19: 174–176.

Hering, M., 1926. *Biologie der Schmetterlinge.* Springer, Berlin.

Herold, M., 1815. *Entwicklungsgeschichte der Schmetterlinge, anatomisch und physiologisch bearbeitet. Atlas des Planches.* Krieger, Cassel und Marburg. Vol. 6.

Heslop-Harrison, J.W., 1935. Courtship and allied problems in insects. *Trans. Soc. brit. Ent.* 2: 115–135.

—— 1946. The abundance of *Pieris brassicae* larvae in Rhum in 1945. *Entomologist* 74: 70.

Hewitt, C.G., 1917. Insect behaviour as a factor in applied entomology. *J. econ. ent. Concord. N.H.* 10: 1.

Hill, J. & W., 1899. Notes from N. Staffordshire in 1898. *Entomologist* 32: 41.

Howarth, T.G., 1973. *South's British Butterflies.* Warne, London.

Hübner, J., 1806. *Sammlung Exot. Schmetterlinges.*

Huggins, H.C., 1952. Migrants in 1951. *Entomologist* 85: 42–43.

Johnson, T., 1878. *Illustrations of the Larvae and Pupae of British Lepidoptera.* Privately published.

—— 1887. *Familiar British Insects.* Privately published.

—— 1906. *A Life History of the British Butterflies.* Vol. 1. Privately published.

—— 1908. *The Butterflies of Essex.* Privately published.

Johnson, W.F., 1927. Lepidoptera in Co. Down, Ireland in 1926. *Entomologist* 60: 114–115.

Jones, A.L., 1960. *Observations on the relationships between feeding, growth and development in lepidopterous larvae.* Ph.D. thesis. University of Leeds.

Junnikkala, E., 1966. Effect of braconid parasitisation in the metabolism of *Pieris brassicae* L. A study of the nitrogenous compounds in the haemolymph at different ages reared in controlled laboratory conditions. *Annls Acad. Sci. fenn.* 100: 1–83.

Kaiser, P., 1949. Histologische Untersuchungen über die Corpora allata und Prothoraxdrusen der Lepidopteren in Bezug auf ihre Funktion. *Wilhelm Roux. Arch. EntwMech. Org.* 144: 99–131.

Karlinsky, A., 1977. *Recherches expérimentales sur les mécanismes endorcrines controlant la vitellogenèse chez un insecte lépidoptère, Pieris brassicae* L. Thèse Docteur Sci. Nat., Université Pierre et Marie Curie.

Kempe, J.E., 1949. Paucity of butterflies in 1948. *Entomologist* 82: 70–71.

Kettlewell, H.B.D., 1931. Collecting 1930 (Lepidoptera). *Entomologist's Rec. J. Var.* 43: 113–115.

Kirby, W.F., 1898. *European Butterflies and Moths.* Cassell, London.

Klein, H.Z., 1932a. Studien zur Oekologie und Epidemiologie der Kohlweisslinge. I. Der Einfluss der Temperatur und Luftfeuchtigkeit auf Entwicklung und Mortalität von *Pieris brassicae* L. *Z. angew. Ent.* 19: 395–448.

—— 1932b. Studien zur Oekologie und Epidomiologie der Kohlweisslinge. II. Zur Bionomie von *Pieris brassicae* L. und Deren Parasit *Microgaster glomeratus* L. *Z. wiss. Insektbiol.* 26: 192–199.

Krzywicki, M., 1962. Key for the identification of Polish insects, Pieridae, Papilionidae. (In Polish) *Polish ent. Soc. Lep. Warsaw* 27: 65–66.

Lal, O.P. & Chandra, J., 1976. Some parasites of cabbage worm *Pieris brassicae* L. (Lepidoptera: Pieridae) from Kulu Valley, Himachal Pradesh. (India). *Curr. Sci.* 45: 766–767.

Linnaeus, C., 1746. *Fauna Svecica Stockholmiae.* Edition 1. p. 244.

—— 1758. *Systema Naturae.* Edition 10. 1: 467.

Lorkovic, Z., 1953. L'accouplement artificielle chez les Lépidoptères et son application dans les recherches sur la fonction de l'appareil génital des insectes. *Physiol. Comp. oecol.* 3: 312–320.

Lundgren, L., 1975. Natural plant chemicals acting as oviposition deterrents on cabbage butterflies (*Pieris brassicae, Pieris rapae, Pieris napi*). *Zool. Scr.* 4: 253–258.

Macdougall, R.S., 1915. Insect pests in 1914. *Trans. highl. agric. Soc. Scotland* 1915.

McLeod, L., 1972. The distribution of insects relating to railway embankments. *Entomologist's Rec. J. Var.* 84: 69–71.

Massee, A.M., 1949. Notes on some interesting insects observed in 1948. *E. Malling Res. Stn Annual Report for 1948.* 102–107.

Mathew, G.F., 1903. Notes on the lepidoptera of the Harwich District 1902. *Entomologist* 36: 140–142.

Mazokhin-Porshniakov, G.A., 1957. Reflecting properties of butterfly wings and role of ultra-violet rays in the vision of insects. *Biofizika* 2: 352–362.

Measures, D.G., 1976. *Bright Wings of Summer.* Cassell, London.

Mellanby, K. & French, R.A., 1958. *Rothamsted Experimental Station, Annual Report for 1957.* 152–153.

Merian, M.S., 1717. *Der Rupsen Begin Voedsel en Wonderbaare.* Valk. Amsterdam.

Merrett, C., 1616. *Pinax Rerum Naturalium Britannicarum, contiens Vegetabilia, Animalia et Fossilia, in hac Insula reperta Inchoatus.*

Meyrick, E., 1895. *Handbook of British Lepidoptera.* Macmillan, London.

—— 1928. *Revised Handbook of British Lepidoptera.* Watkins & Doncaster, London.

Merrifield, F. & Poulton, E.B., 1899. The colour-relation between the pupae of *Papilio machaon, Pieris napi* and many other species, and the surroundings of the larvae preparing to pupate, etc., E. Experiments upon the pupae of *Pieris brassicae. Trans. ent. Soc. London.* 1899. Part IV. 369–433.

Michael, P., 1951. Spring Lepidoptera, 1951. *Entomologist* 84: 186–187.

Ministry of Agriculture, Fisheries and Food (MAFF), 1933. Bulletin No. 66. *Insect Pests of Crops, 1928–1931.*

—— 1936. *Insect Pests of Crops 1932–1934.*

—— 1954. *Report on insect pests of crops in England, Wales 1938–1943.* Technical Bulletin No. 5.

—— 1976. *Cabbage Caterpillars.* Leaflet No. 69.

Mitchell, A.V., 1894. Notes on *Pieris brassicae. Entomologist* 27: 348.

Moiseeva, N.V., Mashkina, L.G., Kalashnikova, G.I. & Ratomskaya, A.A., 1975. The effectiveness of the use of *Trichogramma* for control of certain cabbage pests in the Chu Valley, Kirgizia. *In: Arthropods of Agricultural Importance.* Edited by A.I. Protsenko.

Moore, F., 1904. *Lepidoptera Indica.* Reeve, London.

Moreau, R., 1973. *Recherches sur quelques aspects des phénomènes physiques métaboliques et physiologique qui accompagnent et conditionnent l'expansion des ailes de Lépidoptères.* Thèse Doct. d'État. Sc. Nat. Bordeaux. 145 pp. Ref. C.N.R.S. AO 8458.

—— 1974. Variations de la pression interne au cours de l'émergence et de l'expansion des ailes chez *Bombyx mori* et *Pieris brassicae. J. Insect Physiol.* 20: 1475–1480.

Moreau, R. & Gourdoux, L., 1971. Étude comparative du métabolisme respiratoire au cours de la seconde partie de l'ontogenèse et plus particulièrement pendant l'émergence et l'expansion des ailes chez *Pieris brassicae* et *Tenebrio molitor. C. r. Séance. Soc. Biol.* 169: 953–958.

Moss, J.E., 1933. The natural control of the cabbage caterpillars, *Pieris* spp. *J. Anim. Ecol.* 2: 210–231.

Moufet, T., 1634. *Insectorum Sive Minimorum Animalium Theatrum.* Thos. Cotes, Benjam, Allen.

Mukerji, G.P., 1961. On the biology of "Cabbage White", *Pieris brassicae. J. zool. Soc. India.* 13: 121–127.

Murray, J., 1934. Butterflies in Dumfriesshire in 1933. *Entomologist* 67: 92–93.

Newman, L.H., 1977. *Looking at Butterflies.* Collins.

Nicolson, S.W., 1975. *Osmoregulation, Metamorphosis and the Diuretic Hormone of the Cabbage White Butterfly, Pieris brassicae.* Ph.D. thesis. Cambridge.

—— 1976. Diuresis in the cabbage white butterfly, *Pieris brassicae* water and ion regulation and the role in the hind-gut. *J. Insect Physiol.* 22: 1623–1630.

Obara, Y. & Hidaka, T., 1964. Mating behaviour of the cabbage white, *Pieris rapae crucivora*, I. The "Flutter Response" of resting males to flying males. *Zool. Mag.* 73: 131–135.

—— 1968. The recognition of the female by the male on the basis of ultra-violet reflectance in the white cabbage butterfly, *Pieris rapae crucivora* Boisduval. *Proc. Japan Acad.* 44: 829–832.

Oldaker, F.A., 1902. Notes from Dorking for the season of 1901. *Entomologist* 35: 118–121.

Oldham, C., 1950. *Vegetable Growers Guide.* Lockwood, London.

Ormerod, E.A., 1878. Notes on injurious insects, *Report 1878.* West Newham & Co., London.

Owen, D.F., 1952. Dispersal of *Pieris brassicae. Entomologist* 85: 77.

Oxford Scientific Films, 1976. *Nature's Way, The Butterfly Cycle.* G. Whizzard / André Deutsch.

Paton, C.I., 1944. *Pieris brassicae*: pupation site. *Entomologist* 77: 63.

Petiver, J., 1795. *Rariora Naturae.* Londini.

Pictet, A., 1918a. Observations biologiques sur *Pieris brassicae* en 1917. *Bull. Soc. lépidopt. Genève* 4: 53–66.

—— 1918b. Les éclosions de papillons et la pression barométrique. *Bull. Soc. Lépidopt. Genève* 4: 67–74.

Pollard, E., 1977a. Monitoring butterfly numbers. *R. ent. Soc. Lond. Workshop on "Ecology and Evolution of Butterflies".* December, 1977.

—— 1977b. A method for assessing changes in the abundance of butterflies. *Biol. Cons.* (12) 115–134.

Pollard, E., Hooper, M.D. & Moore, N.W., 1974. *Hedges.* Collins New Naturalist Series, London.

Porritt, G.T., 1917. The season of 1917. *Entomologist's mon. Mag.* 53: 259–260.

Portier, P., 1949. *Biologie des Lépidoptères.* Lechevalier, Paris.

Pott, A.J.H., 1961. Scarcity of the Large White. *Bull. amat. ent. Soc.* 20: 81.

Poulton, E.B., 1890. *The Colours of Animals.* Kegan, Paul, Trench, Trüber & Co. Ltd. London.

—— 1928. Notes on the flight of *Colias croceus* and *Pieris brassicae* near Bere Regis, Dorset. *Proc. ent. Soc. Lond.* 78–79.

Prideaux, R.M., 1913. A few comparative notes of some diurni in the seasons 1912 and 1913. *Entomologist* 46: 324–328.

Pyett, C.A., 1898. Notes on Suffolk lepidoptera in 1897. *Entomologist* 31: 46–47.

Ray, J., 1710. *Historia Insectorum.*

Richards, A.W., 1922. *Colias croceus,* etc. in Derbyshire. *Entomologist* 55: 234.

Richards, O.W., 1940. Biology of *Pieris rapae* with special reference to factors controlling abundance. *J. Anim. Ecol.* 9: 243–288.

Roer, H., 1955. *Über Flug-und Wandergewohnheiten von Pieris brassicae* L. Doktorgrade. University of Bonn.

Rosel, A.J., 1781. *De Naturlyke Historie der Insecten.* Bohn, Amsterdam.

Rothschild, M. & Schoonhoven, L.M., 1977. Assessment of egg load by *Pieris brassicae* (Lepidoptera: Pieridae). *Nature, Lond.* 266: 352–355.

Rothschild, M., Gardiner, B.O.C., Valadon, L.R.G. & Mummery, R.S., 1975. The Large White Butterfly: oviposition cues, carotenoids, and changes of colour. *Proc. R. ent. Soc. Lond.* 40: 13.

Schrank, F.V.P., 1801. *Fauna Bioca.* 2: 152.

Seitz, A., 1929–1932. *The Macrolepidoptera of the World.* The Palaearctic Butterflies, Supplement. Alfred Kernen Verlag, Stuttgart.

Sepp, J.C., 1762. *Nederlandsche Insecten.* Dag-Vlinders, Amsterdam.

Shapiro, V.A.A., 1976. *Apanteles* – a parasite of the Cabbage White Butterfly (in Russian). *Zashch. Rast.* (10) 17–18.

Sheffield Airey Neave, 1940. *Nomenclatur Zoologicus.*

Smith, F.H.N., 1950. The 1950 season. *Entomologist* 83: 239.

Smith, K.M., 1931. *Textbook of Agricultural Entomology.* Cambridge.

Smith, K.M. & Rivers, C.F., 1956. Some viruses affecting insects of economic importance. *Parasitology* 46: 235–242.

South, R., 1899. The nomenclature and arrangement of British butterflies according to various recent authors. *Entomologist* 32: 31–37.

—— 1936. *Butterflies of the British Isles.* Warne, London.

South London Entomological and Natural History Society (SLENHS), 1917. Meeting May 10th. *Entomologist* 50: 143–144.

—— 1953. Meeting November 26th. *Entomologist's mon. Mag.* 89: 72.

Spuler, A., 1908. *Die Schmetterlinge Europas.* Vol. I. Schweizerbartsche Verlag. Stuttgart.

Street, M.C., 1975. *The growth and development of the reproductive system of Pieris brassicae and its control by hormones.* Ph.D. thesis. University of Newcastle upon Tyne.

Swammerdam, J., 1758. *The Book of Nature, of the History of Insects.* Seyffert, London.

Tetley, A.S., 1899. Notes from North Wales: 1898. *Entomologist* 32: 95–97.

Thimann, R.G., 1961. A note on *Pieris brassicae* L. (Lepidoptera). *Bull. amat. ent. Soc.* 20: 100.

Tulloch, B., 1938. Extraordinary abundance of larvae of *Pieris brassicae.* *Entomologist* 72: 266.

Valletta, A., 1954. An unusual pairing between *Pieris* and *Colias* (Lep.). *Entomologist* 87: 38.

—— 1966. The Butterflies of the Maltese Islands. *Entomologist's Rec. J. Var.* 38–42.

Vaughan, J.W., 1898a. *Platyptila tesseradactyla. Entomologist* 31: 139–140.

—— 1898b. The Rhopalocera of the Wye Valley. *Entomologist* 31: 140–141.

Voltukhov, V.A., 1975. Economic evaluation of the chemical and biological methods. (In Russian.) *Zashch. Rast.* (5) 29.

Walker, J.J., 1918. The butterflies of the Oxford District. *Entomologist's mon. Mag.* 54: 246–250.

—— 1924. Oxford butterflies in the season of 1924. *Entomologist's mon. Mag.* 60: 247–249.

Walsh, G.B., 1952. The present status of the butterfly population of the Scarborough District. *Entomologist's mon. Mag.* 88: 185–189.

Warren, B.C.S., 1920. Some records of, and observations on, the flying habit of butterflies when paired. *Entomologist's Rec. J. Var.* 32: 218–223.

—— 1938. The flying habits of butterflies when paired. *Entomologist* 71: 114–117.

Warren, R.G., 1975. *Atlas of the Lepidoptera of Staffordshire.* Staffs. Biol. Recording Scheme.

Westwood, J.D., 1854. *The Butterflies of Great Britain.* W.S. Orr & Co., London.

Wheeler, G., 1918. Paired lepidoptera in flight. *Entomologist's Rec. J. Var.* 30: 152–153.

Wigglesworth, V.B., 1964. *The Life of Insects,* Weidenfeld & Nicolson, London.

Williams, C.B., 1930. *The Migration of Butterflies.* Edinburgh.

Williams, C.B., Cockbill, G.F., Gibbs, M.E. & Downes, J.A., 1942. Studies in the migration of lepidoptera. *Trans. R. ent. Soc. Lond.* 92: 101–283.

Wolfsberger, J., 1966. *Die Macrolepidopterenfauna des Gardaseegebietes.* Verona.

—— 1971. *Die Macrolepidopterenfauna des Monte Baldo in Oberitalien.* Museo Civico di Storia Naturale di Verona.

Wilkes, B., 1747–1749. *The English Moths and Butterflies: together with the plants, flowers and fruits whereon they feed, and are usually found.* Benjamin Wilkes, Fleet Street, London.

Worms, B. de., 1949. The butterflies of London and its surroundings. *Lond. Nat.,* 29: 46–80.

—— 1962. *The Macrolepidoptera of Wiltshire.* Wilts. Arch. nat. hist. Soc.

Further references

Abbott, C.R., 1936. *Entomologist* 69: 69 (early and late imagines).

Adkin, R., 1909. *Entomologist* 42: 320 (abundance at Eastbourne).

Aldrovandi, U., 1602. *De Animalibus Insectis.* Bologna.

Bandermann, F., 1909. *Z. wiss. Insektenbiol.,* 5: 135–136 (relative numbers of Pierids found).

Banner, J.V., 1938. *Entomologist* 72: 168 (late larvae).

—— 1938. *Bull. amat. ent. Soc.* 3: 24 (late larvae).

Baunache, 1926. *Die Kranke Pflanze* 3: 123–125 (description of three Pierids).

Bawden, G., 1958. *Entomologist* 91: 95 (early imago).

Beadle, H.A., 1895. *Entomologist's Rev. J. Var.* 5: 276–283 (abundant in Keswick).

Blunck, H. & Riehm, E., 1958. *Pflanzenschutzberichte.* Frankfurt am Main (general description).

Caland, M., 1921–25. *Ent. Ber., Amst.* 6: 388–396 (pupation of Pieridae).

Cardew, P.A., 1943. *Entomologist* 76: 59–60 (courting with *G. rhamni*).

Ceton, J.C., 1936. *De Levende Natuur* 40: 1–9 (photographs of whites).

—— 1939. *De Levende Natuur* 43: 300–306 (ova batch drawn).

Chapman, T.A., 1917. *Entomologist's mon. Mag.* 53: 196–197 (illustration).

Day, F.M., 1915. *Entomologist's mon. Mag.* 608: 19 (late larvae).

Dimic, A., 1946a. *Riv. Biol.* 38: 170–190 (effect of temperature on pupation).

—— 1946b. *Soc. Ital. Biol. Sper. Bol.* 22: 466–468 (effect of temperature on pupation).

Dixey, F.A., 1915. *Trans. R. ent. Soc. Lond.* 47–48 (courtship).
—— 1919. *Proc. R. ent. Soc. Lond.*, 1918. clii. (nuptial flight).
—— 1932. *Trans. R. ent. Soc. Lond.*, 80: 57–75 (plume scales).
Dowdeswell, W.H., 1936. *Entomologist* 69: 49–53 (abundance).
Eckstein, K., 1930. *Anz. Schädlingsk.* 6: 9–10 (hatching).
Entomological Society of London, 1909. *Entomologist* 42: 324–325 (larval development).
Ford, H.D., 1920. *Entomologist* 53: 139 (scarcity).
Furneaux, W., 1905. *Butterflies and Moths.* Longmans, London.
Garland, G.R., 1896. *Entomologist* 29: 127 (late larvae).
Gray, D., 1978. *Butterflies on my mind.* Angerson & Robertson, Brighton.
Guggisberg, C.A.W., 1945. *Schmetterlinge und Nachtfalter*, Hallwag, Bern.
Guilbot, R., 1975. *Opie Cahiers de Liaison.* (8) 19–25 (general biology).
Harold Smith, B., 1943. *Entomologist* 76: 145 (early appearances).
Harris, M., 1969. Reprinted from 1775. *English Lepidoptera: Aurelian's Pocket Companion.* E.W. Classey, Faringdon (description).
Henneguy, F., 1904. *Les Insectes.* Masson, Paris (general description).
Heslop Harrison, J.W., 1944. *Entomologist's mon. Mag.* 80: 23 (curious behaviour of female).
Higgins, L.R. & Riley, N.D., 1970. *Field Guide to the Butterflies of Britain and Europe.* Collins, London.
Hillis, J.P., 1973. *Ent. Gaz.* 24: 313–314 (abundance).
Hodgson, S.B., 1921. *Entomologist* 54: 173–174 (early imagines).
Holloway, P.H., 1951. *Bull. amat. ent. Soc.* 10: 85 (early imago).
Howard, L.O., 1930. *History of Applied Entomology.* Smithsonian Miscellaneous Coll. 84: 564 (on Aldrovandi's work).
Huggins, H.C., 1956. *Entomologist* 89: 173 (pupation site).
Kidner, A.R., 1942. *Entomologist* 75: 91 (early imago).
Krasucki, A., 1929. *Mem. Inst. Nat. Polon Econ. Rur. Pulaway.* 10: 216–223 (pest in Poland).
Lane, C. & Rothschild, M., 1957. *Entomologist* 90: 303 (early imago).
Lang, H.C., 1884. *Rhopalocera Europae descripta et delineata.* Reeve, London.
Laurence, B.R., 1945. *Entomologist's Rec. J. Var.* 81: 117 (abundance).
Legros, C.V., 1913. *Fabre, Poet of Science.* Unwin, London (biology).
L'Homme, L., 1935. *Catalogue des Lépidoptères de France et de Belge.* Carriol, Par Douelle, Lot (general biology).
Lowther, R.C., 1926. *Entomologist* 59: 172–173 (early imago).
Lucas, W.J., 1894. *Entomologist* 27: 295 (scarcity).
Lyle, G.T., 1914. *Entomologist's Rec. J. Var.* 50: 289 (late imago).
Ma, C.C. & Lui, S.P., 1947. *Fukien agic. J.* 8: 42–53 (ecological notes).
Manchester Entomological Society, 1938a. *Entomologist* 71: 168 (bred specimen).
—— 1938b. *Entomologist* 71: 287–288 (bred series).
Melrose, M.M., 1943. *Entomologist* 76: 82 (early larva).
Miller, F.R., 1956. *Agricultural entomology* (in Czechoslovakian). Akademia, Prague (general biology).
Morley, C., 1917. *Entomologist* 50: 64–66 (parasitism).
Morris, W., 1957. *Entomologist's Rec. J. Var.* 69: 75 (early emergence).
Moufet, T., 1643. *Insectorum Sive Minimorum Animalium Theatrum*, London. (general biology).
Newman, L., 1977. *Looking at Butterflies.* Collins, London.
Petersen, B., 1954. *Ent. Tidskr.* 75: 194–203 (egg laying).

Pictet, A., 1933. *Lambillionea* 33: 89–97 (eclosion).

Rees, A., 1819. *Cyclopaedia or Universal Dictionary of Arts, Sciences, and Literature*. Longmans, Hurst, Rees, Orme, Brown, London. Vol. 26 (description).

Robson, J.P., 1916. *Entomologist* 99: 164 (early eclosions).

Samouelle, G., 1819. *Entomologist's Useful Compendium*. Thomas Boys, London (description).

Sandars, E., 1955. *A Butterfly Book for the Pocket*. Oxford University Press, London (general biology).

Scorer, A.G., 1913. *Entomologists Log Book*. Routedge, London (description).

South, R., 1936. *Butterflies of the British Isles*. Warne, London (description).

South West Yorkshire Entomological Society, 1929. *Entomologist* 62: 119 (life history).

Stanek, V.J., 1974. *Encylopédie Illustrée des Insectes*. Grund, Paris (description).

Stokoe, W.J., 1944. *Caterpillars of British Butterflies*. Warne, London.

Strand, E., 1935. *Lepidopterorum Catalogus*. pp. 65–69 (general biology).

Strawinsky, K., 1928. *Polskie Pismo ent.* 7: 227 (general biology).

Tutt, J.W., 1899. *Natural History of British Lepidoptera*. Sonnenchen, London.

Vorbrodt, G. & Muller Rutz, J., 1912. *Mitt. Schweiz. ent. Ges.* 14: 201–390 (description).

Walrecht, B.J.J.R., 1960. *Levende Natuur* 63: 288 (late pupation).

Walsh, G.B., 1935. *Entomologist's mon. Mag.* 71: 91 (late larvae).

Wood, J.G., 1858. *The Common Object of the Country*. Routledge & Co., London (description).

Woolhouse, J.W., 1929. *Entomologist* 62: 136 (early appearance).

4. Foodplants

"Observe the motions of that common white butterfly which you see flying from herb to herb. Her object is the discovery of a plant that will supply the sustenance appropriated by Providence to her young, upon which to deposit her eggs ... But no: as if aware that this food would be to them poison, she is in search of some plant of the cabbage tribe. But how is she to distinguish it from the surrounding vegetables? She is taught of God!"

From Kirby & Spence, Letter XI: *On the affection of insects for their young* (1858)

Introduction

As an oligophagous insect *P. brassicae* has been accredited with, or has been observed feeding upon, an extraordinarily large number of plants, but by far the most important of these belong to the Cruciferae family. This close association between the butterfly and the Cruciferae is manifest in the highly specialised sensory system of the larva and imago and their responses to chemical substances produced by the plant. Here, it is important to draw attention to Professor L.M. Schoonhoven's work on electrophysiology of *P. brassicae* (Schoonhoven, 1967, 1969a,b, cf. Chapter 10, 12).

Five main plant families, the Cruciferae, Tropaeolaceae, Capparaceae, Resedaceae and the Papilionaceae have been shown here to have been exploited by *P. brassicae* (Table 4.1). These are designated as principal plant families since *P. brassicae* may be found regularly in the wild on any of these species. Eighty-three species of foodplant belong to this first group, the largest number of species belonging to the genus *Brassica*. Wild crucifers play a very important part in supporting first generation larvae

Table 4.1 Summary of foodplants.

Families		Number of species
Principal families		
Capparaceae		4
Cruciferae		60
Papilionaceae		10
Resedaceae		5
Tropaeolaceae		4
	Total	83
Secondary families		
Chenopodiaceae		1
Euphorbiaceae		1
Geraniaceae		1
Liliaceae		1
Phytolaccaceae		1
Polygonaceae		1
Umbelliferae		2
	Total	8
Families from which species are accepted in the laboratory (only when specially treated)		
Compositae		2
Cucurbitaceae		1
Gentianaceae		1
Gramineae		2
Liliaceae		3
Malvaceae		1
Papaveraceae		1
Rosaceae		2
Solanaceae		1
	Total	14

of *P. brassicae* in several countries, but most important of all, many crops suffer severe damage. This latter aspect has been dealt with by country in the chapter on Economic Importance (Chapter 13).

A secondary and in some cases doubtful category of *P. brassicae* foodplants has been made up to include only those species on which *P. brassicae* has been or would be expected to be found occasionally in the wild. These comprise only eight species from seven other plant families (Table 4.1). There are also 15 other species belonging to 10 families including another member of the Umbelliferae (found also in the secondary group) which *P. brassicae* larvae have been induced to eat in the laboratory under unnatural conditions.

The practical implications of rearing larvae on various foodplants is discussed in this chapter and the concept of acceptability is dealt with according to the results of numerous workers who have attempted to elucidate it.

Principal foodplants

Capparaceae

Four foodplants have been documented here, three of which have been noted to have been eaten eagerly, or the larvae found in pest proportions (Table 4.2). One crop attacked by *P. brassicae* larvae which does not seem to be widely recorded elsewhere is *Capparis spinosa* L. (Caper). This has been recorded as a foodplant in Palestine by Buxton (1924), Bodenheimer (1935) and Rothschild *in* Gardiner (1978). *Cleome spinosa* L. was found by Tolman (1955) to be more attractive for oviposition by *P. brassicae* than other species of the family and also other hosts in the Cruciferae and Resedaceae.

Table 4.2 Capparaceae foodplants.

Acceptability ratings (used for subsequent tables in this chapter)
1   "avidly eaten", "completely eaten", "very attractive", "quickly eaten" ...
1p   observed in "pest proportions", "defoliation", "100% attack" ...
2   stated as "foodplant", "eaten", or larvae "found on" ...
3   "fairly readily eaten" ...
4   "gnawed but not readily eaten", "slightly eaten" ...
R   refused
?   doubtful or exceptional records

Specific name	Acceptability rating	Authority
CAPPARACEAE		
Capparis spinosa Linnaeus	1	Verschaffelt, 1910
	1	Plotnikov, 1914
	2	Buxton, 1924
	3	Terofal, 1965
Capparis sp.	2	Kurentzov, 1930
(Caper)		
Steriphana paradoxum Endlicher	1	Verschaffelt, 1910
Cleome spinosa Linnaeus	1	Verschaffelt, 1910
(Spider Flower)	2	Tolman, 1955
	3	Terofal, 1965
Polonisia trachysperma Linnaeus	3	Terofal, 1965

Cruciferae

Brassica

The cabbage, *Brassica oleracea* var. *capitata* Linnaeus is one of the most important foodplants of *P. brassicae* and many species of brassicas and their cultivars are susceptible to attack from this insect (Table 4.3). The present day cabbage is thought to have been specially selected for its different characteristics such as large leaves, buds and heads by civilised man over the last 2000 years, "before Pliny's time", when there were already some wild cabbage forms in cultivation (Bailey, 1927). Before then, cabbage existed wild on cliffs and sea shores, as it does today as an indigenous species of Great Britain.

In the laboratory *P. brassicae* can be reared successfully on many cultivars and hybrids of cabbage (Table 4.3). The next chapter on breeding can be consulted for information on the effect of adding or omitting cabbage from an artificial diet for *P. brassicae* and the effect of feeding different parts of the cabbage on growth is also discussed.

The cultivars of cabbage were tested by Chandra & Lal (1977) for their susceptibility to *P. brassicae* larvae in India. They found that there were no significant differences in the weight and size of larvae at maturity but that both the food consumption and percentage of larvae which developed into imagines varied a great deal. *P. brassicae* larvae ate very little and had low percentages reaching maturity on EC 24856 and Savoy Best of All.

Other cruciferous foodplants

In many regions *P. brassicae* sustains itself for part of the season on wild crucifers. This may be indicative of its evolutionary history when *P. brassicae* had to rely solely on these plants before man selected the present day cultivars. For instance, first generation *P. brassicae* have been recorded in the past and present to develop on wild crucifers in Czechoslovakia (Speyer, 1956a, b), Norway (Spitzer, 1975, pers. comm.), Poland (Starega, 1976, pers. comm.) and west Pamir in USSR (Kadamshoer, 1971); however, in some of these areas later generations of *P. brassicae* larvae feed on *Sinapis arvensis* L., *Raphanus sativus* L., *Matthiola annua* Sweet and others. It has been suggested that in England *P. brassicae* may feed on *Matthiola incana* (L.) R. Brown (Wild Stock) which grows profusely on the sea cliffs on the Isle of Wight (Fearnclough, 1972).

Most of the cruciferous plants listed in Table 4.4 are those which have been tested in the laboratory for their acceptability to the larvae, such as those used by Verschaffelt, Terofal and Dethier. This means that the foodplants have been screened for feeding response of the larvae with

Table 4.3 Brassica foodplants in the wild, or in the laboratory (in the laboratory marked as *).

Specific name	Acceptability rating (see Table 4.2)	Authority
Brassica sp. (Mustard)	2	Theobald, 1909
Brassica oleracea var. *botrytis* Linnaeus (Cabbage)	1p	Zanon, 1919; Buxton, 1924
	1	Terofal, 1965
B. o. var. *capitata* Linnaeus (Cabbage)	1	Terofal, 1965
B. o. var. *caulorapa* Linnaeus (Kohl-rabi)	1	Ormerod, 1882–87; Newman, 1942
	1p	Friederichs, 1931
B. o. var. *gemmifera* Linnaeus (Rape)	1p	Anonymous, 1932; Granville-Clutterbuck, 1941
_B. o._ var. *gongylodes* Linnaeus	2	Kayser, 1975
Brassica napus Linnaeus (Rutabaga, Swede, Turnip)	1p	Granville-Clutterbuck, 1941 Heslop-Harrison, 1946 Terofal, 1965
Brassica nigra (Linnaeus) Koch (Black Mustard)	1	Swinton, 1898; Terofal, 1965
Red Cabbage	1p	Blunck, 1953
	1	Terofal, 1965
	2	Friederichs, 1931
White Cabbage	1	Terofal, 1965
	2	Friederichs, 1931
Savoy	1	Terofal, 1965
	2	Swinton, 1898
*Baby Head	2	Chandra & Lal, 1977
*EC 10109, 24856, 40412	2	Chandra & Lal, 1977
*EC 24855 X EC 10109	2	Chandra & Lal, 1977
*Golden Acre X Baby Head	2	Chandra & Lal, 1977
*Greyhound	1	Feltwell & Rothschild, 1974
*January King	1	Gardiner, 1977, pers. comm.
*Kasmodeskaja	2	Chandra & Lal, 1977
*Louisiana All Year	2	Chandra & Lal, 1977
*Pennstate Ball	2	Chandra & Lal, 1977
*Primo	1	Feltwell & Valadon, 1972, 1974
*Savoy Best of All	2	Chandra & Lal, 1977
*1000 Head Kale	1	Gardiner, 1977, pers. comm.

Table 4.4 Cruciferous foodplants other than *Brassica*.

Foodplant	Acceptability rating (see Table 4.2)	Authority
Alliaria petiolata (Bieberstein)		
Cavara & Grande	1	Terofal, 1965
(Garlic Mustard)	2	Fernando, 1971
Alyssum saxatile Linnaeus	4	Verschaffelt, 1910
(Golden Alison)		
Arabis alpina Linnaeus	1	Terofal, 1965
(Alpine Rocket Cress)	2	Scott, 1942;
		Dethier, 1947
	4	Verschaffelt, 1910
Armoracia rusticana Gaertner,		
Meyer & Scherbius	1	Verschaffelt, 1910;
(= *Cochlearia armoracia* Linnaeus)		Terofal, 1965
(Horse Raddish)	2	Wilson, 1880;
		Friederichs, 1931;
		Dethier, 1947;
		Johannson, 1951;
		Kidner *in*
		Chalmers-Hunt, 1960
Aubretia deltoides De Candolle	4	Verschaffelt, 1910
(Aubretia)		
Barbarea stricta Andrzeiwsky	2	Vepsäläinen, 1968
(Small Flowered Yellow Rocket)	4	Verschaffelt, 1910
Barbarea vulgaris R. Brown	1	Terofal, 1965
(Winter Cress)	2	Dethier, 1947
Biscutella auriculata Linnaeus	2	Terofal, 1965
Biscutella laevigata Linnaeus	2	Terofal, 1965
Bunias orientalis Linnaeus	2	Dethier, 1947;
(Warty Cabbage)		Johannson, 1951
	3	Terofal, 1965
	4	Verschaffelt, 1910
Cakile maritima Scopoli	2	Picard Cambridge,
(Sea Rocket)		1918
		Roer (pers. comm.)
		Speyer, 1956b
Capsella bursa-pastoris		
(Linnaeus) Medicus	4	Verschaffelt, 1910;
(Shepherd's Purse)		Dethier, 1947
N.B. See also Table 4.8, where this species is regarded as being refused		
Cardamine hirsuta Linnaeus	2	Dethier, 1947
(Hairy Bitter Cress)	4	Verschaffelt, 1910
Cardaria (Lepidium) draba		
(Linnaeus) Desvaux	2	Fabre *in* Gautier
(Hoary Cress)		& Riel, 1919

Table 4.4 continued

Foodplant	Acceptability rating (see Table 4.2)	Authority
Cheiranthus cheiri Linnaeus	2	Friederichs, 1931
(Wallflower)	4	Terofal, 1965
Crambe cordifolia Steven	2	Dethier, 1947;
		Vepsäläinen, 1968
	4	Verschaffelt, 1910
Crambe maritima Linnaeus	2	Vepsäläinen, 1968
(Sea Kale)		
Diplotaxis erucoides		
(Linnaeus) De Candolle	1	Terofal, 1965
(White Wall Rocket)		
Diplotaxis muralis		
(Linnaeus) De Candolle	1	Terofal, 1965
(Wall Rocket)		
Diplotaxis tenuifolia		
(Linnaeus) De Candolle	1	Terofal, 1965
(Perennial Wall Rocket)		
Draba incana Linnaeus	2	Vepsäläinen, 1968
(Hoary Whitlow Grass)		
Eruca sativa Miller	2	George, 1927
Erucastrum nasturtiifolium		
(Poir) O.E. Schulz	2	Terofal, 1965
—		
Erysimum hieraciifolium Linnaeus	2	Vepsäläinen, 1968
—		
Erysimum perofskianum		
Fischer & Meyer	4	Verschaffelt, 1910;
—		Dethier, 1947
Hesperis matronalis Linnaeus	1	Terofal, 1965
(Dame's Violet)	2	Dethier, 1947
	4	Verschaffelt, 1910;
		Schwarz, 1948
Hirschfeldia (Sinapis) incana		
(Linnaeus) Lagrèze-Fossat	2	Gautier & Riel, 1919
(Hoary Mustard)		Sandars, 1955
Hornungia petraea		
(Linnaeus) Reichenbach	1	Terofal, 1965
(=*Hutchinsia alpina*)		
(Rock Hutchinsia)		
Iberis amara Linnaeus	2	Terofal, 1965
(Wild Candytuft)		
Iberis gibraltaria Linnaeus	2	Terofal, 1965
—		
Iberis umbellata Linnaeus	2	Terofal, 1965
(Candytuft)		

Table 4.4 continued

Foodplant	Acceptability rating (see Table 4.2)	Authority
Isatis tinctoria Linnaeus (Woad)	1	Sandars, 1955; Terofal, 1965
Kernera saxitilis Reichenbach	2	Terofal, 1965
—		
Lepidium sp.	2	Frohawk, 1934; Sandars, 1955
Lepidium latifolium Linnaeus (Dittander)	2	Anglade *et al.*, 1963
Lepidium sativum Linnaeus (Garden Cress)	2	Terofal, 1965
Lobularia (Clypeola) maritima (Linnaeus) Desvaux (Sweet Alison)	2	Zeller, 1847
Lunaria annua Linnaeus (Honesty)	2	Verity, 1905–11; Hicken, 1942; Monks Wood Data, 1942; Scott, 1942; Allan, 1949
Matthiola sp. (Stock)	1	Borg, 1932
Matthiola annua Sweet (Double Stock)	2	Frohawk, 1927; Holford, 1928
Matthiola incana (Linnaeus) R. Brown (Stock)	2	Terofal, 1965; Bonet, 1976, pers. comm.
Peltaria turkmeria Lipsky	1	Terofal, 1965
—		
Raphanus raphanistrum Linnaeus (Wild Radish/White Charlock)	1	Terofal, 1965
Raphanus sativus Linnaeus (Radish)	2	Theobald, 1909; Buckler, 1928; Frohawk, 1934; Allan, 1949; Sandars, 1955
Rorippa nasturtium-aquaticum (Linnaeus) Hayek (Water Cress)	2	Theobald, 1909; Allen & Selman, 1957
Rorippa sylvestris (Linnaeus) Besser (Creeping Yellow Cress)	2	Terofal, 1965
Sinapis alba Linnaeus (White Mustard)	1	Terofal, 1965
Sinapis arvensis Linnaeus (Charlock)	4	Terofal, 1965; Campbell, 1970

Table 4.4 continued

Foodplant	Acceptability rating (see Table 4.2)	Authority
Sisymbrium officinale (Linnaeus) Scopoli (Hedge Mustard)	2	Dethier, 1947; Terofal, 1965
	4	Verschaffelt, 1910
Sisymbrium sophia Linnaeus (Flixweed)	2	Terofal, 1965
Sisymbrium strictissimum Linnaeus	2	Dethier, 1947
—	4	Verschaffelt, 1910
Thlapsi arvense Linnaeus (Field Penny Cress)	2	Terofal, 1965
Thlapsi rotundifolium Bory & Chaub	2	Terofal, 1965

respect to the presence or absence of feeding stimulants; but not necessarily that they have been reared to maturity on the foodplants. Other foodplants, such as those quoted by Allan (1949) and Sandars (1955) are from books on larval foodplants and field guides respectively and can be relied upon as accurate food sources.

In the laboratory, *P. brassicae* can be reared on *Diplotaxis tenuifolia* (L.), *Sinapis incana* L., *Isatis tinctoria* L., *Raphanus raphinistrum* L., *Lepidium draba* L. and *Sisymbrium officinale* (L.) Scopoli (Fabre *in* Gautier & Riel, 1919), *Tropaeolum* sp. and *Reseda* sp. (Rothschild *et al.*, 1975).

Pieris brassicae larvae reared on certain strains of *Cheiranthus cheiri* L. (Wallflower) always die before the second instar (Rothschild, *loc. cit.*) and this is probably due to the cardiac glycosides found in the leaf, stem and flowers.

In the wild, however, *P. brassicae* have been observed feeding on *Crambe* sp., *Cakile maritima* Scop., *Diplotaxis tenuifolia* L., *Raphanus raphanistrum* L. and *Sinapis incana* L. (Gautier & Riel, 1919).

Many of the foodplants listed in Table 4.4 are cited from the work of Terofal (1965) on acceptability ratings. He rated the plants on a seven point scale: 1 readily accepted, 2 willingly eaten, 3 less willingly eaten, 4 unwillingly eaten, 5 nibbled, 6 tasted and refused, and 7 rejected, but accepted after 40 hours starvation. For the purpose of the present assessment, Terofal's scale 1 and 2 are represented as 1; his 3 and 4 as 2; his 5 as 3; 6 as 4 and his 7 as refused. For those who wish to rear *P. brassicae* larvae on any of the foodplants mentioned by Terofal, it would be best to choose only those under category 1, as listed in Table 4.4 for the greatest chances of success.

Papilionaceae

About 10 species of this family, mostly belonging to the genera *Genista*, *Lathyrus*, *Pisum* and *Vicia* are acceptable to *P. brassicae* larvae, some of them being recommended as foodplants for rearing (Sandars, 1955). It is worth noting that other papilionaceous species, including *Vicia faba*, are refused (Table 4.7). Kornfeld (1935) recorded *P. brassicae* ovipositing on *Glycine soya* in Rumania (Danube Delta) but the larvae caused little damage and moved to other plants.

Resedaceae

Five species of *Reseda* (Mignonette) have been recorded as foodplants of *P. brassicae* and it has been used popularly for rearing larvae (Rothschild *et al.*, 1975) (Table 4.5).

Tropaeolaceae

Members of the Tropaeolaceae are the next most important foodplants of *P. brassicae* after the Cruciferae. There are many accounts in the literature of *Tropaeolum* or Nasturtium as foodplants but use of the latter name is confusing, as it is widely used in Britain as the vernacular name for Tropaeolum. It is also the generic name of Water Cress (*Rorippa nasturtium-aquaticum* (Linnaeus) Hayek (Cruciferae)). References to Nasturtium may not therefore always indicate that the host is *Tropaeolum*.

Table 4.5 Other primary foodplants.

Foodplant	Acceptability rating (see Table 4.2)	Authority
PAPILIONACEAE		
Genista alba Lamark	1	Borg, 1932
Genista tinctoria Linnaeus (Dyer's Greenweed)	2	David & Gardiner, 1953; Sandars, 1955
Glycine soja Sieber & Zuccarini (Soy Bean)	4	Kornfeld, 1936
Lathyrus latifolius Linnaeus (Sweet Pea)	4	Verschaffelt, 1910; Dethier, 1947
Lathyrus sylvestris Linnaeus (Narrow leaved Everlasting Pea)	4	Verschaffelt, 1910; Dethier, 1947
Lathyrus tuberosus Linnaeus	4	Verschaffelt, 1910

Table 4.5 continued

Foodplant	Acceptability rating (see Table 4.2)	Authority
(Earth-nut Pea)	4	Verschaffelt, 1910
Medicago sativa Linnaeus	1p	Plotnikov, 1914
(Lucerne, Alfalfa)		
Pisum sp.	2	Sandars, 1955
(Pea)		
Vicia sp.	2	Sandars, 1955
(Bean)		
Vicia cracca Linnaeus	2	Sandars, 1955
(Tufted Vetch)		
RESEDACEAE		
Caylusea abyssinica		
Fischer & Meyer	3	Terofal
—		
Reseda sp.	2	South, 1936;
(Mignonette)		Morley *in*
		Chalmers-Hunt, 1960
Reseda alba Linnaeus	3	Terofal, 1965
(Upright Mignonette)	4	Verschaffelt, 1910
Reseda lutea Linnaeus	2	Theobald, 1909;
(Wild Mignonette)		Harford, 1927;
		Allan, 1949;
		Sandars, 1955;
		Terofal, 1965
	4	Verschaffelt, 1910
Reseda luteola Linnaeus	1	Thorsteinson, 1953
(Weld/Dyer's Rocket)	2	Allan, 1949
Reseda odorata Linnaeus	1	Thorsteinson, 1953
(Fragrant Mignonette)		
Reseda virgata Linnaeus	4	Verschaffelt, 1910
—		
TROPAEOLACEAE		
Tropaeolum brasiliense Casaretto	4	Terofal, 1965
(Nasturtium — see text)		
Tropaeolum majus Linnaeus	1	Frohawk, 1934;
(Nasturtium)		Thorsteinson, 1953
	2	Wilson, 1880;
		Buxton, 1924;
		Buckler, 1928
	4	South, 1936;
		Sandars, 1955;
		Terofal, 1965
Tropaeolum peregrinum Linnaeus	1	Verschaffelt, 1910
(Canary Creeper)	2	Nicholson, 1939

Unusual foodplants

Chenopodiaceae
Beta (Beet) was recorded as supporting larvae of *P. brassicae* in Palestine by Bodenheimer (1935).

Euphorbiaceae
Rothschild (1978) exhibited larvae of *P. brassicae* at a meeting of the Royal Entomological Society of London which were feeding on *Euphorbia polychroma* Kerner. Her attention had been drawn to the habit of the females of fluttering around this foodplant in the laboratory. The larvae ate the leaves when they were broken.

Geraniaceae
An unusual and possibly suspect foodplant is *Geranium* sp. which was included in the list by Theobald (1909).

Table 4.6 Unusual foodplants.

Specific name	Acceptability rating (see Table 4.2)	Authority
CHENOPODIACEAE		
Beta sp.	2	Bodenheimer, 1935
Beet		
EUPHORBIACEAE		
Euphorbia polychroma Kerner	2	Rothschild, 1978
Spurge		
GERANIACEAE		
Geranium sp.	2	Theobald, 1909
Geranium		
LILIACEAE		
Allium sativum Linnaeus	3	Terofal, 1965;
Garlic		Meeuse, 1973
PHYTOLACCACEAE		
Petiveria alliacea Linnaeus	4	Terofal, 1965
—		
POLYGONACEAE		
Rumex sp.	2	Homer, 1969
—		
UMBELLIFERAE		
? *Daucus carota* Linnaeus	1p	Pictet, 1918
Carrot		
? *Petroselinum crispum* (Miller)		
Nyman	1p	Pictet, 1918

Liliaceae and Phytolaccaceae

Two families on which the chances of rearing *P. brassicae* larvae to maturity are small are *Allium* sp. (Liliaceae) and *Petiveria* sp. (Phytolaccaceae), as these plants are not eagerly eaten (Terofal, 1965).

Polygonaceae

There is only one reference for this family. Homer (1969) reported seeing *P. brassicae* larvae feeding on Dock (*Rumex* sp.). No specific name was given nor was it mentioned whether the larvae developed successfully on the foodplant.

Umbelliferae

Two crops, *Petroselinum* (Parsley) and *Daucus* (Carrot), not mentioned as foodplants by anyone else, were recorded as destroyed during the summer of 1917 in Switzerland as a result of larvae produced from migrants (Pictet, 1918). Minimum numbers of larvae from a dozen fields were recorded as 983 m^2.

Acceptability

The relative acceptability of foodplants to *P. brassicae* larvae has intrigued and baffled scientists for a long time. As early as 1660 John Ray commented on how larvae would readily eat cabbage and consume rape but refused other foodplants (Mickel on Ray, 1973). The chemical affinity of the Cruciferae, Tropaeolaceae, Capparaceae and the Resedaceae, which are the four families with the majority of foodplants of *P. brassicae*, was established by Guignard (1890, 1893) *in* Gautier & Riel (1919). The foodplants of the Pieridae were reviewed by Kurentzov (1930) but he did not include *P. brassicae*; he recorded 16 plant families which had acceptable species and three (Compositae, Gramineae, Hypericaceae) of doubtful credibility.

The most extensive trials or investigations of *P. brassicae* have been carried out by the following workers: Verschaffelt (1910), Gautier & Riel (1919), Dethier (1947), Rainey (1936), Johannson (1951), Kusnezov (1953), Thorsteinson (1953, 1958), Tolman (1955), Terofal (1961, 1965), Fernando (1972), Lundgren (1975), Rothschild *et al.* (1975) and Rothschild (1977). Others have investigated mustard oil glycosides (MOG) as feeding stimulants and their effect on acceptability in the foodplants of *P. brassicae* larvae (Chapter 12).

An account of some of their results follows in chronological order:

Verschaffelt (1910) found that the foodplant preference of *P. brassicae* larvae declined throughout the families Resedaceae (5 spp.),

Tropaeolaceae (2 spp.), Capparaceae (3 spp.) and Papilionaceae (2 spp.), from being readily eaten to being slightly eaten. He experimented with smearing the leaves of foodplants which are normally unacceptable with the juice of normally acceptable foodplants and found that the larvae ate some of the unusual species offered (Table 4.7 – note other results). Other species which were declined he thought contained substances which were toxic or nauseous to the larvae. By smearing MOGs over filter paper, flour and starch, Verschaffelt induced the larvae to feed. He found that the expressed juice of *Bunias orientalis* Linnaeus (Warty Cabbage) (Cruciferae) also induced the larvae to feed on normally rejected leaves.

Rainey (1936) confirmed Verschaffelt's observations in his research at Imperial College, London and attempted to correlate content of MOGs in *P. rapae* foodplants with the distribution of ova found on them. He presented a useful table of 24 cruciferous foodplants together with their MOGs but he experienced some difficulty in extracting some of the oils.

In 1947 Dethier tested 15 species of Cruciferae for palatability to *P. brassicae* larvae, some of which Verschaffelt had tested, and also noted that MOGs were the key to foodplant selection. He found that gregarious larvae could "unquestionably" detect the odours of different foodplants from a range of 5 cm.

Johannson (1951) experimented with the foodplant preference of fifth instar *P. brassicae* larvae to four species: 1. *Brassica oleracea* Linnaeus (Cruciferae), Cabbage; 2. *Armoracia rusticana* Gaertner, Meyer and Scherbius (Cruciferae), Horse Radish; 3. *Bunias orientalis* Linnaeus (Cruc.), Warty Cabbage and 4. *Tropaeolum majus* Linnaeus (Trop.), "Nasturtium". He found that with the exception of larvae reared on the leaves of *A. rusticana* the larvae were conditioned to whichever foodplant they were reared upon initially. Thus larvae reared on *B. oleracea* had a foodplant preference in the order 1,2,3,4; larvae reared on *B. orientalis* had 3,2,1,4, and those from *T. majus* 4,1,2,3. Johannson (1951) believed that larvae could distinguish between four different odours emitted by these plants.

Thorsteinson (1946, 1953) tested more thoroughly the role of MOGs in the foodplant selection of *P. brassicae* larvae and repeated much of Verschaffelt's work. He found that the larvae readily ate the leaves of 19 species of foodplant (Cruc. 16 spp., Trop. 1 sp., Res. 2 spp.) which are known to contain MOGs. When the larvae were offered the leaves of 21 spp. of foodplant which were normally refused and which had been smeared with 3% sinigrin or 3% sinalbin (both MOGs), only 7 spp. were partially eaten. These included *Allium* sp. (Onion), *Calendula* (Marigold), *Cucumis* sp. (Cucumber), *Lactuca* sp. (Lettuce), *Salvia* sp. (Common Salvia), *Papaver* sp. (Poppy) and *Pisum* sp. (Pea) (Table 4.7, see also *Allium* and *Pisum* in Tables 4.5, 4.6).

The attraction of *Cleome spinosa* Linnaeus (Spider Flower) to *P. brassicae* larvae was recorded by Tolman (1955) to be greater than other Capparaceae, Cruciferae and Resedaceae plants offered to it at the same time, although MOG myrosin was found to be present in each.

Terofal (1965) compared the acceptability of 50 foodplants to *P. brassicae* and *P. rapae* larvae and found fairly similar tastes between the two species. On the basis of his seven point scale of acceptability there were only three species which *P. brassicae* larvae refused, *Capsella bursa pastoris* (Linnaeus) Medicus (Shepherd's Purse), *Carica papaya* Linnaeus and *Limnathus douglasii* R. Brown (Table 4.8).

Fernando (1970) demonstrated that *P. brassicae* prefers the foodplant on which it was reared initially. He offered imagines whose larvae had been reared on cabbage a choice of three foodplants on which to oviposit. On the three choices, *B. oleracea* var. *oleracea* Linnaeus, *Allaria petiolata* (Bieberstein) Cavara & Grande and *Tropaeolum* sp. there was 90%, 10% and 0% ovipositing respectively. This conditioning effect is substantiated by Chew (1981) who stated that "there is no evidence for specific conditioning effects in a variety of pierid species (with the exception of *P. brassicae* (L.)) whose food related responses have been studied".

Further work on the effect of mustard oil glycosides used in synthetic diets may also be found in the next chapter.

Thorsteinson (1960) classified *P. brassicae*'s foodplant preference types as Type II – "Feeding and ovipositing are induced by chemical stimuli, one or more of which are extraordinary and occur in some plants but not in others". This was further divided into subtype IIa – "The botanical distinction of the extraordinary stimulus substances is highly correlated with natural plant taxonomic groups."

Table 4.7 Foodplants only eaten under special conditions.

Foodplant	Acceptability rating (see Table 4.2)	Authority
DICOTYLEDONS		
COMPOSITAE		
Calendula sp.	4 with sinigrin	Thorsteinson, 1953
(Marigold)	R with sinalbin	
Lactuca sativa Linnaeus	4 with sinalbin	Thorsteinson, 1953
(Lettuce)	R with sinigrin	
CRUCIFERAE		
Brassica nigra (Linnaeus) Koch	1 with sinalbin	Verschaffelt, 1910
(Black Mustard)		

Table 4.7 continued

Foodplant	Acceptability rating (see Table 4.2)	Authority
CUCURBITACEAE		
Cucumis sativus Linnaeus (Cucumber)	1 with sinigrin 3 with sinalbin	Thorsteinson, 1953
GENTIANACEAE		
Menyanthes trifoliata Linnaeus (Buck Bean)	R with juice of *Bunias orientalis*	Verschaffelt, 1910
MALVACEAE		
Malva sylvestris Linnaeus (Common Mallow)	3 with singrin R with sinalbin	Thorsteinson, 1953
PAPAVERACEAE		
Papaver sp. (Poppy)	4 with singrin R with sinabin	Thorsteinson, 1953
PAPILIONACEAE		
Apios tuberosa Moench —	2 with juice of *Bunias orientalis*	Verschaffelt, 1910
Vicia faba Linnaeus (Broad Bean)	2 in laboratory	David & Gardiner, 1953
	3 in laboratory	Feltwell (unpublished)
ROSACEAE		
Prunus laurocerasus Linnaeus (Cherry Laurel)	R with juice of *Bunias orientalis*	Verschaffelt, 1910
Rosa sp. (Rose)	R. with juice of *Bunias orientalis*	Verschaffelt, 1910
SOLANACEAE		
Petunia sp. (Petunia)	2 but died immediately afterwards	Borg, 1932
UMBELLIFERAE		
Daucus carota Linnaeus (Carrot)	4 (root only)	Verschaffelt, 1910
MONOCOTYLEDONS		
LILIACEAE		
Allium sp.	1p	Kartzov, 1914
Allium azureum Lebedour (Onion)	3	Verschaffelt, 1910
Allium cepa Linnaeus (Onion)	3 (bulb scales only)	Verschaffelt, 1910
Allium porrum Linnaeus (Leek)	3	Verschaffelt, 1910
GRAMINEAE		
Triticum sp. (Wheat)	1 with juice of *Bunias orientalis*	Verschaffelt, 1910
Zea mays Linnaeus (Maize, Corn)	1 with juice of *Bunias orientalis*	Verschaffelt, 1910

Table 4.8 Foodplants always refused.

Specific name	Authority
CARICACEAE	
Carica papaya Linnaeus	Verschaffelt, 1910; Terofal, 1965
COMPOSITAE	
Lactuca sativa Linnaeus	Fabre, 1912
(Lettuce)	
CRUCIFERAE	
Brassica oleracea var. *acephala* Linnaeus	Mathew, 1892
(Cole, Borecole)	
Capsella bursa-pastoris (Linnaeus) Medicus	Terofal, 1965
(Shepherd's Purse)	(see also Table 4.4)
LABIATAE	
Mentha piperita Linnaeus	Terofal, 1965
(Mint)	
Salvia officinalis Linnaeus	Terofal, 1965
LIMNANTHACEAE	
Limnanthes douglasii R. Brown	Terofal, 1965
VALERIANACEAE	
Valerianella locusta (Linnaeus) Betcke	Fabre, 1912
(Corn Salad)	

Meeuse (1973) postulated a "chemical repellent-tolerance hypothermia" to explain feeding in *P. brassicae* larvae. Lourens (1971) showed that thioglucosides which act as repellent substances in certain foodplants make the food unacceptable for *P. brassicae* larvae. It has also been shown on the other hand that isothiocyanates, sulphate and glucose units coupled with a more or less taxon specific radical act as attractants in the Pieridae (cf. Schoonhoven, 1969a, b). Meeuse (*loc. cit.*) believed that divergent chemotaxis must have originated in European gardens as populations of *P. brassicae* live on *Tropaeolum majus* which contains a particular thioglucoside (with $R =$ phenylmethylene or benzyl $C_6H_5CH_2$—) differing from sinigrin which is present in all original foodplants ($R =$ butenyl-1-2 or allyl, $CH_2=CH_3$—CH_3—CH_2—) (cf. Heywood, 1973).

References cited

Allan, P.B.M., 1949. *Larval Foodplants*. Watkins & Doncaster, London.
Allen, M.D. & Selman, I.W., 1957. The response of larvae of the large white butterfly (*Pieris brassicae* L.) to diets of mineral deficient leaves. *Bull. ent. Res.* 48: 229–242.

Anglade, P., Robin, J.C. & Roehrich, R., 1963. Observations de 1963 sur les migrations de la piéride du chou (*Pieris brassicae* L.) Lepidoptera: Pieridae, dans le Sud-Ouest. *Revue Zool. agric. appl.* 10–12: 98–102.

Anonymous, 1932. Plant distribution and pests in Denmark 1932. *Tijdschr. PlZiekt.* 39: 453–511.

Bailey, L.H., 1927. *The Standard Cyclopedia of Horticulture.* I. MacMillan, London.

Blunck, H., 1953. *Tierische Schädlinge an Nutzpflanzen* I. Teil. Lepidopteren und Trichopteren. Berlin.

Bodenheimer, F.S., 1935. *Animal Life in Palestine.* L. Mayer, Jerusalem.

Boldyrev, V.F., 1922. An incident in the life of the cabbage white butterfly. (In Russian.) *Herald N. Dist. Stn. Pl. Prot.* 7.

Borg, P., 1932. *Lepidoptera of Maltese Islands.* Government Printing Press, Valetta, Malta.

—— 1936. Plant pathologist report. *Rpt. Dept. Agric., Malta* 1935: 53–91.

Buckler, W., 1928. *Larvae of British Butterflies and Moths.* I. Butterflies. Ray Society.

Buston, P.A., 1924. Applied entomology in Palestine, being a report to Palestine Government. *Bull. ent. Res.* 14: 289–340.

Campbell, J.L., 1970. Macro-lepidoptera Cannae. Butterflies and moths of the Isle of Canna, Inner Hebrides. *Entomologist's Rec. J. Var.* 82: 1–27.

Chalmers-Hunt, J.M., 1960. Lepidoptera of Kent. *Entomologist's Rec. J. Var.* 72: 18–19.

Chandra, J. & Lal, O.P. 1977. Development and survival of caterpillars of cabbage butterfly, *Pieris brassicae* Linnaeus on some varieties of cabbage. *Ind. J. Ent.* 38: 187–188.

Chew, F.S., 1981. Co-evolution of Pierid butterflies and their cruciferous foodplants II. The distribution of eggs on potential foodplants. *Evolution* (in press).

David, W.A.L. & Gardiner, B.O.C., 1953. Systemic insecticidal action of nicotine and certain other organic bases. *Ann. appl. Biol.* 40: 91–105.

Dethier, V.G., 1947. *Chemical Insect Attractants and Repellents.* H.K. Lewis & Co. Ltd., London.

Fabre, J.H., 1912. *The Life of the Caterpillar.* Hodder & Stoughton, London.

Fearnclough, T.D., 1972. The butterflies of the Isle of Wight. *Entomologist's Rec. J. Var.* 84: 102–109.

Feltwell, J.S.E. & Rothschild, M., 1974. The carotenoids of 38 species of lepidoptera. *J. Zool. Lond.* 114: 441–465.

Feltwell, J.S.E. & Valadon, L.R.G., 1972. Carotenoids of *Pieris brassicae* and of its foodplant. *J. Insect Physiol.* 18: 2203–2215.

—— 1974. Carotenoid changes in *Brassica oleracea* var. *capitata* L. with age in relation to the Large White Butterfly, *Pieris brassicae* L. *J. agric. Sci. Cambs.* 83: 19–26.

Fernando, L.V.S., 1971. Selection and utilisation of different foodplants of *Pieris brassicae.* *Spolia zeylan.* 32: 115–127.

Friederichs, K., 1931. Zur Ökologie des Kohlweisslings (*Pieris brassicae*). *Z. angew. Ent.* 18: 568–581.

Frohawk, F.W., 1927. A new foodplant of *Pieris brassicae. Entomologist* 60: 243.

—— 1934. *British Butterflies.* Ward, Lock & Co. Ltd., London & Melbourne.

Gardiner, B.O.C., 1978. Instar number and pupal colouration in Palestinian *Pieris brassicae. Proc. Trans. Brit. ent. nat. Hist. Soc.* 11: 21–23.

Gautier, C.L. & Riel, P.H., 1919. Sur l'alimentation des chenilles des genres *Pieris* et *Euchloe. C. r. Séanc. Soc. Biol.* 82: 1371–1374.

George, L., 1927. Observations sur la biologie de deux Hyménoptères ento-mophages. *Bull. Soc. Hist. nat. N. Afr.* 18: 55–71.

Granville-Clutterbuck, C., 1941. Notes on lepidoptera at the Lizard in 1920. *Entomologist* 41: 121–123.

Harford, H.C., 1927. *Pontia daplidice* in Malta. *Entomologist* 60: 161–162.

Heslop-Harrison, J.W., 1946. The abundance of *Pieris brassicae* larvae in Rhum in 1945. *Entomologist* 74: 70.

Heywood, V.H., 1973. *Taxonomy and Ecology*. Systematics Association, Academic Press, London & New York.

Hickin, N.E., 1942. *Pieris brassicae* L. (Lep.: Pieridae) larvae feeding on Honesty. *Entomologist's mon. Mag.* 78: 205.

Holford, H.C., 1928. Foodplant of *Pieris brassicae. Entomologist* 61: 10.

Homer, T.J.G., 1969. Communication. *Proc. Brit. ent. Soc.* 2: 36.

Johansson, A.S., 1951. The foodplant preference of the larvae of *Pieris brassicae* L. (Lep.: Pieridae). *Norsk. ent. Tidsskr.* 8: 187–195.

Kadamshoev, M., 1971. Biology of the cabbage butterfly, *Pieris brassicae* L. in Western Pamir. (In Russian.) *Izv. Akad. nauk. Tadkzhik SSR Otd. Biol. nauk.* 2: 89–91.

Kartzov, A.S., 1914. Cultivation of onions, leeks and garlic. (In Russian.) *Mkt. Gdn. Lib. Suppl. Progressive Fruit Growing & Market Gardening in Petrograd.* 1914.

Kayser, H., 1975. Fatty-acid esters of lutein in *Pieris brassicae* fed on natural and artificial diets. *Insect Biochem.* 5: 861–875.

Kirby, W. & Spence, W., 1858. *An Introduction to Entomology.* Longman, Brown, Green, Longmans & Roberts, London.

Kornfeld, A., 1935. Injuries to and diseases of the Soy Bean, so far as hitherto known in Europe. *Z. Pflkrankh. PflPath. PflSchutz.* 45: 577–613.

Kurentzov, A.I., 1970. *Butterflies of the Far East USSR.* Academy of Sciences of USSR, Siberian Branch.

Le Cerf, F., 1972. *Atlas des Lépidoptères de France, Rhopalocères.* Boubée, Paris.

Lourens, J.H., 1971. *Voedselspecialisatie bij Pieridae.* Unpublished MS report. Department of Applied Entomology, University of Amsterdam, The Netherlands.

Lundgren, L., 1975. Natural plant chemicals acting as oviposition deterrents on cabbage butterflies (*Pieris brassicae, Pieris rapae, Pieris napi*). *Zool. Scr.* 4: 253–258.

Mathew, G.F., 1892. Abundance of the larvae of *Pieris brassicae. Entomologist* 25: 287.

Meeuse, A.D.J., 1973. Co-evolution of plants and parasites. *In: Taxonomy and Ecology.* Ed. V.H. Heywood. Systematics Association, Academic Press, London & New York. pp. 289–316.

Mickel, C.E., 1973. John Ray: Indefatigable student of nature. *A. Rev. Ent.* 18: 1–16.

Monks Wood Data Card, 1942. Monks Wood, Huntingdon, England.

—— 1971. Monks Wood, Huntingdon, England.

Newman, L.H., 1942. Note. *Entomologist* 75: 76.

Nicholson, C., 1939. Larvae of *Pieris brassicae* (?) in December. *Entomologist* 72: 111.

Ormerod, E.A., 1882. *Notes and Observations on Injurious Insects during 1882.* Simpkin Marshall, London.

—— 1887. Notes of observations of injurious insects and common crop pests in

1887. Report on observations on injurious insects and common farm pests in 1886. 87–89.

Pickard-Cambridge, A.W., 1918. Notes from North Wales. *Entomologist* 51: 18.

Pictet, A., 1918. Observations biologiques sur *Pieris brassicae* en 1917. *Bull. Soc. lépidopt. Genève.* 4: 53–66.

Plotnikov, V., 1914. Insects injurious to orchards, field-crops and market gardens in Turkestan, – with methods of fighting them. *Turkestan ent. Station, Tashkent* 1914: 122.

Rainey, R.C., 1936. *Observations on insect metabolism.* Diploma thesis, London University (Imperial College).

Rothschild, M., 1978. Exhibit: *Pieris brassicae* larvae eating *Euphorbia polychroma* leaves. *Proc. R. ent. Soc. Lond.* 43: 20.

Rothschild, M. & Schoonhoven, L.M., 1977. Assessment of egg load by *Pieris brassicae* (Lep.: Pieridae). *Nature, Lond.* 266: 352–355.

Rothschild, M., Gardiner, B.O.C., Valadon, L.R.G. & Mummery, R.S., 1975. The large white butterfly: oviposition cues, carotenoids and changes of colour. *Proc. R. ent. Soc. Lond.* 40: 13.

Sandars, E., 1955. *A Butterfly Book for the Pocket.* Oxford University Press, London.

Schoonhoven, L.M., 1967. Chemoreception of mustard oil glycosides in larvae of *Pieris brassicae. Koninkl. Ned. Akad. Wetensch. Proc.* Ser. C. 70: 556–568.

—— 1969a. Amino-acid reception in larvae of *Pieris brassicae* (Lepidoptera). *Nature, Lond.* 221: 1268.

—— 1969b. Gustation and foodplant selection in some lepidopterous larvae. *Entomologia exp. appl.* 12: 555–564.

Schwarz, R., 1948. *Butterflies Diurnal.* Vesmir, Prague.

Scott, H., 1942. *Pieris brassicae* L. (Lepidoptera: Pieridae) feeding on Honesty. *Entomologist's mon. Mag.* 78: 192.

South, R., 1936. *Butterflies of the British Isles.* Warne, London.

Speyer, W., 1956a. Das Verhalten der Kreuzkrote (*Bufo calamita* Laur.) gegenüber den Raupen von *Pieris brassicae* L. *NachrBl. dt. PflSchutzdienst, Berl.* 8: 190.

—— 1956b. Die Bedeutung Schleswig-Holsteins für den Massenwechsel des Grossen Kohlweisslings (*Pieris brassicae* L.) *Faun.-ökol. Mitt.* 7: 17–20.

Swinton, A.H., 1898. Butterflies seen in and around Jerusalem. *Entomologist's mon. Mag.* 34: 181–184.

Terofal, F., 1961. *Zum Problem der Wirtsspezifität bei Pieriden (Lepidoptera). Unter besonderer Berücksichtigung der einheimischen Arten Pieris brassicae L., Pieris rapae L. und Pieris napi L.* Disseration, University of Munich.

—— 1965. Zum Problem der Wirtsspezifität bei Pieriden (Lepidoptera). *Mitt. münch. ent. Ges.* 55–56: 1–76.

Theobald, F.V., 1909. Animals injurious to vegetables. *Jl. S.E. agric. Coll. Wye* 1909: 157–164.

Thorsteinson, A.J., 1946. *The chemotropic response that determines host specificity in an oligophagous insect Plutella maculipennis Curt., Lepidoptera.* Ph.D. thesis. London University (Imperial College). pp. 119 + illustrations.

—— 1953. The chemotactic responses that determine host speciation in an oligophagous insect (*Plutella maculipennis* (Cut.) Lep.) *Can. J. Zool.* 31: 52–72.

—— 1958. The chemotactic influence of plant constituents on feeding by phytophagous insects. *Entomologia expl. appl.* 1: 23–27.

—— 1960. Host selection in phytophagous insects. *A. Rev. Ent.* 5: 193–218.

Tolman, R., 1955. Aberrant food of butterfly larvae. (In Dutch.) *Ent. Ber., Amst.* 15: 459.

Turner, H.J., 1931. The South London Entomological Society. *Entomologist* 64: 96.

Vepsäläinen, K., 1968. Immigration of *Pieris brassicae* L. into Finland in 1966, with general discussion on insect migration. *Ann. ent. fennici.* 34: 223–243.

Verity, R., 1905–1911. *Rhopalocera Palearctica.* Iconographie et description des papillons diurnes de la région palaéarctique. Florence. 2 vols.

Verschaffelt, E., 1910. The cause determining the selection of food in some herbivorous insects. *Proc. Acad. Sci. Amsterdam* 13: 536–542.

Wilson, O.S., 1880. *Larvae of the British Lepidoptera and their Foodplants.* Reeve, London.

Zanon, V., 1919. Horticulture at Benghasi, Tripoli. *Agric. Colon Florence* 13: 154–176.

Zeller, P.C., 1847. Uber die auf einer Reise nach Italien und Sicilien beobachteten Schmetterlingsarten. *Isis Gudiss.* 1: 214–243.

Further references

Auger, R., 1945. *Les plantes de grande culture de la famille des Crucifères.* Thèse Docteur, Université Strasbourg (general).

Böning, K., 1938. *Nachr. Schädbl.* 13: 62–87 (larvae on *Cochlearia* sp.).

Burrows, W.F., 1964. *Entomologist's mon. Mag.* 100: 110 (oviposition on *Cleome spinosa* L.).

Chapman, R.F., 1974. *Feeding in leaf-eating insects.* Oxford Biology Readers (chemoreception of foodplants).

Chittenden, F.J., 1951. *Dictionary of Gardening.* Clarenden, Oxford (larvae on *Tropaeolum* sp.).

Geijskes, D.C. & Doeksen, J., 1947. *Tijdschr. Ent.* 99: 16–34 (larvae on crucifers).

Ghika, G. & Loibl, H., 1925. *Int. ent. Z.* 19: 243–247 (larvae on *C. maritima*).

Hoffmann, F., 1909. *Ent. Z.* 22: 225–227 (larvae on Compositae).

Peregudt, M.F., 1977. *Rastit Resur.* 13: 94–99 (as a pest on *Crambe steveniana* Rupr.).

Rivnay, E., 1962. *Field Crop Pests in the Far East.* Junk, Netherlands. (mustard family foodplants preferred).

Southwood, T.R.E., 1972. *In: Insect/Plant Relationships.* Ed. Royal Entomological Society of London. pp. 3–30 (general interest).

Vogler, W., 1969. *Atalanta* 2: 301–302 (on *Cakile maritima*).

5. Breeding

Introduction

Pieris brassicae has been reared traditionally in the laboratory and in the home, on cabbage, which has often been readily available. However, with the acceptance of *P. brassicae* as a laboratory insect, and with the need for regular supplies of the insect, whether it be for teaching purposes or for screening of insecticides, there has always been a need for the development of synthetic diets. Over the last 20 years there has been much work in the development of insect diets with the consequence that now several hundred recipes have been published (Singh, 1974, 1977; Gardiner, 1978). Gardiner (*loc. cit.*) pointed out that *P. brassicae* has been successfully reared on two types of diet described in his review paper: Ignoffo's and Yamamoto's.

The use of *P. brassicae* as a typical laboratory insect is due principally to the pioneer work of Dr. W.A.L. David and Mr. B.O.C. Gardiner at the University of Cambridge during the 1960s who developed the synthetic diets on which the larvae can feed. This allows the laboratory worker to rear *P. brassicae* throughout the year and thus be independent of the seasonality of supplies from the wild. Moreover, the system provides a continuous supply of *P. brassicae* ova, larvae, pupae and imagines which can be used for experimentation. Many commercial companies have used *P. brassicae* for the screening of insecticides.

There are several advantages in using *P. brassicae* as a laboratory insect. It is a species which can be acquired without too much difficulty in the wild and the fifth instar larva provides usefully large material on which to experiment. Moreover, *P. brassicae* can be reared all the year round on synthetic diets. As a teaching aid, Philip Harris (1974), the biological suppliers, said that "its full potential as a teaching organism has yet to be realised ... suited for showing the effects of environmental factors or diapause and is a useful organism for project work."

On the other hand Lahargue (1952) pointed out that there were three

118

disadvantages in choosing *P. brassicae* as a laboratory insect: first, that its numbers fluctuated from year to year in the wild, second that the larvae were susceptible to disease, and third, that they were very prone to parasitism. Wild collected specimens can sometimes upset laboratory experiments because of high degrees of parasitism (e.g. Poulton, 1890), or infection due to the fungus *Entomophthora sphaerosperma* Fresenius (e.g. Kanervo, 1946). Rearing stocks of *P. brassicae* continuously in the laboratory is liable to increase their susceptibility to disease, thus precluding total reliability as a laboratory insect (Felton *et al.*, 1977). However, this can be alleviated with periodical introduction of stocks from suppliers or from the wild.

Breeding

The choice of rearing room for *P. brassicae* varies enormously with available facilities. For instance, *P. brassicae* may be, and is, reared throughout the year in basement rooms, laboratories, corridors, constant temperature rooms, garden sheds and greenhouses, provided the light regime, temperature and the humidity are controlled.

A number of papers describe the rearing of *P. brassicae* in capitivity and from these a number of useful tips can be gleaned: Lahargue, 1952; David, 1957; David & Gardiner, 1952, 1961a,b, 1962a,b; Lyon, 1960; Grison & Silvestre de Sacy, 1957; Junnikkala, 1966; Chaufaux, 1975.

Cage size is an important factor as this may affect incidence of mating and fecundity (Dusaussoy & Delplanque, 1964). The best cage size found by these workers was $0.50 \times 0.50 \times 0.50$ m and was sufficient to obtain optimum fertility when 20 pairs are introduced. This works out to be 3000 cm^3 per individual of each sex. David & Gardiner (1961a, b) obtained satisfactory results using a cage size of $0.32 \times 0.32 \times 0.37$ m with 12 pairs of insects.

Ovum

In most *P. brassicae* breeding systems, potted cabbage plants are introduced into the mating cage for a few hours each day and females cover these with ova. A suitable cage for ovipositing was figured in Chaufaux (1975) where the plant is raised on a shelf near the top light source. These plants are removed from the cage at regular intervals, labelled with the date of oviposition and placed in a well ventilated room at 20°C. They can then be treated to remove viruses adhering to their shells. After a few days at room temperature, the black headed first instar larvae eat their way out of the egg case which they finally consume. At lower and higher temperatures ova take a much longer or much shorter time to hatch respectively (David & Gardiner, 1962b).

Larva

Larvae can be reared in large cages, such as those sold for locusts, in which several thousand can live, or smaller numbers can be kept in transparent plastic (sandwich) boxes with ventilated lids. In order to keep the relative humidity (RH) constant, the whole room can be regulated or if boxes are used, they can be kept in another box or container; Junnikkala (1966) used glass jars with a little potassium chloride solution in the bottom to maintain an RH at 83% (the larvae were supported above on a perforated shelf). Alternatively, the relative humidity can be controlled by using different concentrations of sulphuric acid and/or water. Larvae can also be reared successfully on muslin supported over water in glass jars. Water has the added advantage of keeping the cabbage fresh (cf. David & Gardiner, 1962a,b). A further satisfactory method is to use 5×2.5 cm glass tubes in which 20 larvae can be kept, or 1 lb (0.45 k) jam jars in which 100 larvae may be reared (Gardiner, 1978).

Many breeders of *P. brassicae* are fastidious about cleaning out cages of old cabbage debris and frass, in the belief that this is an ideal breeding ground for pathogenic organisms. However, Gardiner has successfully maintained his Cambridge strain for over 25 years (i.e. 200 generations) without cleaning out the cages during broods, and without significant outbreak of disease (Gardiner, 1976, pers. comm.). The advantage of rearing larvae in small numbers is that if disease does occur, the infected specimens can be removed easily and there is greater chance that the colony survives. Grison & Silvestre de Sacy (1957) published photographs of the effects of virus and bacterial attacks on *P. brassicae* larvae (cf. Chapter 15). It can be seen therefore that many varying methods are successful and the would-be breeder is advised to choose the method that most suits his individual inclination and/or available time.

Pupa

Adequate pupation sites must be provided in a cage, the larvae tending to choose the corners, roof and sides. If pupae are to be removed for storage, care must be taken in not pulling the cremasters away from the silk bases. Often it is possible to remove several pupae on a large continuous sheet of silk.

Non-diapausing pupae can be successfully stored in a refrigerator at 10°C for up to six weeks to overcome holiday difficulties. Littlewood

(1941) claimed that pupae of *P. brassicae* could be "forced" to emerge earlier by subjecting them to 13–21°C in a damp atmosphere.

Imago

A relatively small cage is adequate for mating and oviposition in *P. brassicae*, although larger sized cages are often used. To prevent imagines escaping, one side fitted with muslin sleeves allows regular maintenance of the colony. Radiant energy provided by the sun is invaluable in inducing imagines to copulate (Dusaussoy & Delplanque, 1964) and therefore a greenhouse is preferable. Artificial light provided by large Tungsten filament bulbs, or in the case of Dusaussoy & Delplanque (*loc. cit.*) 32 fluorescent tubes giving out 5000–8000 Lux, does not always match up to sunlight but can be used successfully. Gardiner, who originally used a 500 W tungsten bulb, now uses natural daylight only (Gardiner, 1978, pers. comm.).

Freshly emerged imagines must be given food, as they reabsorb their yolk after their second day if they are starved (Street, 1975). This is particularly important as the ova continue development after the imago has eclosed. Sugar and honey solutions are most often used, and may be saturated into coloured cotton wool in petri dishes or in artificial flowers.

Artificial flowers are easily made in the laboratory (David & Gardiner, 1952, 1961b; Grison & Silvestre de Sacy, 1957, Figure 5.1).

In their experiments David & Gardiner (*loc. cit.*) found that large blue artificial flowers were more attractive than small ones, and that fresh honey solution was very suitable as food. Fifty percent of imagines which fed on 10% honey solution under normal conditions lived for 18 days. Only a few imagines will feed at artificial flowers and it is necessary to select only these individuals from which to breed, so that greater success will be achieved in later generations.

It is advantageous to have the flowers raised off the ground and near the lights as in the 2 m high cage illustrated by Dusaussoy & Delplanque (1964), or they may be arranged on a supporting alighting board. In any case the blue, red or purple artificial flowers have a hole in the centre in which a small glass tube filled with sugar solution is inserted.

Imagines can live for up to 21 days in the laboratory but much depends on the conditions of temperature, relative humidity and light and food. Further information may be found in Chapter 3 (Life History) and Chapter 6 (Development).

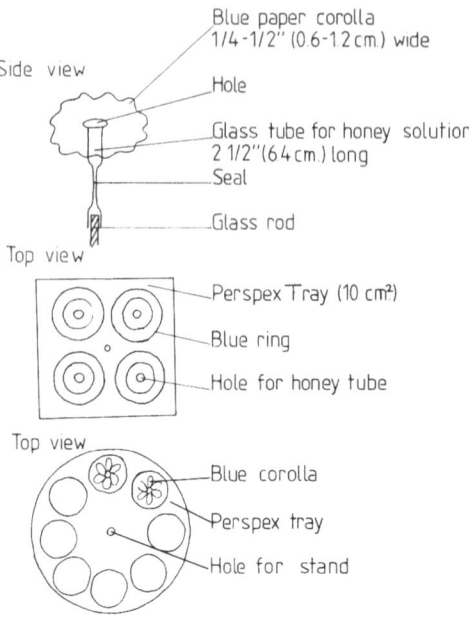

Side view

Blue paper corolla
1/4-1/2" (0.6-1.2 cm) wide

Hole

Glass tube for honey solution
2 1/2"(6.4 cm) long

Seal

Glass rod

Top view

Perspex Tray (10 cm²)

Blue ring

Hole for honey tube

Top view

Blue corolla

Perspex tray

Hole for stand

Fig. 5.1 Artificial flowers
(after David & Gardiner, 1952, 1961b)

Synthetic diets

Introduction

The terms "Artificial", "Synthetic" and "Semi-synthetic" have all been applied to the same diets at some time and it is not strictly accurate to refer to the last two as containing or not containing dried leaf material respectively. A "true synthetic diet" or "defined diet" would contain absolutely no plant substances, only chemicals. However, agar, wheat germ and casein are plant materials and are often included in diets of an artificial nature. An early and unsuccessful attempt to rear *P. rapae* on cellulose filter paper impregnated with cabbage was tried by Rainey (1936).

Semi-synthetic diets and synthetic diets have been used for the culture of lepidoptera for almost 30 years (see the review of artificial diets by Singh, 1977). Semi-synthetic diets were first described and used for the Large White Butterfly by Thorsteinson (1947, 1953, 1958), David & Gardiner (1965b,c 1966a) and Ma (1972), while synthetic diets were first used by David & Gardiner (1966a,b), Akhtar (1966) and Wardojo (1969).

The "synthetic" diets used by David & Gardiner were not entirely synthetic as they contained wheat germ and casein. When these two materials were omitted from the diet all the larvae died.

Synthetic diets have wide applications in research laboratories and allow much scope for investigating effects of different food components, as the basic diet can be easily manipulated to suit the research workers' requirements. Indeed, the diets described by some workers for rearing *P. brassicae* were modified from diets previously described for other species; for example David & Gardiner (1965a,b; 1966b) based their diet on Ignoffos *Trichoplusia ni* (Hübner) (Cabbage Looper) medium, only substituting cabbage leaf powder for cotton leaf powder; and those used for *Plutella maculipennis* (Curtis) (Diamond-black Moth), *Pectinophora gossypiella* (Saunders) (Pink Bollworm), while Ma (1972) based his diet on that used for rearing saturniid moths.

The advantages of the semi-synthetic diet over the conventional method of rearing larvae have been described by David & Gardiner; they are a) useful for insecticidal screening, b) investigating effects of pathogenic organisms on insects as the diet can be sterilised, c) the nutrients of the diet can be defined quantitatively and qualitatively, d) there is less likelihood of infection of the culture, e) the diet can be used throughout the whole year, particularly during the winter months when natural foliage may be in short supply, f) the diet is more convenient for use in laboratories situated in cities, and g) it allows a greater facility for counting and manipulating larvae.

The diets of David & Gardiner (*loc. cit.*) have been used successfully by many workers including Van der Geest (1968), Van der Geest & Sloog-Hoebe (1972), Feltwell (1973) and Rothschild (1975). It has also proved suitable for a wide range of other lepidopterous species even when cabbage flavoured.

Ingredients

The composition and preparation of most synthetic diets is essentially the same and all the ingredients are usually homogenised throughout an inactive carrying substance, such as a suspension of agar, sodium alginate or Carageenen (Table 5.1). Carbohydrates and proteins are provided in the form of sugar, wheat germ and casein, and the fat as linseed oil which also aids in successful eclosion. A salt mix, antibiotics and sterols are usually included. Cabbage leaf powder may or may not be included, but if not a feeding stimulant is often added (David & Gardiner, 1966b). Diets based on that of Yamamota for *Manduca sexta* (L.) and containing yeast powder do not require additional flavouring when used for *Pieris* (Gardiner, 1977, pers. comm.).

Table 5.1 Artificial diet for *P. brassicae*. (after David & Gardiner, 1966b, 1975. To make 7.6 litres, cf. Singh, 1974).

Ingredients	Amount
Water	6.2 l
Agar	180 g
Casein	252 g
Wheat germ	216 g
Wessons salts	72 g
Sugar	252 g
Cellulose	36 g
Potassium hydroxide	36 ml
Choline chloride	72 ml
Methyl-p-benzoate	72 ml
Linseed oil	20 ml
Singrin solution	30 ml
Formalin solution	30 ml
Vitamin mix	37 g
Antibiotic mix	
Vitamins (as above)	5 g
Streptomycin sulphate	10 g
Ascorbic acid	200 g
Aureomycin (25 g/lb)	90 g

Vitamin mix

There are fourteen vitamins which are included in diets of *P. brassicae*, which comprise vitamin A, the vitamin B complex, vitamin C and vitamin E (tocopherols) (Table 5.2).

There has been very little standardisation between workers on the amounts and even the presence of the vitamins included in diet mixes, and a scan through the works of Thorsteinson (1947), David & Gardiner (1965b, 1966a), Akhtar (1966), Jones (1968), Wardojo (1969), Ma (1972), Van der Geest & Slook-Hoebe (1972) and Gardiner (1974) will make this clear. Only Ma (1972) added vitamins C and E with the B complex vitamins in what is called a Vanderzant mixture. Choline chloride, inositol, p-aminobenzoic acid and cyanocobalamine (B_{12}) have not been included in diets by all workers.

One of the problems of using natural products as ingredients for synthetic diets is that they contain vitamins in minute amounts which may upset physiological experiments. It is therefore a necessity to chemically assay each natural product to determine both qualitatively and quantitatively the particular vitamins present. Wheat germ for instance, in the

Table 5.2 Vitamins used in diets of *P. brassicae*.

Vitamin A
Vitamin B Complex
Aneurine hydrochloride
Calcium pantothenate
Choline chloride
Cyanocobalamine (B_{12})
D-biotin
Folic acid
Inositol
Nicotinic acid
p-aminobenzoic acid
Pyridoxine hydrochloride (B_6)
Riboflavin (B_2)
Vitamin C (ascorbic acid)
Vitamin E (tocopherols)

form of "Bemax" contains carotenoids and most of the B complex vitamins (Feltwell & Rothschild, 1974). Ma (1972) used vitamin-free casein which is liable to contain traces of vitamin B in its untreated state.

It is appropriate to mention the physiological works which included details of synthetic diets of Feltwell (1973) on the metabolism of vitamin A and that of Turunen (1976) on vitamin E and lipid synthesis in *P. brassicae*.

Salt mix

Standardisation can be achieved by using Wessons salts (Salt Mix W of N.B.C. Co.) (Table 5.3), although different concentrations of the salt mixture have been used by different workers; for instance, Wardojo (1969) used 0.50 g/100 g of diet, while Ma (1972) and Gardiner (1975) used 9.8 g/litre of diet mixture. Once prepared the salt mixture can be stored indefinitely at room temperature.

Antibiotics

Synthetic diets are subject to contamination from bacterial and fungal agents, less so from protozoan species. Some workers have autoclaved diets in order to sterilise the medium. However, inclusion of antibiotics does not ensure against contamination and diets for *P. brassicae* have to be changed frequently, up to a maximum of about seven day intervals. Fungal and bacterial infection may occur however after a few days.

Table 5.3 Salt mix – Wessons salts.

Salt	Formula	Amount in g
Calcium carbonate	$CaCO_3$	120
di-Potassium hydrogen orthophosphate	K_2HPO_4	129
Calcium hydrogen phosphate	$CaHPO_4$	30
Magnesium sulphate	$MgSO_4$	40.8
Sodium chloride	NaCl	67
Ferric citrate	FeC_6H_2O	11
Manganese sulphate	$MnSO_4$	2
Potassium iodide	KI	0.32
Zinc chloride	$ZnCl_2$	0.1
Copper sulphate	$CuSO_4$	0.12

Freshly made diet can be kept at $-25°C$ for six months without contamination.

Veterinary grade aureomycin is recommended by Gardiner who has also tested another antibiotic called ampicillin (= Penbritin). However, no consistent results were obtained from these experiments to show that either was better, when compared with controls. Aureomycin is responsible for the blue colour of diets especially those supplied to schools. The aureomycin content is probably much higher in this case to allow for the greater risk of contamination by not using autoclaves which are not generally available in schools. Gardiner (1975) recommended that an "antibiotic mix" should be made up using a selection of vitamins, including vitamin C, aureomycin and streptomycin sulphate. This and other dietary ingredients are now marketed commercially by N.B.C. Biochemicals, Miles Road, Cleveland, Ohio, U.S.A. English agents are V. Howe & Co., and Messrs. Harris Biological Supplies, Oldmixon, Weston-Super-Mare, Avon market both ready made and also *P. brassicae* ova. Wardojo (1969) used an "inhibitor" mix which contained ascorbic acid (0.15 g), methyl-p-hydroxybenzoate (0.10 g) and streptomycine sulphate (0.02 g).

Other factors

Linseed oil which contains both linolenic and linoleic acids, was recommended by David & Gardiner to remedy the crippledness which was liable to arise in males reared on a synthetic diet. They trid concentrations of 0.5, 2.0 and 1.0 ml per 379 g of diet and found that the latter concentration results were equivalent to rearing on normal foodplants. It was also shown that domestic corn oil was equally effective as linseed oil (Gardiner, 1977, pers. comm.) and this is now widely used in place of the now very expensive linseed oil. In a recent paper Kastari & Turunen (1977) recommended a supplementing diets for *P. brassicae* with seed oils with added phospholipids to offset the unbalanced nature of lipid composition in the diets.

Preparation

The best method to adopt if the diet is to be made up from raw materials is that of David & Gardiner (1965c, 1966c); if not, a ready mixed commercial diet as supplied by a biological supplier can be made up in minutes by adding water.

If the diet is to be made up from raw materials certain points should be noted. The vitamin mix breaks down at room temperature and must therefore be stored in a refrigerator, so also should the choline chloride solution which is liable to bacterial contamination. The vitamin mix should always be mixed with the diet after it has cooled down. A dispensing apparatus for diet has been figured by David & Gardiner (*loc. cit.*). Care must be exercised in adhering to the diet recipe, as too much water can lead to drowning of small larvae on the surface of the diet.

When cool the tubes containing diet can be covered with a wax or plastic cover and stored at $-25°C$ for many months without deterioration. In an ordinary refrigerator at $3°C$ the diet dries up, cracks and goes a dark colour and is quite unsuitable for larvae.

Rearing tubes

For rearing groups of young larvae David & Gardiner (*loc. cit.*) found glass tubes measuring 30×75 mm containing about 20 first instar and second instar larvae per tube sufficient; while Van der Geest (1968) found 40 to 50 young larvae in each glass tube measuring 2.5×11 cm convenient. David & Gardiner (*loc. cit.*) also used glass tubes measuring 23×50 mm but transferred larger larvae to 1 lb (0.45 kg) jam jars (6×11 cm). In the case of Van der Geest (*loc. cit.*) plastic boxes $19 \times 11 \times 7$ cm covered with 0.5 cm of diet for 25 groups of larvae after the third moult were used. Gardiner (1978, pers. comm.) also advises that losses when using strips of diet in plastic boxes are considerable but in jars negligible. With other lepidoptera such as *Manduca sexta* Linnaeus, the converse applies however.

For rearing larvae individually David & Gardiner (*loc. cit.*) used 5×2.5 cm glass tubes; Wardojo (1969) used glass tubes 2 cm in diameter with a capacity for 17.5 ml of diet or alternatively petri dishes for individual young larvae six to eight hours old. Later these were transferred to plastic boxes $18.5 \times 10 \times 7$ cm.

If the boxes or tubes are turned upside down the frass falls to the bottom and does not foul the diet. Condensation is also a problem but can be overcome by using filter paper early on and wire gauze at later stages. Most workers rear the larvae in laboratories under defined conditions, while Wardojo (1969) used a phytotron for his controlled environment.

Temperature and humidity

Larvae were reared successfully by David & Gardiner (*loc. cit.*) at 20°C and 60% RH. At higher temperatures (25°C) and higher humidity (62–63% RH) development was faster and thought not to have any detrimental effect. However, at low humidities (37–40%) the diet dried up and the larvae did not develop well (David & Gardiner, 1966b).

Effect of normal diet on larval development

Although David & Gardiner (1961a; 1965b,c) recorded that imagines of diet-reared larvae mated normally, laid fertile eggs and 100% success was obtained from larvae placed on this diet, there are some characteristic features of this development which are worth noting.

First, it has always been found that *P. brassicae* larvae which have previously hatched and eaten cabbage leaves will not feed on diet. This is probably due to some kind of imprinting of olfactory cues obtained from the plant. For some unexplained reason however, David & Gardiner found that the subspecies *P. b. cheiranthi* would change over to diet after eating *Tropaeolum* or cabbage first. Certain other lepidoptera will also switch from plant to artificial diet and back again.

Second, the period of larval development is usually increased by one to two days provided the diet is changed every seven days. At longer intervals the larvae take up to five days longer to develop fully and the imagines have more abnormalities.

Third, pupae are not as heavy as normal reared pupae, on average being 0.37 g when compared to controls of 0.39 g (David & Gardiner, 1966b).

In trying to establish which factor was responsible for the increase in larval development David & Gardiner (*loc. cit.*) found that the absence of formaldehyde and aureomycin in separate experiments did not affect the rate of development when compared to the rate on normal diets. Thorsteinson (1947) also found that neither the texture of the diet nor the hydrogen ion concentration appreciably altered the attractiveness of the diet.

Effect of diet without cabbage powder on larval development

David & Gardiner (1966a,b) showed that larvae of *P. brassicae* could be reared on diets lacking cabbage powder but fewer larvae settled on the diet and there was a high proportion of crippled males. Wardojo (1969) also reared larvae on an entirely synthetic diet but instead of finding cripples, he found that females were disinclined to copulate and the imagines died out in the second generation even when 80% were successfully obtained from the first generation and the female ovaries were fully developed. Wardojo also found that larval growth was suboptimal and that this could be improved by heating the diet by autoclaving. Apparently this method may either increase the digestibility of the food by hydrolysing some of the materials, distribute some of the materials more evenly throughout the medium or destroy some undesirable substances. However, this method may also destroy essential vitamins.

Semi-synthetic diets with mustard oils included

Table 5.4 shows that 14 mustard oils have been tested in synthetic diets on *P. brassicae*. Thorsteinson (1947, 1953) was the first to include these compounds into a synthetic diet for the Large White and he found that sinigrin was a most effective glycoside for inducing attractiveness of diet. He also found that pea leaf powder treated with sinigrin gave no significantly different feeding response when compared with the effect of introducing normal leaf powder into the diet. However, mustard oil and unhydrolysed sinigrin at moderate concentrations in the semi-synthetic diet did enhance its attractiveness. When Thorsteinson added the enzyme myrosin, which hydrolyses sinigrin, he found it reduced the attractiveness of the diet. In contrast to the other mustard oil glycosides sinalbin was found to make the diet substantially less attractive than with sinigrin.

David & Gardiner's (1966a) experiments showed that a) larvae reared on cabbage would not change to semi-synthetic diet with or without sinigrin; b) larvae reared on diets containing cabbage powder and then transferred onto diets containing sinigrin increased their rate of feeding, and c) larvae reared on diets without sinigrin increased their feeding rate when transferred onto diets containing sinigrin.

What is learnt from this important work is that when rearing *P. brassicae* the larvae must be placed onto the diet before they have had a chance to nibble at the leaf on which they were laid, otherwise the larvae will not accept the diet. Furthermore, sinigrin may be added to diets in the place of cabbage powder giving almost the same length of larval development time.

Further work by David & Gardiner (1966b) indicated that certain

Table 5.4 Additives used in the diet of *P. brassicae.*

Additive	Authority
Antibiotics	
aureomycin	Gardiner, 1975, pers. comm.
Foodplant	
cabbage	David & Gardiner, 1965a,b; Wardojo, 1969
Glucosides	
sinigrin, sinalbin	Thorsteinson, 1946
glucocoiberin, glucocheirolin,	David & Gardiner, 1966b
glucocoerucin, sinigrin,	
glucocapparin, progoitrin,	
glucoconringiin	
Glucosides as tetramethyl ammonium salts	
glucoconsinigrin	David & Gardiner, 1966b
glucotropaeolin	
Growth promotors	
tryptophan-methylesterhydrochorid	Kayser, 1978, pers. comm.
$(C_{12}H_{15}ClN_2O_2)$	
Minerals	
iron, nitrogen, potassium, phosphorus	Evans, 1938a,b, 1940; Allen, 1954; Allen & Selman, 1957
Polysaccharide	
cellulose	Akhtar, 1966
Tetraacetyl glycosides	
acetylglucocheirolin	David & Gardiner, 1966b
acetylsinigrin,	
acetyl-gluconringiin	
Vitamins	
A (as B-carotene)	Akhtar, 1966
A palmitrate	Wardojo, 1969
E	Turunen, 1976

mustard oil glycosides had specific effects on *P. brassicae* larvae. They tested four glycosides present in cabbage, viz. glucoiberin, glucoerucin, sinigrin and progoitrin, and eight other glycosides and tetraacetyl glycosides not found in cabbage (Table 5.4) on fifth instar larvae. By recording the number of frass pellets dropped as a measure of effectiveness of the feeding stimulants, they found that glycosides were more efficient than the acetyl compounds (see Table II in David & Gardiner, 1966b). One of the glycosides which was effective as a feeding stimulant, but which is not reported from *Brassica* spp. is glucoconringin.

It is interesting to note the action of two glycosides in respect of the foodplant preference of *P. brassicae*; first, that glucocapparin was most effective at low concentrations. This has not been reported from Cruciferae, but it is present in the Capparaceae, which are occasionally taken as a foodplant (Chapter 4). Second, that glucotropaeolin was the most effective glycoside at high concentrations. This is present in *Tropaeolum majus*, one of the most frequently selected foodplants by the imago. The effectiveness of all the glycosides was found to increase with increasing concentration up to $0.333 \ 10^{-3}$ molar. David & Gardiner concluded that a mixture of glycosides might well give a good feeding response for a group of larvae.

Cabbage powder can be incorporated into a purely synthetic diet to make it more acceptable. Feeding stimulants, such as mustard oil glycosides and sinigrin are present in most parts of cabbage and allow the larva, and imago to a lesser extent, to identify their foodplant. Rearing larvae on diets without cabbage powder necessitates adding stimulants. If *P. brassicae* larvae are reared on the central white region of cabbage which is devoid of pigments, the time spent in the larval stage is increased.

Feeding inhibitors

It is known that certain substances have a detrimental effect on growth and development of insects. In the present decade, work has been carried out on the effect on feeding inhibition of *P. brassicae* of inorganic salts, alkaloids, steroids and terpenes; a brief summary of this appears in Chapman (1974) (Table 5.5).

Butterworth & Morgan (1971), while working primarily on the Desert Locust, *Schistocerca gregaria* Forskål, found that the hexanortriterpenoid substance called azadirachtin ($C_{35}H_{44}O_{16}$), which is obtained from the Neen or Nim Tree (*Azadirachta indica* A. Joss), showed only moderate activity against fifth instar larvae of *P. brassicae*. The larvae were offered the azadirachtin on filter paper. No further details of this were given. The use of the azadirachtin as a larvicide is dealt with in Chapter 17.

In 1972 Ma published his extensive work on feeding inhibitors in his thesis. He tested 22 different compounds against *P. brassicae* larvae by using leaf discs. Only nine of these substances gave a strong inhibitory response, and those which were most effective were of an alkaloidal or steroidal nature with a high molecular weight. Ma tested the effect of three substances especially carefully: quinine, strychnine and ecdysone. Strychnine could apparently still be detected by the larvae which had their maxillary organs removed.

If the concentration of ecdysone was increased to $2 \ K \ 10^{-4}$ molar then a 50% drop in the feeding response was noticed (Ma, 1972: Fig. 39, p. 53) and thereafter a slow decrease in feeding inhibition occurred. Schoonhoven & Jermy (1977) noted that berberine, conessine, quinine,

Table 5.5 Feeding inhibitors tested on *P. brassicae*. (data from Butterworth & Morgan, 1971; Ma, 1972; Schoonhoven & Jermy, 1977). Those which gave best results are indicated as *. cf. Ma, 1972, p. 49, Table 10).

COMPOUND	*Quinine hydrochloride
Atropin	*Solanine
Azadirachtin	*Spartein sulphate
Berberine	Sparteine
*Berberin hydrochlorine	Strychnine
Brucine	*Strychnine nitrate
Caffeine	*2-thio uracil
Colchincine	Tomatine
*Conessine	Uric acid
Conline	
d-salicin	*NO REACTION*
*Ecdysone	Amygdalin
*Inokosterone	Betulin
Morine	Coniferin
Morphine hydrochloride	Pilocarpin
Narangin	Quercetin, saturated
Picrotoxin	Salicin
Pilocarpin hydrochloride	Sinigrin
Ponasterone A	Tannic acid
Quinine	Theophylline

sparteine and strychnine caused positive reactions when applied to fifth instar larvae, and azadirachtin caused the larvae to become very agitated and inhibited their drinking.

Eliminating pathogenic organisms

The naturally-occurring pathogenic organisms likely to be a nuisance in breeding *P. brassicae* are viruses and bacteria and to a lesser extent protozoa. Special attention should be given to cleanliness, as infection from the outside, for example with these organisms being brought into the laboratory on locally acquired cabbage leaves, can lead to problems.

A method of eliminating virus from ova and plants is recommended by Mr. B.O.C. Gardiner (1977, pers. comm.): contaminated ova may be washed in a 1% NaOH solution to free them from infection. Carefully remove egg batches from the egg plant with a fine brush and transfer them to a small watch glass. Submerge the ova in 1% NaOH solution for 20 minutes. Then wash the ova thoroughly in distilled water and place carefully on filter papers to dry. The ova may be transferred to an uncontaminated cabbage plant.

Plants destined for oviposition may be carefully wiped with a 1% NaOH solution, left for 20 minutes, then wiped with clean water to decontaminate them. This must be done daily to keep the culture free from virus.

At the end of each generation it is wise to disinfect equipment such as the breeding cages and artificial flowers (Grison & Silvestre de Sacy, 1957). They recommend using a mixture made up from sodium and potassium salts of dioxyquinolein (1 g) and ammonium (10% = 4 cc) made up to 1000 cc with tap water. It should be employed at 0.250 litres per cubic metre. It is often a good idea to maintain a virus free room for breeding, if this is at all possible.

When cultures of the subspecies *Pieris brassicae cheiranthi* (Hübner) are reared on cabbage it is customary for 20% of the larvae to die from virus disease; *P. b. cheiranthi* is usually reared on the Nasturtium (*Tropaeolum* sp.).

Micro-organisms develop very rapidly in synthetic diets held at room temperature and in household type refrigerators, and are usually kept at bay by the inclusion of an antibiotic such as aureomycin in the diet. David *et al.* (1972) recommended enclosing day old ova on diet in fumes of formaldehyde to eliminate the virus.

Current research on P. brassicae

Research into *P. brassicae* is being carried out in at least 24 institutes in Great Britain, Finland, France, Germany (BRD), the Netherlands, Switzerland and the USSR, and cultures of this insect are permanently maintained in all these countries (Table 5.6). This information is based on work which has already been published on *P. brassicae* and it is likely that *P. brassicae* is also reared and researched upon in other countries especially those countries in eastern Europe where *P. brassicae* is still a serious pest. It is particularly important to point out that the origin of a great deal of stock, in the past and present in western Europe, is from Mr. B.O.C. Gardiner's continous cultures at Cambridge. It is also interesting to note that Shell Laboratories, Sittingbourne, Kent, have ceased to use *P. brassicae* as a laboratory test insect since in their opinion, it is no longer regarded as an important pest in western Europe. Stocks of *Pieris rapae* (L.), *P. b. cheiranthi* (Hb.) and the former Cambridge (David & Gardiner) "virus free" strains are maintained at the Glasshouse Crops Research Institute, Littlehampton and in the École Normale Supérieure, in Paris where *P. napi* is also maintained. Van der Geest & Van der Laan (1971) stated that *P. brassicae* is kept in the USA at the Department of Entomology and Parasitology in the University of California, and the

Table 5.6 Current research on *P. brassicae*.

Address	Research and persons concerned
FINLAND University of Helsinki, Department of Physiological Zoology, HELSINKI, Finland 10.	fat metabolism (Turunen)
FRANCE Centre National Recherches Scientifiques (CNRS), Laboratoire de Biologie et Génétique Évolutives, GIF-SUR-YVETTE, 91190	physiology of diapause (Claret) juvenile hormone (L'Hélias)
École Normale Supérieure, Laboratoire de Zoologie, 46, rue d'Ulm, PARIS, 75230	nucleic acid metabolism, hormones, (Descimon, Lafont, Mauchamp, Pennetier)
Institut National de la Recherche Agronomique (INRA), Station de Recherches de Lutte Biologique et de Biocentique, La Minière, VERSAILLES, 78000	granulosis virus (Dusaussoy)
Université de Bordeaux 1, Laboratoire de Physiologie générale, Faculté des Sciences de Bordeaux, Avenue des Facultés, 33-TALENCE	respiration (Dutrieu, Gourdoux, Moreau) pesticide action (Lamy)
Université Claude Bernard-Lyon 1, Département de Biologié Géneral et Appliquée, 43, Boulevard du 11 novembre 1918, 69621 VILLEURBANNE	metabolism (Fourche, Guillet)
Université Pierre et Marie Curie, Paris VI, Laboratoire d'Histophysiologie Fondamentale et Appliquée, 12, rue Cuvier, 75005 PARIS	excretion, physiology (Lhonoré)
Université Pierre et Marie Curie, Paris VI, Laboratoire de Physiologie des Insectes, 7, Quai Saint Bernard, 75230 PARIS	physiology (Chovet, Karlinsky)

Table 5.6 continued

Address	Research and persons concerned
GERMANY Max Planck Institut, 8131 Ehrling-Andrechs, Seeweizen, MUNICH	ultrastructure of eyes, electrophysiology (Kolb)
University of Göttingen, Institute of Phytopathology, Grisebach Str. 6, D34 GÖTTINGEN	hymenopterous parasites (Führer)
University of Ulm, Abteilung für Biologie I., 7900 ULM, Oberer Eselberg	biochemistry of pigments (Kayser)
GREAT BRITAIN Agricultural Research Council, Unit of Invertebrate Chemistry and Physiology, Department of Zoology, Downing Street, CAMBRIDGE, CB2 3EJ	physiological aspects including excretion (Gardiner, Maddrell)
Ashton Wold, PETERBOROUGH, PE8 5LZ	colouration, behaviour, general biology (Rothschild)
Glasshouse Crops Research Institute, LITTLEHAMPTON, Sussex, BN16 3PU	virus and bacterial assay and field trials (Burges, Payne)
ICI, Jealott's Hill Research Station, BRACKNELL, Berkshire, RG12 6EY	screening insecticides (Ruscoe)
University of London, Imperial College, Silwood Park, ASCOT, Berkshire	post-graduate working stock
University of London, Wye College, Department of Biology, WYE, Kent, TN25 5AH	hymenopterous parasites (Abdulla, Copland)

135

Table 5.6 continued

Address	Research and persons concerned
University of Reading, Department of Microbiology, London Road, READING, RG1 5AQ	comparative study of granulosis viruses (Hunter, unpublished)
May & Baker Ltd., ONGAR, Essex	screening insecticides (Twinn)

Places which used to maintain P. brassicae cultures until recently

Chesterton Park Research Station, SAFFRON WALDEN, Essex	17 year stock (McCleod)
Shell Laboratories, Sittingbourne Research Centre, SITTINGBOURNE, Kent, ME9 8AG	screening insecticides (Felton, Watkinson)
Unit of Invertebrate Virology, Commonwealth Forestry Institute, South Parks Road OXFORD, OX1 3RB	insect viruses (Rivers)

THE NETHERLANDS

Agricultural University, Entomology Department, WAGENINGEN	electrophysiology (Ma, Schoonhoven)
Laboratory for Experimental Entomology, University of Amsterdam, Kruislaan 3502, AMSTERDAM	bacterial control (Van der Geest)
Philips-Duphar B.V., S'GRAVELAND	screening insecticides
Laboratorium voor Virologie, Landbouwhogeschool, Binnenhaven 11, WAGENINGEN	granulosis virus in culture

SWITZERLAND

Entomologisches Institut, Eidg. Technische Hochschule, Universitätstrasse 2, ZURICH	granulosis virus in culture

USSR

Institute of Zoology & Botany, Academy of Sciences of Estonia SSR, TALLINN	(T. Hansen)

USDA Agricultural Research Station, Beltsville; also in Canada at the Forestry Research Laboratory, Quebec, however, this is likely to be in error for *P. rapae* (L.). *P. brassicae* is also used extensively in schools, colleges and universities for teaching purposes, as it is one of the insects specified in the Nuffield Biology Teaching Project.

References cited

Akhtar, M., 1966. *Studies on the nutrition and feeding of the larvae of Large White Butterfly, Pieris brassicae (L.)*. Ph.D. thesis. University of Bangor, U.K.

Allen, M.D., 1954. *The effect of plant mineral nutrition on leaf-eating insects*. M.Sc. thesis. University of London (Wye College).

Allen, M.D. & Selman, I.W., 1957. The response of the Large White Butterfly (*Pieris brassicae* L.) to diets of mineral deficient leaves. *Bull. ent. Res.* 48: 229–242.

Butterworth, J.H. & Morgan, E.D., 1971. Investigation of the locust feeding inhibition of seeds of the Neem Tree, *Azadirachta indica. J. Insect Physiol.* 17: 969–977.

Chapman, R.F., 1974. *Feeding in Leaf-eating Insects*. Oxford Biology Readers.

Chaufaux, J., 1975. *Le matériel d'élevage des insectes*. OPIE, Cahiers de Liaison. No. 8. 11–18.

David, W.A.L., 1957. Breeding *Pieris brassicae* (L.) and *Apanteles glomeratus* L. as experimental insects. *Z. PflKrankh. PflPath. PflSchutz.* 64: 572–577.

David, W.A.L. & Gardiner, B.O.C., 1952. Laboratory breeding of *Pieris brassicae* and *Apanteles glomeratus* L. *Proc. R. ent. Soc. Lond.* (A) 27: 54–56.

—— 1961a. The mating behaviour of *Pieris brassicae* (L.) in a laboratory culture. *Bull. ent. Res.* 52: 263–280.

—— 1961b. Feeding behaviour of adults of *Pieris brassicae* L. in a laboratory culture. *Bull. ent. Res.* 52: 741–762.

—— 1962a. Oviposition and the hatching of the eggs of *Pieris brassicae* L. in a laboratory culture. *Bull. ent. Soc.* 53: 91–109.

—— 1962b. Observations on the larvae and pupae of *Pieris brassicae* (L.) in a laboratory culture. *Bull. ent. Res.* 53: 417–436.

—— 1965a. Resistance of *Pieris brassicae* (L.) to granulosis virus and the virulence of the virus from different host races. *J. Invert. Path.* 7: 285–290.

—— 1965b. The incidence of granulosis deaths in susceptible and resistant *Pieris brassicae* (L.) larvae following changes of population, density, food and temperature. *J. Invert. Path.* 7: 347–355.

—— 1965c. Rearing *Pieris brassicae* L. larvae on semi-synthetic diet. *Nature, Lond.* 207: 882–883.

—— 1966a. The effect of sinigrin on the feeding of *Pieris brassicae* L. larvae transferred from various diets. *Entomologia expl. appl.* 9: 95–98.

—— 1966b. Mustard oil glycosides as feeding stimulants for *Pieris brassicae* larvae in a semi-synthetic diet. *Entomologia expl. appl.* 9: 247–255.

—— 1966c. Rearing *Pieris brassicae* L. on semi-synthetic diets with and without cabbage. *Bull. ent. Res.* 56: 581–593.

David, W.A.L., Ellaby, S. & Taylor, G., 1972. The fumigation action of Formaldehyde incorporated in a semi-synthetic diet on the granulosis virus of *Pieris brassicae* and its evaporation from the diet. *J. Invert. Path.* 19: 76–82.

Dusaussoy, G. & Delplanque, A., 1964. L'élevage de *Pieris brassicae* L. en toutes saisons: accouplement et ponte en conditions artificielles. *Rev. path. veg. ent. Afr. fr.* 43: 119–134.

Evans, A.C., 1938a. *The biology and physiology of Lucilia sericata, Pieris brassicae and Brevicoryne brassicae.* Ph.D. thesis. University of London (Rothamsted Experimental Station).

—— 1938b. Physiological relationships between insects and their host plants. I. Effects of the chemical compositions of the plant on reproduction and production of winged forms in *Brevicoryne brassicae* L. (Aphididae). *Ann. appl. Biol.* 25: 558–572.

—— 1940. The utilisation of food by certain lepidopterous larvae. *Trans. R. ent. Soc. Lond.* 89: 13–22.

Felton, J.C., Watkinson, I. & Whitehead, S.E., 1977. The provision of test insects for continuous screening programmes. *In: Crop Protection Agents.* Ed. by N.R. McFarlane, pp. 531–540.

Feltwell, J.S.E., 1973. *The metabolism of carotenoids in the Large White Butterfly (Pieris brassicae (L.)), and its foodplant, cabbage (Brassica oleracea var. capitata).* Ph.D. thesis. University of London (Royal Holloway College).

Feltwell, J.S.E. & Rothschild, M., 1974. The carotenoids of 38 species of lepidoptera. *J. Zool. Lond.* 174: 441–465.

Gardiner, B.O.C., 1974. Food for caterpillars. *Nat. Sci. Schools* 12: 8–10.

—— 1975. *The Large White Butterfly, Pieris brassicae.* Philip Harris Biologicals Ltd.

—— 1978. The preparation and use of artificial diets for rearing insects. *Entomologist's Rec. J. Var.* 90: 181–184, 267–270, 287–291.

Grison, P. & Silvestre de Sacy, R., 1957. L'élevage de *Pieris brassicae* L. pour les essais de traitements microbiologiques. *Ann. Épiphyt.* 7: 661–674.

Harris, P., 1974. *Catalogue.* Oldmixon.

Jones, R.R., 1968. Personal communication *in* Ziegler, I. & Harmsen, R. *The biology of pteridines in insects. Adv. Insect. Physiol.* 6: 139–203.

Junnikkala, E., 1966. Effect of braconid parasitization on the nitrogen metabolism of *Pieris brassicae* L. A study of the nitrogenous compounds in the haemolymph of larvae at different ages reared in controlled laboratory conditions. *Annls Acad. Sci. fennici* 100: 1–83.

Kanervo, V., 1946. Sporadic observations concerning diseases in certain species of insects. *Ann. ent. fennici* 11: 218–227.

Kastari, T. & Turunen, S., 1977. Lipid utilisation in *Pieris brassicae* reared on meriodic and natural diets: implications for dietary improvement. *Entomologia expl. appl.* 22: 71–80.

Lahargue, J., 1952. Élevage complet (de l'oeuf à l'oeuf) de la piéride du chou en captivité. *Procès verbaux Soc. Linn. Bordeaux* 95: 61–66.

Littlewood, F., 1941. On rearing lepidoptera. *Entomologist* 74: 161–165, 270–274.

Lyon, J.P., 1960. *L'élevage de Pieris brassicae* L. Rapport Interne 1960. C. r. activité de Labor. Lutte Biol. et Biocènotique de la Minière. 1960–1961.

Ma, W.C., 1972. *Dynamics of feeding responses in Pieris brassicae L. as a function of chemosensory input: a behavioural, ultrastructural and electrophysiological study.* Doctoral thesis. Agricultural University, Wageningen.

Poulton, E.B., 1890. *The Colours of Animals.* Kegan, Paul, Trench, Trübner & Co. Ltd.

Rainey, R.C., 1936. *Observations on insect metabolism.* Diploma of Imperial College. University of London.

Rothschild, M., 1975. Remarks on carotenoids in the evolution of signals. *In: Co-*

evolution of Animals and Plants. Ed. by L.E. Gilbert & P.H. Raven. pp. 20–51. University of Texas Press, 1975.

Schoonhoven, L.M. & Jermy, T., 1977. A behavioural and electrophysiological analysis of insect feeding deterrents. *In: Crop Protection Agents – their biological evaluation.* Ed. by N.R. McFarlane. pp. 133–146.

Singh, P., 1974. *Artificial diets for insects and mites.* New Zealand Department of Science and Industrial Research, Auckland. Bulletin No. 24.

—— 1977. *Artificial diets for insects, mites and spiders.* IFI/Plenum, New York, Washington and London.

Street, M.L., 1975. *The growth and development of the reproductive system of Pieris brassicae and their control by hormones.* Ph.D. thesis. University of Newcastle on Tyne.

Thorsteinson, A.J., 1947. *The chemotrophic responses that determine host specificty in an oligophagous insect, Plutella maculipennis Curt. Lepidoptera.* pp. 119 + illust. & Table. Ph.D. thesis. University of London (Imperial College).

—— 1953. The chemostatic responses that determine host specificity in an oligophagous insect (*Plutella maculipennis* (Curt.) Lep.). *Can. J. Zool.* 31: 52–72.

—— 1958. The chemotactic influence of plant constituents on feeding by phytophagous insects. *Entomologia expl. appl.* 1: 23–27.

Turunen, S., 1976. Vitamin E: effect on lipid synthesis and accumulation of linolenate in *Pieris brassicae. Ann. zool. fennici* 13: 148–152.

Van der Geest, L.P.S., 1968. Effects of diets on the hameolymph proteins of larvae of *Pieris brassicae. J. Insect Physiol.* 14: 537–542.

Van der Geest, L.P.S. & Sloog-Hoebe, A.A.M., 1972. Effects of various vitamins on the development of *Pieris brassicae* larvae. *Medd. Faculteit Landbouw Gent.* 37: 713–715.

Van der Geest, L.P.S. & Van der Laan, P.A., 1971. Insect pathogens available for distribution. *In: Microbial Control of Insects and Mites.* Ed. by H.D. Burges & N.W. Hussey. Academic Press, London. pp. 733–739.

Wardojo, S., 1969. Artificial diet without crude plant material for two oligophagous leaf feeders. *Entomologia expl. appl.* 12: 698–702.

6. Development

Introduction

Two aspects of development are dealt with in this chapter, the development of the imaginal discs, and the development of the larva. The latter section has been given special attention as scientists have always been keen to know how extrinsic factors affect the duration of larval development. This is particularly important in controlled laboratory experiments.

Other aspects of development may be found in the chapter on physiology (Chapter 8), where for instance changes in respiratory quotient, muscle and nerve development in the larva, pupa and imago are discussed. In addition, changes in biochemistry during development may be found in Chapter 10.

Attention must be drawn to the important work carried out by scientists on various biochemical, anatomical and physiological changes which occur during development of *P. brassicae*. Under the heading of biochemistry René Lafont and his colleagues at the École Normale Supérieure in Paris studied the metabolism of proteins, nucleic acids, pteridines and nitrogen excretory products during the fifth instar, prepupa and imago of *P. brassicae* and have expressed much of this data in a series of 21 graphs which show the fluctuations which occur (Lafont *et al.*, 1975, 1977; Mauchamp & Lafont, 1975). Growth and differentiation have been drawn up in the form of a table (Table 6.1).

Anatomical changes in the intersegmental muscles of *P. brassicae* during the fifth instar larva and pupa were studied by Auber-Thomay & Srihari (1973) and development of the central nervous system in larvae, pupae and imagines by Heywood (1965). Oogenesis was studied by Ali (1973).

One aspect of development of *P. brassicae* which has not received much attention is the growth of pleuropodia. Roonwal (1936) reported that they are found in *Pieris* as evaginate structures (see also Matsuda, 1976). According to Wigglesworth (1967) these are paired glandular organs with

Table 6.1 Main developmental events during the metamorphosis of *P. brassicae* (after Lafont, *et al.*, 1975).

Developmental areas	Life cycle			
	fifth instar	*prepupa*	*pupa*	*imago*
larval tissues	growth	histolysis	—	—
imaginal wing discs	slow growth	very rapid growth	differential growth; scale development; pigmentation by pteridines	cell degeneration
fat body	growth	growth and storage	little or no growth; release of most compounds	renewal of excretory compounds

141

varying function, not usually found in the lepidoptera, which are homologous with the appendages of the first abdominal segment. Little if any work on pleuropodia seems to have been carried out on *P. brassicae.*

SECTION A   IMAGINAL DISCS

The arrangement of the imaginal discs (buds) *in situ* in the anterior of the larva of *P. brassicae* has been illustrated by Gonin (1894) and their development has been shown in a series of six diagrams by Mercier *in* Portier (1949). Eassa (1953) examined their development in the head and more recently Cals (1970) studied the development of the thoracic imaginal discs in all larval instars. René Lafont (1970a,b, *et al.*, 1977) studied the metabolism of nucleic acids in the discs of the larva and Lavenseau & Surlève-Bazeille (1976) studied the structure of the pupal discs. The first signs of the imaginal discs may be seen during the larval stage but they become prominent during the pupal stage. Work has been carried out on the development of the wings, legs, antennae and compound eyes from imaginal discs.

Eassa (1953) examined the development of the imaginal discs in the head of *P. brassicae* and in particular followed the development of the head appendages, e.g. the antennae, maxillae, labium, as well as the labrum, clypeus, epicranium and mandibles from the larva through to the imago. His work is well illustrated on the development of larval maxillae and antennae, as well and a plate showing sections through larval appendages is given. Earlier work by Gonin (1894, cf. Packard, 1898) did not include details of antennal development with consequent changes in form of the head capsule or on the development of the epicranium, both of which were enlarged upon by Eassa (1953).

Most of the head appendages arise as rudiments closely associated with their larval counterparts. Thus there are few major changes in head development from the larva to the imago (Eassa, 1953). These are a) the position of the antenna changes dramatically from being ventral in the larva to dorsal in the imago; the antennae develop behind a membrane through which they eventually burst (see figure 3 *in* Eassa, 1953); b) the imaginal head capsule is greatly extended by additions from the antennal peripodial membrane; c) a pupal crest is formed in which the clypeus of the imago develops; and d) the imaginal epicranium develops as a novel pupal structure.

Larva

Cals (1970) showed that the imaginal discs of the larva are little differentiated in the first two instars, but start to change in the third instar and in the fourth instar when the nerves begin to be differentiated.

Two phases of mitosis, corresponding to secretion of juvenile hormone and an important synthesis of nucleic acids and proteins, occur in the imaginal discs of the fifth instar (Lafont, 1970a). Lafont's work involved incorporating labelled thymidine and uridine. Lafont et al. (1977) demonstrated that imaginal disc development in the fifth instar occurred in a series of steps which required the presence of ecdysone, however the ecdysone peaks (at 44% and 70% of the larval duration) did not seem to be associated with these steps.

Chang-Whan (1959) studied the post-embryonic development of the larval true leg up to the development of the imaginal leg. He found that in the epidermis between segment two and three of the larva there was active cell division soon after the final ecdysis, and that it received a sufficient oxygen supply from early on. This region Chang-Whan thought to be a differentiation centre which possibly absorbs more hormone than other parts.

Pupa

Lafont (1970b) found that he could interfere with the normal formation of scales on the wing by incorporating a compound called 5-fluorouracil (5-FU). This he injected into seven day old pupae (when there are maximal amounts of nucleic acid in the wings) and found that the synthesis of scales on the wing was inhibited to the extent that only the sockets were left. Lafont (loc. cit.) published a plate depicting some of these relatively scaleless insects which look transparent with pronounced venation. A few of the melanin containing scales are left particularly in the apical region of the forewing.

5-FU interferes with nucleic acid synthesis, affecting the formation of thymidine monophosphate (TMP), by inhibiting thymidinekinase. This results in accumulations of thymidine triphosphate (TTP) in the cells which blocks the functioning of nucleic acid polymerase.

Necrosis of larval characteristics starts shortly after pupa formation (Lavenseau & Surlève-Bazeille, 1976). These workers published a plate of a parenchyma cell of P. brassicae being broken down and in it one can see masses of ribosomes ($\times 21,000$ magnification). A diagram of the structure of the pupal hindwing disc was also published, showing a peripheral "mucron" around the margin of the wings.

SECTION B LARVAL DEVELOPMENT

So much work has been carried out specifically on the growth and development of larvae of *P. brassicae* that the following section has been prepared to outline the main biological changes which occur and the extrinsic factors which affect their development. For intrinsic control of development see Chapter 9 (Hormones).

a) Weight relationships

The increase in wet weight through the larval instars follows a typical sigmoid curve, the greatest increase occurring in the fifth instar (Figure 6.1a, Srihari, 1972; Srihari & Gahukar, 1975). This wet weight then falls off to less than half the larval weight in the imago (cf. Figure 6.1j). How- ever, when the weight is expressed as percentage dry weight the increase is seen to be put on earlier in the fifth instar (Figure 6.1b); expressed logarithmically the fresh weight increases steeply during the larval life (Figure 6.1c), and if expressed as an increase in log. dry weight, most is put on steadily up to the pupa stage (Figure 6.1d).

The relative amount of weight put on in each instar can be expressed as a value of "k", a constant of weight increase such that $L_1 = 0.85$; $L_2 = 0.80$; $L_3 = 0.85$; $L_4 = 0.81$; $L_5 = 0.68$ (Srihari, 1972). According to Lafont *et al.* (1975) the wet weight of the larva increases about five times during the fifth instar.

Lafont *et al.* (1975) demonstrated that there were two distinct periods of growth in the fifth instar larva; a) a rapid growth phase during the first half of the intermoult and b) slow weight decrease in the prepupa. The relative growth rate (weight increase per day ÷ average weight) also showed a typical sigmoid curve (Figure 6.1f).

Different parts of the body studied by Lafont *et al.* (1975) grow at different rates – the imaginal discs grow slowly during the first two days, then quickly before pupation; the fat body grows continuously, its dry weight increasing fifty times during the intermoult, and the "carcass" comprising the larval tissues and imaginal discs, but not the wings, increases in weight rapidly during the first two days, reaches a plateau and then falls towards pupation (Figure 6.1g,h,i).

After eclosion the larval weight decreases enormously through water loss via the meconium (Figure 6.1e). Nicolson (1976) found that if the anus of imagines was plugged the rate of water loss did not continue so quickly.

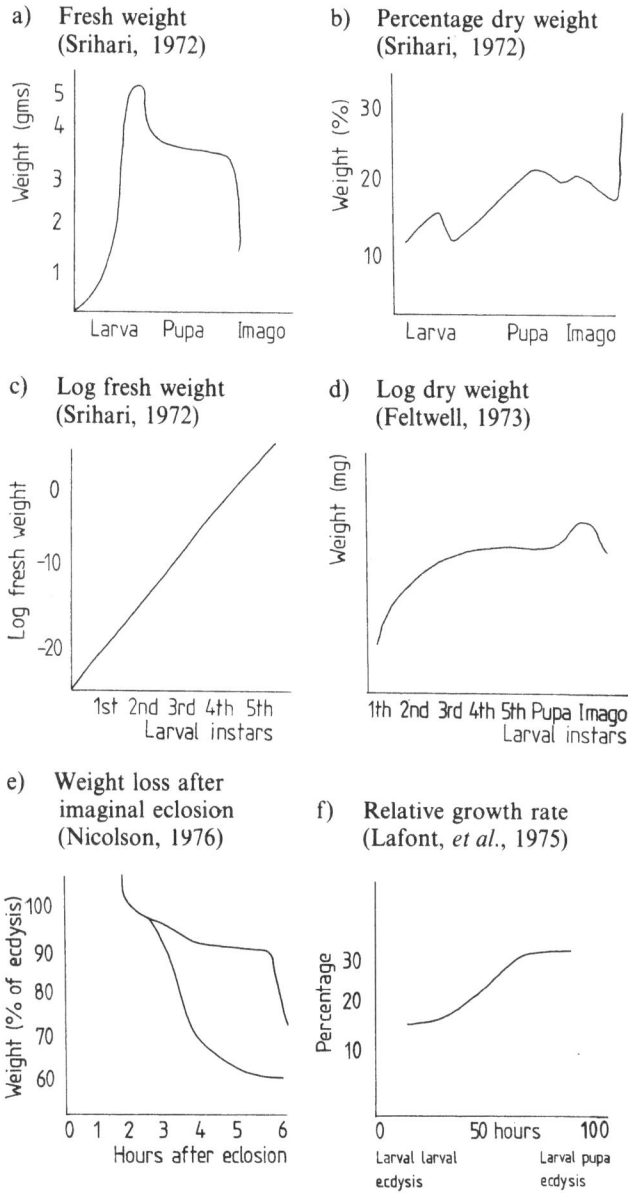

a) Fresh weight
   (Srihari, 1972)

b) Percentage dry weight
   (Srihari, 1972)

c) Log fresh weight
   (Srihari, 1972)

d) Log dry weight
   (Feltwell, 1973)

e) Weight loss after
   imaginal eclosion
   (Nicolson, 1976)

f) Relative growth rate
   (Lafont, *et al.*, 1975)

Fig. 6.1 Weight relationships

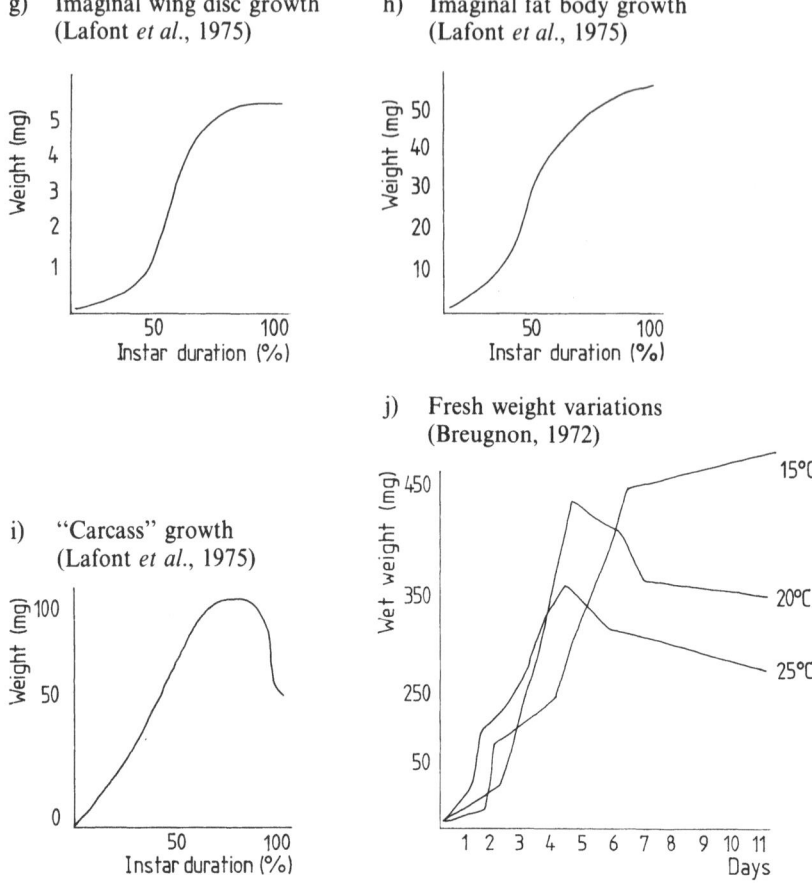

g) Imaginal wing disc growth
(Lafont *et al.*, 1975)

h) Imaginal fat body growth
(Lafont *et al.*, 1975)

i) "Carcass" growth
(Lafont *et al.*, 1975)

j) Fresh weight variations
(Breugnon, 1972)

Fig. 6.1 continued

b) Size relationships

Larval length. The change in length throughout the larval stages is also directly proportional (Figure 6.2b). At every instar the total length of the mesothorax and first abdominal segment decreases with the corresponding increase in width. After this they increase up to the end of the instar (Srihari, 1972). Pictet (1922) showed that size varies with temperature, the colder the laboratory, the shorter the larvae.

Head capsule. The size of the head capsule of the larva is a useful means of calculating age, as size and age of the larva are directly proportional (Figure 6.2a). The variation in head capsule width for each instar has been published by Srihari (1972) and Eassa (1963).

a) Growth of larval head capsule
    (Klein, 1932; David & Gardiner, 1962b)

b) Growth of larval length
    (Srihari, 1972)

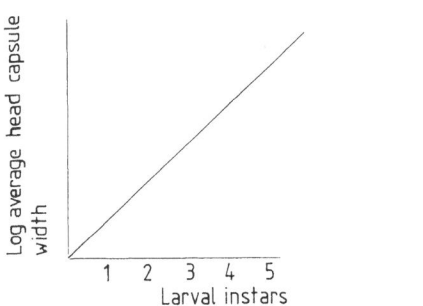
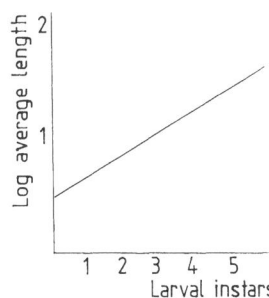

Fig. 6.2 Size relationships

c) Food consumption, utilisation and energetics

A number of workers have studied food energetics in larvae of *P. brassicae*: Chlodny (1967), David & Gardiner (1962a,b), Srihari (1970), Feltwell (1973), Feltwell & Valadon (1974), Srihari & Gahukar (1975), Fourche *et al.* (1977) and Ripa (1978). Many different coefficients and indices have been used to give information on development (Table 6.2).

Voracity, production output, assimilation (EAD) and digestibility coefficients were used by Chlodny (*loc. cit.*), and consumption index, relative growth rate, efficiency of conversion index and utilisation (CUDA) (= digestibility *in* Chlodny) by Srihari (*loc. cit.*). Fourche (*loc. cit.*) made measurements of metabolic heat and water loss, oxygen consumption and utilisation of glycogen and fats in non-diapausing and diapausing pupae of *P. brassicae*.

Chlodny's (1967) figures showed that the L_5 larvae consume about 11 times their maximum body weight and 85% of the food consumed in the life of the larva is consumed in the L_4 and L_5 stages. Most of the food ingested by the larva occurs in the final instar but this is accounted for in part by the longer duration of the stage. During L_3 there is a seven-fold increase in food consumption and this is accredited to a change in the larva's feeding habit, i.e. that it is then able to bite across the whole lamina of the leaf (Figure 6.3a).

The voracity of the larva is highest in the L_3 instar (Chlodny, *loc. cit.*) (Table 6.3), but Evans (1940) who studied food utilisation and requirements of the L_5 larva recorded it 40% higher at 1.504 than that of Chlodny (0.96). This is, however, explained by the fact that Evans used larvae of a lower average weight.

Srihari's (*loc. cit.*) results showed that the CUDA, the EAD, and the ETI indices fluctuated during the L_5 and L_4 stages and finally increased

Table 6.2 Indices calculated for *P. brassicae.*

Indices according to Chlodny (1967)

$$\text{Voracity} = \frac{\text{Daily consumption}}{\text{Body weight}}$$

$$\text{Production Output} = \frac{\text{Production}}{\text{Consumption}}$$

$$\text{Assimilation} = \frac{\text{Amount of food ingested}}{\text{Amount of food voided and excreted}}$$

$$\text{Digestibility} = \frac{\text{Weight of food consumed - weight of faeces}}{\text{weight of food consumed}} \times 100$$

Indices according to Srihari & Gahukar (1975)

CUDA

$$\begin{array}{l}\text{Coefficient of utilisation} = \\ (? = \text{Digestibility})\end{array} \frac{\text{weight of food eaten - weight of frass}}{\text{weight of food used}} \times 100$$

EAD

$$\begin{array}{l}\text{Efficiency of assimilation} \\ \text{expressed according to} = \\ \text{body weight}\end{array} \frac{\text{gain in weight}}{\text{weight of food eaten - weight of frass}} \times 100$$

ETI

$$\text{Efficiency of assimilation} = \frac{\text{gain in weight}}{\text{weight of food used}} \times 100$$

CI

$$\text{consumption index} = \frac{\text{fresh or dry weight of food eaten}}{\text{Time (days)} \times \text{mean weight (net or dry) of animal during feeding time}}$$

CR

$$\text{Relative growth rate} = \frac{\text{wet or dry gain in weight during feeding time}}{\text{duration of feeding (days)} \times \text{wet or dry gain in weight during feeding time}}$$

rapidly at the end of the final instar (Figure 6.3b,c). On the other hand the CI and the CR decreased at the end of L_5, the CI the more so (Figure 6.3b,c).

The relative growth rate of larvae through the last three instars rises and falls with the efficiency of conversion of ingested food to body matter (ECI), and the efficiency of conversion of ingested food to body matter (ECD). The approximate digestibility (AD) varied from instar to instar and did not follow the pattern of any other index. The percentage of dry weight of larvae increased with the number of instars.

Table 6.3 Larval energetics (data selected from Chlodny, 1967).

Instar	Voracity	Require-ment for food %	Biomass pro-duction %	Pro-duction output %	Food assimi-lated per larva (g)	per instar	Assimi-lation of energy	Digesti-bility
1	1.99	0.73	1.05	0.127	6.024	16.665	0.81	0.797
2	1.33	1.30	3.04	0.202	8.976	24.326	1.18	0.595
3	2.30	13.07	13.32	0.090	83.712	307.537	14.92	0.753
4	1.85	35.53	30.20	0.074	167.376	763.329	37.04	0.691
5	0.96	49.36	52.38	0.093	158.952	948.775	46.04	0.615

Ripa (1978) developed a method for measuring rate of frass production, which is related to food consumption, by marking frass with a phenol red dye. The dye is dissolved in distilled water with 0.2% Teepol to increase wetting power, then it is applied to leaf discs and evaporated. Larvae are allowed to feed on the discs and their frass is collected (see Ripa for apparatus designed for quantitative frass collection). Recovery of the dye is measured in a spectrophotometer.

Sudah (1970 thesis) believed that the available techniques used by different workers (he compared the results of four workers) were not sufficiently accurate for meaningful coefficients to be used. He studied the physical characteristics of the plant with respect to *P. brassicae* larvae (L_1 and L_5) feeding on it and made the following points:

First instar larvae were restricted to the intervein lamina of young and old leaves usually near the margin, while fifth instar larvae preferred the margin. Intervein areas of young leaves had significantly greater water content than similar areas in old leaves. No differences were detected in the feeding rates of young or old larvae on young or old intact leaves, when measured in leaf area eaten or dry weight of leaf eaten. Larvae fed on old leaves produced significantly heavier faeces (in dry weight) than those on young leaves, and the increase in body weight of fifth instar larvae was significantly greater on old than young leaves. Larvae tended to prefer cut leaf surfaces to intact ones, possibly due to chemical attractants released, but he did not find any evidence that saliva or other pheromones attracted larvae to cut areas.

d) Frass drop frequency

The frass drop frequency is the number of defecations during a particular time interval (Friden, 1958) and can be used to indicate the particular acceptability of a foodplant or diet. Friden stated that third generation larvae have a lower frass drop frequency than those approaching pupal

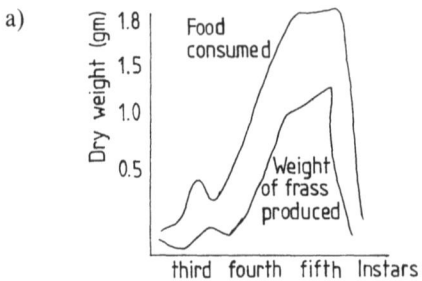

a)

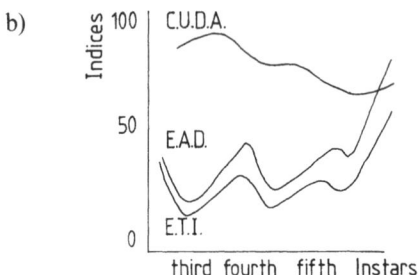

b)

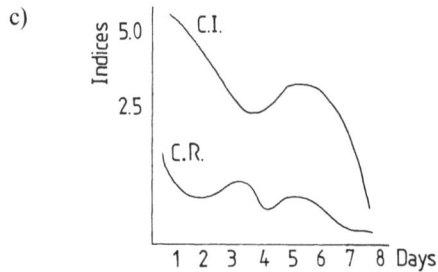

c)

Fig. 6.3 Variations in indices through third, fourth and fifth instar larva of *P. brassicae*
(from Srihari, 1970)

diapause. He also showed in *P. rapae* that there was a higher frass drop frequency in the L_{4-5} larvae than in the L_{2-3}.

There are, however, too many variables which make the use of frass drop frequency an unreliable method for determining food consumption, although some compensation could be achieved by having controlled rearing and feeding conditions. For instance, David & Gardiner (1966) demonstrated that increasing concentrations of mustard oil glycosides in synthetic diets increased frass drop frequency; Sudah (1970 – thesis) found that dry weight of frass varied according to the age of the leaf eaten, older leaves resulting in heavier frass; and Srihari (1970) found that the dry weight and the number of frass pellets varied greatly from one instar to the next.

About one third of the energy of all ingested food is lost in the frass (Tables 6.3, 6.4, 6.5), or more accurately, 66% of the ingested food is assimilated (Feltwell & Valadon, 1974).

In experiments on the amounts of the carotenoid pigment β-carotene ingested and lost in the frass, Feltwell & Valadon (1974) found that over 90% of it was retained, while 1,165.25 g/g was lost in the frass and 11,999.98 g/g ingested.

e) Effect of light

Evans (1938a,b) found that larvae reared in shade increased in weight more slowly and pupated later than those fed on plants grown in normal light. This reaction was probably due, he thought, to a lack of carbohydrate in the shade grown plants. Effects of light on diapause is explained in detail in Chapter 8.

Table 6.4 Assimilation of food by P. brassicae larvae (from Chlodny, 1967).

Details	Figures
Average consumption of fresh cabbage by one larva throughout its life at 20°C	5,4869 g
Equivalent in dry weight	0.7151 g
Conversion to calorific value	3121.22 calories
Calorific value of frass voided during life	1060.59 calories
Energy consumed	2060.63 calories

Table 6.5 Frass production by P. brassicae (adapted after Chlodny, 1967).

Stage	Faeces voided (% of sum total of whole larval life)	Faeces voided (calories)
1	0.57	6.106
2	1.55	16.554
3	9.39	100.075
4	32.43	345.730
5	55.55	592.180

f) Effect of temperature

Many people have investigated the effect of temperature on rate of development of *P. brassicae*, experiments having been made between the temperatures 10–28°C (Klein, 1932a); 15°, 18–20°, 25°C (Grison & Silvestre de Sacy, 1957); 10–35°C (Kozhantchikov, 1935) 12.5–34°C (David & Gardiner, 1966), at 15°, 20° and 25°C (Breugnon, 1972).

Essentially, there is an increase in rate of development up to 27.5°C followed by a decrease (Jones, 1960), the optimum temperature for development being between 20–26°C (Maercks, 1934) (Figure 6.4). Reared at normal laboratory temperature (20°C) the total development time for larva and pupa is 31.5 days compared to 21 days at 25°C and 64 days at 15°C (Breugnon, 1972) (Figure 6.1j, Table 6.6). Her figure of 18 days for the complete development of the larva at 20°C agrees accurately with that of David & Gardiner (1962b) and Klein (1932a). The constancy of these results indicates that a precise temperature sensitive mechanism

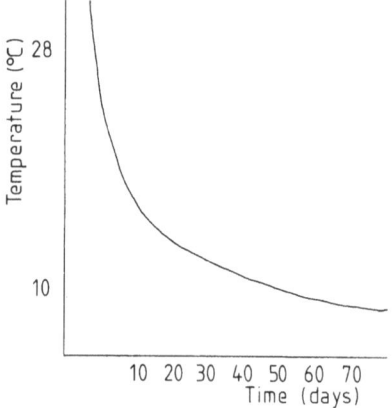

Fig. 6.4 Larval duration with temperature
(after Klein, 1932b)

Table 6.6 Duration of larval development at various temperatures (from Breugnon, 1972).

Rearing temperature (°C)	Length of larval instars (days)					Total days	Duration of pupa	Total of larva and pupa
	1	2	3	4	5			
15	7	8	7	7	11	40	24	64
20	4	3	3	3	6	18	12.5	31.5
25	2	2	2	2	5	13	7.5	21

must be present; it is not known at present how the larvae respond to temperature but it was found by Klein (*loc. cit.*) that they are sensitive to a change of $\frac{1}{10}-\frac{3}{10}°C$.

g) Temperature limitations in the wild

P. brassicae is restricted in its world range by temperature. According to Bonnemaison (1965) the threshold temperature of *P. brassicae* is 10°C but Klein (1932a) stated that it was limited in its world range between the temperatures of 18–28°C and Kozhantchikov (1936) showed the world distribution of *P. brassicae* to be confined between the summer and winter isotherms of 28°C (July) and 20°C (Jan.), respectively.

Cold temperatures, including frosts, can increase the time of larval development for over two months. Pictet (1922) demonstrated in the field that when larvae were exposed to temperature between −3°C and +12°C they took up to 75 days to develop, whereas the control larvae, reared at 20°C, developed in 22 days. The larvae were, however, smaller in size. Pupae of *P. brassicae* can withstand temperatures down to −22.5°C (Bonnemaison, 1965).

High temperatures such as those experienced in the summer of 1947 in Europe were responsible for considerable control of *P. brassicae*, as many ova were lost by overheating (Blunck, 1950).

h) Effects of population density

Long (1953, 1955) found that crowded larvae developed about seven days more quickly than those reared individually (30.5 and 37.6 days respectively), and there appeared to be no differences in colour markings or patterns attributable to these different regimes. David & Gardiner (1962b), however, found that there was less of a difference in duration of development between crowded and solitary larvae.

Other characteristics of crowded larvae were that they were of a smaller size and weight, had a larger ratio of forewing and hindleg femur, had a higher fat content and the imagines had less fat content, imagines had a higher glycogen content, the larvae consumed less food and imagines lived longer (Long & Zaher, 1957). They suggested that the larvae of crowded and solitary reared larvae have different degrees of food utilisation, but in a later paper they stated that there was no apparent effect on the wing pattern or colouration (Long & Zaher, 1958). However, crowding does tend to decrease the area of the hindwing more than the forewing (Long, 1959).

i) Effect of humidity

Larvae of *P. brassicae* appear to like a fairly moist atmosphere, at least for part of their development, and are not found in arid areas. One is reminded of the southerly limit of *P. brassicae* along the north African coast and around the coast of the eastern end of the Mediterranean. Klein (1932a,b) stated that the geographical limits of the Large White are restricted between the 13°C and 28°C isotherms when the humidity is 60%. Imagines can exist in areas below 60% but they are not stimulated to oviposit. It has also been reported that in Israel when the humidity drops below 70% larval mortality is high. This may often happen when the Khamsin wind blows. It was stated by Klein (1932b) that low humidities and high temperatures reduced the number of larval instars but this has not been substantiated. As far as is known, *P. brassicae* has 5 instars regardless of extrinsic factors.

Jones (1960) showed that the percentage weight loss of the pupa in both sexes was substantially higher – about 20% – when they were kept at low humidity (RH 0–10), than at high humidity (RH 90–100) when it fell to 2.5%.

j) Effect of food

The chemical constitution of the foodplant would appear to be of prime importance in the development of the larva. Availability of the foodplant is critical; when other factors are constant the development period will progress at the "normal" rate when adequate food is available, but when food is limited the larval period increases and there is a likelihood that dwarf specimens may be produced. Poor quality food also results in the larvae eating more food (Evans, 1940).

Experimental evidence has shown that different foodplants have different effects on the growth of the larva. Fernando (1971) showed that there was a direct relationship between larval growth and time of development in *P. brassicae* when reared on Brussels sprouts (*Brassica oleracea* L.) as compared with those reared on Jack in the Hedge, *Alliaria petiolata* (Bieb.) Cavara & Grande and *Tropaeolum* sp. The growth curve for Brussels sprouts-reared larvae is seen in Figure 6.5a. On the other hand a typical sigmoid curve has been obtained for larvae reared on cauliflower (Allen, 1954, Allen & Selman, 1957) and on cabbage (Type Primo) (Feltwell, 1973) Figure 6.5b. The differences in growth rates experienced by larvae reared on different foodplants may well be a reflection of the relative acceptability of the foodplant in terms of the chemical constitution.

a) On Brussels sprouts
   (after Fernando, 1971)

b) On cauliflower and cabbage
   (after Allen & Selman, 1957;
   Feltwell, 1973)

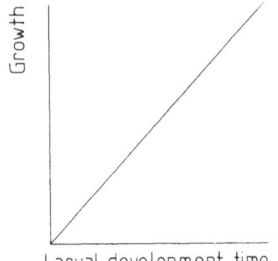
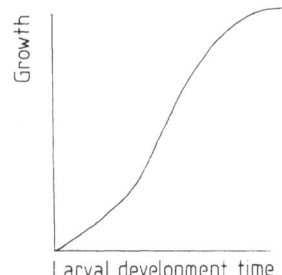

Fig. 6.5 Effect of food on development

Vitamins may also affect the growth of *P. brassicae* larvae. When larvae are reared on the central white leaves of cabbage the length of the larval development time is increased and there may also be some mortality (Feltwell, 1973). This may be due to the lack of vitamin A or carotenoid pigments in the plant tissue which are essential for growth, development and vision in insects.

The effects of a deficiency of various elements in the food of *P. brassicae* has been investigated thoroughly (Evans, 1938a,b, 1940; Allen, 1954; Allen & Selman, 1957) and the elements which have received attention are nitrogen, potassium, phosphorus and iron. Allen (1954) studied the effects of mineral deficient leaves of watercress (*Rorippa nasturtium-aquaticum* (L.) Hayek) and other crucifers on *P. brassicae* larvae, while Allen & Selman (1957) studied those of cauliflower and turnips. Their results for each of the species studied were much the same for each of the following criteria: average larval weight, maximum individual larval weight, length of larval life, relative growth rate, mortality, delay in pupation, pupal weights and body lengths of imagines.

With all foodplants a deficiency of nitrogen and iron resulted in a reduction of the relative growth rate and an increase in the mortality of the larvae (Allen, 1954; Allen & Selman, 1957). The latter authors wondered whether the mortality was due to starvation or a lack of essential minerals in the diets. They also found that a deficiency in all of the minerals reduced the length of the larval life and caused a delay in pupation.

A deficiency in potassium resulted in a reduction in maximum individual larval weight, mean larval weights and an increase in the length of larval life (Allen, 1954).

A deficiency of iron also reduced individual maximum weights as well as reducing the relative growth rate, pupal weight and body length of imagines (Allen, 1954). She also found that there was no apparent difference in the coefficient of utilisation of food (weight of food utilised ÷ weight of food consumed) of larvae reared on food deficient in iron and controls. Evans (1940) showed that the coefficient of utilisation of food for *P. brassicae* larvae was relatively high at 80% utilisation of chemical constituents (in mg/g/larva/day) but that it falls during the first instar.

Details of effects of carbohydrate and protein deficiency on growth of *P. brassicae* larvae are very limited. However, work done by Evans (1938a,b) showed that poor quality food which had been grown in the dark and which was presumed to be lacking in carbohydrates, indicated that the larvae were not utilising their food as efficiently as the control on normal food.

The effects of mustard oil glycosides on the development of larvae on synthetic diets may be seen in the preceding chapter.

References cited

Ali, F. 1973. Post-embryonic changes in central nervous system and perilemma of *Pieris brassicae* (L.) (Lepidoptera: Pieridae). *Trans. R. ent. Soc. Lond.* 124: 463–498.

Allen, M.D., 1954. *The Effects of Plant Mineral Nutrition on Leaf-eating Insects.* M.Sc. thesis. University of London (Wye College).

Allen, M.D. & Selman, I.W., 1957. The response of larvae of the Large White butterfly (*Pieris brassicae* L.) to diets of mineral deficient leaves. *Bull. ent. Res.* 48: 229–242.

Auber-Thomay, M. & Srihari, T., 1973. Évolution ultrastructurale des fibres musculaires intersegmentaires chez *Pieris brassicae* (L.) pendant le dernier stade larvaire et la nymphose. *J. Microsc.* 17: 27–36.

Blunck, H., 1950. Zur Kenntnis des Massenwechsels von *Pieris brassicae* L. mit besonderer Berücksichtigung des Dürrejahres 1947. *Z. angew. Ent.* 32: 141–171.

Bonnemaison, L., 1965. Insect pests of crucifers and their control. *A. Rev. Ent.* 10: 233–256.

Breugnon, M.M., 1972. Étude de quelques caractères du developpement postembryonnaire de *Pieris brassicae* à trois conditions thermiques différentes. *Ann. Soc. ent. Fr.* 8: 461–473.

Cals, P., 1970. Discontinuité morphogénétique dans la differentiation postembryonnaire des ailes des Lépidoptères: *Bombyx mori* L. (Bombycidae) et *Pieris brassicae* L. (Pieridae). *Note. C. r. Acad. Sci. Paris* 270: 2666–2669.

Chang-Whan, K., 1959. The differentiation centre inducing the development from larval to adult leg in *Pieris brassicae* (Lepidoptera). *J. embryo. exp. Morph.* 7: 572–582.

Chlodny, J., 1967. The energetics of the development of cabbage white, *Pieris brassicae* L. (Lepidoptera). *Ekol. Pol. Ser.* A 15: 553–561.

David, W.A.L. & Gardiner, B.O.C., 1962a. Oviposition and the hatching of eggs of *Pieris brassicae* L. in a laboratory culture. *Bull. ent. Res.* 53: 91–109.

—— 1962b. Observations on the larvae and pupae of *Pieris brassicae* (L.) in a laboratory culture. *Bull. ent. Res.* 53: 417–436.

—— 1966. Mustard oil glycosides as feeding stimulants for *Pieris brassicae* larvae in a semi-synthetic diets. *Entomologia exp. appl.* 9: 247–255.

Eassa, Y.E.E., 1953. The development of imaginal buds in the head of *Pieris brassicae* Linn. (Lepidoptera). *Trans. R. ent. Soc. Lond.* 104: 39–50.

—— 1963. Metamorphosis of the cranial capsule and its appendages in the Cabbage butterfly, *Pieris brassicae. Ann. ent. Soc. Am.* 56: 510–521.

Evans, A.C., 1938a. *The Biology and Physiology of Lucilia sericata, Pieris brassicae and Brevicoryne brassicae.* Ph.D. thesis. Rothamsted Experimental Station.

—— 1938b. Physiological relationships between insects and their host plants I. Effects of the chemical composition of the plant on reproduction and production of winged forms in *Brevicoryne brassicae* L. (Aphididae). *Ann. appl. Biol.* 25: 558–572.

—— 1940. The utilisation of food by certain lepidopterous larvae. *Trans. R. ent. Soc. Lond.* 89: 13–22.

Feltwell, J.S.E., 1973. *The Metabolism of Carotenoids in Pieris brassicae and its foodplant, Brassica oleracea var. capitata* (Cabbage). Ph.D. thesis. University of London (Royal Holloway College).

Feltwell, J.S.E. & Valadon, L.R.G., 1974. Carotenoid changes in *Brassica oleracea* var. *capitata* L. with age in relation to the Large White Butterfly, *Pieris brassicae* L. *J. agric. Sci. Cambs.* 83: 19–26.

Fernando, J.V.S., 1971. Selection and utilisation of different food plants by *Pieris brassicae. Spolia Zeylanica* 32: 115–127.

Fourche, J., Guillet, C., Calvez, B. & Bosquet, G., 1977. Le métabolisme énergétique des nymphes de *Pieris brassicae* (Lepidoptera) au cours de la métamorphose et de la diapause. Essai d'établissement d'un bilan. *Ann. zool. ecol. Anim.* 9: 51–61.

Friden, F., 1958. *Frass Drop Frequency in Lepidoptera.* Almqvist & Wiksells Boktryckeri Ab. Uppsala, Sweden.

Gonin, J., 1894. Recherches sur la métamorphose des Lépidoptères de la formation des appendices imaginaux dans la chenille du *Pieris brassicae. Bull. Soc. vand. Sci. nat.* 30: 89–139.

Grison, P. & Silvestre de Sacy, R., 1957. L'élevage de *Pieris brassicae* L. pour les essais de traitements microbiologiques. *Ann. Épiphyt.* 7: 661–674.

Heywood, R.B., 1965. Changes occurring in the central nervous system of *Pieris brassicae* L. (Lepidoptera) during metamorphosis. *J. Insect Physiol.* 11: 413–430.

Jones, A.L., 1960. *Observations on the relationships between feeding, growth and development in lepidopterous larvae.* Ph.D. thesis. University of Leeds.

Klein, H.Z., 1932a. Studien zur Oekologie und Epidemiologie der Kohlweisslinge I. Der Einfluss der Temperatur und Luftfeuchtigkeit auf Entwicklung und Mortalität von *Pieris brassicae* L. *Z. angew. Ent.* 19: 395–448.

—— 1932b. Studien zur Oekologie und Epidemiologie der Kohlweisslinge II. Zur Bionomie von *Pieris brassicae* L. und deren Parasit *Microgaster glomeratus* L. *Z. wiss. Insek. Biol.* 26: 192–199.

Kozhantchikov, I., 1936. The role of ecological factors in the distribution of *Pieris brassicae* L. (In Russian.) *Plant Protection* 11: 30–57.

Lafont, R., 1970a. L'évolution des acides nucleiques dans les disques imaginaux alaires de *Pieris brassicae* au cours de cinquieme stade, *C. r. Acad. Sci. Paris* D 270: 1599–1602.

—— 1970b. Étude du développement des ébauches alaires de la chrysalide de

Pieris brassicae en presence de 5-fluorocil. *C. r. Acad. Sci. Paris* 271: 2186–2189.

Lafont, R., Mauchamp, B., Boulay, G. & Tarroux, P., 1975. Developmental studies in *Pieris brassicae* (Lepidoptera). I. Growth of various tissues during the last larval instar. *Comp. Biochem. Physiol.* 518: 439–444.

Lafont, R., Mauchamp, B., Blais, C. & Pennetier, J.L., 1977. Ecdysones and imaginal disc development during the last larval instar of *Pieris brassicae. J. Insect Physiol.* 23: 277–283.

Lavenseau, L. & Surlève-Bazeille, J.E., 1976. Necrose cellulaire et morphogènese alaire chez quelques lépidoptères. *Bull. Soc. zool. Fr.* 101: 69–74.

Long, D.B., 1953. Effects of population density on larvae of lepidoptera. *Trans. R. ent. Soc. Lond.* 104: 544–585.

—— 1955. Observations on sub-social behaviour in two species of lepidopterous larvae, *Pieris brassicae* L. and *Plusia gamma* L. *Trans. R. ent. Soc. Lond.* 106: 421–437.

—— 1959. Observations on adult weight and wing area in *Plusia gamma* L. and *Pieris brassicae* L. in relation to larval population density. *Ent. exp. appl. Amst.* 2: 241–248.

Long, D.B. & Zaher, M.A., 1957. Effects of population density on lepidoptera. *Rothamsted Experimental Station, Harpenden*, Report for 1956. p. 157.

—— 1958. Effect of larval population density on the adult morphology of two species of lepidoptera, *Plusia gamma* L. and *Pieris brassicae* L. *Ent. exp. appl. Amst.* 1: 161–173.

Maercks, H., 1934. Untersuchungen zur Ökologie des Kohlweisslings (*Pieris brassicae* L.), die Temperaturreaktionen und das Feuchtigkeitoptimum. *Z. Morph. Ökol. Tiere.* 28: 692–721.

Matsuda, R., 1976. *Morphology and Evolution of the Insect Abdomen*. Pergamon Press, Oxford.

Mauchamp, B. & Lafont, R., 1975. Development studies in *Pieris brassicae* (Lepidoptera: Pieridae). II. A study of nitrogenous excretion during the last larval instar. *Comp. Biochem. Physiol.* 51 B: 445–449.

Nicolson, S.W., 1976. Diuresis in the Cabbage White Butterfly, *Pieris brassicae*: fluid secretion by the Malpighian tubules. *J. Insect Physiol.* 22: 1347–1356.

Roonwal, M.L., 1936. Studies on the embryology of the African Migratory Locust. *Phil. Trans. R. Soc. Lond.* 226: 391–421.

Packard, A.S., 1898. *A Textbook of Entomology*. New York.

Pictet, A., 1922. Recherches sur l'hibernation de *Pieris brassicae* à l'état de chenille. *Bull. Soc. lépidopt. Gèneve* 5: 47–57.

Portier, P., 1949. *Biologie des Lépidoptères*. Lechavalier, Paris.

Ratzenhofer, M., 1953. Studien über die Gewichtsänderungen bei der Entwicklung des grossen Kohlweisslings. *S. B. mat-nat. Kl.* 1: 162.

Ripa, R., 1978. *Studies on the susceptibility of Pieris brassicae (L.) to a granulosis virus*. Ph.D. thesis. University of London (Imperial College).

Srihari, T., 1970. Étude quantitative de la consommation et de l'utilisation de la nourriture au cours de la croissance larvaire de *Pieris brassicae. Ann. Soc. ent. Fr. (N.S.)* 6: 1003–1014.

—— 1972. Observations sur le poids et la taille au cours de la croissance et de la métamorphose chez *Pieris brassicae* L. *Ann. Soc. ent. (N.S.)* 8: 359–376.

Srihari, T. & Gahukar, R.T., 1975. The influence of juvenile hormone on food consumption in the last larval instar of *Pieris brassicae* L. (Lepidoptera: Pieridae). *Bull. Soc. zool. Fr.* 100: 327–333.

Street, M.L., 1975. *The growth and development of the reproductive system of Pieris brassicae L. and their control by hormones.* Ph.D. thesis. University of Newcastle upon Tyne.

Sudah, M.I.E., 1970. *Some effects of larval age and leaf age on the feeding behaviour of Pieris brassicae* L. (Lepidoptera: Pieridae). M.Sc. thesis. Bangor University.

Wigglesworth, V.B., 1967. *The Principles of Insect Physiology.* Methuen & Co. Ltd., London.

Zylberberg, L., 1965. Variation de la teneur en glycogène après irradiation des cellules de la paroi de testicule chez *Pieris brassicae. J. Microsc.* 4: 172.

Further references

Auel, H., 1902. *Allg. Z. Ent.* 7: 113 (development over several generations).

Clas, P., 1970. *C. r. Acad. Sci. Paris* Ser. D 270: 2666–2669 (post-embryonic development of the wings).

Heimbach, F., 1973. Diplomabeit. University of Cologne. (effect of temperature on the development of the larva).

House, H.L., 1965. *In: The Physiology of the Insecta.* Ed. by M. Rockstein. Academic Press, London. pp. 769–813 (insect nutrition).

Jones, A.L., 1956. *Leeds phil. lit. Soc. sci. Sect. Proc.* 7: 8 (rate of development with temperature).

Lal, O.P. & Chandra, J., 1977. *Indian J. Ent.* 37: 310–311 (sex differentiation).

Lavenseau, L., 1973. *Deuxième Séminaire Diff. Cell. Insectes.* École Normale Superieure 1–20 (problem of wing development).

Lipp, C., 1957. *Biol. Zbl.* 76: 681–700 (cell differentiation in wings).

Peters, T.M. & Barbosa, P., 1977. *A. Rev. Ent.* 22: 431–450 (factors affecting development rate).

Stepanova, L.A., 1962. *Rev. ent. USSR* 41: 721–736.

Vats, L.K., Singh, J.S. & Yadava, P.S., 1977. *Agro-Ecosyst.* 3: 303–312 (food energy budget).

Waldbauer, G.P., 1968. *Adv. Insect Physiol.* 5: 229–288 (food utilisation).

Wardzinski, K., 1938. *Z. angew. Ent.* 25: 478–486 (population density).

Williams, C.B., 1953. *Rothamsted Exp. Stn Rep. 1953.* p. 118 (population density).

7. Morphology and anatomy

General introduction

The first part of this chapter on morphology brings together aspects of gross morphology, body ratios, features of the cuticle and associated tissues, genitalia, ocelli and compound eyes in a logical sequence from ovum to imago. While preparing this chapter the author was surprised to find that there is such a paucity of information on what would seem to be everyday aspects of pierid morphology, such as basic descriptions of the silk girdle, cremaster and larval legs, the three pairs of legs in the imago and of the wings and antennae. It has been assumed perhaps that most of these descriptions have been published in the past thus leading them to be overlooked.

In the second part of this chapter on anatomy the major tissues of the insect, the fat body and the haemolymph, have been dealt with as separate sections. All the glands of the body, excluding those of the endocrine system, have been designated a separate section; this follows the part on the alimentary canal with which there is much in common.

Four major systems, the digestive, reproductive, nervous and muscular system have also been given separate sections. Locomotion, which comprises much work on the musculature and anatomy of flight and walking, has been appended to the section on muscles.

Separate chapters have been allocated to the endocrine system (Chapter 9), the biochemistry of the cells (Chapter 10) and the sense organs associated with the cuticle (Chapter 12).

For further information reference should be made to Portier's (1949) book on *The Biology of Lepidoptera*, Teotia & Pathak's (1957) anatomy of the lepidopteron *Enarmonia pseudonectis* Meyr., Ehrlich's (1958) study of the anatomy of *Danaus plexippus* L. and Yagi & Koyama's (1963) book on *The Compound Eye of Lepidoptera.*

SECTION A. MORPHOLOGY

Ovum

The ova of many lepidoptera have been described and illustrated by Ewald Döring (1955) in *Zur Morphologie der Schmetterlingseier*, and it contains illustrations of the ova of the three pierids, *P. brassicae*, *P. rapae* and *P. napi*. One of the earliest accounts of the ovum of *P. brassicae* is that of John Swammerdam (1758) in his book with René Réaumur called *A History of an animal in an animal*, while later accounts and figures may be found in Sepp (1762), Westwood (1854), and Sarlet (1949a,b) (cf. also Hinton, 1981).

Swammerdam (1758) noted that the ovum of *P. brassicae* has 15 small longitudinal ridges which have regular cross grooves or channels; these are well seen in the diagrams of Döring (*loc. cit.*) who has indicated 16 longitudinal ridges. The average height of an ovum is 1.40 mm, while its diameter is 0.58 mm based on a sample of 100 ova (Bartell, 1966). He also figured lateral and apical aspects of the ovum of *P. brassicae* as well as a section through the chorion. Henking (1890) published an illustration of a section through the upper pole of the ovum of *P. brassicae* at the point of fertilisation. It was pointed out by Wigglesworth (1967) that polyspermy seemed to be the normal case in *P. brassicae* as in other insects, that is many spermatozoa enter the ovum. The figure published by Wigglesworth (1967) shows that the *P. brassicae* ovum has a typical yolk, micropyle, submicropylar region and chorion.

Much of the yellow colouring substance of the ovum is contained in the ovum case and this has been identified as carotenoid by Feltwell (1973).

Larva

Head and appendages

Eassa (1963) described the metamorphosis of the cranial capsule of *P. brassicae* from the larva to the imago. He described the origins of the clypeus and frons, parietals, genae and postgenae in the light of misinterpretations made earlier by others, and readers are therefore referred to his work (see his pp. 510–514) for historical information. The structure of the head is shown in Figure 7.1.

The labrum is a broad structure which articulates with the conjuntiva and is continuous with the epipharynx below. The mandibles articulate dorsally and ventrally with the cranium and there are two apodemes at their bases, one inner and one outer. The mandibles are shovel-like with

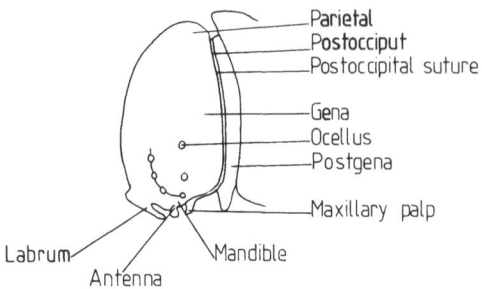

Fig. 7.1 Lateral view of fifth instar head
(modified from Eassa, 1963)

four major "teeth" (other smaller teeth are present), and when closed, the right mandible usually overlaps the left.

The labium lies ventrally on the head and is divided into two parts, the postmentum and prementum (see Eassa, 1963, figure 1d, ventral view of the head). The maxillae consist of three parts; the cardo, the stipes which are fused to the labium, and a terminal lobe which bears the galea and maxillary palps. Eassa (1963) discussed disagreements amongst other authors over the homologies of the maxillae. Sensory apparatus on the labium, maxillae and labrum is described in Chapter 12.

Ocelli

P. brassicae larvae possess 12 ocelli each of which is about 90–100 μm in diameter, and which stands proud of the surface by up to 70 μm (Eassa, 1963; Angersbach, 1970; El-Dakroury, 1972-thesis, Figure 12.1). Six ocelli are situated on each side of the head in the form of an ellipse and all point in different directions. The only ocellus which is different from the others is number three which has three corneas instead of one. The function of the ocelli is dealt with in Chapter 12.

Thorax

As far as the author is aware no specific and thorough descriptive account has been made of the thorax of *P. brassicae*, and particularly of the three pairs of true legs.

Abdomen: Spiracles

There are nine pairs of spiracles in the larva, one relatively large pair on the first true leg segment, one each on segments four and five, one each on

162

the four proleg segments and one on segments 10 and 11. In the pupa there are only six pairs of spiracles (cf. Jackson, 1890).

The spiracles are oval in shape, heavily chitinised around the margin, and occluded by between 27–30 projecting invaginations, each of which contains several hairs. There appear to be at least two levels of the projecting hairs.

A study of the tracheal system of *P. brassicae* appears to be wanting but for comparative purposes the reader is referred to Teotia & Pathak (1957) and Ehrlich (1958).

Larval cuticle

The cuticle of the fifth instar larva is transparent in a few areas along the dorsal and lateral surfaces, these areas being most noticeable when a larva moves. During wandering there is a transparent area behind the head at the junction of the first segment. As the larva moves it exposes a folded triangle of transparent cuticle at either side. This is especially noticeable when the head moves from one side to the other during silk-laying, i.e. when the head moves to the right, the left triangle is exposed and vice versa.

Furthermore there is an area of clear cuticle on the side of the body at the junction of segments 2 and 3, positioned on segment 2. These areas of clear cuticle appear whitish green as they show up the interior of the larval body. In some cases the nerves and muscles are visible inside. As far as the author is aware the function of these transparent areas has not been investigated, but there could be some connection with photoreception.

The cuticle on the dorsal surface of the fifth instar is folded from segment two posteriorly and there are about six folds per segment. Between the folds internal views of the body can be seen as the cuticle is also opaque in these areas. The width of the folds varies according to the relative stretching during locomotion.

The larva possesses areas of opaque cuticle on the head which are thought to allow light to react on the brain. The significance of these receptor areas is discussed under larval vision in Chapter 12.

Chaetotaxy

The chaetotaxy of lepidopterous setae was studied in depth by Hinton (1946) who noted that in *P. brassicae* larvae the head has a relatively large surface area clothed with long setae and a relatively small area from which these setae are absent. His figure of the first instar larva indicates the arrangements of the long setae on the cranium. It is worth pointing out for comparative purposes that the chaetotaxy of another lepidopteron,

163

Enarmonia pseudonectis Meyr. (Eucosmidae) was studied by Teotia & Pathak (1957). Setae on the compound eye do not appear to have been studied in *P. brassicae*, but in *Pieris rapae crucivora* more setae were found on females (2.6 per 100 facets) than on males (1.8 per 100 facets) (Yagi & Koyama, 1963).

A number of sense organs have been found on the cuticle of larvae of *P. brassicae*; tactile hairs (Portier, 1949), and sensilla styloconica (Ma, 1972). Younger larvae are covered in setae which are longer in proportion to their body width than those of older larvae. The fact that the hymenopterous parasite, *Apanteles glomeratus*, selects first instar larvae in which to oviposit may in part be due to the lack of obstruction of these hairs since in the second instar the setae are equal in length to the width of the larvae. However, in the final instar the setae are equal in length only to half the body width. There are two types of setae in the fifth instar: a) long curved white setae originating from the large black areas, and b) small and large black setae which originate from the edge of the black markings and other small black spots (Feltwell, unpublished).

Microstructure

The light microscopy work of Kayser-Wegmann (1976a,b) shows that the cuticle of the larva during the fourth and fifth instar and the cuticle of the freshly formed pupa are essentially different in structure as well as pigment distribution. The larval cuticle is traversed by epidermal cell processes connected by fibrils which act possibly as supporters. The pupal cuticle, however, is traversed by fibril bundles (1–20 fibres/bundle) which transport lipid-containing material to the outside. The distribution of black pigment in the cuticle also differs, being present in the epicuticle of the larvae and in the exocuticle of the pupa. During the larval–pupal ecdysis no processes of the epidermal cells were found to extend into the cuticle as has been shown in other species, even though sections of cuticle were taken at two hourly intervals through the ecdysis. Only microvilli were present.

Tubercles

The only person who seems to have studied these is Forbes (1909) who compared the tubercles of *P. brassicae*, *P. rapae* and *P. daplidice*.

Forbes' method involved cutting off a complete segment from which drawings were then made showing distribution of spots, hairs and tubercles. The only figures given of *P. brassicae* are of the metathorax, the "middle abdominal segment" at two different stages, and a tubercle. Forbes traced the development of *P. brassicae* tubercles through larval life and made the following points: a) in the first instar the tubercles are twice as high and two-thirds the diameter of those in *P. daplidice*; b) in the second instar the cuticle is pale greenish yellow with black dots and pale tubercles. The stigmatal tubercles are small with minute tubercles laterally and subdorsally; c) in the third instar some large and acute tubercles are present with white setae; d) in the fourth instar tubercle number three is twice as big as number two or one and the latter is four times as large as the one directly above number three. An exceedingly large tubercle is present laterally on the thorax. In the fifth instar there is little change from above.

Pupa

Head

There are certain radical changes which take place during metamorphosis between the larva and the imago, for instance on the head the ocelli disappear and are replaced by compound eyes. Other larval parts which disappear, diminish or disintegrate are the hypodermal furrows caused by part of the epistomal ridge, the mandibles and mandibular muscles, the parietals, antefrons, maxillary palps, conjunctiva, postocciput and spinnerets. Overall, the head is compressed anteroposteriorly giving a shorter dorsoventral axis than before (see Eassa, 1967, for plates of pupal head).

Thorax

Like the thorax of the larva, that of the pupa also appears to have been overlooked and a full description of the three pairs of legs and two pairs of wings is wanting.

Abdomen: Genitalia

Jackson (1890) made extensive studies on the genitalia of pupae of *P. brassicae* as well as other species, and noted that in most species the features were consistent. However, in *P. brassicae* the features are not consistent and Jackson said that he was going to publish the inconsistencies at a later date. (The author has been unable to trace them).

P. brassicae pupae clearly show sexual differences, the female has two genital openings and is altogether much wider than the male (Figure 7.2). A photograph showing the ventral groove running anterior-posteriorly along the length of the 10th segment of the male was illustrated in Dusassoy & Delplanque (1964). Gérard Chovet's (1974) thesis on the male genitalia is worth studying, although it is mostly on that of the imago.

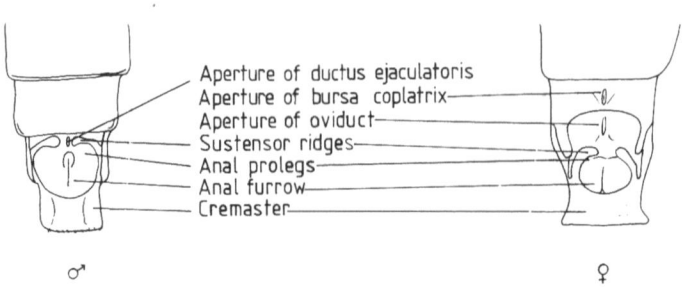

Fig. 7.2 Sexual differences in genitalia of pupae of *P. brassicae* (modified from Jackson (1890) and Richards & Davies (1977))

Cuticle: Pupal spines

Johansson (1959) recorded a particular morphological structure on pupae he collected near Oslo by which he maintained he was able to differentiate between diapausing and non-diapausing pupae. Apparently only his non-diapausing pupae had a small spine on the dorso-lateral surface of the third abdominal segment. He looked at 141 non-diapausing pupae and 462 diapausing pupae and on no occasion did a diapausing pupae have a spine. However, 15% of the non-diapausing pupae had features which fell between those morphological groups. An histological investigation of the spine showed that it was derived from a protrusion of the epidermis, while the cuticle remained the same thickness below.

In a note communicated by Green (1927) mention was made of a translucent green pupa found attached to a bramble leaf which had a "moderately long, slender spine, taking the place of the stout conical process that normally occurs on each side of the body, near the outer angle of the wing". This specimen was figured beside a normal specimen. Main (1937) noted that of 46 pupae which resulted from larvae of *P. brassicae* collected in Epping Forest, near London in 1895 the first 16 to hatch all came from pupae with a particularly large spine.

A recent study by Feltwell (unpublished) on some 239 pupae with spines collected in France (Cévennes, 30440) may indicate the effects of environmental factors such as abundance of predators (in this case lizards) influencing the presence of spines. Such an unusual occurrence of spines was also witnessed in Morocco by Allcard but here it may be that small birds influence the production of spines (Allcard, 1979, pers. comm.).

Imago

Head

The head of an insect is essentially an integrated sensory apparatus with provision for the acceptance of food. In the case of the imago of *P. brassicae* the head, on first inspection, appears to be densely covered with long setae, but beneath this lies the coiled proboscis enclosed within the labrum, the compound eyes, and on top of the head a pair of sensitive antennae. First, the different parts of the cranium will be dealt with here, followed by the other parts.

The head of the imago of *P. brassicae* has been described by Eassa (1963). The clypeus is the "face" of the imago; it is limited laterally by the genae and ventrally by a furrow next to the labrum. It serves as an attachment place for the muscles and increases in size as a hypodermal cone or crest projects anteriorly. The frons is the area between the bases of the antennae and is limited dorsally by the epicranium. The epicranium is a new part of the imago formed from a growth of the hypodermis posterior to the pupal frons and results in a constriction of the occipital foramen. The postocciput is a narrow area which passes along the occipital foramen and is limited by the postoccipital suture and ending at the two tentorial pits. The tentoria on the anterior tentorial areas are inflections of the cuticle which arise from the two tentorial pits just above the mandibles and extend posteriorly. They are thick, as they also serve for insertion of muscles. There is a transverse tentorial bar which extends across the posterior tentorial areas.

Antennae

The antennae are situated in a socket which is lateral to the frons and above the clypeus. The basal segment is large, the second ring-shaped, and the third is the largest of the three. All the other segments are nearly the same length except those at the distal end. The 13-segmented antennae are about 16 mm in length (Mukerji, 1961; Eassa, 1963) and are knobbed at the distal end in the characteristic lepidopteran manner and are scattered with white and black scales which point proximally and give a banded appearance. Jordan (1898) noted that the formation of the club in the Pierinae offers a good characteristic for genera distinctions. An analysis of the antennal length of 25 males and females by the author (unpublished) showed that there is no apparent sexual difference, viz. males: av. 14.95 mm (range 11.5–16.0), females: av. 14.73 (range 12.0–16.5).

Jordan (1898) investigated the structure of the antennae of many lepidopteran families and found that the general characteristic of all pierids was that the dorsal side was never without scaling, and that the fine sense hairs and bristles are often specialised. He published a small figure of an antennal joint of *P. brassicae* at high power magnification and stated that the last joint was not scaled dorsally and that there was a ventral unscaled stripe down the stalk. According to Jordan (*loc. cit.*) the pierids have such specialised antennae that they have definitely not given rise to other lepidoptera, their nearest "precursor" being the Erycinidae.

In his study of the antennae of 180 lepidoptera, Dethier (1941) stated that he examined those of *P. brassicae* and *P. rapae*, as well as five other members of the Pieridae. In his paper there are many details on musculature, innervation and histology of antennae. One of the characteristics of the pierids is that the second segment of the antenna is short; it is more squat than some of the other pierids studied, and the hairs on the antennae were $4\frac{1}{2}$ times as long as the segmental length (see Dethier's figure 75.8). For an account of the function of the antennae the reader is referred to Chapter 12.

Compound eyes and ommatidia

The most complete analysis of the compound eye of lepidoptera was carried out by Yagi & Koyama (1963) who investigated species from 39 families. Among these was *P. rapae crucivora* Boisduval from which much information, particularly on ratios, may be found. The authors found that the lesser the ratio of antennal length/wing expansion the larger that of the eye size/antennal length. Interfacial hair was examined in 27 pierid species and in *Pieris melete* Ménétriés more hairs were found in females than males. Ommatidia, pseudopupils and pigment migration were also studied.

In a recent study of the compound eyes of 15 insect species, Frantsevitch & Pichka (1976) found that in *P. brassicae* as in other anthropophiles, the binocular zone enclosed 20–25% of the facets in each eye. "Middle sized insects had from 2000–9000 ommatidia". They carried out their work using ophthalmological methods, such as observation of the pseudopupil and the glow from the ommatidia illuminated from the inside.

It appears that information is wanting on the compound eyes of *P. brassicae*.

Proboscis

The proboscis is made up of two parts: the maxillary galea which fit together by means of exocuticular hooks, and dorsal overlapping plates, thus enclosing the food canal. This double barrelled tube is between

16–18 mm long (Müller, 1883; Portier, 1949) but may be as short as 13 mm (Mukerji, 1961 in India).

Eastham & Eassa (1954) made an extensive study of the proboscis of *P. brassicae*, particularly its musculature and the functioning of the sucking pumps. Extension was thought to be brought about by a combination of four factors: a) the flexible endocuticle and rigid exocuticle, b) muscle action, c) internal skeleton of longitudinal septa, and d) haemocoelic turgidity; and not solely by the action of intrinsic muscles as was previously thought (Eastham & Eassa, 1954). Coiling was thought due to the elasticity of the endocuticular dorsal bar which lies in the upper wall of the food tube. However, two papers published in 1971 re-examined Eastham & Eassa's work and threw new light on the workings of the proboscis. Bänziger (1971) disclaimed that proboscis extension occurs by a) muscles inside the tube causing a dorsal transverse convexity; b) direct muscle action or c) elasticity; instead he found by carrying out many experiments that blood-pressure causes extension. Hepburn (1971) found resilin in the dorsal bar of the proboscis which provides sufficient cuticular elasticity for initial coiling.

Thorax: Venation

Early development of venation can be seen in the figures published on the imaginal wing discs of *P. brassicae* by Spuler (1908) and Cals (1970). These show a typical peripheral vein which travels around the outside distal region of the wing and joins up with the radiating arms of the main veins (cf. Grote, 1900; Ford, 1945).

Wing patterns of pierids have been illustrated by Niculescu (1963) with two other species for comparison, while figures and details of wing venation have been published by Graham-Smith & Graham-Smith (1930), Portier (1949) and Eitschberger (1969).

Scales

The wings of *P. brassicae* are covered with rows of small scales which are socketed into the wing structure by little pegs (Feltwell & Rothschild, 1974, plate a, b; cf. Verity, 1905–11) for blocks of scales *in situ* and comparison of *P. brassicae* scales with those of *P. b. chariclea*). In a study of over 300 contour scales from the wing of a normal male *P. brassicae* (Feltwell, unpublished) eight types of scale were found: a) broad with four rounded "serrations" at the distal end; b) long and thin with a rounded end; c) very short and wide with five rounded serrations; d) rectangular and wide with three round serrations; e) thin with two round serrations; f) long with three round serrations; g) short and wide with four smooth serrations. The

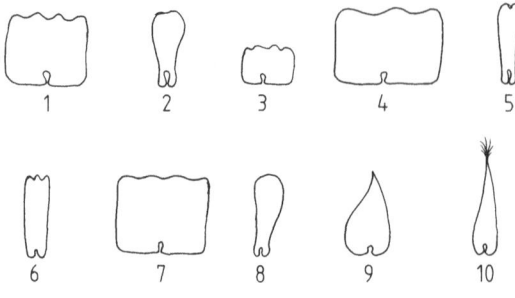

Fig. 7.3 Scale types
(Feltwell, unpublished)

1 Broad with four rounded serrations; 23–31 ridges
2 Long and thin with rounded end; 26–30 ridges
3 Very short and wide with five rounded serrations; 27–29 ridges
4 Rectangular and wide with three rounded serrations; 27 ridges
5 Thin with two rounded serrations, 19 ridges
6 Long with three rounded serrations; 25–31 ridges
7 Short and wide with four smooth serrations; 31 ridges
8 Quama lubrica (after Verity, 1905–11)
9 Quama lubrica cordiformis (after Verity, 1905–11)
10 Androconial scale

number of ridges running longitudinally for each scale type appeared to be fairly constant with a range in 357 scales of 23–31 ridges per scale. If these scales were investigated in any further detail other types may be found and sexual differences may be apparent. Verity (*loc. cit.*) made drawings of two types (Figure 7.3).

Scales help to impart colour to the insect's wing. Yagi (1956) found that "there were great differences between males and females in the form of pterin pigment masses in the scales which formed the ground colour of the fore-wings". Feltwell & Rothschild (1974, plate 1b) recorded green-blue pigment around the sockets on the wings of *P. brassicae ab. coerulea* and they further discussed how colouration was produced with the interaction of pterin and bile pigments, haemolymph colouration, the "impressionist effect" of Wigglesworth (1928), and the mixing effect. Mayer (1896) investigated the development of scales in the wings of pierids and stated that the pigment flowed into the scales' lumen before it evaporated on the inside.

Wings: Surface area

Long (1959) calculated the wing surface area of *P. brassicae* using pieces of tracing paper, and also calculated the wing loadings in his studies on the effect of crowding of larvae on imago size. He found that the wing area varied between 1537–1841 mm^2; the hindwings being the largest at 795–965 mm^2 and the forewings being 741–876 mm^2.

The wing loading measurements, which are critical for flight, were for males and females respectively (sample 115) 11.6–12.7 and 11.6–12.0 mg/cm^2. Wing loadings between 8.0–12.1 mg/cm^2 have been quoted by Hoff (1919); Portier & de Rorthays (1926); Pieron (1927); Osborne (1951) and Müllerhoff (1930) in Banks (1930). The wide range of figures may have been due to differences in larval breeding conditions, seasons or imaginal age.

Imaginal size

It is clear that imaginal size is controlled not only by the intrinsic factors as in ab. *minor* (cf. Chapter 1) but by extrinsic factors which include the availability of food, crowding and possibly by decreased temperature. However, Allen (1954) found that imaginal size was not affected by rearing on diets deficient in sodium, potassium, or phosphorus.

Small size can be induced in the laboratory by two methods: a) limiting the food during the facultative feeding time of the larva (Street, 1975) or by starving (Gardiner, pers. comm.); or b) crowding the larvae in groups at 125 or 500 per square foot (0.37 m^2) (Street, 1975; Gardiner, 1976, pers. comm.).

Small size of imagines was observed in the wild by Pictet (1922) who found small specimens produced from larvae which had been feeding from October until· January in France when the temperature was between −9°– +10°C, with frosts. It has been suggested by Auel (1937) that wing length is shortened in very dry years.

Large wingspan can be induced experimentally by decreasing the density of larvae to 100 per square foot (0.37 m^2) (Gardiner 1976, pers. comm.). In the wild large specimens have been recorded from France by Rowland-Brown (1907), Higgins (1930) and also in Greece (Kalávryta by Rosa (1924).

The generally accepted distance for wing measurement of butterflies is twice the distance from the apex of the forewing to the centre of the thorax in millimetres. Frohawk (1934), often quoted as a source of such information, stated that the wing length of an average male *P. brassicae* should be 63 mm and an average female 76 mm (Table 7.1). Meyrick (1928) however stated that imagines (sexes not stated) vary between 60–76 mm. Frohawk's figure is suspected of being a misprint by Gardiner (1963) who found the male/female ratio 58:63 mm from specimens from his own collection. Bowden (1966) compared the data of Frohawk (*loc. cit.*) with specimens from Gardiner's cultures, and with specimens of his own from various countries and showed that the female is generally the

Table 7.1 Wing length of *P. brassicae* in mm, sample size if known in brackets.

Notes on specimens	Males	Females	Authority
England			
Normal	63.5	76.2	Frohawk, 1934
Cambridge stock			
1951	53	58	David & Gardiner, 1952
Cambridge stock			
1960	58.5	57.9	David & Gardiner, 1961
from collection	58	63	Gardiner, 1963
F_1 from wild			
caught F	58	63	Gardiner, 1963
Sussex (Hartfield)			
wild	64.9 (2)	65 (2)	Feltwell, unpublished
Kent (Sutton			
Valence) wild	67.46 (3)	67.4 (3)	Feltwell, unpublished
Kent (Sevenoaks)			
bred from egg batch	59.58 (11)	58.36 (11)	Feltwell, unpublished
Germany		60–65	Blunck, 1953
Poland		66	Strawinski, 1929
France			
St. Martial			
30440 wild	68.8 (2)	69.16 (4)	Feltwell, unpublished
Switzerland			
Stein	70.00 (1)	—	Feltwell, unpublished
India			
Agra	57		Mukerji, 1961
	(sex and number not stated)		
Aberrations			
ab. *minor*	37 (not given)	38 (not given)	Gardiner, 1963
ab. *coerulea*			
bred (source			
Oxford labs.)	65.91 (7)	65.91 (7)	Feltwell, unpublished
race *cheiranthi*	60–65 (6)	65.70 (6)	Gardiner, 1963

larger of the sexes. This was particularly evident with specimens from Corsica, Britain and for *P. b. cheiranthi* from the Canary Islands.

The combined data on wing size of imagines from many sources compiled in Table 7.1 also show that the female is usually the larger sex, being larger in six cases opposed to two cases of males being larger. There were also two cases where the sexes were on average the same size. The mean size of males to females is 61.74:63.89; males on occasions have been recorded as large as 70 mm for a Swiss male (Feltwell, unpublished), just less than the maximum for an English female as recorded by Frohawk. This latter figure of 76.2 mm for females does appear to be rather excessive.

Abdomen: Genitalia

The genitalia of *P. brassicae* imagines have been illustrated by Verity (1905–10), Drohsinn (1933), Beirne (1942), Lagnel (1966), Chovet (1970) and Higgins (1975), cf. Ogata *et al.*, 1957. Genitalia of the subspecies *P. b. cheiranthi* (Hübner) were figured by Kudrna (1973) and those of *P. b. nepalensis* by Bernardi (1958).

The most recent description of the genitalia of *P. b. brassicae* (Linnaeus) is that of Higgins (*loc. cit.*): "Uncus short, slender, tegumen about twice as long as uncus; valve with deep incision or short process near inferior posterior angle; penis robust, almost straight, larger than valve, caecum short, trochanter well developed, shaft with slightly shorter, wider, posterior notch emphasised; penis dorsal hump larger. These characters most marked in specimens from Madeira, least obvious in a specimen from the Azores, scarcely distinct from continental *P. brassicae*."

Lagnel (1966) studied the genitalia of the three pierid species, *P. brassicae*, *P. deota* de Nicéville and *P. brassicoides* Guérin, and found that the distinguishing feature for *P. brassicae* was the anterior vaginal lobes of the female. The distal point of the male valva is too variable to represent a good specific character. Lagnel (*loc. cit.*) found that *P. brassicae* was allopatric with *P. deota*, and found that there was an overall difficulty in distinguishing the species on their genitalia because of the extreme amount of geographical variation. He examined *P. brassicae* specimens from Ireland, Canaries and the Yunnan. Valva of *P. brassicae* from Morocco were illustrated.

Sex ratios

There are more accounts of females being more abundant than males in the literature (eight cases out of 12) giving an overall ratio of female to males as 1.26:1 (Table 7.2). The most thorough investigation of sex ratios was made by David & Gardiner (1952) who, for three consecutive years, analysed the contents of eight cages comprising in all some 2,687 insects and found that there were more females than males, the ratios varying from 0.84–1.45, the average ratio being 1.05.

The work of Graham-Smith and Graham-Smith (1930) showed otherwise, that there were more males than females in a sample of 907 insects. Males were also found to be more abundant than females by Church (1957, 1958) but here small samples were taken. Little comparison with allied species on sex ratios appears to have been made but work by Tulluch (1941) showed that the ratio in *P. rapae* was 3:1 compared with 1:4 in *P. brassicae* for males to females. Here again the samples were small.

Table 7.2 Sex ratios of *P. brassicae*.

Numbers counted		Ratios		Site or origin	Authority
Males	Females	Males	Females		
a) Where females predominate					
10	18	1	1.8	Harpenden, on migration	Williams et al., 1942
9	12	1	1.3	Austria	Church, 1958
1	4	1	4	Wales	Tulluch, 1941
113	312	1	2.76	Great Britain (?)	Williams, 1958
		0.85	1.35	Cambridge, in laboratory	David & Gardiner, 1961
2,687 insects of both sexes excess of females over males by 9% at eclosion				(?)	Buparai, 1968
11	12	1	1.09	Sevenoaks, bred from ova	Feltwell (unpublished)
43	47	1	1.09	Biological suppliers (probably Gardiner's strain)	Feltwell (unpublished)
b) Where males predominate					
455	452	1	0.99	Aberdeen	Graham-Smith & Graham-Smith, 1930
33	24	1.37	1	Berwick	Church, 1957
25	9	2.78	1	Berwick	Church, 1957
4	0	4	0	Berwick	Church, 1958
Totals					
704	890	1	1.26		

Church (1957) found that in one locality over two years (1955 & 1956) the sex ratio doubled in favour of the males. However, his samples were also small and no explanation for this occurrence was given. He believed there was a tendency for equal numbers of each sex with a decrease in latitude, his localities being Berwick-on-Tweed (Berwickshire) (55°N) and Stubal, Austria (47°N).

It appears that some unknown factors may be responsible for changes in the sex ratio. However, Allen (1954) found that the sex ratios are not affected when larvae are reared on diets deficient in nitrogen, potassium or phosphorus. Genetic control of the sex ratio is suggested by Buparai's observation that at eclosion there was an excess of females by 9% (Buparai, 1968). Williams (1958) showed that there is always a greater number of females in a migratory stream, often as many as three times the number of males. During the immigration of 1940 Williams counted 18 females to 10 males at Harpenden (Williams *et al.*, 1942).

Cuticle

Cuticular setae: Ultrastructure

Ma & Schoonhoven (1973) investigated the setae on the tarsi of *P. brassicae* with scanning electron microscopy (SEM) and transmission TEM. They found two types of setae: the tactile hairs and bristles which they called "A-type" structures, and the smaller "B-type" trichoid sensilla. The "A-type" structures included the very small microtrichia which cover the surface of the cuticle.

The "B-type" hairs are approximately 55 μm by 5.5 μm with a pore at the distal end which is 0.6 μm in diameter. There are more "B-type" setae on the tarsi of the females (263 to 197 – average of four specimens) than on males. "A-type" structures were present in about equal numbers on the mid- and hind-tarsi. The "B-type" structures are innervated by 5 bipolar neurones which at their distal end are divided into an inner and outer segment by a ciliary structure. The cilia have nine peripheral sets of double tubules.

SECTION B. ANATOMY

Fat body

Larva and pupa

The presence of the yellow coloured fat body in the larva of *P. brassicae* cannot really be overlooked when a larva is dissected, as it occupies a large proportion of the haemocoel. Pardi (1939) who made sections ($\times 750$ magnification) through the abdominal fat body of *P. brassicae* pupae noted the presence of chromolipids (cf. Nakahara, 1917). Descriptions of the fat body of *P. brassicae* appear in a paper by Chippendale & Kilby (1969) who studied the relationship between the proteins of the haemolymph and fat body during development (Figure 7.4).

In the first instar (1 day old) larva the fat body cells typically have large nuclei and there are many mitochondria in the cytoplasm. In the larva and prepupae lipid droplets, proteinic globules and urate granules appeared and mitochondria are again present (Lhonoré, unpublished electron-micrographs $\times 400$–13,800) Plate III.

During the fourth instar larval stage the fat body is made up of thin sheets which surround the alimentary canal and are known as the perivisceral fat body. There are also dorsolateral and superficial parts of the fat body. At this stage the fat body comprises about 8% of the larva's dry weight.

In the fifth instar there is always one type of cell but these differ in their inclusion. The fat cells in the two dorsolateral sheets are filled with prominent yellow fat vacuoles and in the late prepupal stage pro-teinaceous granules appear. The two parts of the fat body, one yellow the other green, are visible to the naked eye at this stage. In non-diapausing larvae, most of the ventrolateral sheets are used up indicating that the energy reserves were utilised for imaginal differentiation. At the prepupal stage the fat body has reached about 27% of the body weight. It is of interest to note that during the course of larval development the total body weight and length changes mark out a typical sigmoid growth curve (Srihari, 1972).

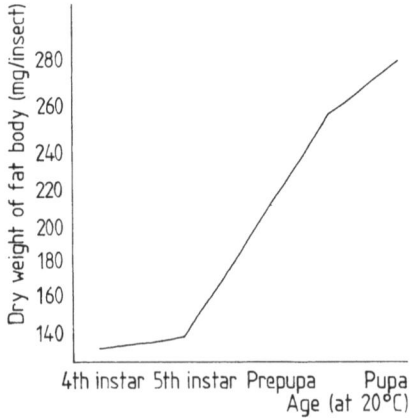

Fig. 7.4 Fat body development in *P. brassicae* (from Chippendale & Kilby, 1969)

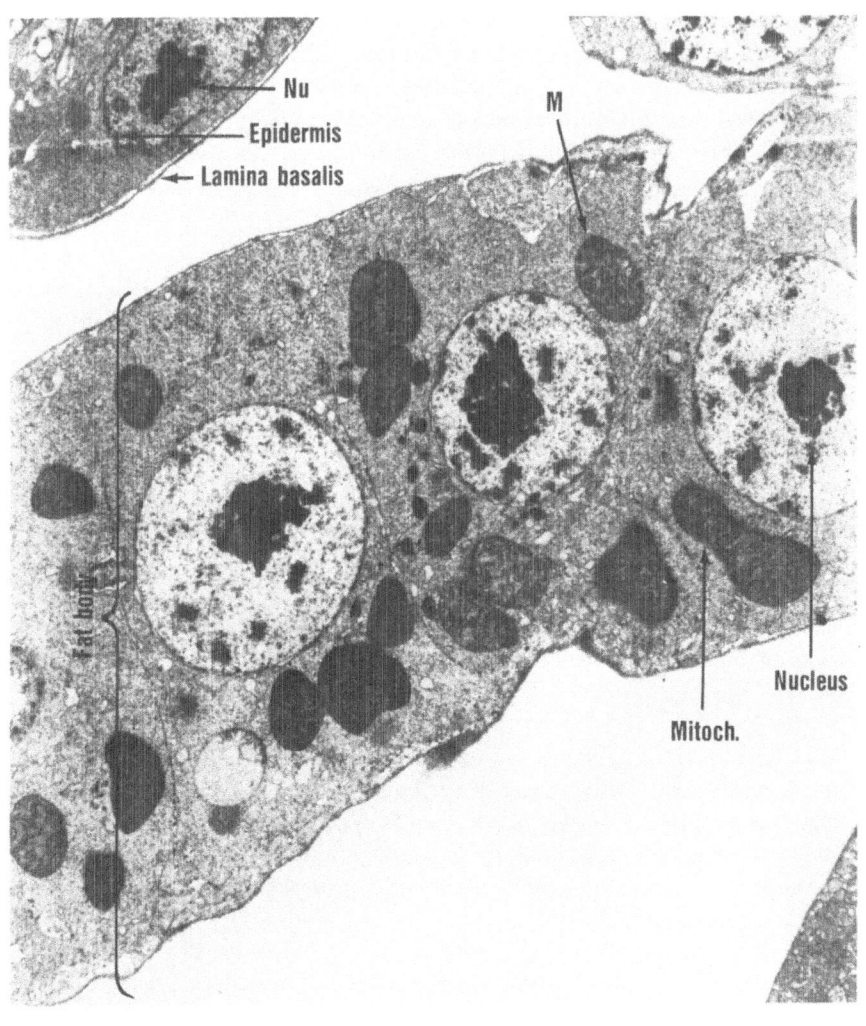

Plate III Section through fat body of one-day-old first instar larva of *P. brassicae* (× 5000)
(Courtesy of Jacques and Denise Lhonoré, Université Pierre et Marie Curie, Paris)

Chippendale & Kilby (1969) showed that glycogen, lipid and protein are deposited in the fat body cells from the early fifth instar larva up to the fresh pupa; the amounts of lipid increase by a factor of 18. Those workers also showed that the fat body makes about 5.5 mg gain of protein during this time, most of it coming from the haemolymph. This was established by incorporating labelled leucine into the tissues. Free amino nitrogen also increased in the fat body from 0.01 to 0.07 mg/insect, while it decreased in the haemolymph during the same period.

The development of the fat body and its relationship with the synchronous development of other organs in the body has been dealt with by Lafont *et al.* (1975) in Chapter 6, while the aspects of its biochemistry, such as its protein content may be found in Chapter 10. Keeley (1978) reviewed work on the endocrine regulation of fat body development, function and metamorphosis and emphasised the importance of the fat body as a tissue for intermediary metabolism.

The yellow colouring substances, carotenoids, which are derived from the food, are present in the fat body (Feltwell, 1978) but presumably are unequally distributed throughout the different areas of the fat body, as Jackson (1890) noted that the colour varied in the fifth instar larva from white through opaque yellow to green.

Imago

Lepidoptera exhibit two types of fat body cell metamorphosis. Either immature fat body cells remain intact and persist into the imago with little change, or mitosis occurs in the larva and new fat body cells are synthesised in the imago from precursor cells (cf. Keeley, 1978). *P. brassicae* comes into the second group as considerable change takes place in the fat body cells.

Karlinsky & Poulaert (1971) studied the development of the fat body during the first 10 days of imaginal life in fed and unfed specimens as well as those ligatured post-cephalically, and took as their parameters size of fat cell and nucleus, and amounts of glycogen, DNA, protein inclusion and fat.

In normal specimens the size of the fat body nucleus decreases from $12\text{--}15\,\mu$ in the prepupa and pupa to $7\text{--}8\,\mu$ in the imagines of both sexes. No detectable difference was found in the ligatured specimens compared to a steady increase through development in controls. Glycogen stores decreased from pupa to imago (more in males) in both controls and ligatured insects.

The most striking effect of ligaturing was that increased amounts of protein inclusions occurred six days after ligaturing, thus supporting the theory that earlier larval characteristics were being retained i.e. that hormone was not getting through from any of the endocrine glands or cells in the head.

Lhonoré (1980) found lipid and protein inclusions which contain small birefringent bodies 0.5–2.0 μ in diameter in the fat body of the imago. More numerous in the female were inclusions of potassium urate with calcium, and traces of magnesium, sodium or chloride ions. Twelve days after eclosion the adipocytes became particularly rich in calcium and potassium.

Reproductive system

Ovaries: Structure

The ovaries of *P. brassicae* were first illustrated in colour by Herold (1815, fig. xv and xxxiii) showing that the tips of the ovaries were yellow. Jackson (1890) gave a very detailed account of the structure and development of the ovaries, based on the previous work of Herold (1815).

The female genital apparatus comprises the common oviduct, the bursa copulatrix, the seminal receptacle, paired accessory glands and sebaceous gland, all of which are situated near the dorsal midline of the abdomen. These parts increase in size during the first four days after pupal ecdysis and are complete before ecdysis of the imago (Street, 1975).

After copulation one to four spermatophores may have been deposited in the bursa copulatrix. Each spermatophore has nine partition walls containing spermatozoids – and the wall is mineralised for an unknown function (Lhonoré, 1978, pers. comm.). Karlinsky (1963a,b) published figures of the ovaries of a newly eclosed imago compared with that of a six day old imago showing the ripe ova. After 48 hours of imaginal life 20–30 ova are ripe and by the sixth day 50–60 ova are ready.

Oogenesis

Oogenesis in *P. brassicae* is imaginal (postmetabolic), i.e. the ova mature only after imaginal eclosion. A mature female may lay up to 750 ova, although this will vary according to several conditions (see section on oviposition in Chapter 3).

Benz (1970a) studied aspects of the diet and physiology which affected oogenesis. Indirect stimulation of the ova is brought about by drinking water or sugar solution (Figure 7.5) and mating, and perhaps by temperature, and it is thought that they activate the corpora allata which releases the juvenile hormone thus stimulating ova maturation (Benz, 1970b).

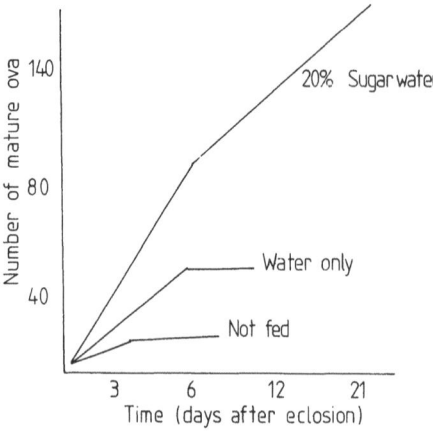

Fig. 7.5 Influence of feeding on oogenesis
(after Benz, 1970a)

Unless spermatozoa reach the receptaculum seminis oogenesis cannot be stimulated experimentally by mating, or by the presence of a spermatophore or a secretion of the male. A negative feedback mechanism regulating diapause was proposed by Benz (*loc. cit.*). Earlier work carried out by Karlinsky (1962, 1963a,b), involving the extirpation of the corpora allata from late fifth instar larvae and fresh female imagines, demonstrated that oogenesis could not take place. Similarly, implantation of the corpora allata into a neck-ligatured female caused oogenesis to continue.

Testes: Structure

These are paired structures situated dorsally between segments six and seven and are well supplied with tracheols (see Herold, 1815, tables V and VI). The testes are paired during the larval stage, but fuse just before the pupa is formed when they take on a clear rose colour. Several layers of cells were seen by Lhonoré (1980) who noted red, brown and clear yellow colourations as well. The yellow external layer contained birefringent crystals of potassium urate associated with xanthine and calcium. Ribosomes and "rosettes" of glycogen were also present. The internal layer contained more pigments, calcium and magnesium phosphate and xanthin. Two types of inclusion bodies were also found by Lhonoré (1978): clear agranular vacuoles containing a homogenous floccular material devoid of pigments, and long (0.2–0.3 μ) needle-like "osmiophiles". In transverse section the testis shows a typical arrangement of secretory cells, follicular and collecting ducts (Figure 7.6).

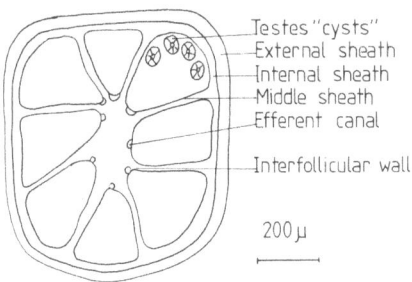

Testes "cysts"
External sheath
Internal sheath
Middle sheath
Efferent canal

Interfollicular wall

200 μ

Fig. 7.6 Transverse section through testis of *P. brassicae* (after Lhonoré, 1978)

Junnikkala (1966) showed that testes of larvae parasitised with *Apanteles glomeratus* (L.) stayed the initial small size and remained colourless instead of changing into the colour typical of normal larvae.

Spermatogenesis

The development of the mitochondria during spermatogenesis was investigated by André (1959) who published a series of electronmicrographs illustrating various stages. Mature spermatozoa are present before pupal ecdysis but are only released two days after imaginal eclosion (Street, 1975).

Zylberberg (1965a,b) found that there were two types of spermatozoa, a typical and an "apyrenes" type in *P. brassicae*. She investigated the effect of temperature on the development of spermatozoa and reared larvae at 30°C (at 25°C the insects were too "fragile", and at higher temperatures too few imagines hatched). The resultant imagines were crossed with specimens which had been reared at 20°C and the percentage of sterile ova consequently laid was noted. On comparison with controls, males reared at 30°C produced more of the "apyrenes" types of spermatozoa which were responsible for a decrease in fecundity.

Small mitochondria with typical structure are found in the spermatogonia, but these increase in size and number during the development of the spermatocytes, when they develop a clear matrix which pushes the cristae to the external membrane (André, 1959). They fuse end to end and are called canalicular chondrioconts at this stage. Other "resistant" mitochondria do not change like this. In the spermatid the chondrioconts pass through a number of stages termed the radiating stage, the entwined stage, the onion stage, the chromophobic stage, the loaf stage and the bowel stage.

181

Two types of mitochondria were also found in follicular cells of the testes of *P. brassicae*, normal mitochondria 0.5 μ in diameter and up to 3 μ in length, and long mitochondria, up to 3–5 μ in length, which show a lamellar or para-cristalline structure (Zylberberg, 1965a, see her six excellent plates). The lamellae are 80–100 Å. The follicular cells were found by Zylberberg (*loc. cit.*) to be rich in glycogen and lipids.

Zylberberg (1969) made light and electron microscopic investigation of double spermatogenesis in *P. brassicae*. She found that morphological and chemical transformation in the nuclei of typical spermatids were synchronous. A proteinaceous Acrosomia complex was shown only to form in the typical line of spermatozoa, and the structure of the flagellum in both types was shown to contain the usual $9+2$ fibres with 9 secondary fibres, 9 external ones (ribs) each of diameter 300 Å. Adenosine triphosphatase was found in the flagellae.

Chromosome number

The chromosome number of *P. brassicae* is n = 15 (Beliaieff, 1930; Bigger, 1960; Doncaster, 1912a,b; Robinson, 1971). More recently the late Dr. Bigger (1975, 1976) developed a technique for air drying chromosomes and has fully documented the karyotypes, having investigated meiotic and mitotic divisions of *P. brassicae* (Plate IV).

The chromosomes characteristically have G band patterns and are further characterised by having two unusual features: a) primary constrictions on the mitotic chromosomes and b) satellites on the acrocentric chromosomes number 7 and 13. Altogether the 15 chromosomes comprise: "7 acrocentric, 5 sub-metacentric and 2 metacentric autosomes and a sub-metacentric sex chromosome which is heteromorphic."

Eleven of the 15 chromosomes of *P. brassicae* are similar to those of *P. napi* (Bigger, 1976), and have similar lengths. As *P. brassicae* has less chromosome material than *P. rapae* and *P. napi*, this lends support to Kudrna's (1973a,b) suggestion that *P. brassicae* should be separated from those other two species.

The number of chromosomes in the male *P. brassicae* can be altered by irradiating with X-rays. Bauer (1967) subjected males to 6000 R. units and produced 19.9% lethal zygotes to 7.7% mortality in controls. Of the fifth instar larvae 64.9% contained in their spermatocytes 1–4 heterozygous translocation rings or chains consisting of 5–14 chromosomes.

Phylogenetically *P. brassicae* is similar to its subspecies *P. b. azorensis* Rebel and *P. b. nepalensis* Gray (Federley, 1942; Maeki & Ae, 1966) in possessing 15 chromosomes, and Federley (1936) also pointed out the analogy of the different chromosomes of *P. brassicae* with those of *Argynnis ino* Rottemburg, compared with other species.

182

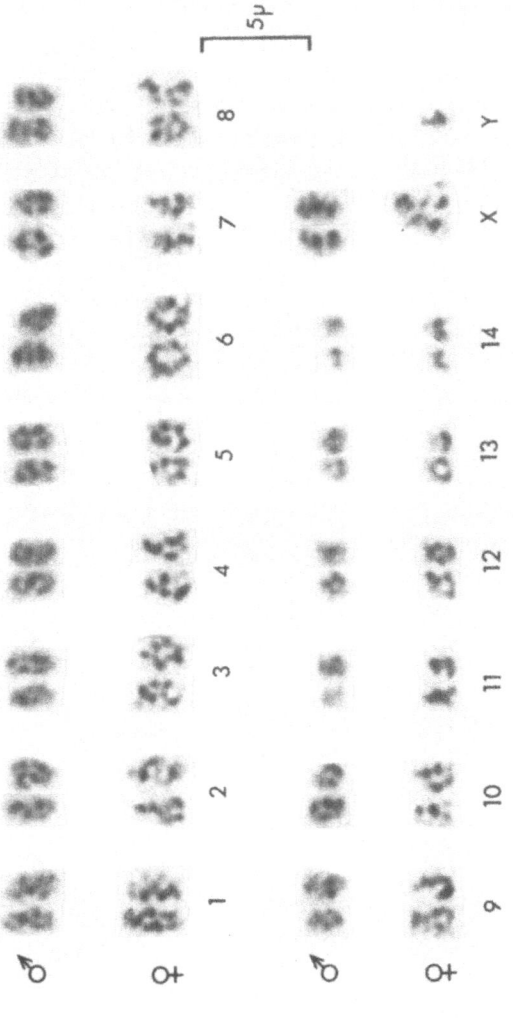

Plate IV Karyotypes of *P. brassicae* made from G-banded chromosomes of a spermatogenial mitosis (♂) and oogonial mitosis (♀) where 2n = 30
(Courtesy of the late Trevor Bigger, Harwell)

Digestive system

Alimentary canal: Structure

P. brassicae larvae have a typical insect gut, that is, with fore-, mid- and hind-gut regions together with a crop, six Malpighian tubules and rectal caecum (Figure 7.7). A comparative study of the alimentary canal of 57 species of lepidoptera representing six families and including six Pieridae species, of which *Pieris rapae crucivora* Boisduval and *Pieris melete aglaope* Motscholsky were included, was published by Homma (1954). Dauberschmidt's (1933) study of the morphology of alimentary canals includes some diagrams of *P. brassicae*.

An histological study of the epithelia of the mid-gut of *P. brassicae* larvae is found in the work on parasitism by Lartschenko (1933) and on the effect of insecticides on the guts of several insect species by Pilat (1935), and more recently by Martouret *et al.* (1965) and Plantevin & Nardon (1970). Pilat found that the gut of *P. brassicae* larvae was similar to that of *Lymantria dispar* (L.) (Gypsy moth) in that there were similar columnar uniserial epithelia with marginal fringes interspersed with smaller calyciform type cells (see Pilat *loc. cit.* figure 26 – L.S. of mid-gut epithelia). The chromatin of the nuclei was composed of finer granules and the nucleolus was smaller. The protoplasm did not contain any oxyphil granules and the nuclei were not divided.

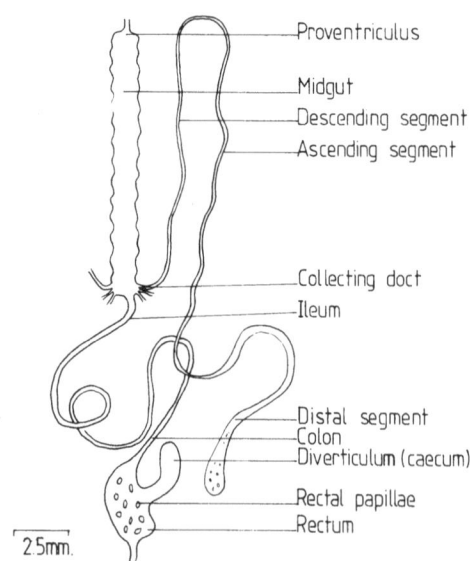

Fig. 7.7 Alimentary canal with Malpighian tubules
(after Lhonoré, 1976)

Two types of cell were described from the mesenteron of 48 hour old larvae by Martouret (*loc. cit.*) and he published 14 photomicrographs of sections through gut cells. Type I cells containing vacuoles typically are contracted at their bases, while Type II cells have much larger vacuoles.

A more recent study of the gut of *P. brassicae* was carried out by Plantevin & Nardon (1970). They compared the development of *P. brassicae* gut with that of *Galleria mellonella* (Linnaeus) and showed that there were three types of cells in the mesenteron villi and microvilli: cylindrical cells, goblet cells (=calyciform cells) and basal cells (Figure 7.8). During the prepupal stage the goblet cells disappear and the cylindrical cells increase in length to about 30–40 μ. The basal cells play the usual part in cell replacement. Differentiation of the cells is thought to be brought about by a "double genetic system". Biochemically the cylindrical cells were found to have irregular deposits of glycogen, as well as proteins including essential amino acids and acidic monosaccharides located in the apices of the villi. No mucopolysaccharides were found in the goblet cells but glycoproteins were present.

Merocrine cells play a part in excretion by expelling mucopolysaccharide grains from the apices of cylindrical cells. These grains are held in small vesicles. Apocrine excretion occurs in two ways, first the end of the villus breaks down thus releasing into the gut lumen protein and mucopolysaccharides which are contained within very small vesicles. Second, by a process of budding, where a small bud of protein forms at the distal end of the cylindrical cells, and then eventually falls off into the lumen.

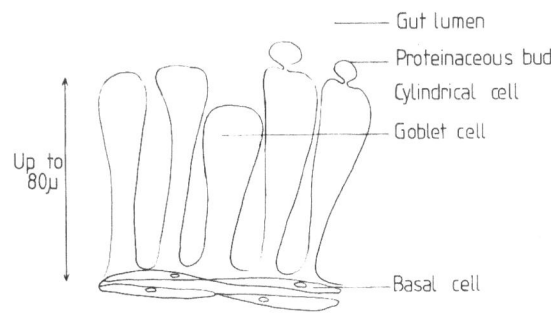

Fig. 7.8 Villi of mesenteron of *P. brassicae*

Malpighian tubules

The Malpighian tubules of *P. brassicae* were originally described by Henson (1932, 1937) but have recently received new attention with the anatomical and physiological studies of Lhonoré (1976) in Paris and of Nicolson (1976a,b) in Cambridge, England.

There are six Malpighian tubules arranged in a cryptonephric complex around the rectum of the larva (Nicolson, 1976a,b). Each is swollen at the distal end and runs back to the mid-gut where it is looped onto the outside in descending and ascending parts; at the proximal end the tubules empty into the distal end of the mid-gut (Figure 7.7).

According to Nicolson each tube of the imago is made up of three parts, the thick upper region, the translucent middle region and the looped lower region, all of which make up a length of about 40 mm. Lhonoré (*loc. cit.*) divided the tubules into two functional parts, the distal part which appears to accumulate minerals (phosphate, magnesium, manganese) and a proximal part which comprises two cellular types, the excretory cells and the flat cells. In the lumen of the tubes are found pigmented cells containing potassium.

Lhonoré's (*loc. cit.*) work contains many figures of histological sections through all regions of the tubules at electronmicroscope resolution and shows much of the detail of the cell types (Plate V). A transverse section through a Malpighian tubule of *P. brassicae* has also been published by Weiser (1977) and in it one can see infestation by the spores of a protozoan.

Glands

Introduction

Apart from endocrine glands, which are described in Chapter 9, there are a number of other glands which have received attention in *P. brassicae* (Table 7.3). The information included here on the Verson's glands which is kindly supplied by Dr. Jacques Lhonoré (Paris) is original and has not been published before. It will be seen that much is still to be learnt about some of the glands and particularly of their functions; thus a complete study of the glands of *P. brassicae* is still necessary and would serve as an ideal research topic.

Earlier work by Swammerdam, Malpighi, Lyonet and Herold must not be overlooked because of their intricate and often lengthy descriptions of morphological and anatomical details of lepidopterous larvae. Gonin (1894) also investigated at some length the formation of the imaginal appendages of *P. brassicae* and mentioned some of the earlier work. Recently a review on the fine structure of insect epidermal glands by Noirot & Quennedey (1974) gives details of exocrine glands.

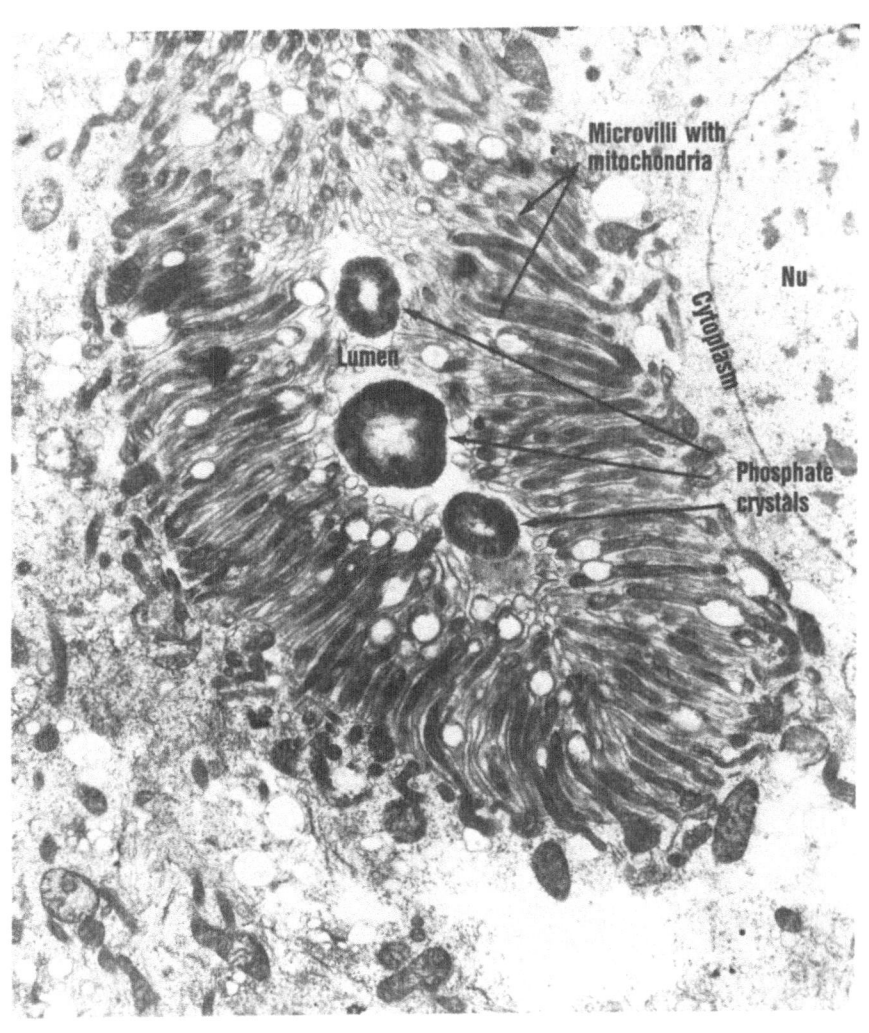

Plate V Section through Malpighian tubule of fifth instar larva of *P. brassicae* (430 g), showing rising part of free tubule (×5200)
(Courtesy of Jacques and Denise Lhonoré, Université Pierre et Marie Curie, Paris)

Table 7.3 Glands of *P. brassicae* (excluding endocrine and certain digestive glands which are dealt with elsewhere).

Stage	Gland	Comments/Function
Larva	Gland of Filippi	Proteinaceous secretions, uncertain function
	Salivary glands	Digestive function
	Silk glands	Silk production
	Unknown glands	Associated with salivary glands
	Verson's glands	Uncertain function
Imago	Abdominal glands	? pheromone marker (postulated)
	Andoconia	Modified scales producing pheromone
	Accessory glands ⎫	
	Annex glands ⎬ All part of the male genitalia	
	Mesospermatic duct ⎪	
	Seminal vesicles ⎭	
	Bursa copulatrix ⎫	
	Spermatheca gland ⎬ All part of the female genitalia	
	Colleteral gland ⎭	
	Proboscis glands	Lubrication of proboscis

Gland of Filippi

This gland, which is otherwise referred to as Lyonet's gland or accessory gland in other lepidopterous larvae (cf. Waku & Sumimoto, 1974), is found close to the spinning duct of the silk glands to which it communicates by a small duct. Srihari (1972) published a drawing which included details of a pair of these filiform glands in *P. brassicae* which extended along either side of the prothoracic glands. No mention of these was made in the text.

The function of these glands remains obscure today, although they produce proteinaceous secretions including mucoidin and are involved in the transfer of water and ions (Lhonoré, 1978, pers. comm.). The glands have also been thought to produce a substance which "cements" the two strands of silk together before it is extruded from the spinning duct (Wigglesworth, 1967).

Salivary glands

These consist of two long tubes, folded upon themselves and divided into three segments and were described from *P. brassicae* larvae in the thesis of Srihari (1971, fig. 1). They are smaller but of the same type as *Manduca sexta* (Linnaeus) (Lhonoré, 1978, unpublished results).

Silk glands and unknown glands

The silk glands are a modified pair of labial glands situated along the ventral anterior part of the larval body (cf. Bordas, 1971), and are made up of a secretory region, reservoir, canal and silk gland. The histology of the distal secretory segment and reservoir were studied by Lhonoré (1980) (Plate VI) who noted that the cubical epidermal cells (25–40 μ high) had large chromatin nuclei. Calcium phosphate was particularly noticeable in the third instar and fibroin and sericin were present in the lumen. The glands produce a secretion which passes through a press which squeezes the two fibres together into one thread (Wigglesworth, 1967) and become functional in the second instar (cf. glands of Filippi). The spinnerets are situated just posterior to the mouthparts (see Spuler, 1908).

Allegret (1956) illustrated the dissected silk glands of *P. brassicae* in his thesis on lepidopterous silk glands, and also found some hitherto undescribed glands with unknown functions. An account of the biochemistry of silk is given in Chapter 10.

Verson's gland

Verson's glands are found in all lepidopterous larvae and have been known for nearly 90 years. Typically in *P. brassicae* there are 15 pairs situated in the latero-dorsal position, two pairs in each thoracic segment, seven pairs in the first seven abdominal tergites, and two pairs in the eighth segment (Lhonoré, 1978, unpublished results (Plate VII and Figure 7.9)).

Each gland is made up of three cells – the secretory cell (300–1800 μ long in the fifth instar), the reservoir cell and the canal cell, the former being the largest. Glucosides, lipids, proteins and minerals are not found in the

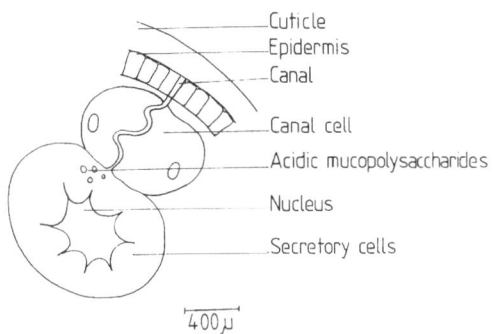

Fig. 7.9 Verson's gland
(Lhonoré, 1978, pers. comm.)

189

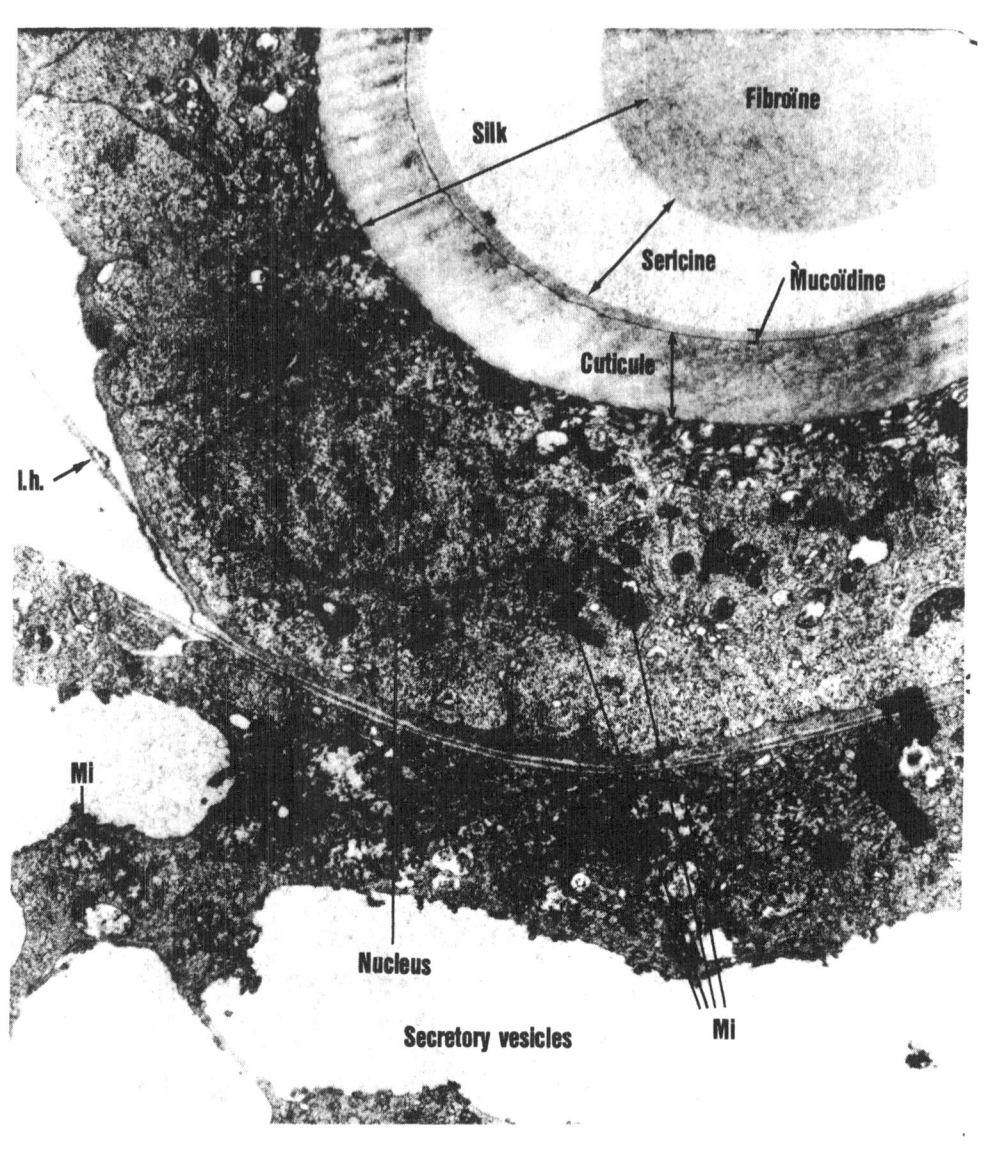

Plate VI Section through canal of silk gland (×5600)
(Courtesy of Jacques and Denise Lhonoré, Université Pierre et Marie Curie, Paris)

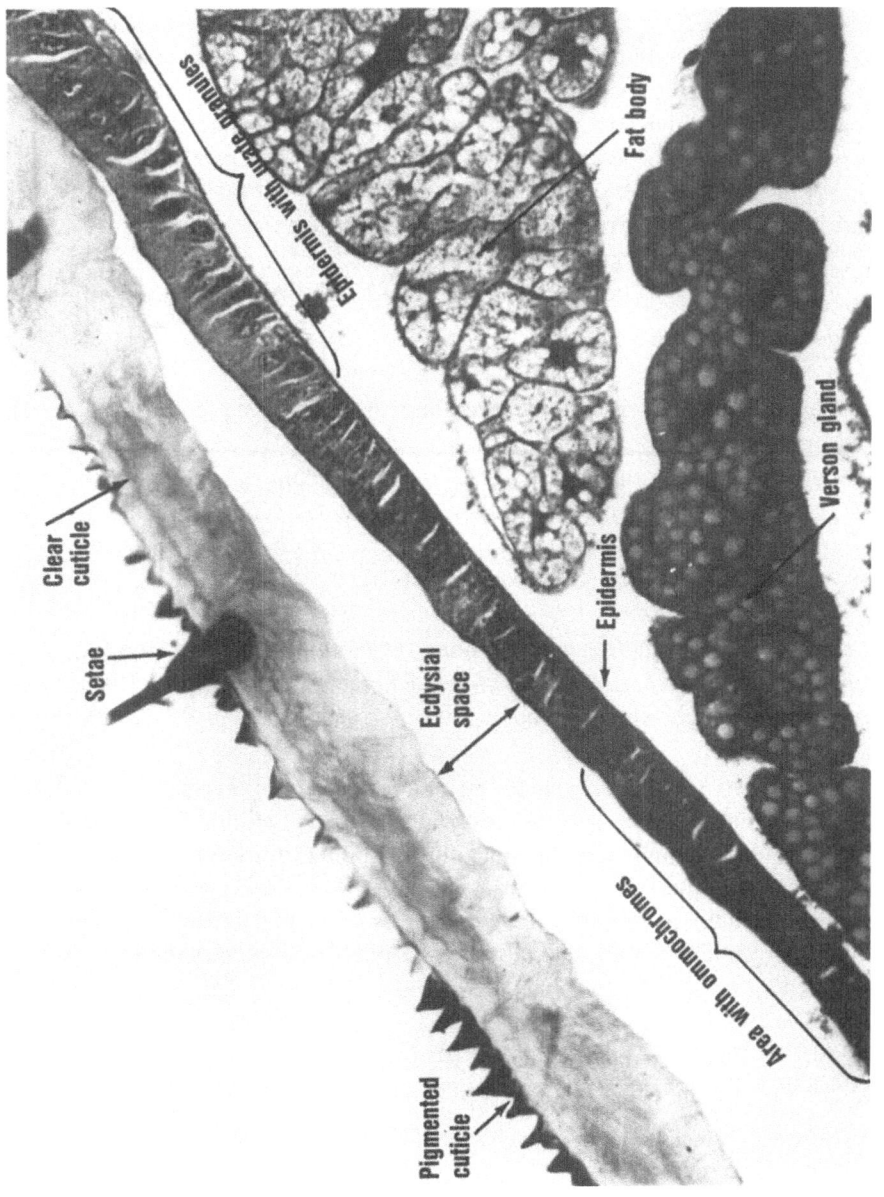

Plate VII Section through cuticle of fifth instar larva of *P. brassicae* (430 g) showing Verson's gland (Courtesy of Jacques and Denise Lhonoré, Université Pierre et Marie Curie, Paris)

191

cytoplasm but during the prepupa small grains of acidic mucopoly-saccharides appear. The glands become large and very active in the fifth instar but disappear in the pupa.

The precise role of the Verson's glands is still undetermined. It has been established that they do not participate directly in excretion (Lhonoré, 1978, pers. comm.). Barbier (1971) found that Verson's glands contribute to the stability of the base of the endocuticle and it is thought that a factor in the secreted fluid stimulates the metabolism of the epidermal cells particularly in the processes of pigmentation. In *Diataraxia* larvae the glands produce a secretion which is poured over the surface of the wax and helps in the formation of the cement layer (Way *in* Wigglesworth, 1967).

Glands of the imago

Abdominal glands. Rothschild believed that the ovipositing females may well produce a pheromone-like substance from a gland in the tip of their abdomen which they may use as a marker (Rothschild & Schoonhoven, 1977) (see also Chapter 4). It is noteworthy that Rothschild & Schoonhoven (*loc. cit.*) also demonstrated that ova give off an odour which was thought might originate in the ova.

Androconia. The structure and function of the modified scales on the wing called androconia are dealt with in Chapter 12 in respect of the pheromones which they produce.

Annex glands. The annex glands are associated with the male genital apparatus but have not yet been designated with a function. Transverse sections of these glands were figured coincidentally to Lhonoré's study of Malpighian tubules in *P. brassicae* (1976, 1980) and were shown to contain calcium, potassium and phosphorus probably localised in the birefringent microcrystals and calcium phosphate and magnesium in the secretion.

Proboscis glands. Eastham & Eassa (1955) identified about 150 pairs of unicellular pear-shaped gland cells which lie in the anterior dorsal angle of the haemocoel of each galea of the proboscis. The occurrence of these glands was not known elsewhere in the lepidoptera by these authors. As the glands are in a position in which they would be subjected to pressure it was suggested that they functioned either in sealing up the gaps between the dorsal plates, or in lubricating the movement of the plates as they slide over each other.

Blood system

Haemolymph constituents

Information on the haemolymph of *P. brassicae* comes from a variety of sources: on resistance to parasitism and pathogens (Lartschenko, 1933), on bacterial studies (Ratcliffe, 1975, Misyalunene, 1976a,b) and on defense reactions (Gysels, 1975). Breugnon & Le Berre (1976) carried out extensive investigations into the cells of the haemolymph. It is fairly evident from the literature that there has not been any consistency in the classification of the various cells of the haemolymph; there have been four classifications of *P. brassicae* cells according to size and structure, three of which were published within four years of each other.

Klaudia Lartschenko (1933) investigated the resistance of *P. brassicae* larvae to the hymenopterous parasite *Hyposotor ebenina* (Gravenhorst) with particular attention to the effect on the mesenchyme cells, and noted the presence of four different cells: young proleucocytes and amaebocytes (6–8 μ in diameter), phagocytes (8–12 μ) and cells of "another" type with much plasma and a small nucleus.

More recently Ratcliffe (1975) recorded five types of haemocytes in *P. brassicae in vitro*: a) spherule cells which are the smallest cells present and are the most common (10 μ in diameter). Each contains 5–20 inclusions (1–3 μ in diameter) and after a short time *in vitro* they aggregate into groups of 2–20$^+$ cells. These cells are not like the granular cells typical of other species' which are larger and with more inclusions. They contain mucosubstances; b) plasmocytes – these are the largest cells (10–25 μ in diameter) with fanlike pseudopodia *in vitro* and are less prone to aggregation (see Ratcliffe, 1975 for diagrams of these); c) small numbers of prohemocytes, granular cells and oenocytes are also present.

Misyalunene's (1975, 1976a,b) study of the haemolymph of *P. brassicae* also recognised five types of cell which he classified as: proleucocytes, macronucleocytes, micronucleocytes, enocytes and phagocytes. In *P. brassicae* imagines there were only micronucleocytes and phagocytes. Both serum and haemocytes were rich in protein, polysaccharide and lipids, which are presumably food reserves.

Breugnon & Le Berre (1976) drew attention to the variability in the shape of the plasmocytes and observed several morphological types. Other classes of haemocytes found by these workers were prohaemocytes, coagulocytes, spherulé cells and granulocytes (see also their diagrams). Mean density of the cells was found to increase during the fourth instar and prenymphal stage due to multiplication of prohaemocytes.

Haemolymph volume

This may be measured either by the difference between body weight before and after removal of haemolymph and adjusting for specific gravity (Chippendale & Kilby, 1969), or by using coloured dyes (Breugnon & Le Berre, 1976; cf. references in Turunen & Junnikkala, 1974).

A decrease in the haemolymph volume has been recorded for *P. brassicae* at the larval–pupal ecdysis (Chippendale & Kilby, 1969; Turunen & Junnikkala, *loc. cit.*; Nicolson, 1976a,b,c). This decrease is from $114 \pm 2\mu$l in the prepupa to $74 \pm 4 \mu$l in a freshly ecdysed pupa (Chippendale & Kilby, *loc. cit.*).

Turunen & Junnikkala (*loc. cit.*) made quantitative measurements of the haemolymph volume of male and female larvae during the wandering stage but found no significant differences between them. The average haemolymph volume for a larva weighing 451 mg was found to be 143 μl using a sample of 30.

A direct relationship exists between larval haemolymph volume and body weight (Figure 7.10).

The haemolymph volume was shown to increase from 20 microlitres to 160 microlitres from the beginning of the fourth instar up to the end of the fifth instar (Breugnon & Le Berre, 1976). The rate of increase of the haemolymph volume was also shown to match the increase in fresh weight of the larva.

Preventing blackening of the haemolymph

If haemolymph is kept exposed to the air for any length of time it readily oxidises and turns black. In order to prevent blackening several methods

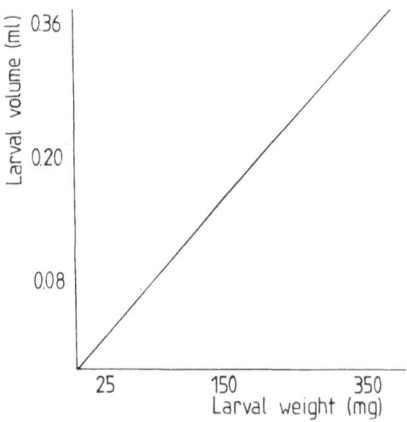

Fig. 7.10 Haemolymph volume by weight (after Miah, 1976)

have been employed: keep the haemolymph below 20°C or add EDTA (Post, 1972), add 10^{-4} M sodium diethyldithiocarbamate (Post, 1972) or phenylthiourea (van der Geest, 1968), use neutral cyanide solution (0.015 M KCN and 0.135 M NaCl) (Junnikkala, 1966), or collect haemolymph in a tube with chilled water deproteinised with either tungstic acid or by adding ethanol to a final concentration of 80% (Howden & Kilby, 1961).

Heart

The heart of the larva of *P. brassicae* is typically a tube which runs along the inner dorsal surface of the cuticle and which can be seen to pump haemolymph forwards along its length. Lartschenko (1933) figured a transverse section through the heart of the larva and it can be seen that it is rather flat-topped and is relatively deep in the dorso-ventral plane. Radiating filaments lead away ventrally and presumably link up with the pericardial membrane. As far as is known there has been no detailed study of the distribution of ostia, valves, septa and chambers or of the circulatory system of *P. brassicae*.

Central nervous system

Introduction

Particular attention has been paid to the brain and neurosecretion in *P. brassicae*; however, there are a number of specific nerves which have been described and these have been found to facilitate tactile, gustatory, visual and muscular systems. From Ali's (1967, 1973, 1974) detailed work on the central nervous system it is possible to identify 14 major nerves in *P. brassicae* (Table 7.4).

Several workers have studied the development of the brain of *P. brassicae* (Hanström, 1925; Heywood, 1965; Satija & Chopra, 1969, 1970; Ali, 1967, 1973, 1974; Srihari, 1972; Strausfeld, 1976). Ehnbom, 1948 investigated the central and sympathetic nervous systems of over 100 species of lepidoptera including *P. napi*.

Brain: Structure

Typically the brain of *P. brassicae* is made up of three major parts, the protocerebrum, the deutocerebrum and the tritocerebrum, bounded on the outside by a neurilemma made up itself of two parts, an outer non-cellular layer termed the neural lamellae, and an inner cellular perinerual layer, and a cortex containing ganglion cells, giant cells, globuli cells and

Table 7.4 Major nerves of *P. brassicae* with innervations (from Ali, 1973, 1974).

Stage of life cycle	Nerve	Innervation
(L=larva; I=imago)		
L (from sub-		
oesophageal ganglion)	Peripheral I	Mandibular segments
L	Peripheral II	Maxillary segments
L	Peripheral III	Labial segments
I (from brain)	Antennal (motor)	Extrinsic antennal muscles and muscles of first antennal segment
	Antennal (sensory)	Various sensillae of antennae
L,I	Labrofrontal	Labrum, frontal ganglia
L	Ocellar	Ocelli
I	Tegumentary (dorsal)	Small chaetosema just internal to compound eyes
I	Mandibular	Mandibular area
I	Galea	Tentorio-maxillary muscles
I	Maxillary	Tentorio-maxillary muscles
I	Labial	Labium
I	Optic	Ommatidia
Sympathetic nervous system		
L,I	Median Recurrent nerve	

cortical cells. The protocerebrum consists of protocerebral lobes, corpora pedunculata, organs of the central complex, corpora ventralis, pons cerebralis, and commisures joining the two halves of the brain (Satija & Chopra, 1970).

Behind the brain is the suboesophageal ganglion, three thoracic ganglia and eight abdominal ganglia in the larva. Each ganglion is linked by double connectives. The structure of the suboesophageal ganglion of the larva, prepupa and young pupa is very much the same in that it is a bilobed organ lying beneath the brain. Leading off from the sub-oesophageal ganglion (SOG) are a pair of glomerula bodies which extend from the middle of the ganglion to near the posterior end. It is interesting to note that Rehm (1955) stated that the suboesophageal ganglia were absent in *P. brassicae*.

According to the classification of insect brains by Strausfeld (1976) that of *P. brassicae* comes under category II, which is "two simple bipolar neurones from lobula to an optic focus on the ventro-lateral process".

Development

Hanström (1925) studied the brain of *P. brassicae* as a typical holometa-bolous insect, and made serial sections through larvae and imagines in an attempt to relate structure with function. He noted particularly the great changes that take place in the post-embryonic development of the optic ganglia.

Satija & Chopra (1970) showed that there were three major changes which take place in the post-embryonic development of the central nervous system; first, the length of the whole nerve cord shortens, second, some of the ganglia fuse and third, the optic ganglia increase in size dramatically in the imago. Many of the following details come from the illustrated work of these two workers.

The neurilemma is hardly distinguished in the first two instars but gradually thickens through the next three instars and remains constant in size in the pupa. The extent of the cortex diminishes during development as the size of the medulla increases. Both giant cells and globuli cells are clearly seen in the fifth instar.

In the first instar larva, the protocerebrum divides into three lobes and is connected to the deutocerebrum. However, this connection disappears at the end of the first instar and the tricerebrum decreases in size until in the pupal stage it is insignificant.

The suboesophageal ganglia changes in shape considerably in the imago where it fuses with the brain and the neurones become concentrated on the ventral and lateral sides of the ganglia (Srihari, 1972). The olfactory glomeruli develop quite late and appear first in the 72 hour pupa. At this time the imaginal antennal nerve grows into the lobe.

Two long papers on the development of the central nervous system (CNS) of *P. brassicae* were published by Firdausia Ali (1973, 1974), who studied at Imperial College, London. In his first paper he reviewed the earlier literature on CNS development in all insects and stressed the point that most work had been done on the brain rather than the ventral ganglia.

Central to his particular study Ali (1973) found that during development of the larva the length and number of the ganglia became reduced. Essentially the SOG fuses with the brain, and the first and second abdominal ganglia fuse with the second and third thoracic ganglia, and the sixth abdominal ganglion fuses with the seventh and eighth.

Ali traced the development of each type of glial cell and presented data on the size of the nuclei with age and also gives diagrams of serial transverse sections through the ganglia during all stages of metamorphosis.

The three types of neurones which occur in *P. brassicae* were found to increase in size with development; medium sized neurones increased from 4.4 μ at first instar to 8.3 μ in the imago. Similarly, large sized neurones increased from 6.0 to 12.5 μ and associated neurones increased from 4.0 to 5.3 μ. In the larva, the neurones were found to have two additional types of cells, beaker cells and optic ganglion cells.

The imaginal and larval brain differ only in the structure of the optic lobes and olfactory glomeruli. The optic lobes are principally associated with the compound eyes, and are made up of the following: a) ommatidia, b) basement membrane of postretinal fibres and tracheae, c) nerve bundles layer, d) lamina ganglions, e) external chiasma, f) medulla externa, g) internal chiasma, and h) medulla interna. The ommatidia have also been discussed in Chapter 12.

Corpora pedunculata (CP)

The most noticeable feature of the brain of *P. brassicae* is the corpora pedunculata (Ali, 1974). In the fifth instar this is a paired structure which lies on either side of the brain and occupies most of the protocerebrum. On each side are three beakers or calyces, one lies anterior and slightly dorsolaterally, and the other two are close together, one posteriorly and the other dorsal in position. Ali (1974) provided a very detailed account of the development of the CP together with diagrams of serial sections taken through the head. The CP of the larva and imago are very much the same in structure.

There are seven well defined connectives of the CP with the brain in the imago (Ali, 1974): a) transverse processes branching into calyces; b) olfactory nerve tracts (possibly); c) inner olfactory globular tracts (probably); d) nerve tracts from calyses to deutocerebrum; e) strong tracts to accessory lobes; f) tract to medulla interim and g) fibres connecting posterior calyces to suboesophageal ganglion. In lepidoptera and dictyoptera the CP is said to be well developed, but in *Pieris* and *Blaberis* the CP/Central Body Ratio is low (Howse & Williams, 1969).

Neurosecretion and other roles

In the study of neurosecretion activity of 23 species of lepidoptera (not including *P. brassicae*) Hincks (1971) found that on average there were about eight to ten types of neurosecretory cells in each species. As these were all found in the brain, and there was an apparent absence of these cells in the ventral ganglia Hincks thought that there had been complete cephalisation in all the species he studied. It is perhaps worth bearing in mind that Ali (1973) stated that not a great deal of work had been done on the ventral ganglia.

Ratios of certain areas of the brain are always useful for interpreting behaviour. Professor Grassé (1975) noted that the ratio of the optic lobes to the olfactory lobes of *P. brassicae* was high (80–2.3%) and this was therefore in keeping with diurnal insects. All non-social insects have low ratios of their mushroom bodies to their central bodies, and *Pieris* as well as *Mantis* sp. and *Chrysops* sp. are no exception (Howse, 1974, 1975).

Ali (1974) believed that the haemocytes were important in the breakdown of the neural lamella in the pupal stage, while he thought that the perineural cells were responsible for maintaining ionic relationships between nerves and haemolymph.

Although Kono (1975) studied neurosecretion in *P. b. crucivora*, much of his important work may be very relevant to *P. brassicae*.

Kono studied neurosecretion in three types of cells of the neurosecretory type-II cells which are present in each brain hemisphere (males were used only); these comprise two small cells, six large cells and two others which have vacuolated endoplasmic reticulum. From ultrastructural changes he found a secretory cycle which was essentially the same whether fifth instar larvae were reared in long or short day conditions. The cycle was as follows: day one there was reduced secretory activity; day two formation of organelles necessary for synthesis of secretory materials during rest of development; day three active synthesis and secretion of secretory material, and reversion to a reduced level of cell activity after the end of feeding.

Nerves of the feeding mechanism

While investigating the feeding mechanism of adult *P. brassicae* Eastham & Eassa (1955) noted the presence of three nerves. As the mandibles are very much reduced in the imago, it was not surprising to find that the mandibular nerve had disappeared. It originates from either side of the anterior surface of the SOG and innervates the retractor muscles, stipital muscles and up to the probosis tip. The labial nerve originates from the SOG and innervates the labial glands. Two other unnamed nerves arise near the SOG and innervate a) the sucking pump and b) the muscles of the sucking pump.

Sympathetic nervous system

This was studied by Ali (1974) in relation to the anterior region of the central nervous system, but not in any great detail. He showed that there was a median frontal ganglion which is joined by a bilateral labiofrontal connection to the tritocerebrum. A median recurrent nerve originated from the frontal ganglion. The corpora cardiaca innervates the dorsal vessel and sent nerves to the corpora allata. Attention must be drawn to the excellent review of Mme Raabe (1979), in which there are correlations on the neurohormonal systems of many insects in the Lepidoptera (but not including *P. brassicae*), Dictyoptera, Phasmida, Coleoptera, Diptera, Hymenoptera and Hemiptera.

Muscular system

Introduction

A small number of workers have given quite detailed information on the muscles of *P. brassicae*, ranging from descriptive accounts of the central nervous system (Heywood, 1965; Eassa, 1967) to accounts of more specialised regions – segmental muscles of the fifth instar larva (Auber-Thomay & Srihari, 1973) and muscles of the head appendages and cephalic stomadaeum of the prepupa and pupa (Eassa, 1967). Eastham & Eassa (1955) investigated the muscles of the proboscis and how proboscis extension is effected.

Flight muscles have received more recent attention, although no one appears to have published a descriptive account of their arrangement, particularly of the vertical, longitudinal and direct flight muscles of the thorax. The Italians Camatini & Saita (1969) made an extensive study of the flight muscles of *P. brassicae* and both Donnellan & Beechey (1969) and Njio & Peik (1977) investigated the effect of certain inorganic ions in the electrophysiology of the muscles.

It may be useful to refer to the works of Ehrlich & Ehrlich (1962, 1963) who studied the head, thorax and abdominal musculature of 76 species of butterfly including two *Pieris* species, for comparative information on *P. brassicae*.

Arrangement

Forty-seven muscles have so far been identified in *P. brassicae* and these are associated with the abdomen of the larva, the head, thorax, abdomen, the proboscis, and sucking pump of the imago (Table 7.5).

Diagrams of the arrangements of muscles *in situ*, often with accounts of origins and insertions of muscles, have been published of the ventral longitudinal muscles of the prepupa (Heywood, 1965) and of the head showing muscles of the antennae and proboscis (Eastham & Eassa, *loc. cit.*). Details of the wing joints of *P. brassicae* were given by Petersen (1966).

Larval muscles

Srihari (1972) distinguished between dorsal and ventral muscles of the fifth instar larva, which are nearly all longitudinal and intersegmental. The muscles in the thorax, particularly the prothorax, were found to be very complex. Lateral muscles are found between the two series of ventral muscles.

Table 7.5 Muscles of *P. brassicae* (several sources).

LARVA	**IMAGO** (*continued*)
Brain muscles	*Sucking pump muscles*
Dorsal muscles	a) controlling entry and exit of
Mandibular muscles	fluids
Suspensory muscles of oesophagus	Labral compressor muscles
Ventral muscles (lateral and central)	Transverse muscles
	Other muscle fibres
	b) the compressors
PUPA	Diagonal muscle bands
Central muscles	Lateral longitudinal fibres
Dorsal muscles	Longitudinal muscles
Flight muscles	Transverse muscle of the
Lateral muscles	posterior stipital muscle
Neck muscles	c) the dilators
Ventral muscles	Anterior cibarial dilators
	Lateral cibarial dilators
	Posterior pharyngeal dilators
IMAGO	*Abdomen*
Head	Circular muscles of the ejaculatory
Antennary muscles	bulb
Anterior stipital adductor muscles	Lateral muscles
Cranial adductor muscle of stipes	Valve muscles
Posterior tentorial stipital adductor	Ventral muscles
muscle	*Muscles of the genitalia*
Proboscis	Fultura ⎫
Elevator muscles of the galea base	Penis ⎪
Dorsal and ventral elevator muscles	Uncus ⎬ male
Primary oblique muscles	Valves ⎭
Proximal elevators of proboscis	Anal papillae ⎫ female
Retractor muscles	lateral lobes ⎬ ovipositor
Secondary oblique muscles	vaginal lobes ⎭
Stipital muscles	

During the first three days of the pupal life most of the larval muscles degenerate, and are replaced by the imaginal muscles. Some of the dorsal, ventral and lateral muscles remain until the first few days of imaginal life.

Auber-Thomay & Srihari (1973) investigated the ultrastructure of the ventral intersegmental muscles of the first abdominal segment of the fifth instar larva and pupa of *P. brassicae*. During the first two days the abdominal muscles fibres are 0.7–1.4 mm in length with a basement membrane 1500–2000 Å thick, innervated with trachea and nerves. Mitochondria were found on the myofibrils. The variation in length of the myosin filaments during development may be seen in Figure 7.11.

Auber-Thomay & Srihari (*loc. cit.*) found that there were three distinct phases in development of the myofibrils: a) a growth phase which occurs when the myofilaments lengthen and the number of sarcomeres increase,

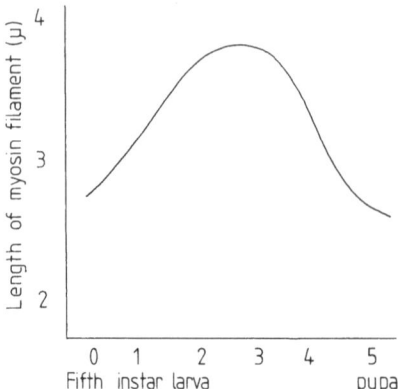

Fig. 7.11 Length of myosin filaments in longitudinal muscles of first abdominal segment of *P. brassicae* larvae
(after Auber-Thomay & Srihari, 1973)

b) a decreasing phase which occurs at the beginning of the pupal stage, in which the fibres shorten and the sarcomeres finally supercontract, and c) a degeneration phase which occurs at the end of the pupal stage, when fibrillar material undergoes lysis, the thick filaments disappear first, the Z discs last and microtubules appear in the whole sarcoplasm and lengthen. Simultaneously, the transverse system swells and forms large vacuoles which are probably filled by degraded fibrillar material from the muscle cells.

The ventral longitudinal muscles of the prepupa, which originate in the second thoracic segment and which run through to the ninth abdominal segment have been figured by Heywood (1965, figures 1 & 2) who was in fact investigating nerves at the time. Diagonal muscles of the thorax are also shown in the prepupa and imago.

Ultrastructure

The ultrastructure of *P. brassicae* muscles does not appear to be any different from that of other insect muscles, in that it is made up of myofibrils, sarcoplasm, sarcoplasmic reticulum, H, I and Z bands, with mitochondria with basement membranes and glycogen granules (Camatini & Saita, 1969). These workers compared the muscles of *P. brassicae* with the moth *Triphaena pronuba* L. and showed that the ratio of the thin/thick filaments was greater, indicating that it was related to a different flutter frequency. Electronmicrographs of the longitudinal muscles of the imago have also been published by Njio & Piek (1977).

Electronmicrographs of mitochondria in intersegmental muscles of *P.*

brassicae were published by Auber-Thomay & Srihari (1973). From the biochemical point of view Donnellan & Beechey (1969) looked at the oxidation of glycerol-1-phosphate in flight muscle and Njio & Piek (1977) investigated the concentrations of sodium and potassium in the same.

Locomotion

Fifth instar larvae

When the larva is moving quite fast searching for food, waves of "contraction" originate from the last segment and visibly move forward up to segment six (or the segment with the first pair of prolegs). These waves originate about every three quarters of a second, even before the previous wave has reached the anterior end. Anterior to segment six the strength of the waves is dulled and is seen as a "touch reaction" across the truelegs. The "touch reaction" is also seen in the prolegs. As one leg moves forwards it touches the posterior part of the leg in front which is then stimulated to move forward itself, thus touching another.

While searching for food the larva can cover about 30 cm per minute. During these exploratory movements the larva will often stop and raise its head and body as far back as segment six, and reach up from side to side as if looking for another possible place to wander. On arrival at food the larva often stops immediately and feeds voraciously at the first contact point with the food.

Imago: Wing beating

P. brassicae is recorded by Wigglesworth (1967) and Chovet (1977) to beat its wings 9–12 times a second but Camatini & Saita (1969) say that they have observed the muscles to contract at a frequency of 6 times a second corresponding to wing beats. On migration *P. brassicae* has been recorded as flying between 1.6–10 metres per second (Table 11.1). Demoll (1918) experimented on *P. brassicae* by progressively cutting off more surface area of the hind-wing but this did not hinder flight horizontally even when the hindwings were cut off completely. He assumed therefore that the forewing has a greater degree of specialisation for flying than the hindwing.

Unlike other butterfly species *P. brassicae* does not glide but has a scarcely visible glide phase at each wing beat (Nachtigall, 1976). It does however, hover and Weis-Fogh (1976) has observed the "clap" motion at the top and bottom of the horizontal stroke. Air is sucked between the forewings during the upstroke and a vortex is created between the hindwings during the downstroke in *P. brassicae* (as in other butterflies) Ellington "Horizon" BBC, January 1978). The rate of flap in *P. brassicae* is equivalent to 400–500 Reynolds numbers in mid-stroke, with lower levels at start and finish (Weis-Fogh, 1976).

Electrophysiology of flight

Physiological aspects of flight were studied by Donnellan & Beechey (1969) who extracted flight muscle mitochondria and submitochondrial particles of *P. brassicae*, after which they subjected them to different inorganic ions. At high concentrations of glucose-1-phosphate the requirement for calcium, strontium and magnesium followed a normal Michaelis-Menton pathway.

Njio & Piek (1977) studied the distribution of sodium ions in the transverse tubular system (TTS) of *P. brassicae* flight muscles, and using various chemicals which precipitated the sodium salt in the cell they were able to examine the distribution of these ions. Sodium was found in the mitochondria, on the sarcolemma, on the sarcoplasmic reticulum, in and out of the lamina basalis and in the myofibril. Evidence suggested that there was more sodium in the TTS lumen than outside the plasma membrane.

References cited

Ali, F.A., 1967. *The postembryonic development of the central nervous system in Pieris brassicae (Lepidoptera: Pieridae)*. Ph.D. thesis. University of London (Imperial College).
—— 1973. Post-embryonic changes in the central nervous system and perilemma of *Pieris brassicae* (L.) (Lepidoptera: Pieridae). *Trans. R. ent. Soc. Lond.* (A) 124: 463–498.
—— 1974. Structure and metamorphosis of the brain and suboesophageal ganglion of *Pieris brassicae* (L.) (Lepidoptera: Pieridae). *Trans. R. ent. Soc. Lond.* 125: 363–412.
Allegret, P., 1956. *Étude des glandes séricigènes des larves de Lépidoptères. Leur rôle dans la physiologie du développement*. Thèse Doctorat Sciences naturelles, Paris. pp. 345.
Allen, M.D., 1954. *The effect of plant mineral nutrition on leaf-eating insects*. M.Sc. thesis. University of London (Wye College).
André, J., 1959. Étude au microscope électronique de l'évolution du chondriome pendant la spermatogenèse du papillon du chou, *Pieris brassicae. Ann. Sci. nat. zool. biol. Anim.* 1: 283–307.
Angersbach, D., 1970. *Die Bedeutung der Einfallsrichtung und der Wellenlänge des Lichtes bei der Melanisierung der Kohlweisslingspuppe, Pieris brassicae* L. Dissertation. University of Giessen.
Auber-Thomay, M. & Srihari, T., 1973. Évolution ultrastructural des fibres

musculaires intersegmentaires chez *Pieris brassicae* (L.) pendant le dernier stade larvaire et la nymphose. *J. Microsc.* 17: 27–36.

Auel, H., 1937. The influence of weather in the size of *Pieris brassicae*. *Z. angew. Ent.* 23: 596–602.

Banks, E., 1930. The relation of weight to wing area in the flight of animals. *J. Malay. br. asiat. Soc.* 8: 334–360.

Bänziger, H., 1971. Extension and coiling of the lepidopterous proboscis and new interpretation of the blood-pressure theory. *Mitt. Schweiz. ent. Ges.* 43: 225–239.

Barbier, R., 1971. *Recherches sur la morphogenèse tegumentaire d'un insecte holométabole: Galleria mellonella L. (Lepidoptera: Pyralidae).* Thèse de Doctorat de Science. Université de Rennes.

Bartell, R.J., 1966. *Studies on the water loss from the eggs of lepidoptera.* Diploma. University of London (Imperial College).

Bauer, H., 1967. Kinetic organisation of lepidopteran chromosomes. *Chromosoma* 22: 101–125.

Beirne, B.P., 1942. Notes on the morphology and taxonomy of the genitalia of the British Rhopalocera. *Entomologist* 75: 211–216.

Beliaieff, N.K. 1930. Die Chromosomenkomplexe und ihre Beziehung zur Phylogenie bei den Lepidopteren. *Z. Ind. Anst. Bererbgsl.* 54: 369–399.

Benz, G., 1970a. L'influence des stimuli externes sur la gamétogenèse des insectes. *Coll. int. natn. Recherches Sci.* (189).

—— 1970b. Stimulation of oogenesis in *Pieris brassicae* by juvenile hormone derivative tarnesenic acid ethyl ether. *Experimentia* 26 (9) 1012.

Bernardi, G., 1958. Taxonomie et zoogeographie de *Talbotia naganum* Moore. *Rev. fr. Ent.* 25: 125–128.

Bigger, T.R.L., 1960. Chromosome numbers in Lepidoptera. *Ent. Gaz.* 11: 149–152; 12: 85–89.

—— 1975. Karyotypes of some lepidopteran chromosomes and changes in their holokinetic organisation as revealed by new cytological techniques. *Cytologia* 41: 713–726.

—— 1976. Karyotypes of three species of lepidoptera including an investigation of B-chromosomes in *Pieris. Cytologia* 41: 261–282.

Blunck, H., 1953. *Tierische Schädlinge an Nutzpflanzen.* Lepidopteren und Trichopteren. Berlin. p. 2.

Bordas, L., 1971. L'appareil digestif et les tubes de Malpighi des larves des Lépidoptères. *Ann. Sci. nat. 9th Series. Zool.* 14: 191–263.

Bounhniol, J.J., 1938. Recherches expérimentales sur le déterminisme de la métamorphose chez les Lépidoptères. *Bull. biol. Fr. Belg. Suppl.* 24: 1–199.

Bowden, S.R., 1966. "Irregular" diapause in *Pieris. Proc. S. Lond. ent. nat. Hist. Soc.* 67: 67–68.

Breugnon, M-M. & Le Berre, J.R., 1976. Fluctuations of the haemocyte formula and haemolymph volume in the caterpillar of *Pieris brassicae* L. *Ann. zool. ecol. Anim.* 8: 1–12.

Buparai, J.S., 1968. Observations on the sex ratio in *Pieris. Res. Bull. (NS) Punjab Univ.* 19: 1–2.

Cals, P., 1970. Discontinuité morphogénétique dans la differentiation postembryonnaire des ailes des Lépidoptères: *Bombyx mori* L. (Bombycidae) et *Pieris brassicae* L. (Pieridae). Note. *C. . Acad. Sci. Paris* 270: 2666–2669.

Camatini, M. & Saita, A., 1969. Analysis of the ultrastructure of the flight muscles of *Noctua pronuba* and *Pieris brassicae. Rend. Sc. fis. mat. e nat. Lincei.* 46: 745–752.

Chippendale, G.M. & Kilby, B.A., 1969. Relationship between the proteins of the haemolymph and fat body during development of *Pieris brassicae*. *J. Insect Physiol.* 15: 905–926.

Chovet, G., 1970. Analyse du déclenchement de la copulation chez *Pieris brassicae* (Lepidoptera: Pieridae). *C. r. Acad. Sci. Paris* 270: 2832–2835.

—— 1974. *Méchanismes d'accouplement, structure et fonctionnement de l'appareil reproducteur mâle de Pieris brassicae (L.) (Lepidoptera: Pieridae)*. 1–121. Thèse, Docteur 3 ème cycle. Université de Paris VI.

—— 1977. Les stimulus visuels dans le déclenchment de la parade nuptiale chez *Pieris brassicae* L. (Lepidoptera: Pieridae). Nouvelle interprétation de certains comportements grégaires. *C. r. Acad. Sci. Paris* 284: 2127–2130.

Church, H.F., 1957. Sex ratios in the Pierinae. *Entomologist* 90: 19–20.

—— 1958. Sex ratios in the Pierinae. *Entomologist* 91: 46–48.

Cretschmar, M., 1920. Besprechung über die Farbe der Puppen der *Pieris brassicae* und die Schutzfärbung. *Int. ent. Z.* 14: 125–127.

Dauberschmidt, K., 1933. Vergleichende Morphologie des Lepidopterendarmes und seiner Anhänge. *Z. angew. Ent.* 20: 204–267.

David, W.A.L. & Gardiner, B.O.C., 1952. Laboratory breeding of *Pieris brassicae* L. and *Apanteles glomeratus* L. *Proc. R. ent. Soc. Lond.* (A) 27: 54–56.

—— 1961. The mating behaviour of *Pieris brassicae* (L.) in a laboratory culture. *Bull. ent. Res.* 52: 263–280.

Demoll, R., 1918. *Der Flug der Insekten und der Vögel*. Fischer, Jena.

Dethier, V.G., 1941. The antennae of lepidopterous larvae. *Bull. Mus. comp. zool. Harvard* 87: 455–507.

Doncaster, L., 1912a. The chromosomes in the oogenesis and spermatogenesis of *Pieris brassicae* and in the oogenesis of *Abraxas grossulariata*. *J. Genet.* 2: 189–200.

—— 1912b. Note on the chromosomes in oogenesis and spermatogenesis of the white butterfly, *Pieris brassicae*. *Proc. Camb. phil. Soc. math. phys. Sci.* 16: 491–492.

Donnellan, J.F. & Beechey, R.B., 1969. Factors affecting the oxidation of glycerol-1-phosphate by insect flight muscles mitochondria. *J. Insect Physiol.* 15: 367.

Döring, E., 1955. *Zur Morphologie der Schmetterlingseier*. Akademie-Verlag, Berlin.

Drohsinn, J., 1933. Über Art- und Rassenunterschiede der männlichen Kopulationsapparate von Pieriden (Lepidopteren). Inaugural Dissertation. Halle, Wittenberg; und *Ent. Rundschau* 50: pp. 135.

Dusaussoy, G. & Delplanque, A., 1964. L'élevage de *Pieris brassicae* L. en toutes saisons: accouplement et ponte en conditions artificielles. *Rev. path. veg. ent. ag.·i. Fr.* 43: 119–134.

Eassa, Y.E.E., 1953. The development of imaginal buds in the head of *Pieris brassicae* Linn. (Lepidoptera). *Trans. R. ent. Soc. Lond.* 104: 39–50.

—— 1963. Metamorphosis of the cranial capsule and its appendages in the Cabbage Butterfly, *Pieris brassicae*. *Ann. ent. Soc. Amer.* 56: 510–521.

—— 1967. Musculature of the head appendages and cephalic stomodaeum of the prepupae and pupa of *Pieris brassicae*. *Bull. ent. Soc. Egypt* 51: 63–70.

Eastham, L.E.S. & Eassa, Y.E.E., 1955. Feeding mechanism of the butterfly, *Pieris brassicae* L. *Phil. Trans.* B. 239: 1–43.

Ehnbom, K., 1948. Studies on the central and sympathetic nervous system and some sense organs in the head of neuropteroid insects. *Opusc. ent. Supp.* 8: 4–162.

Ehrlich, P.R., 1958. The integumental anatomy of the Monarch Butterfly *Danaus plexippus* L. (Lepidoptera: Danaidae). *Kansas Univ. sci. Bull.* 38: 1315–1349.

Ehrlich, P.R. & Ehrlich, A.H., 1962. The head musculature of the butterflies (Lepidoptera: Papilionoidea). *Microentomology* 25: 1–89.

—— 1963. The thoracic and basal abdominal musculature of the butterflies (Lepidoptera: Papilionoidea). *Microentomology* 25: 91–126.

Eitschberger, U., 1969. Die Unterscheidungsmerkmale der europäischen Arten der Gattung *Pieris* Schrank. *Atalanta* 2: 211–223.

El-Dakroury, M.S.I., 1972. *A study of the larval photoreceptors of Pieris brassicae.* Ph.D. thesis. University of London (Queen Mary College).

Federley, H., 1936. Sex linked hereditary cancer in lepidopterous larvae. *Hereditas* 22: 193–216.

—— 1938. Chromosomenzahlen finnländischer Lepidopteren I. Rhopalocera. *Hereditas* 24: 397–464.

—— 1942. Chromosomenzahlen von vier Tagfaltern von Ozeanischen Inseln. *Hereditas* 28: 493–495.

Feltwell, J.S.E., 1973. *The Metabolism of carotenoids in Pieris brassicae and in its foodplant, Brassica oleracea var. capitata (Cabbage).* Ph.D. thesis. University of London (Royal Holloway College).

—— 1978. Carotenoids in the class Insecta. *In: Plant and Animal Evolution.* Ed. G. Harborne. Academic Press, London and New York. 277–307.

Feltwell, J.S.E. & Rothschild, M., 1974. The carotenoids of 38 species of lepidoptera. *J. Zool. Lond.* 114: 441–465.

Forbes, W.T.M., 1909. On certain Pierid caterpillars. *Psyche* 16: 69–75.

Ford, E.B., 1945. *Butterflies.* Collins, London.

Frantsevitch, L.I. & Pichka, V.E., 1976. The size of the binocular zone of the visual field in insects. (In Russian.) *Zh. evol. Biokhim. Fiziol.* 12: 461–465.

Frohawk, F.W., 1934. *British Butterflies.* Ward, Lock & Co. Ltd. London & Melbourne.

Frühstorfer, H., 1908. Neues über die Genitalorgane der Pieriden. *Ent. Z.* 22: 198–202.

Gardiner, B.O.C., 1963. Genetic and environmental variation in *Pieris brassicae.* *J. Res. Lepid.* 2: 127–136.

Gonin, J., 1894. Recherches sur la métamorphose des Lépidoptères (de la formation des appendices imaginaux dans la chenille du *Pieris brassicae*). *Bull. Soc. vand. Sci. nat.* 30: 89–139.

Graham-Smith, G.S. & Graham-Smith, W., 1930. *Pieris brassicae* L. with special reference to aberrations from Aberdeenshire. *Entomologist's Rec. J. Var.* 41: 157–161, 173–180; 42: 17–22.

Grassé, P.P., 1975. *Traité de Zoologie.* Masson, Paris. Vol. 8.

Green, E.E., 1927. An abnormal pupa of *Pieris brassicae.* *Proc. ent. Soc. Lond.* 2: 86.

Grote, A.R., 1900. The descent of the Pierids. *Proc. Am. phil. Soc.,* 39: 5–67.

Gysels, H., 1975. Electrophoretical and histochemical investigations on some noxious Lepidoptera and on the impact of pesticides upon *Ephemera dancia*, a water-dwelling, innocuous mayfly (Ephemeroptera: Ephemeridae). *Acta Zool. Pathol. Antwerp* 62: 129–141.

Hanström, B., 1925. Comparison between the brain of the caterpillar and the imago in *Pieris brassicae.* *Ent. Tidskr.* 46: 43–52.

Henking, H., 1890. Untersuchung über die ersten Entwicklungsvorgänge in Einem der Insekten. *Z. wiss. Zool.* 49: 503–564.

Henson, H., 1932. Development of the alimentary canal in *Pieris brassicae* and the endodermal origin of the Malpighian tubules of insects. *Q. Jl. microsc. Sci.* 1932: 283–305.

—— 1937. The structure and post-embryonic development of *Vanessa urticae* (Lepidoptera). II The larval Malpighian tubules. *Proc. zool. Soc. Lond.* B. 107: 161–174.

Hepburn, H.R., 1971. Proboscis extension and recoil in lepidoptera. *J. Insect Physiol.* 17: 637–656.

Herold, M., 1815. *Entwicklungsgeschichte der Schmetterlinge, anatomisch und physiologisch bearbeitet. Atlas des Planches.* Krieger, Cassel & Marburg.

Heywood, R.B., 1965. Changes occurring in the central nervous system of *Pieris brassicae* L. (Lepidoptera) during metamorphosis. *J. Insect Physiol.* 11: 413–430.

Higgins, L.H., 1930. A lepidopterological excursion to Piedmont. *Entomologist* 63: 125–127.

—— 1975. *The classification of European Butterflies.* Collins, London.

Hincks, C.F., 1971. A comparative survey of the neurosecretory cells occurring in the adult brain of several species of lepidoptera. *J. ent.* (A) 1: 13–26.

Hinton, H.E., 1946. On the homology and nomenclature of the setae of lepidopterous larvae, with some notes on the phylogeny of the lepidoptera. *Trans. R. ent. Soc. Lond.* 97: 1–37.

—— 1981. *The biology of insect eggs.* Pergamon Press, London.

Hoff, W., 1919. Der Flug der Insekten und der Vögel. *Naturwissenschaften* 7: 159–162.

Homma, T., 1954. A comparative study of the alimentary canal in butterflies, with special reference to their systematic relationships. *J. Fac. Sci. Hokkaido Univ. Series.* VI. Zool. 12: 40–60.

Howden, G.F. & Kilby, B.A., 1961. Biochemical studies on insect haemolmph. II. The nature of the reducing material present. *J. Insect Physiol.* 6: 85–95.

Howse, P.E., 1974. Design and function on the insect brain. *In: Insect Behaviour.* Ed. L. Barton-Brown. Springer, Berlin. 180–194.

—— 1975. Brain structure and behaviour in insects. *A. Rev. ent.* 20: 359–379.

Howse, P.E. & Williams, J.L.D., 1969. The brains of social insects in relation to behaviour. *Proc. VI. Cong. IUSSI. Bern.* 59–64.

Jackson, W.H., 1890. IV. Studies in the morphology of the lepidoptera. Part I. *Trans. Linn. Soc. Lond.* 4: 143–176.

Johansson, A., 1959. Diapause and pupal morphology and colour in *Pieris brassicae.* L. (Lepidoptera: Pieridae). *Norsk. ent. Tidskr.* 9: 90–86.

Jordan, K., 1898. Contributions to the morphology of lepidoptera. *Nov. Zool.* 5: 381–382.

Junnikkala, E., 1966. Effect of braconid parasitization on the nitrogen metabolism of *Pieris brassicae* L. A study of the nitrogenous compounds in the haemolymph of larvae at different ages reared in controlled laboratory conditions. *Ann. Acad. Sci. Fenn.* 100: 1–83.

Karlinsky, A., 1962. Effets de l'ablation des corpora allata larvaires sur le développement ovarien de *Pieris brassicae* L. *C. r. Acad. Sci. Paris* 255: 191–193.

—— 1963a. Effects de l'ablation des corpora allata imaginaux sur le développement ovarien de *Pieris brassicae* L. (Lepidoptera). *C. r. Acad. Sci. Paris* 255: 191–193.

—— 1963b. Effects de l'ablation des corpora allata imaginaux sur développement ovarien de *Pieris brassicae* L. *C. r. Acad. Sci. Paris* 256: 4101–4103.

Karlinsky, A. & Poulaert, J., 1971. The changes in fat body during the imaginal life of *Pieris brassicae*. *Bull. Soc. zool. Fr.* 96: 453–466.

Kayser-Wegmann, I., 1976a. Differences in black pigmentation in lepidopteran cuticles as revealed by light and electron microscopy. *Cell. Tiss. Res.* 171: 513–521.

—— 1976b. Ultrastructural differences between larval and pupal cuticles of *Pieris brassicae* (Lepidoptera). *Protoplasma* 90: 319–331.

Keeley, L.L., 1978. Enocrine regulation of fat body development. *A. Rev. Ent.* 23: 329–352.

Kono, Y., 1975. Daily changes of neurosecretory Type II cell structure of *Pieris* larvae entrained by short and long days. *J. Insect Physiol.* 21: 249–264.

Kudrna, O., 1973. On the status of *Pieris cheiranthi* Hübner (Lepidoptera: Pieridae). *Ent. Gaz.* 24: 299–304.

—— 1973. *Artogeia* Verity, 1947, General Revision for *Papilio napi* Linnaeus (Lep: Pieridae). *Ent. Gaz.* 25: 9–12.

Lafont, R., Mauchamp, B., Boulay, G. & Tarroux, P., 1975. Development studies in *Pieris brassicae* (Lepidoptera: Pieridae) I. Growth of various tissues during the last larval instar. *Comp. biochem. physiol.* 51B: 439–444.

Lagnel, M., 1966. Note sur l'armure génitale mâle et femelle du sous-genre, *Pieris* Schrank. *Bull. Soc. ent. Fr.* 71: 91–94.

Lartschenko, K., 1933. Definite reaction to parasites of larvae of *L. stricticalis* and *Pieris brassicae*. *Z. Parasitenk.* 5: 679–707.

Lhonoré, J., 1976. Morphological and histochemical data on the imaginal Malpighian tubules of *Pieris brassicae* (L.) (Lepidoptera). *Ann. Sci. nat. zool. biol. Anim.* 18: 275–294.

—— 1978. Données histophysiologiques sur les accumulations minérales et puriques. *La Cellule* 72: 1–54.

Long, D.B., 1959. Observations on adult weight and wing area in *Plusia gamma* L. and *Pieris brassicae* L. in relation to larval population density. *Entomologia expl. appl.* 2: 241–248.

Ma, W.C., 1972. *Dynamics of feeding responses in Pieris brassicae L. as a function of chemosensory input: a behavioural, ultrastructural and electrophysiological study*. Ph.D. thesis. Wageningen.

Ma, W.C. & Schoonhoven, L.M., 1973. Tarsal contact chemosensory hairs of the large white butterfly, *Pieris brassicae* and their possible role in oviposition. *Entomol. exp. appl.* 16: 343–357.

Maeki, K. & Ae, S.A., 1966. A chromosome study of 28 species of Himalayan butterflies (Papilionidae, Pieridae). *Spec. Bull. lep. Soc. Japan* 107–120.

Main, H., 1937. Some points on the pupae of the large white butterfly. *Amat. ent. Soc.* 2: 45.

Martouret, D., Lhoste, J., Roche, A., 1965. Action sur le mesenteron de *Pieris brassicae* L. de la toxine de l'inclusion parasporale de *Bacillus thuringiensis* Berliner. *Entomophaga* 10: 349–365.

Mayer, A.G., 1896. The development of the wing scales and their pigment in butterflies and moths. *Bull. Mus. comp. zool. Harv.* 29: 209–236.

Meyrick, E., 1928. *Revised Handbook of British Lepidoptera*. Watkins & Doncaster, London.

Miah, M.A.H., 1976. *Some effects of gamma radiation in the different stadia of Pieris brassicae* L. Ph.D. thesis. University of London (Imperial College).

Minnich, D.E., 1924. The olfactory sense of the cabbage butterfly. *J. exp. Zool.* 39: 339–356.

Misyalunene, I.S., 1975. The morphology of the haemocytes in larvae of *Pieris brassicae* (Lepidoptera: Pieridae). (In Russian.) *Tsitologiya* 17: 647–652.

—— 1976a. Morphology and functional activity of cabbage butterfly haemocytes on various stages of postembryonic development (in Russian). *Liet. Tsr. Mokslu. Akad. Darb. Ser. C. biol. Mokslai* 2: 99–110.

—— 1976b. Changes in the morphology and proportions of the various types of cells of the hameolymph in larvae of the cabbage white butterfly during infection with Entobakterin. (In Russian.) *Tsitologiya* 18: 1220–1225.

Mukerji, G.P., 1961. On the biology of "cabbage white" *Pieris brassicae. J. Zool. Soc. India* 13: 121–127.

Müller, H., 1883. *The Fertilisation of Flowers*. Translated and edited by D'Arcy Thompson, London.

Nachtigall, W., 1976. Wing movements and the generation of aerodynamic forces by some medium-sized insects. *In: Insect Flight*. Ed. R.C. Rainey. *R. ent. Soc. Lond. Symp*. Blackwells, Oxford. 31–47.

Nakahara, W., 1917. Studies of amitosis; its physiological relation in the adipose cells of insects and its probable significance. *J. Morph*. 30: 483–526.

Nicolson, S.W., 1976a. Diuresis in the cabbage white butterfly, *Pieris brassicae*: fluid secretion by the Malpighian tubules. *J. Insect Physiol*. 22: 1347–1356.

—— 1976b. Diuresis in the cabbage white butterfly, *Pieris brassicae*, water and ion regulation and the role in the hind-gut. *J. Insect Physiol*. 22: 1623–1630.

—— 1976c. The hormonal control of diuresis in the cabbage butterfly, *Pieris brassicae. J. exp. Biol*. 65: 565–575.

Niculescu, E.V., 1963. Pieridae of Rumania. (In Rumanian.) *Fauna pop. Rom. Bucarest*. Insecta II.

Njio, K.D. & Piek, T., 1977. Localisation of sodium and potassium ions in a flight muscle of *Pieris brassicae. J. Insect Physiol*. 23: 919–929.

Noirot, C. & Quennedey, A., 1974. Fine structure of insect epidermal glands. *A. Rev. Ent*. 19: 61.

Novikoff, M., 1931. Untersuchungen über die Komplexaugen von Lepidopteren. *Z. wiss. Zool*. 138: 1–67.

Ogata, M., Okada, Y., Okagaki, H. & Sibatani, A., 1957. Male genitalia of lepidoptera: morphology and nomenclature III. Appendages pertaining to the 10th somite. *Ann. ent. Soc. Amer*. 50: 237–246.

Osborne, M.F., 1951. Aerodynamics of flapping flight with application to insects. *J. exp. Biol*. 28: 221–245.

Pardi, L., 1939. I corpi grassi degli insetti. *Redia* 25: 87–288.

Peterson, J., 1960. *Morphological and behavioural studies in resting position in lepidoptera*. Ph.D. thesis. University of London.

Pictet, A., 1922. Recherches sur l'hibernation de *Pieris brassicae* à l'état de chenille. *Bull. Soc. lépidopt. Genève* 5: 47–57.

Pieron, H., 1927. De la loi qui relie la surface des ailes au poids des individus dans une même espèce animale, et de quelques problèmes concernant la vol des insectes. *C. r. Acad. Sci. Paris* 184: 239–241.

Pilat, M., 1935. Action of insecticides on the intestinal tube of insects. *Bull. ent. Res*. 26: 165–180.

Plantevin, G. & Nardon, P., 1970. Histologie et activité sécrétoire de l'intestin moyen des larvaires de *Pieris brassicae* et *Galleria mellonella*. Evolution au cours de la mue larvaires et de la nymphose chez *Galleria mellonella. Ann. zool. ecol. Anim*. 2: 25–50.

Portier, P., 1949. *Biologie des Lépidoptères*. Lechavalier, Paris.

Portier, P. & Rorthays, D., 1926. Recherches sur la charge supportée par les ailes des lépidoptères de divers familles. *C. r. Acad. Sci. Paris* 183: 1126–1129.

Post, L.C. 1972. Bursicon: its effect on tyrosine permeation into insects. *Biochim. biophys. Acta* 290: 424–428.

Raabe, M., 1979. Les neurohormones des insectes. *In: Ecole Normale Supérieure Publications de Laboratoire de Zoologie*. No. 14. pp. 1–135.

Ratcliffe, N.A., 1975. Spherule cell test particle interactions in monolayer cultures of *Pieris brassicae* hemocytes. *J. Invert. Pathol.* 26: 217–223.

Rehm, M., 1955. Morphologische und histologische Untersuchungen an neurosekretorischen Zellen von Schmetterlingen. *Z. Zellforsch.* 42: 19–58.

Richards, O.W. & Davies, R.G., 1977. *Imms General Textbook of Entomology*. Chapman & Hall, London.

Robinson, R., 1971. *Lepidoptera Genetics*. Pergamon Press, Oxford & New York.

Rosa, A.F., 1924. A flying visit to Greece. *Entomologist* 57: 265–273.

Rothschild, M. & Schoonhoven, L.M., 1977. Assessment of egg load by *Pieris brassicae* (Lepidoptera: Pieridae). *Nature, Lond.* 266: 352–355.

Rowland-Brown, H., 1907. Butterflies observed during a short tour in southern France in May, 1907. *Entomologist* 40: 149–153.

Sarlet, L., 1949a. Eggs of Pierids. *Lambillionea* 49: 108.

—— 1949b. Iconographie des oeufs des Lépidoptères (Faunes de la Belgique). *Lambillionea* 53: 54–63.

Satija, R.C. & Chopra, P., 1969. Optic ganglia and oesophageal diverticulum in *Pieris brassicae. Res. Bull. Punjab Univ.* 20: 413–421.

—— 1970. Morphology and medullary organs of the brain of *Pieris brassicae. Res. Bull. Punjab Univ.* 21: 1–2, 27–34.

Sepp, J.C., 1762. *Nederlandsche Insecten*. Dag-Vlinders, Amsterdam.

Spuler, A., 1908. *Die Schmetterlinge Europas*. Vol. 1. Schweizerbartsche Verlag, Stuttgart.

Srihari, T., 1972. Anatomie des systèmes musculaire et nerveux et leur évolution au cours du développement post-embryonaire de *Pieris brassicae* (Lepidoptera: Pieridae). *Soc. zool. fr. Bull.* 97: 133–147.

—— 1973. *Croissance, évolution musculaire et actions hormonales au terme de développement post-embryonnaire.* Thèse de Doctorat des sciences naturelles. Université de Paris VI.

Strawinsky, K., 1929. Bielinek kapustnik -*Pieris brassicae* (Biologja oraz zwalczanie). *Bull. ent. Pologne* 8: 227–248.

Stekol'nikov, A.A., 1967. Phylogenetic relationships within the Rhopalocera on the basis of the functional morphology of the genital apparatus. (In Russian.) *Ent. Rev. USSR* 46: 1–11.

Strausfeld, N.J., 1976. *Atlas of an Insect Brain*. Springer-Verlag, Berlin.

Strausfeld, N.J. & Blest, A.D., 1970. Golgi studies on insects. Part 1. The optic lobes of lepidoptera. *Phil. Trans.* B 258: 81–134.

Street, M.L., 1975. *The growth and development of the reproductive system of Pieris brassicae L. and their control by hormones*. Ph.D. thesis. University of Newcastle upon Tyne.

Swammerdam, J., 1758. *The Book of Nature, of the history of insects, A history of an animal in an animal*. Seyffert, C.G., London.

Teotia, T.P.S. & Pathak, M.D., 1957. The anatomy of the larva of *Enarmonia pseudonectis* Meyr. (Eucosmidae: Lepidoptera). *Ann. Zool. Warsaw* 2: 65–85.

Tulluch, J.B.G., 1941. Mass movements of *Pieris brassicae* and *Pieris rapae.*

Entomologist 74: 32–35.

Turunen, S. & Junnikkala, E., 1974. Haemolymph fatty acids and lipid content in larval *Pieris brassicae* L. (Lepidoptera: Pieridae). *Ann. ent. fennici.* 40: 145–149.

Verity, R., 1905–1911. *Rhopalocera Palearctica. Iconographie et descriptions des papillons diurnes de la region palaéarctique.* Florence.

Waku, Y. & Sumimoto, K.I., 1974. Ultrastructure of Lyonet's gland in the silkworm (*Bombyx mori*). *J. Morph.* 142: 165–186.

Weis-Fogh, T., 1976. Energetics and aerodynamics of flapping flight: a synthesis. *In: Insect Flight.* Ed. R.C. Rainey. Royal Entomological Society of London Symposium. Blackwells, London. pp. 48–72.

Weiser, J., 1977. *An Atlas of Insect Diseases.* 2nd edition. Dr. W. Junk. The Hague, Netherlands.

Westwood, J.D., 1854. *The Butterflies of Great Britain.* W.S. Orr & Co., Paternoster Row, London.

Wigglesworth, V.B., 1928. Impressionist colouring among Lepidoptera. *Proc. ent. Soc. Lond.,* 3: 4.

—— 1945. Transpiration through the cuticle of insects. *J. exp. Biol.* 21: 97–114.

—— 1957. The physiology of insect cuticle. *A Rev. Ent.* 2: 37–54.

—— 1967. *The Principles of Insect Physiology.* Methuen & Co., Ltd., London.

Williams, C.B., 1958. *Insect Migration.* Collins, London. New Naturalist Series.

Williams, C.B., Cockbill, G.F., Gibbs, M.C. & Downes, J.A., 1942. Studies in the migration of lepidoptera. *Trans. R. ent. Soc. Lond.* 92: 101–283.

Yagi, N., 1956. Electron microscopical studies on the form of pterin pigments in the scale of Pieridae with reference to their bearing on systematics. *New Entomol.* 4: 1–20.

Yagi, N. & Koyama, N., 1963. *The compound eye of lepidoptera.* Shinko Press & Co., Tokyo, Japan. pp. 319.

Zylberberg, L., 1965a. Modifications intramitochondriales dans les cellules des follicules testiculaires de *Pieris brassicae. Ann. Sci. Nat. Zool.* 12 series. 7: 209–212.

—— 1965b. Action de la temperature sur la spermatogenèse de *Pieris brassicae* (Lepidoptera). Note. Effects of temperature on spermatogenesis in *Pieris brassicae. Acad. Sci. Comp. R.* 260: 3765–3767.

—— 1969. Contribution à l'étude de la double spermatogenèse chez un lépidoptère. (*Pieris brassicae*). *Annls. Sci. Nat.* 11: 569–626.

8. Physiology

Colouration of the larva and pupa

Introduction

For over a hundred years, the control of colouration in *P. brassicae* larvae and pupae has baffled scientists and such workers as Wood (1867), Meldola (1873), Poulton (1887a,b, 1890, 1982), Merrifield & Poulton (1899), Dürken (1916), Brecher (1917), Cretschmar (1920), Bertolini (1924, 1927), Gardiner (1965), Wigglesworth (1970), Bückmann (1971), Angersbach (1970, 1975, *et al.* 1971), Rothschild *et al.* (1975a,b), Kayser-Wegmann (1975) and Kayser & Angersbach (1974, 1975) have sought to elucidate the factors involved.

The main feature of *P. brassicae* colouration which scientists have been interested in is the variable colour of the larvae and pupae. The aim of the scientists has been to elucidate the control mechanism which effect the colour changes.

Many other criteria have been investigated: a) at what stage are the larvae sensitive to light? b) are the ocelli involved? c) are there photo-sensitisers involved (pigments), and if so where are they located? d) does the background or its texture have any effect on the colouration of larvae? e) does the content of the food affect colouration? f) to what spectrum of light is the larva sensitive? g) which part of the integument, if at all, is sensitive to light? and h) how does the system operate, are hormones involved?

Some of these questions can be answered immediately, others still require much attention. It would appear that the critical time for light reception is at the end of larval life, and that the ocelli are definitely involved in photoreception. Carotenoid pigments have been suspected for a long time and recently Seuge & Veith (1976) demonstrated the photoreceptor function of β-carotene and lutein in the brain of fifth instar larvae of *P. brassicae*. Response to background often plays an important part in colour control, while there is no evidence which shows that textures of the substrate plays any part in colour determination. (This is unlike other species such as *Papilio polytes* Linnaeus, cf. Smith, 1976.) The consensus of opinion is that pupae are background-matching but some strains are heterochromic.

It is also worth bearing in mind before considering the factors which control colouration in more detail, that *P. brassicae* is atypical in many respects of its behaviour and physiology but particularly in the following respects.

First, that the diapausing pupae of *P. brassicae* are green, that is at least the Cambridge strain, while most overwintering cryptic British butterfly pupae are brown (Gardiner, 1974).

Second, that the pupae display heterochromy (background-contrasting), while many other species display homochromy (Rothschild *et al.*, 1975a).

Third, that *P. brassicae* pupae are responsive to the entire range of wavelength of light from ultra-violet to infra-red, while most insects react to wavelengths between 400–550 nm (Kayser & Angersbach, 1975).

Factors involved

In some populations genetical effects are more obvious than environmental effects; the reverse may also be true. Most work has been carried out on environmental effects of colouration and this is dealt with under food, background, light and temperature effects.

Gardiner (1963) gave unequivocal evidence to show that at least one colour form of the pupae, namely the golden colour form, was due to a recessive gene. He also assumed that the aberration *coerulea* was due to an autosomal recessive gene as it breeds true.

Gardiner (1963) described how variations in relative humidity temperature, at starvation and crowding in the laboratory can all affect the colouration of the pupa to produce some of the different aberrations hitherto described.

Broadly speaking, whatever species of foodplant is eaten by the larva in the wild or laboratory, the colouration of the imago is not changed. This is true for *P. brassicae* and other insect species (Prach, 1920), and implies that colouration is genetically fixed.

It is from the plant that *P. brassicae* larvae exogenously obtain some white pteridine and yellow carotenoid pigments, while it makes for itself more white pteridines, blue bile pigments and ommochromes. It is through a combination of these pigments that overall colour patterns are created.

Originally it was thought that the chlorophyll pigments were responsible for the green colouration in larvae but they are now widely regarded to be broken down in the gut. The green content of the gut sometimes provides the green colour of the larva. Poulton (*loc. cit.*) found that his larvae reared on midribs of cabbage, died and this can now be explained as due to lack of the growth-promoting carotenoid pigments, which produce vitamin A.

If larvae are reared on entirely synthetic diets, i.e. lacking plant material and/or carotenoids, they produce blue or turquoise pupae (Gardiner, 1974). This is due to the visual effect of seeing the blue bile pigments in the integument, which together with the yellow carotenoids usually produce the green colour due to the Tyndall Effect. Pupae indistinguishably coloured from controls can be produced by adding lutein to the diet (Rothschild *et al.*, 1975a,b, cf. 1977, Rothschild, 1975). In the absence of carotenoids pupae failed to match the background. An optimum of approximately 207 mg of xanthophyll per 100 g diet is the critical level below which carotenoids are excluded from the epidermis and cuticle.

The influence of background in pupal colouration is something which has fascinated scientists and all those involved with the subject have enclosed larvae in boxes and subjected them to a wide array of coloured lights, papers, gauzes, hessian, glass and towels, under different conditions (Table 8.1).

Gardiner (1963) found that the background and photoperiod just prior to pupation had no effect on colouration of the pupa. He also found (David & Gardiner, 1962) that pupae respond to background when diapausing and non-diapausing larvae were fed on artificial diet without carotenoids or vitamin A.

Rothschild *et al.* (1975a) made extensive tests using different backgrounds and found that there was no visible response to background by the pupae when reared under the following conditions: "the caterpillars or prepupae are exposed to long or short hours of daylight, dim or bright illumination, or reared on smooth-surfaced artificial diet, or on rough-surfaced artificial diet in containers covered with coarse dark brown hessian, in deep shade, smooth light-coloured natural wood, in bright summer sunshine, rough white bath towelling in winter sunshine, or in glass jam jars in various conditions". Later however, Rothschild *et al.* (1975b) found that larvae did respond to background when they were reared on carotenoid free artificial diet. This indicated that xanthophylls were possibly mediators in the process of heterochromy. Kayser & Angersbach (1974) inferred at the end of their paper how important heterochromy is in the life of *P. brassicae* by stating that pupae placed against a red background produce a maximum contrast of bright green.

Wood (1867) found different results in his diapausing pupae in the wild which possibly indicated that background has some important effect upon the colouration of the pupae (Table 8.1). He explained that the different

Table 8.1 Colour of pupae reared under different conditions.

Conditions	Pupal colour	Authority
a) *Effect of foodplant*		
larvae ate green or etiolated cabbage	green or brown respectively	Poulton, 1892
b) *Effect of background colouration*		
pupation "under a vine"	orange	Wood, 1867
dark corner with very subdued light	green	
on tarred fence	darkest black	
on white surface	almost albino	
on dark fences	"darker than those on walls"	Meldola, 1873
Larvae reared in:		
black boxes	"largest amount of pigment"	Poulton, 1887b
dark red boxes	"largest amount of pigment"	
orange boxes	"smallest amount of pigment"	
pale yellow boxes	"more pigment present"	
green boxes	"more pigment present"	
pale blue/green boxes	almost red	
dark blue boxes	almost red	
yellow or orange surrounds	green	Poulton, 1890
black or white surrounds	dark and light respectively	
all colours except yellow or orange	dark (more or less)	
boxes lined with black or red paper	dark	Poulton, 1892
boxes lined with yellow or orange paper	green	
red glass in front of yellow or orange paper	green	Poulton, 1892
yellow glass in front of white or orange paper	black	
green glass in front of non-reflecting background	green	
green paper, gauze, cabbage leaves	whitish yellow	Merrifield & Poulton, 1899
green glass	whitish yellow to green	
yellow glass	green	
orange glass	whitish yellow	

216

Table 8.1 continued

Conditions	Pupal colour	Authority
Effect of background colouration — continued		
yellow or orange		
paper	green	Dürken, 1916
white paper	lightest	
yellow paper	green	
black paper	darkest	
light grey, darker		Cretschmar, 1920
grey, black, blue/green,		
yellow/green,		
orange or purple	all "medium" colours	
black background	darkly pigmented	
green or orange, with		Wigglesworth, 1970
more incident light	pure green	
cylinders lined		
with different		
coloured tissues:		
orange, yellow, brown	green	Smith, 1976
red, blue, white and		
acetate tissue	white	
green	confusing results	

c) *Effect of different light wavelengths*

Larvae reared in:		
darkness	green with black flecks	
diffuse light		
(from woods)	more green	
exposed to sunlight	varied between yellow,	
behind glass	grey and green	
warm rays	more green and have	Steiner, 1930
	much weaker black	
	pigment	
cold conditions	very variable: yellow,	
behind glass	green and very darkly	
	pigmented	
artificial light	green and grey/yellow	
blue light, especially		
ultra-violet	very melanised	Brecher, 1938

d) *Effect of painting over ocelli*

Poulton (1887b) observed the same results as mentioned earlier when he blinded the larvae.

ocelli painted with		
dark blue paint	dark	Brecher, 1917
ocelli painted with		
yellow paint	light	

Table 8.1 continued

Conditions	Pupal colour	Authority
e) *Effect of photoperiod*		
Larvae reared on cabbage in 12 h photoperiod (i.e. diapausing)	green	Gardiner, 1974
Larvae reared on cabbage in 18 h photoperiod (i.e. non-diapausing)	brown, including yellow, ochreous and grey	
Larvae reared on cabbage in 18 h photoperiod (i.e. non-diapausing)	silvery white, grey, green/brown, plus extensive blotching and freckling with dark and brown or black	Rothschild *et al.*, 1975a
f) *Effect of photoperiod and background*		
16 h photoperiod given during pupal development on orange background	green	Smith, 1976
8 h photoperiod given during larval development on red background	green	
g) *Effect of coloured or non-coloured constituents in the diet*		
Larvae reared on artificial diet lacking cabbage, virtually devoid of carotenoids and no vitamin A	turquoise blue, both diapausing and non-diapausing	David & Gardiner, 1962b
Larvae reared on artificial diet devoid of plant material	blue or turquoise	Gardiner, 1974
Larvae reared on artificial diet with β-carotene added	brown	Gardiner, 1974
artificial diet virtually lacking carotenoids but with vitamin A added	turquoise blue	Rothschild *et al.*, 1975b
artificial diet but with xanthophyll added, mainly lutein	indistinguishable from controls	

218

Table 8.1 continued

Conditions	Pupal colour	Authority
h) *Effect of photoperiod and foodplant or diet* 18 h photoperiod during larval development on:		
diet with 1.7% dried green cabbage diet with 1.7% dried white cabbage *Manduca sexta* diet wheatgerm added fresh green cabbage fresh white cabbage	blue, pale blue, white	Gardiner, 1974
12 h photoperiod during larval development on:		
diet with 1.7% dried green cabbage diet with 1.7% dried white cabbage fresh green cabbage	green, blue, turquoise	Gardiner, 1974
5 experiments: with non-diapausing larvae fed artificial diet incorporating wheatgerm, different xanthophyll extracts, etc.	turquoise blue, silver grey, green	Rothschild *et al.*, 1975b

colour forms of the pupae were due to the different colour backgrounds on which he found them. For instance, white pupae he found had pupated on a white wall; black pupae on a tarred fence and green pupae under a vine.

It is regretted that experiments on the colouration of *P. brassicae* since Wood's time are lacking in observations made in the field, because Ohtaki (1960), who worked on the colouration of a related species, *Pieris rapae crucivora*, found that results in the laboratory yielded much less uniform results than observations made in the field.

Wood (1867) originally thought that the "skin" of the pupa was photographically sensitive to light for a few hours after shedding the skin and that many colours could be produced depending on the background. Poulton (1887a,b) disproved Wood's sensitivity theory and suggested that the amount of pigment in the cuticle of *P. brassicae* was influenced by the spectral composition of the light incident on the larva immediately before pupation (critical period), an observation which has been substantiated today. Poulton (1903) *in* Kennedy (1961), believed that control of pigmentation was due to a genetic switch, which was triggered by visual stimulus. However, since then a great deal of work has been carried out and there is still controversy, indeed Raphael Meldola (1873) stated that "the particular conditions under which this photographic sensitivity is acquired have not yet been fully investigated". However, despite recent work it is still not apposite.

Claret (1966a,b) showed that the larvae were sensitive to light after the third moult, when the hard black head capsule was replaced by a clypeus, containing a yellow triangle which admits light (see Chapter 7). He found that if the brain of a fourth instar larva was implanted into the fourth instar larva abdomen of a larva reared under short day conditions (i.e. diapausing) it will respond to illumination again, i.e. six hours of light acting on the brain produced a hormone which prevents diapause, while eight hours of light acting on the brain induces diapause. However, Gardiner (1974) found that the time of sensitivity to light was just prior to pupation. Rothschild (1980, pers. comm.) suggests that both the larva and pupa respond to the light cues.

The yellow triangle described by Claret (1966a,b) on the clypeus of the larva and which was thought to be responsive to light, was further investigated by Angersbach (1970), who found an additional light receptor located dorsally in the head. He reported that this area was capable of discriminating between different spectral ranges and could perceive brightness contrast. It may be recalled that Kayser & Angersbach (1975) recorded that fifth instar larvae of *P. brassicae* were unusually sensitive to all wavelengths of light between the ultra-violet and the infra-red.

With non-diapausing larvae reared on artificial diet lacking cabbage, Rothschild *et al.* (1975a) found that there was no response to background when larvae or prepupae were exposed to long or short hours of daylight, dim or bright illumination using different backgrounds as mentioned earlier.

Gardiner (1974) however, found that the position of pupation and the incident light had a significant influence on the production of green or brown pupae. Normally-fed larvae which pupated around stems, where more incident light was reaching the cuticle of the larvae, went green, while those pupating on dark non-transparent surfaces went brown.

Gardiner (1974) found that diapausing pupae of the Cambridge strain were always green and non-diapausing pupae were always brown. However, Johansson (1959) found the opposite results; there was a tendency for more pale or green specimens in non-diapausing pupae than those with diapause. He looked at 599 specimens. This difference may be explained by strain differences as the author has also seen brown diapausing pupae in cultures made up from wild specimens.

Non-diapausing Palestinian pupae of *P. brassicae* were found to be noticeably different in colouration to non-diapausing pupae of the Cambridge strain (Gardiner, 1978). When freshly formed the pupae were a pale mauve colour with a very pale straw coloured dorsal keel, and the mauve colour gradually faded resulting in a silvery grey colour of the pupae.

Rearing temperature of the larvae can affect colouration. Cherchi (1952) found in her study of gregariousness of *P. brassicae* larvae and the factors which affected colouration, that larvae reared at 25–28°C were mostly brown. Those bred in the wild have large spots which are not fused together. David & Gardiner (1962b) showed that larvae reared at a lower temperature (12.5°C) had more intense black markings than those reared at 28°C.

Location of colour in the integument

Pteridine containing pigment granules are found in the interlaminar layer between the wing surfaces (Hopkins, 1895a) and in the wing scales (Baylis, 1924), and bile pigments are present in the scale sockets also in ab. *coerulea* (Feltwell & Rothschild, 1974). Melanin is present in the epicuticle of the larva and the exocuticle of the pupa (Kayser-Wegmann, 1976a,b) and carotenoids are present in the haemolymph and most of the integument (Feltwell, 1978), as well as being present in the brain (Seuge & Veith, 1976). For further information on individual pigments, the reader is referred to Chapter 10 Biochemistry.

Earlier this century, Cretschmar (1920) tried to find out how the different colour forms of the pupae were produced, and he came to the conclusion that microscopic vertical canals, which had small hairs inside them, were capable of shrinkage and expansion. Colour, he thought, was determined by the relative amounts of coloured flecks and the degree of shrinkage of the canals. Thus the green of the pupa was formed from green and black flecks when the canals shrank. More recently, a light microscopic study of the integument of both larvae and pupae revealed larval cuticle traversed by epidermal cell processes and pupal cuticle with microvilli and fibre bundles (Kayser-Wegmann, 1976a,b).

Pupal colour forms

Poulton (1892) recognised three colour types of pupae: a) the "normal form" with the ground colour either grey, yellow, green or orange – the grey being due to many black pigment spots in depressions. There were three subdivisions of the normal form, α, darkest forms having a grey-green, orange, yellow or white ground colour, β, intermediate forms with light ground colour, and γ, lightest coloured pupae, which had a grey ground colour; b) pupae with a white-yellow ground colour; the γ subdivision passes into this form, and c) a more abnormal form which had a bluish green ground colour. This latter form is reminiscent of those fed on artificial diet.

Hormones released from the head region of the fifth instar larva of *P. brassicae* were suspected of controlling melanisation. This was concluded from ligature experiments conducted by Oltmer (1968). Kayser & Angersbach (1974) further thought that a hormone from the brain was responsible for melanin and bile pigment synthesis and that both a melanin-stimulating and a melanin-inhibiting hormone were present (Kayser-Wegmann, 1975). Some hormonal control over ommochrome distribution was thought to take place by Rothschild *et al.* (1975a,b). The amount of juvenile hormone in the haemolymph determines the green colouration of *Pieris* pupae (see Chapter 9 Hormones).

Diapause

Historical

Much has been written on diapause in *P. brassicae* by Russians, Germans and French. One of the earliest workers in this field was the Russian Danilevskii who pioneered modern studies of insect photoperiodism and also investigated the effect of temperature on photoperiodic response in *P. brassicae* (Danilevskii, 1948, 1950, 1957, 1960; Danilevskii & Geispits, 1948; Danilevskii & Goryshin, 1960). He summarised his work in two books *Photoperiod and Seasonal development of Insects* (1961) and *Photoperiodic Adaptions in Insects and Acari* (1968). Following up on this work was another Russian co-worker Goryshin (1963, 1964a,b) and the Germans Bünning & Joerrens (1959, 1960, 1962, 1963) were also investigating the effects of different wavelengths and quantity of light on diapause induction, while another Russian, Maslennikova (1958, 1959, 1960a,b, 1961, 1968, 1970a,b, 1971; Maslennikova & Mustafayeva, 1971; Maslennikova & Chernysh, 1973) was concerned with the different effects of latitude and temperature on photoperiodic responses in *P. brassicae*. Maslennikova also became involved in the hormones regulating diapause and the effects of these hormones on various parasitic hymenoptera. Goryshin & Kozlova (1967) and Goryshin & Tyschenko (1968) were also involved in studying endogenous rhythms of diapause in different geographical areas of the USSR.

In the 1960s and up to the present a great deal of physiological work has come from the French. Studies on the effects of folic acid on growth, tumour development and genetic change were carried out by L'Hélias (1959a,b, 1960, 1961, 1966, 1969, 1972, 1973, 1975). Investigations into spectral sensitivity and time of sensitivity of the larva and effects on parasites were also undertaken by Claret (1966a,b, 1968, 1969, 1971a,b, 1972, 1973) and by Claret & Carton (1975). Vuillaume and her colleagues were responsible for a number of papers on sensitivity of the larva and the involvement of the bile pigments in photoreception (Vuillaume, 1976; Vuillaume & Seuge, 1972; Vuillaume & Berkaloff, 1974; Vuillaume *et al.*, 1970, 1971, 1972a,b, 1973, 1974a,b,c; but also see Seuge, 1973 and Seuge *et al.*, 1972a,b).

A recent book by Saunders (1976) *Insect Clocks* gives an account of some of the early work on *P. brassicae* and an important paper by Ushatinskaya (1976) classifies and discusses all the various types of classification of diapause which have been applied to insects.

Introduction

Diapause may be found in any stage of the life history of insects but in *P. brassicae*, as in other species of *Pieris* and closely related genera, it occurs only in the pupal stage. However, the larval stage is directly involved in the reception and measurement of the extrinsic factors which initiate diapause. According to recent classification of diapause *P. brassicae* would have a typical hiemopaustic diapause as this occurs during the winter months (Ushatinskaya, 1976).

Throughout most of its range in the northern hemisphere *P. brassicae* experiences temperate, mediterranean, continental and asiatic type climates which have relatively long scotoperiods during the winter months and relatively long photoperiods during the warm summers. The subspecies *P. b. cheiranthi* (Hübner) on the Canaries experiences Atlantic type of climate. It would be interesting to investigate the seasonality of *P. brassicae* in Chile, as it has a similar climate to that of the Lebanon.

Adaptations to hot and cold conditions

a) *Larva.* From the abundance of records of larvae found throughout the winter months in Great Britain, particularly during mild winters, it is clear that *P. brassicae* larvae have some means of keeping their metabolism ticking over at a slow rate. Pictet (1918) stated that larvae could "hibernate" for three months during the winter without food as long as the temperature did not drop below 0°C. This suggests that their basal metabolic rate can be turned down under these circumstances. Pictet (1922) further noted that, of larvae collected in the mild October of 1917, several died when the temperature fluctuated from −5°C to +7°C.

A drop in temperature appears to have two other effects on mature larvae. The first is stunting growth of the larvae; Russel (1936) noted that in the second week of December he came across larvae which, in the final instar were well below the average length as quoted by Frohawk (1934)

(36 mm instead of 41.3 mm), and Burton (1941) also found that larvae at the end of March were 2–3 mm smaller than usual. The second effect is to increase the length of time spent in the larval stage, i.e. prolongation of growth processes. Larvae found by Burton (*loc. cit.*) were first discovered already half grown in January and by March 30th they had become fully grown.

Small individuals of larvae are rather common in spring but may also be seen in other broods. It would be an interesting exercise to see whether there was any correlation between seasonal variation in size of imagines and size of larva. Certainly there are cases of small imagines seen in the spring (in Italy and Malta) cf. Chapters 1, 7.

b) *Pupa.* Pupae of *P. brassicae* can undoubtedly survive very great drops in winter temperature coupled with frost. In Scandinavia, Hansen & Merivee (1971) reported that the critical temperature for *P. brassicae* pupae, beyond which they succumb, was −22.3°C. They go on to say that in Estonia (NW USSR, 58°N), there should be 100% mortality because temperatures below the critical are experienced there. The large number of imagines in this region in the summer must be explained by some other reason – which is probably the arrival of immigrants from the south. Réamur, *in* Kirby & Spence (1858), stated that pupae of *P. brassicae* could withstand 15–16°F. Danilevsky (1961) gave a figure of −22.5°C for the supercooling point of *P. brassicae* pupae from Leningrad, where it only has one generation per year. The lethal temperatures in Norway are between −24.6°C and −26.5°C (Sømme, 1967). Kozhantchnikov (1936), working in the USSR, stated that *P. brassicae* must survive −30°C to −35°C as these severe frosts have been experienced often.

One of the best examples of pupae enduring severe cold is reported by Egerton Collier (1941) who found 200 healthy pupae in England which experienced 25 degrees of frost and were embedded in ice for 12 days.

Embedding in ice for relatively short periods does not appear to affect the pupae or imago and it is interesting to note that climbers in Austria and Switzerland have revived imagines embedded in ice at 3100–3410 m (Mellows, 1924). Many records exist of larvae, mostly fifth instar larvae, being found feeding on plants such as brocolli, sprouts, kale and cabbage in the middle of winter (from November to February) (Campbell, 1953a,b; Newman, 1942). These reports come from many parts of the British Isles including Abergavenny, South Shields, Middlesbrough, Bexley, Sutton and Rayleigh where 5–8 degrees of frost have been experienced when the larvae have been found. Young larvae in the autumn are likely to be killed off by any first severe frost, as these stages are more susceptible to drop in temperature (Littlewood, 1941).

In Lebanon, the summer brood of *P. brassicae* is thought to aestivate in the pupal stage as a means of overcoming the hottest months of the year, although the author did not elaborate (Larsen, 1974).

c) *Imago*. Both Opheim (1958) and Nordström (1951) reported the distribution of *P. brassicae* above the Arctic circle (67°N) in Scandinavia beyond 70°N and bordering onto the Arctic Sea. Extremes of temperature are experienced in these continental regions and Haig-Thomas (1938) reported *P. brassicae* common round the Lappland town of Maalselval in the summer when the temperature was 95°F in the shade after a bad winter. It may be recalled from Chapter 2 that *P. brassicae* has been recorded from the Orkneys many times and from the Shetlands; odd specimens have occurred in the Faroes and northern Iceland. There is a paper by Warner (1920) which questions whether *P. brassicae* imagines can actually hibernate, as he came across imagines apparently resting in the curtains of this house. The year of 1920 was in fact one of the coldest summers on record this century, so it seems likely that Warner's account is unreliable.

The oppressive heat in Arabia is said to exclude *P. brassicae* (Talhouk, 1969), and in the Lebanon *P. brassicae* is restricted to the coastal regions (Larsen, 1974).

Diapause flexibility to latitudes

Maslennikova studied diapause response of *P. brassicae* and its parasites at different geographical areas from 40°N to 60°N. The responses of both host and parasite (*A. glomeratus* (Linnaeus) and *Pteromalus puparum* (Linnaeus) varied independently with latitude, temperature and humidity, and was assumed to be genetically controlled (Maslennikova, 1958, 1959). The photoperiodic response of *A. glomeratus* was found to be much more pronounced than in *P. brassicae* which indicated that it had evolved this system independently of its host.

The two regions studied by Maslennikova, Sukhumi (43°N) and Leningrad (60°N), were used in later experiments together with samples from Belgorod (50°N), Yerevan (40°N) and Kuba (41°N) (Maslennikova & Mustafayeva, 1971). These workers found that the northern populations of *P. brassicae* had a significantly higher diapause threshold at 16 hours photoperiod than southern populations near the Caucasus, which responded to a threshold of 12 hours. *A. glomeratus* also showed similar flexibility to suit the environment and its host in these different areas.

Another feature which is possibly related to local geographical or rearing conditions is that in some areas pupae are obtained which hatch out much later than their own kin but in a shorter time than those which diapause. This phenomenon has been termed irregular diapause. Such was the case with pupae brought back from Corsica (Bowden, 1966).

Characteristics

a) *Morphological*

1) Diapausing pupae tend to be light green in colour as opposed to the silvery-green or brown if non-diapausing (Gardiner, 1974, 1979). The green colour is quite distinct in Gardiner's specimens and can be separated easily from the other forms. However, this effect is not experienced everywhere, and indeed, the colour of most pierines is brown when in the diapausing state and green in the non-diapausing state which accords with the natural seasonal distribution of colours.

2) A pair of spines in the ventral mid-line position of non-diapausing pupae is said to be present by Johansson (1959), but absent in diapausing ones. Further details on this have been presented in Chapter 7 under the section on pupal spines.

b) *Physiological*

Larva. Physiologically, there are a number of different points which have been observed between diapausing and non-diapausing larvae and pupae of *P. brassicae*. This is not surprising as diapausing organisms would be expected to have certain physiological adaptations to enable them to cope with the different conditions.

In comparison with non-diapausing larvae, diapausing larvae have:

1) an increased frass drop frequency, indicating that they consume more food (Friden, 1958).

2) a phase of ecdysone secretion immediately after the larval–pupal moult (Calvez, 1976). (It is worth pointing out that diapausing larvae of *P. b. crucivora* have a greater secretory activity than non-diapausing larvae (Kono, 1975).

3) often more carotenoid pigments (Rothschild *et al.*, 1975a,b).

4) more trehalose reserves (Dutrieu & Moreau, 1970).

5) a rate of respiration which decreases very rapidly after the larval–pupal moult and then remains steady while ATP and ADP do not increase in the same ratio (Guillet, 1976a,b).

6) a decreased susceptibility to sub-lethal amounts of insecticides (Champ, 1958).

7) a slightly longer larval life (16.6 days) as opposed to 14.83 days in non-diapausing larvae (Gardiner, 1979).

8) a tendency to pupate on roofs of cages rather than on sides (Gardiner, 1979).

9) a tendency to escape more often than non-diapausing ones (Gardiner, 1979).

226

Pupa. Diapausing pupae in comparison with non-diapausing pupae have:
1) more sorbitol, which was suggested may help in withstanding the cold (Sømme, 1967).
2) a higher total lipid content than diapausing pupae (Fourche *et al.*, 1977).
3) a lower respiratory rate (Fourche *et al.*, 1977).
and tend to be:
4) 6% heavier than non-diapausing pupae (402 gm/1000 pupae as opposed to 347 gm/1000 pupae of non-diapausing pupae) (Gardiner, 1979).

c) *Biochemical adaptation*

There are two distinct ways in which insects can overcome low temperatures; first, an insect may supercool its body fluids by including various compounds such as the polyols sorbitol and inositol or sugars such as trehaloses or even amino acids; or second, it may be physiologically resistant to ice crystals which form in the body fluids.

Sømme (1967) showed that the carbohydrate sorbitol increased in concentration in diapausing pupae of *P. brassicae* up to 90 days after pupation (Figure 10.1) and suggested that the increase of this solute in the haemolymph aided the pupa in withstanding cold. In other insect species glycerol acts as an "anti-freeze" but this has not been found in *P. brassicae* Sømme (*loc. cit.*) also found that trehalose, lactate alanine and glucose were increased in concentration with supercooling. Pupae were stored at room temperature for two weeks and then transferred to 2°C.

Ushatinskaya (1976) pointed out in her discussion of cold resistance in insects that in many insects large amounts of glycerol are found in the haemolymph during diapause but that it is resynthesised into glycogen prior to termination of diapause. Other compounds such as sorbitol, glucose, unsaturated fatty acids and lactic acids all increase in concentration throughout the winter, thus helping in supercooling mechanisms.

Beck (1976) used his own mathematical Dual System Theory to compute the photoperiodic features of diapause in a number of insects, including *P. brassicae*. His theory is based on the postulation that there are "two interacting biochemical systems whose kinetics are regulated by temporally spaced photoperiodic signals". The first stage, termed (S) system, begins at the onset of darkness and results in the build up of the concentration of the product of the system (S) four hours later. The product is then metabolised away. During the (S) system there is a time when continued development of the insect into either diapause or non-diapause is determined. This is termed the Determination Gate and is followed by the second system (P). Beck (1976) used data already published on *P. brassicae* photoperiodic responses to determine the number of constants and thresholds in both the (S) and (P) systems.

It has been claimed by several workers that many metabolic processes cease during diapause but this view is challenged by the evidence that Fourche et al. (1980) have uncovered through their work on P. brassicae. They showed that if diapausing pupae were held at 5°C for two months considerable physiological modifications take place, for instance the respiration rate increased by 45% after two months (cf. Fourche, 1977a,b). Diapausing pupae held under these conditions had a higher concentration of haemolymph proteins and at ecdysis were richer in fats (6.20 mg/100 fresh weight, to 4.23 mg/100 mg in non-diapausing pupae), and they also accumulate sorbitol (Sømme, 1967). The nucleotides pools also decrease in diapausing pupae.

How is diapause initiated?

It has been shown earlier that the amount of photoperiod, how and when it is delivered, the wavelength of light and the temperature are the critical factors in the determination of diapause in P. brassicae.

"A few days" in the life of the third and fourth instar larva is the time when the light hitting the integument is quantitatively measured and the physiological response of diapause is either turned on or off (Claret, 1966b). However, the way in which light is quantitatively measured has not yet been established. Light penetrating the integument is presumably absorbed by the pterobilin and carotenoid pigments.

Unsuccessful attempts to induce diapause in P. brassicae using fluorinated mescaline, an hallucinogenic drug derivative of tyrosine, were carried out by Aranda & Vuillaume (1977).

Photoperiod/Scotoperiod

Way et al. (1949) gave the first account of suppressing diapause when they increased the photoperiod to 16 hours per day and decreased diapause by 10–20%. If larvae are reared with less than 16 hours per day they are switched into diapause. Other methods can be used to prevent diapause, by subjecting larvae to two hour light breaks making a 6:8:8 light/regime (Bünning, 1969), or by giving a 8:8:1:7 light/dark regime (Dumortier, 1973, pers. comm.). Barker et al. (1963) have found that virtually instantaneous flash photoperiods, correctly timed, will inhibit diapause in P. rapae.

Temperature also influences the percentage of diapause. As early as 1934 Maercks (1934) showed that diapause could be suppressed by an increase of temperature.

Rearing larvae in complete darkness completely reduces the numbers going into diapause (Claret, 1972) as long as the temperature is above the critical level of 22°C (Danilevskii, 1968). On a 12:12 light/dark (L/D) photoperiod there is no diapause above 29°C. However, recent work has shown that if larvae are exposed for the whole of their lives to a 24 hour thermoperiodic cycle in constant darkness, individuals of *P. brassicae* may produce non-diapausing pupae according to the cold and warm sequences (Dumortier & Brunnarius, 1977a,b).

Wavelength of light

Several workers have tested the effect of different wavelengths of light on diapause induction in *P. brassicae* (Geispits, 1953, 1957; Bünning & Joerrens, 1960; Claret, 1972; Vuillaume *et al.*, 1974a,b,c; Vuillaume, 1976).

Colours from many parts of the visible spectrum have been shown to induce diapause: green, violet, and red (in order of decreasing ability) (Geispits, 1957); blue, yellow and white (Claret, 1972) and red (Bünning & Joerrens, 1960; Vuillaume, 1976). Vuillaume *et al.* (1974a,b,c) showed that UV (260–360 nm), violet (400–440), green (510–560) and yellow (550–640) produced 80% diapause, while blue light (430–480) had a feeble effect.

Seuge *et al.* (1972a,b) and Vuillaume *et al.* (1974) studied the effects of white, blue, green, red and yellow light at different photoperiods and found that they all induced greatest percentage diapause at nine hours photoperiod per day, with the exception of white light which peaked at seven hours photoperiod.

Red light and reversal of diapause

Red light may induce or inhibit diapause according to the wavelength used (Vuillaume, 1976; Claret, 1972). Red light and dark red may act antagonistically under the same conditions, red light giving only 4% diapause, opposed to 93% by dark red light (Vuillaume *et al.*, 1974). The way in which red light acts on larval tissues has not yet been explained but it has been established that the temperature of the larva does not increase (Seuge *et al.*, 1972a; Vuillaume *et al.*, 1974).

Bünning & Joerrens (1960) found that critical amounts of red light administered to larvae at different stages in the photoperiod/scotoperiod stages could reverse what would normally take place. For instance, with a 10:14 L/D regime most (60%) of the pupae would go into diapause but this could be reduced to 15% by giving nine hours or red light before the dark period. Conversely, a 16:8 L/D regime would normally give a high percentage of non-diapausing pupae (98%) but if four hours of red light were given before the dark period the percentage of diapausing pupae could be increased to 95%.

Photoreception by pterobilin

The bile pigment, pterobilin, extracted from *P. brassicae* (Ruediger *et al.*, Chapter 10), was thought to play a part in photoreception by Vuillaume *et al.* (1971, 1974). It has an absorption spectrum which has two maxima, one in the blue region and the other larger one in the red, although it absorbs light in the ultra-violet, green and yellow regions.

Red light (630–670 nm) is strongly absorbed by pterobilin and accelerates larval development and reduces mortality and diapause to about zero (Vuillaume *et al.*, 1971, 1974; Veith *et al.* 1974).

Hormone regulation of diapause

Novak (1975, p. 324) points out that it had been claimed that a "specific inhibitory factor" inactivated by the cold "participates" in diapause-induction-in diapause determined *P. brassicae* pupae. However, in a related species *P. r. crucivora* induction of diapause is believed to be dependent upon a daily secretory activity of the larva (Kono, 1975). Kono (*loc. cit.*) demonstrated that SD reared larvae had a greater secretory activity than LD reared larvae. It has been shown that diapause-induced *P. brassicae* larvae are resistant to chilling, and pupae would continue to develop even when their brains had been extirpated at the end of diapause (Maslennikova, 1968, 1970a,b, 1971; Bielozevov, 1962 *in* Novak, 1975; Claret, 1966; McDaniel & Berry, 1967).

L'Hélias (1975) demonstrated the presence of juvenile hormone (JH) in diapausing pupae of *P. brassicae* by using cyclic AMP phosphodiesterase, and more recently Guillet & Fourche (1976) found that when diapausing pupae were injected with α-ecdysone, breakdown of diapause resulted.

In her discussion on diapause Ushatinskaya (1976) believed that these earlier workers were concerned too much with the effects of exogenous factors on diapause, not enough on possible endogenous mechanisms. She claimed that the process of oxybiosis which occurs during diapause appeared to be under more direct control by endodermal regulators than an oxybiosis, although no experimental support was forthcoming. She said that neurosecretory and hormonal factors bring about either stimulation or inhibition of certain links of metabolism, and that their activity determines the wide spectrum of physiological rest which she divided into sleep, oligophase, diapause and superpause. Kind (1976) stated that when pupae of *P. brassicae* are decerebrated they can resume development only after reactivation at 10°C (Maslennikova, 1970a,b), but that if this occurred after 24 hours from the onset of the pupal stage this did not take place (Chernysh, 1973).

Kind (*loc. cit.*) also showed by extirpation experiments of the heads of

230

pupae and by cooling that the reaction of the prothoracic glands differs amongst different geographical populations of *P. brassicae*; those from the south of USSR respond to higher temperatures. Resumption of development after diapause is brought about by a deficit of hormone from the brain. It is interesting to note that diapause was also reported to be broken by chilling in *Pieris* (Daniliveski, 1965, *in* Bowden, 1966).

Excretion

Introduction

Excretory products are removed from the body in four ways: from the six Malpighian tubules which empty into the alimentary canal, the meconium which is expelled at eclosion, the exuviae which are lost at each larval ecdysis and via the spiracles. Elimination of carbon dioxide via the spiracles is dealt with later in this chapter under respiration.

Some of the excretory products such as uric acid are not removed from the body but are stored in the wings and the fat body of the imago. Although frass is not technically an excretory product, it does contain excretory products produced by the Malpighian tubules and is thus dealt with here.

Excretory products

The excretory products of *P. brassicae* include water, inorganic ions, carbon dioxide, ommochrome, carotenoid and pterin pigments, nitrogenous compounds, phospholipids, ketocatechol and arterenone, protein and amino acids, fats and mucopolysaccharides (Table 8.2).

Excretion in the Malpighian tubules

Potassium and sodium ions are always found in the Malpighian tubules and potassium is always in greater concentration (28 mM/l), as it is actively secreted into the tubules, than sodium (7 mM/l) (Ramsey, 1953; Craig, 1960; Nicolson, 1976a,b,c). Nicolson (*loc. cit.*) found that the osmotic pressure of the tubules did not vary significantly from one region to another.

Malpighian tubules of *P. brassicae* can maintain a salt balance under heavy loading (Ramsay, 1976) and those of the larvae, unlike the imago, can exhibit high rates of nicotine excretion (about × 12 normal) (Maddrell & Gardiner, 1976).

Table 8.2 Excretory product and sites of excretion in *P. brassicae.*

Excretory tissue/organs	Excretory product (UA = uric acid)	Authority
a) *Excretory products removed from the body*		
Exuviae (larvae)	carotenoids	Feltwell & Valadon, 1972
Exuviae (pupae)	neutral ketocatechols	Anderson & Barrett, 1971
Frass	UA, kynurenine	Harmsen, 1966a,b
Frass	Fats (various)	Turunen, 1973b
Frass	Ommochromes, UA	Mauchamp & Lafont, 1975
Frass	carotenoids	Feltwell, 1978
Gut	proteins, mucopolysaccharides	Plantevin & Nardon, 1970
Malpighian tubules	sodium, potassium	Nicolson, 1975
Meconium	kynurenine, isoxanthopterin, leucopterin, urea	Harmsen, 1966a,b
Meconium	carotenoids	Feltwell, 1978
Meconium	sodium, potassium	Nicolson, 1976b
b) *Excretory products stored in the body*		
Body (male)	Adenosine	Lafont, 1972
Body	UA	Wigglesworth, 1925
Fat body (urate cells)	UA, allantoic acid, pterin	Lafont *et al.*, 1976a,b
Fat body	xanthine	Harmsen, 1966b
Wings	leucopterin, isoxanthopterin, UA	Harmsen, 1966a,b Maddrell, 1971 Lafont, 1972
Wings (trichogen cells)	UA, allantoin, pterin	Lafont *et al.*, 1976a,b

Diuresis

Nicolson (*loc. cit.*) recorded a high rate of diuresis at the pupal-imago ecdysis, the first account of such a high rate for a non-bloodsucking insect. She estimated the flow of fluid from all tubules as 420 nl/min. for the first three hours after eclosion, giving an average of 70 nl/min. from each tubule. This loss of fluid caused the imago to loose 40% of its weight during the same period, most of this coming from the decrease in the haemolymph volume which decreased by 74%. The pupal exuviae and the meconium accounted for 8% drop in weight.

Using excised Malpighian tubules *in vitro* Nicolson showed that fluid secretion increased with a rise in temperature in a similar way in which *Rhodnius* tubules react under the same conditions. The optimum pH for secretion in *P. brassicae* was 6.2, unlike responses recorded for *Calliphora* sp. (Diptera: Calliphoridae) and *Carausius* sp. (Phasmida: Phasmidae).

Nicolson's experimental results with cyclic AMP (adenosine 3′,5′-mono phosphate), which is a mediator of diuresis, led her to suggest that the fluid secretion in *P. brassicae* is under control of a diuretic hormone. She found that excised Malpighian tubules increased their secretion 20 fold from 4–5 nl/min. to 30–100 nl/min. when stimulated with cyclic AMP.

Meconium

Meconium is an excretory fluid produced at imaginal eclosion and contains the following substances: allantoic acid, carotenoids, the pteridines kynurenine, isoxanthopterin and leucopterin, sodium, potassium, urea and uric acid (Table 8.2).

There are two phases of meconium secretion (Lafont & Pennetier, 1975). During the first stage, termed "meconium-1", ommochromes and all of the uric acid are excreted, while during "meconium-2" allantoic acid and a colourless fluid are excreted.

Harmsen (1966a,b) showed that very little pteridine pigment is excreted via the meconium, in fact only 2% of that contained in the pupa is lost in the imago. Kynurenine is the highest pteridine lost in the meconium at 50 μg/insect, while leucopterin and isoxanthopterin are 6.5 and 0.4 μg/insect. David (1965) found that granulosis virus was not excreted in the meconium.

Some of the meconium is reabsorbed by the imago and may be involved chemically in wing expansion (Moreau, 1973, cf. Chapter 3).

Wings and fat body

Mauchamp & Lafont (1975) investigated UA metabolism in larvae, pupae and imagines, and Lafont *et al.* (1976a,b) studied fluctuations of UA, allantoic acid and pterin excretory products in *P. brassicae*. Both maintained that the fat body was a most important organ in which UA was synthesised and stored. Mauchamp & Lafont (*loc. cit.*) showed this by incorporating labelled guanine into the larvae and finding it later in the fat body of the prepupa. They also showed that UA together with ommochromes were present in the red frass pellets produced by fully fed larvae and that the UA content of the frass may represent 14% of the dry weight. In young larvae most of the UA is present in the integument associated with protein, while in later life most eventually is found in the fat body. The details of the enzymes associated with UA breakdown investigated by these two workers can be found in the section on enzymes.

Lafont *et al.* (*loc. cit.*) showed that UA and allantoic acid made up 2.5–3 mg (of dry body weight) and that pterin made up 1 mg in the body of *P. brassicae*. Adenosine was also shown by Lafont to make up 0.5 mg of the fat body of imagines three days after eclosion.

Wings were shown by these workers to be the principal site of pteridine synthesis during the pharate adult stage and acted as storage areas for these excretory products. In addition wings were shown to act as excretory organs which accumulated compounds released by the fat body.

During pupal-imago development there are three distinct phases of excretory metabolism (Lafont *et al.*, 1976a,b): a) for 0–50% of the time little nitrogen is excreted and the total UA content rises slowly; b) 50–85% synthesis of pterins (isoxanthopterin and leucopterin) and some degradation of UA to allantoic acid occurs; and c) 85–100% synthesis of UA in the fat body starts again and extends to the young imago. In the male the fat body accumulated adenosine (Figure 8.1).

Total nitrogen excretion

The nitrogen content of meconium was calculated to be 164 μg, comprising 12 μg of total pteridine content (average representing 4 μg of nitrogen, the average purine content of 456 μg representing 153 μg of nitrogen, while the content of kynurenine at 50 μg represented only 7 μg of nitrogen (Harmsen, 1966a,b)

When Harmsen (*loc. cit.*) tried another method (Ultra-micro Kjeldahl technique) of determining the nitrogen content of meconium he found a higher result of 241 μg. Thus the nitrogen content must come from another source besides pteridines, purines and kynurenine. Chromatography of meconium and subsequent spraying with dimethyl-amino benzaldehyde showed the presence of urea which was determined to be about 12 μg.

In the wings and body 549 μg of pteridines are stored and this represents 197 μg of nitrogen. The high content of UA and xanthine (2960 μg) represents 987 μg of nitrogen (Harmsen, 1966a,b).

Frass

Frass contains carotenoids, pteridines, uric acid, fats, ommochromes and presumably from its colour chlorophyll or its breakdown products (Table 8.2).

a) *Carotenoids*. *P. brassicae* larvae expel from their bodies 11.5% of their total intake of β-carotene during their larval life, which is equivalent to about 1,1165 μg/g of β-carotene (Feltwell & Valadon, 1972). The qualitative and quantitative nature of the carotenoids found in the frass will vary considerably with the foodplant source and with the time of the year

a) Uric acid content of fat body (*)
   and carcass (+) during larval life
   (after Mauchamp & Lafont, 1975)

b) Concentration of uricase and
   allantoinase in larvae
   (after Mauchamp & Lafont, 1975)

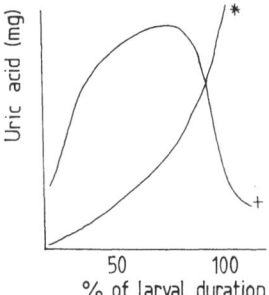

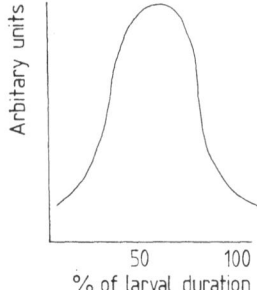

c) Concentration of GD & XDH in larvae
   (after Mauchamp & Lafont, 1975)

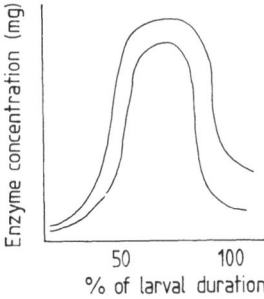

Fig. 8.1 Uric acid metabolism

(Feltwell & Valadon, 1974). Nine carotenoids were found in the frass when the larvae were feeding on cabbage.

b) *Pteridine and uric acid.* In the fifth instar larva about 1 mg of UA is lost every 24 hours (Harmsen, 1966a,b), and this is accompanied by kynurenine excretion at 100 μg/24 hours. In the prepupal stage the amount of UA lost in the frass decreases rapidly and then the level remains constant at 1.5–2 mg during the pupal stage. At imaginal eclosion 20% of the UA is excreted, while the rest is stored in the body.

Water metabolism

Water represents such a high percentage of the body weight of all organisms (50–90% in insects, up to 70% in leaves) that its significance in biochemical reactions must not be underestimated. Indeed Harmsen (1966a) mentioned the very important point that when assessing physio-

235

logical data it is essential to know the water content of the insect body at all times during its metamorphosis. Thus a water-content curve must be prepared; and this is especially important when one macerates insects or tissues in aqueous solution as so often happens in biochemical analysis.

The water content of an insect body can vary according to the source of food. For instance Junnikkala (1969) stated that the fifth instar larva of *P. brassicae* reared on fresh leaves contained 82.4% water, whereas those reared on artificial diet had less at 76.5%. Even intraspecifically the water content of an individual can vary by as much as 20% (Harmsen, 1966a). However, a more reliable parameter which can be used is the wet weight/dry weight ratio which varies less than 1.3% (Harmsen, *loc. cit.* cf. Strogaya, 1961, 1962).

One of the greatest dangers facing an insect is desiccation, although it is provided with a thick outer integument which helps to counteract this. Water loss can occur, however, at ecdysis when the new integument is being formed and water would tend to be lost straight through the integument and spiracles. Nicolson (1976a) studied the water loss of freshly emerged imagines and found that when the anus was plugged with wax water loss, presumably from the integument, was only 1%. *Pieris* larvae do not take in water through their mouths directly but obtain their water via the metabolic digestion of water-rich leaves. *Pieris* imagines have probosci for the imbition of water.

The integument of *Pieris* serves as a barrier against water loss by providing two main layers, one on the surface of the cement and the other on the outer surface of the cuticulum (Beament, 1958, 1959a,b; Locke, 1964). The structure of the integument varies with age; for instance, in young pupae of *Pieris* there is a primary wax layer and no cement layer, but in mature pupae there is a primary wax layer covered with cement and a second wax layer covering the cement (Ebeling, 1964). Some protection against desiccation is brought about by epicuticular lipids. Those of *P. brassicae* have a transition point of 30°C, i.e. the temperature at which (lower than melting point) the state of the lipid changes to an increase in the carbon-carbon chain and other properties. Previously Wigglesworth (1945) had recorded "critical temperatures" for many insect species above which a great increase in evaporation of water would occur. That of *Pieris* larvae was about 40 C, whereas *Agrotis* larvae was 10°C. (Figure 8.2 cf. Edney, 1957.)

Beament (*loc. cit.*) believed that the outer layers of the epidermis are capable of modifying the lipid layer on the surface. Thus, when cockroaches' grease was placed on *Pieris* wing membranes at 22°C it took six hours for the permeability of the wing to water to drop from 3.8 mg/cm^2/hr to 0.35 in 100 hrs. Beament thought that perhaps an orientated and compressed monolayer of molecules had been formed.

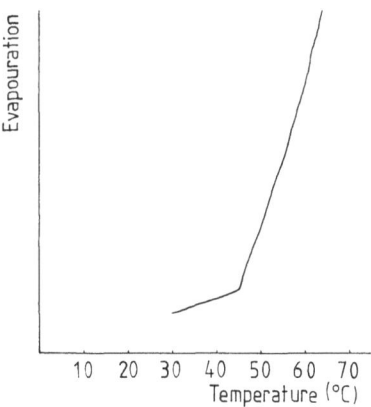

Fig. 8.2 Evaporation from *Pieris* larvae
(after Wigglesworth, 1945)

Stobbart & Shaw (1964) found that *P. brassicae* larvae were twice as susceptible to desiccation as pupae of the same weight when measured as a percentage of initial weight per day at 20°C and 60% RH.

In the course of a study on water loss from lepidopteron ova Bartell (1966–thesis) showed that ova laid in compact batches have a slightly higher survival potential than those laid singly. At relative humidities of 95% or above there was a greater incidence of fungus attack.

In the larva the water loss may be considerable at eclosion as Nicolson (1976b) showed; the osmolarity of the haemolymph falls from 750 mOsm to 150 mOsm two hours later (Nicolson, 1976b).

When performing ligaturing experiments on larvae and pupae it is always necessary to minimise water evaporation from the cut surfaces. The method used by Harmsen (1966a,b) on *P. brassicae* larvae is therefore of particular interest. He used a mixture of low melting point wax, or a smear made up from pyroxylin and castor oil dissolved in ethanol, ethyl acetate, butanol and amyl acetate ("New Skin") to cover up underlying patches of exposed cuticle. When three of four wings are removed, the pupae stays alive until eclosion.

Hydrogen ion concentration

a) *Alimentary canal.* Ripa (1978) made an extensive study of the pH of nine regions of the alimentary canal of fourth and fifth instar larvae of *P. brassicae* and innovated a method for measuring pH values which involved freezing the larvae in liquid hydrogen, cutting them up into sections, dissecting out the gut and placing these between two micro-electrodes. The characteristic pattern of pH through the gut was thus found to be slightly alkaline in the crop and increasing up to the central mid-gut, whereafter it decreased (Figure 8.3). Martouret *et al.* (1965) recorded slightly lower pH values for the oesophagus and mesenteron.

b) *Haemolymph.* The pH of larvae haemolymph of *P. brassicae* was found to be 6.83 (range 6.70–6.90) (Martouret *et al.*, 1965).

Respiration

Several workers from France, Great Britain and the USSR have studied respiration in *P. brassicae*, and most have published their results in the last seven years. (Champ, 1958; Dutrieu & Moreau, 1970; Breugnon, 1972; Samedov, 1974; Moreau, 1973, 1974; Moreau & Gordeaux, 1971a,b; Moreau *et al.*, 1974, 1975, 1977; Miah, 1976; and at the Université Claude Bernard, Lyon, Guillet, 1976a,b; Guillet & Fourche, 1976 and Fourche, 1977a,b).

Measurement of gaseous exchange

Measurement of oxygen consumed and carbon dioxide liberated has been calculated with data obtained from a simple Warburg flask (120 cc) with attached manometer, as was used by Champ (*loc. cit.*); the carbon dioxide was calculated from the amount absorbed by the potassium hydroxide in a "Gilson differential respirometer" (Moreau *et al.*, 1977), or in a specially devised piece of apparatus where the insect is imprisoned underwater in a container to which sensitive equipment is attached (Fourche, unpublished).

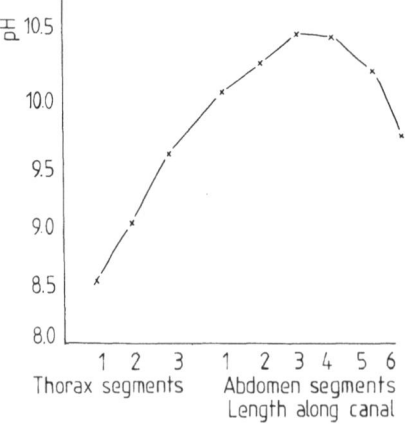

Fig. 8.3 pH of alimentary canal of fifth instar larva (96 hours old) (from Ripa, 1978)

Gaseous exchange in the larva

Samedov (1974) reported that third instar larvae of *P. brassicae* have a bicyclic type of diurnal rhythm of oxygen (O_2) consumption during the day, and that fifth instar larvae have a monocyclic rhythm at night.

In the course of research into the effects of gamma radiation on *P. brassicae*, Miah (1976) investigated oxygen consumption of third instar larvae. His data showed that heavier larvae did not use more oxygen during respiration than lighter larvae, the oxygen consumed being between 0.80–1.40 μl/g/hour.

Breugnon (1972) studied gaseous exchange in fourth and fifth instar larvae at three different temperatures $-15°$, $20°$, and $25°C$, and observed the following: that slightly more oxygen is consumed at the end of an instar than at the start; that the fourth instar larvae need 4.3 mm^3 O_2 at ecdysis and 11.75 mm^3 O_2 immediately afterwards; that fifth instar larvae consume 19.77 mm^3 O_2 after ecdysis and 25.10 mm^3 O_2 when they change into a prepupa. At these different temperatures the amount of oxygen consumed by a fifth instar larva during one hour varies little; at $15°C$:111,33 mm^3, at $20°C$:154,39 mm^3, and at $25°C$:164,16 mm^3. The amount of carbon dioxide liberated during larval life was recorded as varying according to the amount of oxygen consumed (Breugnon, 1972).

Gaseous exchange in the prepupa and pupa

Before formation of the prepupa, oxygen consumption falls rapidly, as much as between 33–91 mm^3 O_2 per individual per hour (Breugnon, 1972,

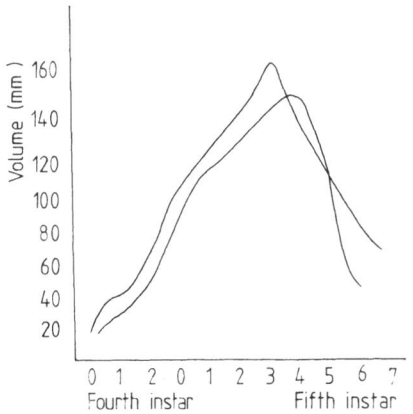

Fig. 8.4 Gaseous exchange in fourth and fifth instar larvae (after Breugnon, 1972)

cf. graphs of Guillet, 1976a,b; Moreau & Gourdoux, 1971a,b). This continues in the pupa and the rate at which this stage consumed oxygen at the three different temperatures was found to be 1 mm^3 O_2, 8 mm^3 O_2, and 13 mm^3 O_2 per day at 15°, 20° and 25°C respectively (Breugnon, *loc. cit.*) (Figures 8.4, 5, 6). Champ's figures (1958) for late pupae at 25°C were 5–10 μmlO_2/g/minute, dropping to 5 μml/g/minute at eclosion. The total amount of oxygen consumed per gram of tissue for the period 100 minutes before emergence to 300 minutes after was 3111 mm^3.

Guillet (1976a,b) showed that the respiration rate of non-diapausing pupae, measured in μl O_2/h in 100 mg of wet weight larva, followed a typical "U" shaped curve, but in diapausing pupae it decreased to a low level. This was confirmed by Fourche *et al.*, 1977a,b (cf. Guillet & Fourche, 1976).

Respiration in non-diapausing pupae was found to follow a sigmoid curve at all temperatures between 5–35°C, but this was only shown in diapausing pupae immediately after moulting. Fourche (1977) assumed that the velocity of one respiratory enzyme was the limiting step, not the access to substrates.

Gaseous exchange in the imago

This was recorded as 8–10 μml/g/minute in freshly emerged imagines (Champ, 1958).

Respiratory quotient (RQ)

The RQ of fourth and fifth instar larvae at different temperatures was shown to vary considerably (Breugnon, 1972). At 15°C, highest RQs were

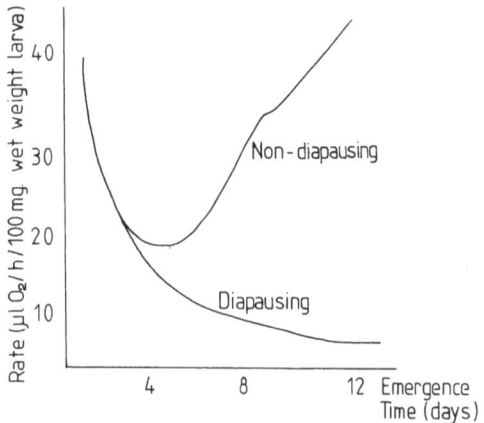

Fig. 8.5 Respiratory rate of *P. brassicae* pupae (after Guillet, 1976a,b)

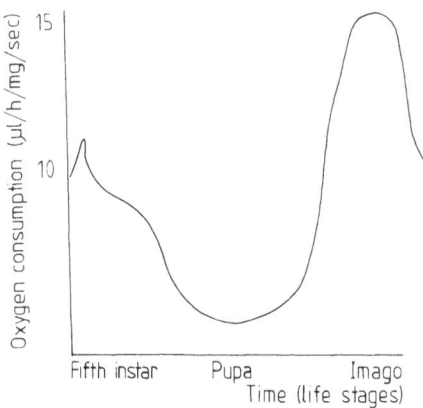

Fig. 8.6 Oxygen consumption in *P. brassicae*
(after Moreau & Gourdoux, 1971a,b)

recorded during the fourth instar when it attained 1.3, whereas at 20°C
and 25°C, a maximum of 1,1 was attained in a six day fifth instar larva.

Later Moreau *et al.* (1977) studied the RQ of prepupae, pupae and
imagines of *P. brassicae* at three different temperatures and showed that
with an increase in temperature, the RQ tended to increase. This was
partially explained from the results of analysing the proportions of
evolved carbon dioxide after the incorporation of labelled carbon dioxide.
Their work showed that at higher temperatures, pentoses were utilised
more. At 20°C pupae had an RQ of 1,1, prepupae 0.83 and imagines 0.77.
It is interesting to note that RQs of male and female larvae of all weights
were not recorded above 1 (Figure 8.7).

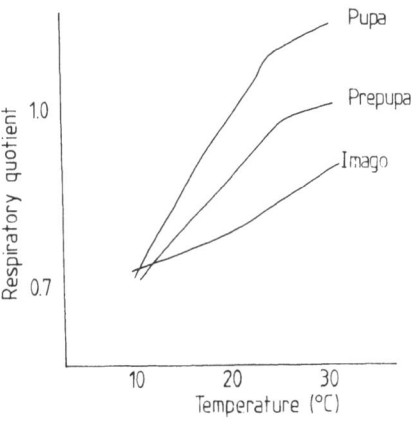

Fig. 8.7 Variation of respiratory quotient with temperature
(after Moreau *et al.*, 1977)

241

Tissue respiration

The rates of tissue respiration were measured in terms of cytochrome oxidase (CO) and succinate dehydrogenase (SDG) activities throughout larval and pupal life of *P. brassicae* by Tikhonravova (1973). Enzymatic activities were measured by spectrophotometric analysis. Three measurements were made in each instar.

The level of CO was found to be highest in youngest larvae and this may possibly be explained by the high oxygen debt incurred during egg life when there is a lack of oxygen. This had been found to be the case in the Chinese Oak Moth (*Antheraea pernyi* Guerin-Ménétville). The pattern of CO followed the same pattern through each instar, that is there was an increase at first, followed by a peak during feeding and then a decrease occurred at the end (Figure 8.8). The great fluctuations of the CO during each instar were accounted for by the "complex morphological rearrangements", such as moulting of trachea and replacement of cuticle which take place.

The concentration of SDG throughout larval life was more or less opposite to that of CO. It increased during larval feeding when respiration was high and during moulting when gaseous interchange is reduced and oxygen entrance is impeded.

Adenosine triphosphate (ATP)

The amount of ATP in the larva increases exponentially with weight and development, 0.26 nmoles per larva at hatching to 700–740 nmoles in fifth

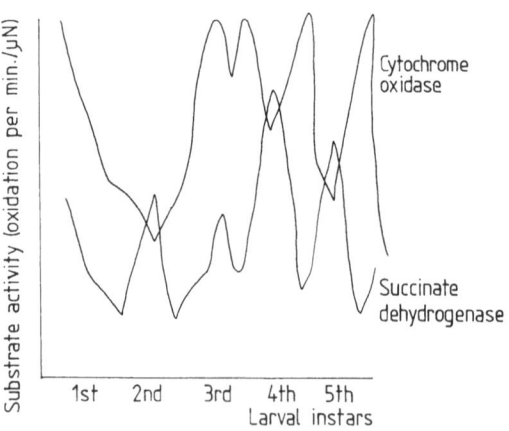

Fig. 8.8 Respiratory-enzyme activity during larval life of *P. brassicae* (after Tikhonravova, 1973)

instar larvae whether reared in short or long day conditions (Guillet, 1976, Figure 8.9).

The amount of ATP and ADP in non-diapausing pupae follows a characteristic "U" shaped curve, the level of ATP falling during the first two days of pupal life from 165 to 99 nmoles/100 mg of wet weight larva (Guillet, 1976a,b). In diapausing pupae the level of ATP and ADP declines steadily. Guillet (*loc. cit.*) followed the metabolism of ATP during six months of pupal diapause and noted that the level dropped by a half (from 64 to 35 in 188 days). The results of Guillet's work on ATP and respiration in pupae showed that in both diapausing and non-diapausing pupae the rate of respiration always dropped more quickly than that of ATP.

Moreau and workers made extensive investigations into the respiratory substrates of *P. brassicae*. In his research work Moreau (1973) established that glycogen, trehalose and fats are metabolised during pupal and imaginal life. By injecting labelled glucose into *P. brassicae* (Table 10.10) and by analysing the respired carbon dioxide, he proved the presence of an active pentose cycle (Moreau *et al.*, 1974). On amputation of the wings the cycle was not very important. Small amounts of trehalose were found in non-diapausing pupae (Dutrieu & Moreau, 1970), compared to large amounts in diapausing pupae.

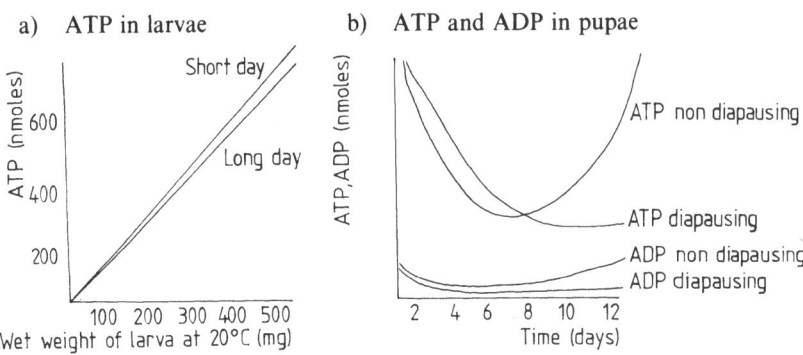

Fig. 8.9 ATP in *P. brassicae* larvae and pupae
(after Guillet, 1976a,b)

References cited

Andersen, S.O. & Barrett, F.M., 1971. The isolation of the ketocatechols from insect cuticle and their possible role in sclerotization. *J. Insect Physiol.* 17: 69–83.

Angersbach, D., 1970. *Die Bedeutung der Einfallsrichtung und der Wellenlänge des Lichtes bei der Melanisierung der Kohlweisslingspuppe von Pieris brassicae* L. Dissertation. University of Giessen.

—— 1975. The direction of incident light and its perception in the control of pupal melanisation in *Pieris brassicae*. *J. Insect Physiol.* 21: 1691–1696.

Angersbach, D. & Kayser, H., 1971. Wavelength dependence of light controlled pupal pigmentation. *Sond. Naturwiss.* 11: 571–572.

Aranda, G. & Vuillaume, M., 1977. Étude du métabolisme de mescaline fluorée par *Pieris brassicae* (Insecta, Lépidoptèra). *Experimentia* 33: 261–262.

Barker, R.J., Mayer, A. & Cohen, C.F., 1963. Photoperiod effects in *Pieris rapae*. *Ann. ent. Soc. Am.* 56: 292–294.

Bartell, R.J., 1966. *Studies on the water loss from the eggs of lepidoptera*. Diploma. University of London (Imperial College).

Baylis, H.A., 1924. Colour production in lepidoptera. *Entomologist* 57: 2–6, 78–82.

Beament, J.W.L., 1958. The effect of temperature on the waterproofing mechanism of an insect. *J. exp. Biol.* 35: 494–519.

—— 1959a. The waterproofing mechanism in arthropods. I. The effects of temperature on cuticle permeability in terrestrial insects and ticks. *J. exp. Biol.* 36: 391–422.

—— 1959b. The active transport and passive movement of water in insects. *Adv. Ins. Physiol.* 2: 67–130.

Beament, J.W.L. & Lal, R., 1957. Penetration through the egg-shell of *Pieris brassicae* L. *Bull. ent. Res.* 48: 109–125.

Beck, S.D., 1976. Photoperiodic determination of insect development and diapause. *J. comp. Physiol.* A 107: 97–111.

Bertolini, F., 1924. Richerche sulla struttura del tegumento delle larve di *Pieris brassicae*. *Redia* 15: 69–71.

—— 1927. Biological and histological investigation on *Pieris brassicae*. *Redia* 16: 29–39.

Bowden, S.R., 1966. "Irregular" diapause in *Pieris*. *Trans. Br. ent. nat. hist. Soc.* 1966: 67–68.

Brecher, L., 1917. Die Puppenfärbungen des Kohlweisslings, *Pieris brassicae* L. I. Teil: Beschreibung des Polymorphismus. II. Teil: Prüfung des Lichteinflusses. III. Teil: Chemie der Farbtypen. *Wilhelm Roux Arch. entwMech. Org.* 43: 88–221.

—— 1938. Der Weg der Farbanpassung bei Schmetterlingspuppen vom Rezeptor bis zum Effektor. *Biol. Gen.* 14: 212–237.

Breugnon, M.M. 1972. Étude de quelques caractères du développement post-embryonnaire de *Pieris brassicae* à trois conditions thermiques différentes. *Ann. Soc. ent. Fr.* (N.S.) 8: 461–473.

Bückmann, D., 1971. Melanisierungsverlauf und Melanisierungshemmung bei der Kohlweisslingspuppe *Pieris brassicae* L. *Wilhelm Roux Arch. EntwMech. Org.*

Bünning, E., 1969. Common features of photoperiodism in plants and animals. *Photochem. Photobiol.* 9: 219–228.

Bünning, E. & Joerrens, G., 1959. Versuche zur photoperiodischen Diapause-Induktion bei *Pieris brassicae* L. *Naturwissenschaften* 46: 518–519.

—— 1960. Tagesperiodische antagonistische Schwankungen der Blauviolett- und Gelbrot-Empfindlichkeit als Grundlage der photoperiodischen Diapause-Induktion bei *Pieris brassicae*. *Z. Naturf.* 15: 205–213.

—— 1962. Versuche über den Zeitmessvorgang bei der photoperiodischen Diapause-Induktion von *Pieris brassicae*. *Z. Naturf.* 17: 57–61.

—— 1963. Rot-und Infrarotwirkungen auf die Diapause-Induktion bei *Pieris brassicae*. *Z. Naturf.* 18: 324–327.

Burton, R.J., 1941. *Pieris brassicae* larvae overwintering. *Entomologist* 79: 194.

Calvez, B., 1976. Taux d'ecdysone circulante aux derniers âges larvaires et induction de la diapause nymphale chez *Pieris brassicae*. *C. r. Acad. Sci. Paris* D. 282: 1367–1370.

Campbell, J.L., 1953a. *Pieris brassicae* larvae in November and December. *Entomologist* 86: 51.

—— 1953b. *Pieris brassicae* larvae feeding in November. *Entomologist* 86: 107.

Champ, B.R., 1958. *The effect of sub-lethal dosages of insecticides on parasitism in Pieris brassicae*. Ph.D. thesis. University of London (Imperial College).

Chapman, T.A., 1909. Temperature experiments with pupae of *Pieris brassicae*. *Trans. ent. Soc.* 57–58.

Cherchi, M.A., 1952. Gregarismo artificiale in "*Pieris brassicae*". *Boll. Mus. Inst. Biol. Univ. Geneva* 24: 42–54.

Chernysh, S.I., 1973. Materials of the third conference of young scientists of the Scientific-Research Institute of Leningrad State University. (In Russian.) Leningrad, 1973. p. 124.

Claret, J., 1966a. Recherche du centre photorécepteur lors de l'induction de la diapause chez *Pieris brassicae* L. *C. r. Acad. Sci. Paris* 262: 1464–1465.

—— 1966b. Mise en évidence du rôle photorécepteur du cerveau dans l'induction de la diapause, chez *Pieris brassicae*. *Ann. Endocr.* 27: 311–320.

—— 1968. Modifications physiologiques provoquées pars photopériode pendant la pédisponce chez la chenille de *Pieris brassicae*. *C. r. Acad. Sci. Paris* 266: 1156–1159.

—— 1969. Influence de la photopériod sur les variations des proteins de l'hemolymph de *Pieris brassicae* avant l'entrée en diapause. *C. r. hebd. Séance. Acad. Sci., Paris* 268: 3126–3129.

—— 1971a. Période de sensibilité des chenilles de *Pieris brassicae* à la photopériode controllant la diapause. *C. r. Acad. Sci. Paris* 174: 1055–1058.

—— 1971b. *La diapause facultative de Pieris brassicae recherchée du centre photorécepteur et étude de l'action de la photopériode sur le métabolisme larvaire.* Thèse Doctorat, Paris. pp. 109.

—— 1972. Sensibilité spectrale des chenilles de *Pieris brassicae* (L.) lors de l'induction photopériodiques de la diapause. *C. r. Acad. Sci. Paris* 274: 1055–1058, 1727–1730.

—— 1973. La levée photopériodique de la diapause nymphale de *Pieris brassicae* (L.). *C. r. Acad. Sci. Paris* D 277: 733–735.

Claret, J. & Carton, Y., 1975. The effect of host species on diapause in *Pimpla instigator* (Hymenoptera: Ichneumonidae). *C. r. hebd. Séanc. Acad. Sci., Paris.* 281: 279–282.

Craig, R., 1960. The physiology of excretion in the insect. *A. Rev. Ent.* 5: 53–68.

Cretschmar, M., 1920. Besprechung über die Farbe der Puppen der *Pieris brassicae* und die Schutzfärbung. *Int. ent. Z.* 14: 125–127.

Danilevskii, A.S., 1948. The photoperiodic reaction of insects in conditions of artificial light. *Rep. Acad. Sci. USSR* 60: 481–484.

—— 1950. The temperature conditions of reactivation of diapausing stages of insects. *Trans. nat. hist. Soc.* 70: 90–107.

—— 1957. Photoperiodism as a factor in the formation of geographical races of insects. *Ent. Obozr.* 36: 6–27.

—— 1960. *In: Ontogeny of Insects.* Ed. I. Hardy. Academic Press, London.

—— 1961. *Photoperiodism and Seasonal Development of Insects.* English translation. Oliver & Boyd, Edinburgh & London.

—— 1968. *Photoperiodic Adaptations in Insects and Acari.* Leningrad University Press.

Danilevskii, A.S. & Geispits, K.F., 1948. The effect of the daily periodicity of light on seasonal cycles of insects. *Rep. Acad. Sci. USSR* 59: 237–240.

Danilevskii, A.S. & Goryshin, N.I., 1960. The relation between temperature and light conditions in regulating diapause in insects. *Trans. Peterhof Biol. Inst. Lsu* 18: 147–168.

David, W.A.L., 1962. Observations on the larvae and pupae of *Pieris brassicae* (L.) in a laboratory culture. *Bull. ent. Res.* 53: 417–436.

—— 1965. The granulosis virus of *Pieris brassicae* L. in relation to natural limilations and biological control. *Ann. appl. Biol.* 56: 331–334.

Dumortier, B. & Brunnarius, J., 1977a. L'information thermopériodique et l'induction de la diapause chez *Pieris brassicae*. Note. *C. r. Acad. Sci. Paris* D 284: 957–960.

—— 1977b. Existence d'une composante circadienne dans l'induction thermopériodique de la diapause chez *Pieris brassicae* L. Note. *C. r. Acad. Sci. Paris* D 285: 361–364.

Dürken, B., 1916. Über die Wirkung verschiedenfarbiger Umgebung auf die Variation von Schmetterlingspuppen. Versuche an *Pieris brassicae*. *Z. wiss. Zool.* 116: 587–626.

Dutrieu, J. & Moreau, R., 1970. Étude comparative de l'évolution taux de glycogène et de trehalose au cours du développement d'insectes et de la diapause chez Pieris *brassicae*. *C. r. Séanc. Soc. Biol.* 164: 1999–2003.

Ebeling, W., 1964. The permeability of insect cuticle. *In: The Physiology of the Class Insecta* III. Ed. M. Rockstein, Academic Press, London, pp. 507–556.

Edney, E.B., 1957. *The Water Relations of Terrestrial Arthropods.* Cambridge University Press, Cambridge.

Egerton Collier, A., 1941. *Pieris brassicae,* 1939 and 1940. *Entomologist* 74: 70.

Feltwell, J.S.E., 1978. Carotenoids in the class Insecta. *In: Plant and Animal Evolution.* Ed. G. Harborne. Academic Press, London & New York.

Feltwell, J.S.E. & Rothschild, M., 1974. Carotenoids in 38 species of lepidoptera. *J. Zool.* 174: 441–465.

Feltwell, J.S.E. & Valadon, L.R.G., 1972. Carotenoids of *Pieris brassicae* and of its foodplant. *J. Insect Physiol.* 18: 2203–2215.

—— 1974. Carotenoid changes in *Brassica oleracea* var. *capitata* in relation to the large white butterfly, *Pieris brassicae* L. *J. agri. Sci. Cambs.* 83: 19–26.

Fourche, J., 1977a. Influence de la température sur la respiration des nymphes de *Pieris brassicae* (Lépidoptère) diapausantes et non diapausantes. *C. r. hebd. Séanc. Acad. Sci. Paris* 284 D: 1693–1696.

—— 1977b. The influence of temperature on respiration of diapausing pupae of *Pieris brassicae* (Lepidoptera). *J. Thermal Biol.* 2: 163–172.

Fourche, J., Bosquet, G., Guillet, C. & Calvez, B., 1980. Cold acclimatization during the wintering of diapausing pupae of *Pieris brassicae* (Lepidoptera). *Comp. biochem. Physiol.* (in press).

Fourche, J., Guillet, C., Calvez, B. & Bosquet, G., 1977. Le métabolisme énergétique des nymphes de *Pieris brassicae* (Lépidoptère) au cours de la métamorphose et de la diapause. Essai d'établissement d'un bilan. *Ann. zool. ecol. anim.* 9: 51–61.

Friden, F., 1958. *Frass Drop Frequency in Lepidoptera.* Almqvist & Wiksells Boktryckeri, Uppsala, Sweden.

Frowhawk, F.W., 1934. *British Butterflies.* Ward, Lock & Co. Ltd., London & Melbourne.

246

Gardiner, B.O.C., 1963. Genetic and environmental variation in *Pieris brassicae*. *J. Res. Lepid.* 2: 127–136.

—— 1965. Hybrids between typical *Pieris brassicae* L. and race *Cheiranthi* Hübner. *Proc. XII Int. Cong. Ent. London.* 1964. p. 261.

—— 1974. Observations on green pupae in *Papilio machaon* and *Pieris brassicae* L. *Wilhelm Roux Arch. entwMech. Org.* 176: 13–22.

—— 1978. Instar number and pupal colouration in Palestinian *Pieris brassicae*. *Proc. & Trans. Br. ent. nat. Hist. Soc.* 11: 21–23.

—— 1979. Talk at workshop on the Biology of the Pieridae held at the Royal Entomological Society, London on 17th October, 1979.

Geispits, K.F., 1953. The reaction of univoltine lepidoptera to day-length. *Ent. Obozr.* 33: 17–31.

—— 1957. The mechanism of acceptance of light stimuli in the photoperiodic reaction of lepidoptera larvae. *Zool. Zh.* 36: 548–560.

Goryshin, N.I., 1963. The photoperiodic reaction of insects to periods of light and darkness of different durations. (In Russian.) *Rev. ent. URSS* 42: 22–28.

—— 1964a. Influence of diurnal light and temperature rhythms on diapause in lepidoptera. *Rev. ent. URSS* 43: 43–46.

—— 1964b. The effect of daily rhythms of light and temperature on the occurrence of diapause in Lepidoptera. (In Russian.) *Ent. Obozr.* 43: 86–93.

Goryshin, N.I. & Kozlova, R.N., 1967. Thermoperiodism as a factor in the development of insects (in Russian). *Zhur. obshch. Biol.* 28: 278–288.

Goryshin, N.I. & Tyschenko, V.P., 1968. Physiological mechanism of photoperiodic reaction and the problem of endogenous rhythms. (In Russian.) *In: Photoperiodic Adaptations in Insects and Acari*. Ed. A.S. Danilevskii, Leningrad University Press. pp. 192–269.

Guillet, C., 1976a. Le métabolisme de diapause chez *Pieris brassicae*. Les Nucléotides adényliques. *Année Biol.* 15: 77–90.

—— 1976b. Le métabolisme de diapause chez *Pieris brassicae*. Les nucleotides adényliques. *Bull. Biol. Fr. Belg.* 110: 31–44.

Guillet, C. & Fourche, J., 1976. Les modifications de métablisme energétique lors de la levée de diapause par injection d'α-ecdysone chez *Pieris brassicae*. *Bull. biol. Fr. Belg.* 110: 31–44.

Haig-Thomas, P., 1938. Arctic butterflies. *Entomologist* 71: 1–5.

Hansen, T. & Merivee, E., 1971. Cold-hardiness of the butterflies *Pieris brassicae* L. and *Pieris rapae* (L.). *Eesti Nsv. Tead. Akad. Toim.* 20: 298–303.

Harmsen, R., 1966a. Identification of fluorescing and ultra-violet absorbing substances in *Pieris brassicae* L. *J. Insect Physiol.* 12: 23–30.

—— 1966b. A quantitative study of the pteridines in *Pieris brassicae* L. during post embryonic development. *J. Insect Physiol.* 12: 9–22.

Hopkins, F.G., 1895a. The pigments of the Pieridae: A contribution to the study of excretory substances which function in ornament. *Phil. Trans.* 186: 661–682.

—— 1895b. The pigments of the Pieridae: A contribution to the study of excretory substances which function in ornament. *Entomologist* 28: 1–2.

Johansson, A., 1959. Diapause and pupal morphology and colour in *Pieris brassicae* L. (Lepidoptera: Pieridae). *Norsk. ent. Tidsskr.* 9: 90–86.

Junnikkala, E., 1969. Effect of a semi-synthetic diet on the level of the main nitrogenous compounds in the hemolymph of larvae of *Pieris brassicae* L. *Annl. acad. sci. Fenn.* 155: 1–10.

Kayser, H. & Angersbach, D., 1974. Action spectra for light controlled pupal pigmentation in *Pieris brassicae* melanisation and level of bile pigment endocrine control. *J. Insect Physiol.* 20: 2277–2286.

—— 1975. Dose effects in light controlled pupal melanisation in *Pieris brassicae*: specificites to spectral ranges. *J. Insect Physiol.* 21: 589–594.

Kayser-Wegmann, I., 1975. Untersuchungen zur Photobiologie und Endokrinologie der Farbmodifikationen bei der Kohlweisslingspuppe *Pieris brassicae*. Zeitverlauf der sensiblen und kritischen Phasen. *J. Insect. Physiol.* 21: 1065–1072.

—— 1976a. Differences in black pigmentation in lepidopteran cuticles as revealed by light and electron microscopy. *Cell. Tiss. Res.* 171: 513–521.

—— 1976b. Ultrastructural differences between larval and pupal cuticles of *Pieris brassicae* (Lepidoptera). *Protoplasma* 90: 319–331.

Kennedy, J.S., 1961. *Insect Polymorphism*. Symposium of the Royal Entomological Society of London. No. 1. 115 pp.

Kind, T.V., 1976. Endocrine regulation of the reactivation of diapausing pupae of the cabbage worm *Pieris brassicae*. (In Russian.) *Dokl. Akad. nauk. SSSR* 229: 1266–1269.

Kirby, W. & Spence, W.L., 1858. *An Introduction to Entomology*. Longman, Brown, Green, Longmans & Roberts, London.

Kono, Y., 1975. Daily changes of neurosecretory Type II cell structure of *Pieris* larvae entrained by short and long days. *J. Insect Physiol.* 21: 249–264.

Kozhanchnikov, I., 1936. The role of ecological factors in the distribution of *Pieris brassicae* L. (in Russian). *Plant Prot.* 2: 30–57.

Lafont, R., 1972. Modalités de l'excrétion azotée (Ptérines et acide urique) chez les lépidoptères Pieridae au cours de la vie nymphale. *Bull. Soc. zool. Fr.* 97: 401–411.

Lafont, R., Mauchamp, B., Pennetier, J.-L., Tarroux, P. & Blais, C., 1976a. Biochemical correlations during metamorphosis in *Pieris brassicae*. *Insect Biochem.* 6: 97–103.

Lafont, R., Mauchamp, B., Tarroux, P. & Blais, C., 1976b. Biochemical parameters of imaginal wing discs development in *Pieris brassicae*. *Abstract from Insect Biochem. Symp. Hamburg. August 1976.* p. 649.

Lafont, R. & Pennetier, J.L., 1975. Uric acid metabolism during pupal–adult development of *Pieris brassicae* L. *J. Insect Physiol.* 21: 1323–1336.

Larsen, T.B., 1974. *The Butterflies of the Lebanon*. C.N.R.S., Lebanon.

L'Hélias, C.L., 1959a. Facteur inducteur de tumeur provoquée par l'acide folique chez *Pieris brassicae* en état de diapause. *Année. Biol.* 63: 237–247.

—— 1959b. Purification partielle du facteur viral induisant les tumeurs provoquées artificiellement chez *Pieris brassicae* (Insecta: Lépidoptère). *C. r. Acad. Sci. Paris* 240: 3646–3648.

—— 1960. Tumour induction factor provoked by folic acid in *Pieris brassicae* during the diapause. *Folia biol. Praha* 6: 310–318.

—— 1961. Transmission génétique du facteur inducteur de tumeur de *Pieris brassicae* à la drosophile. *C. r. Acad. Sci. Paris* 252: 2015–2016.

—— 1966. Induction of tissue disorders in insects by altering the equilibrium between pterinic growth factors and growth hormones, and the mechanism of this induction. *Annl. Endocr.* 27: 343–352.

—— 1969. Altération de transmission des information génétiques chez la drosophile par injection d'extraits de chrysalides diapausantes (*Pieris brassicae*) traités par des pterines. *Abst. Gen. comp. Endrocr.* 13: 510.

—— 1972. Mutations et altérations dans la transmission des informations génétiques chez la drosophile par injections d'extraite de chrysalide diapausantes (*Pieris brassicae*) traités par des ptérines. *Mutation Res.* 14: 207–224.

—— 1973. Mutations et altérations dans la transmission des informations génétiques chez la drosophile par injection d'extraits de chrysalides diapausantes

(*Pieris brassicae*) traités par des ptérines. II. variations dans l'expression des gènes par addition d'hormones de croissance au traitement. *Mutation Res.* 20: 53–66.

—— 1975. Hormones juveniles de *Pieris brassicae* diapausante et mutations. *Ann. Épiphyt.* 36: 63–85.

Littlewood, E., 1941. On rearing lepidoptera. *Entomologist* 74: 161–165, 270–274.

Locke, M., 1964. The Structure and Formation of the integument in insects. *In: The Physiology of the Insecta.* III. Ed. M. Rockstein. Academic Press, London. pp. 379–470.

McDaniel, E.N. & Berry, S.I., 1967. Activation of the prothoracic glands of *Antheria polyphenus. Nature, Lond.* 214: 1032–1034.

Maddrell, S.H.P., 1971. Mechanisms of insect excretory systems. *Adv. Ins. Physiol.* 8: 199–331.

Maddrell, S.H.P. & Gardiner, B.O.C., 1976. Excretion of alkaloids by Malpighian tubules of insects. *J. exp. Biol.* 64: 267–281.

Maercks, H., 1934. Untersuchungen zur Ökologie des Kohlweisslings (*Pieris brassicae* L.), Die Temperaturreaktionen und das Feuchtigkeitoptimum. *Z. Morph. Ökol. Tiere* 28: 692–721.

Martouret, D., Lhoste, J. & Roche, A., 1965. Action sur le mésentéron de *Pieris brassicae* L. de la toxine de l'inclusion parasporale de *Bacillus thuringiensis* Berliner. *Entomophaga* 10: 349–365.

Maslennikova, V.A., 1958. The conditions determining diapause in the parasitic hymenopteron *Apanteles glomeratus* L. (Braconidae) and *Pteromalus puparum* (Chalcidae). (In Russian.) *Ent. Obozr.* 37: 538–545.

—— 1959. The relation between the seasonal cycles of geographical populations of *Apanteles glomeratus* L. and its host *Pieris brassicae* L. (In Russian.) *Ent. Obozr.* 38: 517–522.

—— 1960a. Relationship of the seasonal development of *Apanteles glomeratus* to that of host *Pieris brassicae* in different geographical populations. (In Russian.) *Ent. Obozr.* 38: 463–467.

—— 1960b. *The conditions which determine synchronism of the seasonal cycles of parasitic insects and their hosts.* Thesis, University of Leningrad. Leningrad University Press.

—— 1968. The hormonal mechanism of regulation of pupal diapause in *Pieris brassicae* L. (In Russian.) *Ent. Obozr.* 47: 429–439.

—— 1970a. Hormonal regulation of diapause in *Pieris brassicae*. (In Russian.) *C. r. Acad. Sci. USSR* 192: 942–945.

—— 1970b. Hormonal regulation of diapause in *Pieris brassicae* L. *Dokl. (Proc.) Ac. Sci. USSR* 192: 412–414.

—— 1971. An investigation of the hormonal mechanism of diapause in *Pieris brassicae* by use of the parasite *Pteromalus puparum* as test subject. *All Union ent. Soc. Leningrad* 1: 415.

Maslennikova, V.A. & Chernysh, S.I., 1973. Effect of ecdysterone on determination of diapause of *Pteromalus puparum* L. (In Russian.) *Dokl. Akad. nauk. SSSR. Ser. Biol.* 213: 480–482.

Maslennikova, V.A. & Mustafayeva, T.M., 1971. An analysis of photoperiodic adaptations in geographic populations of *Apanteles glomeratus* L. (Hymenoptera, Braconidae) and *Pieris brassicae* L. (Lepidoptera: Pieridae). *Ent. Rev. URSS.* 50: 281–284.

Mauchamp, B. & Lafont, R., 1975. Development studies in *Pieris brassicae* (Lepidoptera: Pieridae). II. A study of nitrogenous excretion during the last larval instar. *Comp. biochem. Physiol.* 51B: 445–449.

Meldola, R., 1873. On a certain class of cases of variable protective colouring in insects. *Zool. Soc. Proc.* p. 153.

Mellows, C., 1924. Altitudes at which lepidoptera occur. *Entomologist* 57: 90–91.

Merrifield, F. & Poulton, E.B., 1899. The colour-relationship between the pupae of *Papilion machaon, Pieris napi* and many other species, and the surroundings of the larvae preparing to pupate, etc., E. Experiments upon the pupae of *Pieris brassicae. Trans. ent. Soc. Lond.* 1899: 369–433.

Miah, M.A.H., 1976. *Some effects of gamma radiation on the different stadia of Pieris brassicae* L. Ph.D. thesis. University of London (Imperial College).

Moreau, R., 1973. *Recherches sur quelques aspects des phénoménes physiques métaboliques et physiologiques qui accompagnent et conditionnent l'expansion des ailes de Lépidoptères.* Thèse Doctorat d'état Sci. nat. Université de Bordeaux. 145 pp. Ref. CNRS AO 8458.

—— 1974. Variations de la pression interne au cours de l'émergence et de l'expansion des ailes chez *Bombyx mori* et *Pieris brassicae. J. Insect Physiol.* 20: 1475–1480.

Moreau, R., Castex, C. & Lamy, M., 1975. Examen preliminaire de quelques aspects des effects métaboliques d'un nouvel insecticide synthese chez deux insectes nuisibles, *Pieris brassicae* et *Thaumetopoea pityocampa. Ann. zool. ecol. Am.* 7: 161.

Moreau, R., Dutrieu, J. & Olivier, D., 1974. Degradation du glucose par le cycle des pentoses au cours de la fin de l'ontogenèse chez *Pieris brassicae* normal et aptere. *C. r. Séanc. Soc. Biol.* 168: 1285–1288.

Moreau, R. & Gourdoux, L., 1971a. Étude comparative du métabolisme respiratoire au cours de la seconde partie de l'ontogenèse et plus particulièrement pendant l'émergence et l'expansion des ailes chez *Pieris brassicae* et *Tenebrio molitor. C. r. Soc. Biol.* 169: 953–958.

—— 1971b. Étude comparative du métabolisme respiratoire au cours de la seconde partie de l'ontogènese et plus particulièrment pendant l'emergence des ailes de *Pieris brassicae* et *Tenebrio molitor. Comp. r. hebd. Séanc. Acad. Sci. Paris* 273: 2302–2305.

Moreau, R., Gourdoux, L. & Dutrieu, J., 1977. Utilisation comparée du cycle des pentoses en fonction des variations thermiques chez deux lépidoptères, *Bombyx mori* L. et *Pieris brassicae* L. *Comp. biochem. physiol.* B. 56: 175–180.

Newman, L.H., 1942. Note. *Entomologist* 75: 76.

Nicolson, S.W., 1975. *Osmoregulation, metabolism and the diuretic hormone of the cabbage white butterfly, Pieris brassicae.* Ph.D. thesis. University of Cambridge.

—— 1976a. Diuresis in the cabbage white butterfly, *Pieris brassicae*: fluid secretion by the Malpighian tubules. *J. Insect Physiol.* 22: 1347–1356.

—— 1976b. Diuresis in the cabbage white butterfly, *Pieris brassicae*: water and ion regulation and the role in the hind-gut. *J. Insect Physiol.* 22: 1623–1630.

—— 1976c. The hormonal control of diuresis in the cabbage white butterfly, *Pieris brassicae. J. exp. Biol.* 65: 565–575.

Nordström, F., 1951. *Pieris brassicae* on migration. *Ent. Tidskr.* 72: 79–80.

Novak, V.J.A., 1975. *Insect Hormones.* John Wiley & Sons, New York.

Ohtaki, T., 1960. Humoral control of pupal colouration in the cabbage white butterfly, *Pieris rapae crucivora. Annot. zool. Japan* 33: 97–103.

Oltmer, A., 1968. Die Steuerung des Melanineinbaus in das Farbmuster der Kohlweisslingspuppe *Pieris brassicae* L. *Willhelm. Roux. Arch. entwMech. Org.* 160: 401–427.

Opheim, M., 1958. *Catalogue of the lepidoptera of Norway,* Part 1. Oslo.

Pictet, A., 1918. Observations biologiques sur *Pieris brassicae* en 1917. *Bull. Soc. lépidopt. Genève.* 4: 53–66.

—— 1922. Recherches sur l'hibernation de *Pieris brassicae* a l'état de che chenille. *Bull. Soc. lépidopt. Genève* 5: 47–57.

Plantevin, G. & Nardon, P., 1970. Histologie et activité secrétoire de l'intestin moyen des larves de *Pieris brassicae* et *Galleria mellonella*. Évolution au cours de la mue larvaire et de la nymphose chez *Galleria mellonella*. *Ann. zool. ecol. anim.* 2: 25–50.

Poulton, E.B., 1887a. An enquiry into the cause and extent of a special colour-relation between certain exposed lepidopterous pupae and the surfaces which immediately surround them. *Phil. Trans. R. Soc. Lond.* B. 178: 311–441.

—— 1887b. An Inquiry into the cause and extent of a special colour relation between exposed lepidopterous pupae and the surfaces which immediately surround them. *Proc. R. Soc. Lond.* 42: 94–110.

—— 1890. *The Colours of Animals.* Kegan, Paul, Trench, Trübner & Co. Ltd., London.

—— 1892. Further experiments upon the colour relation between certain lepidopterous larvae, pupae, cocoons and imagines and their surroundings. *Trans. ent. Soc. Lond.* Part 4: 293–487.

Prach, (–), 1920. Besprechung über die Farbe der Puppen der *Pieris brassicae* und die Unterschiede von den Erwachsenen *Int. ent. Z.* 14: 144.

Ramsey, J.A., 1953. Active transport of potassium by the Malpighian tubules of insects. *J. exp. Biol.* 30: 358–369.

—— 1976. The rectal complex in the larrose of lepidoptera. *Phil. Trans. R. Soc.* B. 274: 203–226.

Ripa, R., 1978. *Studies of the susceptibility of Pieris brassicae (L.) to a granulosis virus.* Ph.D. thesis. University of London (Imperial College).

Rothschild, M., 1975. Remarks on carotenoids in the evolution of signals. *In: Coevolution of Animals and Plants.* Ed. L.E. Gilbert and P.H. Raven. University of Texas Press. pp. 20–51.

Rothschild, M., Gardiner, B.O.C., Valadon, L.R.G., & Mummery, R.S., 1975a. The large white butterfly: oviposition cues, carotenoids and changes of colour. *Proc. R. ent. Soc. Lond.* 40: 13.

—— 1975b. Lack of response to background colour in *Pieris brassicae* pupae reared on carotenoid-free diet. *Nature, Lond.* 254: 592–594.

Rothschild, M., Valadon, L.R.G., Mummery, R.S. & Gardiner, B.O.C., 1975c. Storage of carotenoids in diapausing pupae of *Pieris brassicae* L. and *Pieris napi* L. and their influence on larval colouration (Lepidoptera: Pieridae). *Nature, Lond.* 254: 592.

Rothschild, M., Valadon, L.R.G. & Mummery, R.S., 1977. Carotenoids of the pupae of the large white butterfly (*Pieris brassicae*) and the small white butterfly (*Pieris rapae*). *J. Zool. Lond.* 181: 323–339.

Russel, G.M., 1936. An unusually late brood of *Pieris brassicae* larvae. *Entomologist* 49: 44–45.

Samedov, A.N., 1974. Respiration during locomotion and metabolic activity in *Pieris brassicae* larvae and *Barathra brassicae* larvae. (In Russian.) *Zool. Zh.* 53: 188–197.

—— 1976. Diurnal rhythms of gas exchange in the cabbage butterfly (*Pieris brassicae*). (In Russian.) *Zool. Zh.* 55: 1838–1842.

Saunders, D.S., 1976. *Insect Clocks.* Pergamon Press, Oxford.

Seuge, J., 1973. Recherche du rôle des stemmates dans la perception de la photopériode et l'induction de la diapause, chez *Pieris brassicae*. *Bull. Soc. zool. fr.* 98: 435–440.

Seuge, J. & Veith, K., 1976. Diapause de *Pieris brassicae*: rôle des photorecepteurs, étude des carotenoides cerebraux. *J. Insect Physiol.* 22: 1229–1235.

Seuge, J., Vuillaume, M. & Bergerard, J., 1974. Pigment tegumentaire, photo-reception et diapause chez *Pieris brassicae*: recherche de la sensilibité de chacun des stades larvaires aux variations photopériodiques et photoniques. *Arch. zool. exp. gen.* 115: 77–91.

Seuge, J., Vuillaume, M., Biache, G. & Bergerard, J. 1972. Photopériode et pigment tegumentaire vert des chenilles de *Pieris brassicae* conditionement de la diapause par les lumières de lougeurs differentes. *C. r. Acad. Sci.* 274: 2526–2529.

Seuge, J., Vuillaume, M., Jacques, R. & Bergerard, J., 1976. Recherches des variations lumineuses responsables de l'induction de la diapause nymphale chez *Pieris brassicae*. *C. r. Séanc. Soc. Biol.* 166: 526–530.

Smith, A., 1976. *Environmental factors influencing pupal colour determination in some rhopaloceran lepidoptera.* Ph.D. thesis. University of Liverpool.

Sømme, L., 1967. The effect of temperature and anoxia on haemolymph composition and supercooling in the overwintering insects. *J. Insect Physiol.* 13: 805–814.

Steiner (-)., 1930. Licht und Insektenentwicklung. *Z. Bakt.* (2) 77.

Stobbart, R.H. & Shaw, J., 1964. Salt and water balance: excretion. *In: Physiology of the Insecta.* III. Ed. M. Rockstein. Academic Press, London. pp. 189–258.

Strogaya, G.M., 1961. Peculiar features of fat and water balance in the individual development of *Aporia crataegi* and *Pieris brassicae* as a form of adaptation to the surrounding medium. *Akad. nauk. SSSR Dok.* 139: 474–477.

—— 1962. Variation de la quantité d'eau et de la quantité des corps gras contenus dans l'organisme des papillons de la piéride de l'aubergine et de la piéride du chou au cours de leur développement individuel (d'après les résultats de l'analyse biochimique). *J. Zool. Acad. Sci. USSR* 41: 92–100.

Talhouk, A.M., 1969. *Insects and mites injurious to crops in Middle Eastern countries.* Monographien zur Angew. ent. No. 21. Verlag Paul Parey, Hamburg & Berlin.

Tikhonravova, N.M., 1973. Respiratory characteristics of the cabbage butterfly during different stages of larval development. *Sov. J. Ecol.* 3: 320–323.

Turunen, S., 1973a. Utilisation of fatty acids by *Pieris brassicae* reared on artificial and natural diets. *J. Insect Physiol.* 19: 1999–2009.

—— 1973b. Role of labelled dietary fatty acids and acetate in phospholipids during metamorphosis of *Pieris brassicae*. *J. Insect Physiol.* 19: 2327–2340.

—— 1974a. Notes on lipid requirement of a phytophagous lepidopteran. *Ann. ent. fennici.* 40: 151–155.

—— 1974b. Lipid utilisation in adult *Pieris brassicae* with special attention to the role of linolenic acid. *J. Insect Physiol.* 20: 1257–1269.

Ushatinskaya, R.S., 1976. Insect dormancy and its classification. *Zool. Jb. Syst.* 103: 76–97.

Veith, K., Vuillaume, M., Seuge, J., Biache, G. & Bergerard, J., 1974. Existe-t-il une action particulière des differentes du spectre visible sur la physiologie des chrysalides et prénymphes du *Pieris brassicae* conditionnées à la diapause. *Experimentia* 30: 152–153.

Vuillaume, M., 1976. Rôle de l'intensité lumineuse rouge (630–670 nm) dans le determinisme photopériodique de la diapause nymphale de *Pieris brassicae*. *J. Insect Physiol.* 22: 1053–1056.

Vuillaume, M. & Berkaloff, 1974. LSD treatment of *Pieris brassicae* and consequences on the progeny. *Nature, Lond.,* 251: 314–315.

Vuillaume, M. & Seuge, J., 1972. Action de psychodysleptiques (LSD et Mescaline) sur la diapause nymphale de *Pieris brassicae*. *Bull. Biol.* 116: 285–289.

Vuillaume, M., Choussy, M. & Barbier, M., 1970. Pigments tetrapyrroliques verts et bleus des ailes de lépidoptères pterolbilin et neopterobilin. *Bull. Soc. zool. Fr.* 95: 19–28.

Vuillaume, M., Seuge, J. & Bergerard, J., 1971. Photopériode et pigment tégumentaire vert des chenilles de *Pieris brassicae*: conditionnement de la diapause. *C. r. Acad. Sci. Paris* D 273: 1608–1610.

Vuillaume, M., Seuge, J., Jacques, R. & Bergerard, J., 1972b. Diapause nymphale de *Pieris brassicae*: analyse de l'influence des radiations du spectre visible de la lumière sur ce phénomène. *C. r. Soc. Biol.* 166: 541–543.

Vuillaume, M., Biache, G. & Bergerard, J., 1972a. Photopériode et pigment tégumentaire vert des chenille de *Pieris brassicae*: conditionnement de la diapause par des lumières de longuers d'onde différentes. *C. r. Acad. Sci. Paris* 274: 2526–2529.

—— 1973. Induction de la diapause chez *Pieris brassicae*: étude de l'importance relative de la scotophase et de la photophase en lumière blanche et colorée suivant plusieurs rythmes (Circadien ou non) et même en l'abscence de tout rythme. *Arch. zool. exp. Gen.* 114: 653–666.

—— 1974a. Pigment, photoreception and nymphal diapause in *Pieris brassicae*. *Int. J. Chronobiol.* 2: 181–188.

—— 1974b. Analyses récentes sur la photoréception pigmentaires chez *Pieris brassicae*. 99th Congress des Soc. Savantes, Besançon.

Vuillaume, V.K., Seuge, M., Biache, J. & Bergerard, J., 1974. Do different wavelengths of the visible spectrum exert a special action on the physiology of the chrysalids and perinymphs of *Pieris brassicae* conditioned to diapause? *Experimentia* 30: 151–152.

Warner, G.U., 1920. Note on *Pieris brassicae*. *Entomologist* 53: 209.

Way, M.J., Hopkins, B. & Smith, P.M., 1949. Photoperiodism and diapause in insects. *Nature, Lond.* 164: 615.

Wigglesworth, V.B., 1925. Uric acid in Pieridae: a quantitative study. *Proc. R. Soc.* 97: 149–155.

—— 1945. Transpiration through the cuticle of insects. *J. exp. Biol.* 21: 97–114.

—— 1970. *Insect Hormones.* Oliver & Boyd, Edinburgh.

Wood, T.W., 1867. Remarks in the colouration of chrysalides. *Proc. ent. Soc. London* 99–101.

Further references

Akhmedov, R.M., 1967. *Trudy Inst. zool. Baku* 26: 184–190 (photoperiodic reactions in Tashkent).

Auber, J., 1967. *C. r. Acad. Sci. Paris* 264: 621–624 (muscle ultrastructure).

Barker, R.J., 1963. *Experimentia* 19: 1–3 (diapause inhibition).

Barton-Browne, L., 1975. *Adv. Insect Physiol.* 11: 1–116 (regulatory mechanisms in insect feeding).

Beck, S.D., 1968. *Insect Photoperiodism.* Academic Press, London (diapause).

Becker, E., 1937. *Z. Physiol. Chem.* 246: 177–180 (pterins).

Beliaev, M.M., 1946. *Zool. Zhur.* 25: 403–410 (warning colouration).

Brecher, L., 1918. *Wilhelm. Roux. Arch. EntwMech. Org.* 45: 273 (pupal colouration).

—— 1921. *Wilhelm. Roux. Arch. EntwMech. Org.* 48: 1 (pupal colouration).

Bursell, E., 1967. *J. Insect Physiol.* 4: 33–67 (excretion).

Chapman, R.F., 1974. *Bull. Ent. Res.* 64: 339–363 (chemical inhibition).

Comstock, J.H., 1918. *The wings of Insects.* Comstock Publishing Co., New York.

Doskocil, J., 1954. *Cestoslov. Spolecnost. Zool. Vest.* 18: 139–145 (effect of light on the duration of diapause).

Dürken, B., 1923. *Arch. Mikr. Anat.* 99: 222–389 (pupal colouration).

Dutrieu, J. & Moreau, R., 1970. *C. r. Soc. Biol.* 164: 1999–2003 (glycogen and trehalose levels).

Fox, D.L., 1953. *Animal Biochromes and Structural Colours.* Cambridge University Press.

Fox, H.M. & Vevers, H.G., 1960. *The Nature of Animal Colours.* Sidgwick & Jackson.

Friederichs, K., 1934. *S.B. Naturf. Ges. Rostock* 4: 18–29 (light and darkness effects on development).

Friederichs, K. & Steiner, P., 1930. *Zbl. Bakt.* 80: 71–77 (light effects on development).

Geispits, K.F., Sapozhnikova, F.D. & Simonenko, N.P., 1974. *In: Photoperiodism in animals and plants.* Proc. Symp. Leningrad 1974 (regulation of diapause).

Janisch, E. & Maercks, H., 1933. *Z. Morph. Ökol. Tiere* 26: 372–384 (effect of light on development).

Jones, R.R., 1971–1972. *Study of aspects of excretion in the larvae of Pieris brassicae* L. Ph.D. thesis. University of Cambridge (Caius College).

Kalmes, R. & Lepinay, J.P., 1976. *Bull. Ecol.* 7: 105–112 (energy transfer between *P. instigator* and *P. brassicae*).

Lees, A.D., 1955. *The Physiology of Diapause in Arthropods.* Cambridge University Press, London.

—— 1956. *A. Rev. Ent.* 1: 1–16 (biochemistry of diapause).

Lockey, K.H., 1957. *A study of insect cuticle by surface chemical techniques.* Ph.D. thesis. University of London (Imperial College).

Lofts, B., 1970. *Animal Photoperiodism.* Arnold, London.

Meijere, J.C.H. de, 1919. *Tijdschr. Ent.* 61: 57–75 (colour development).

Peter, K., 1930. *Biol. Zbl.* 50: 19–25 (protective colouration).

Ramsey, J.A., 1953. *J. exp. Biol.* 30: 358–369 (active transport of potassium).

Rotman, M.N., 1936. *Izv. Kurs. Prikl. Zool. Leningrad* 6: 2–14 (gaseous exchange).

Sevastopolo, D.G., 1948. *Proc. R. ent. Soc. Lond.* A. 23: 93–95 (pupal colouration).

Stauden-Mayer, T. & Stellwaag, F., 1940. *Z. angew. Ent.* 26: 589–607 (pH of gut).

Tyshchenko, V.P., 1976. *In: Photoperiodism in animals and plants.* Proc. Symp. Leningrad. USSR, 1974 (photoperiodic control of diapause).

Ushatinskaya, R.S., 1952. *Izvest. Akad. Nauk. SSSR. Ser. Biol.* 1952: 101–114 (physiological aspects of diapause).

Wilde, J. de, 1962. *A. Rev. Ent.* 7: 1–26 (review of photoperiodism).

—— 1964. *In: The Physiology of the Insecta.* Ed. M. Rockstein. Academic Press, London. 9–90 (reproduction).

—— 1965. *Arch. Anat. Microsc. Morph. exp.* 54: 547–564 (photoperiodic control).

Williams, C.B., 1950. *Nature, Lond.* 166: 1035 (phase colouration in larvae).

Zaslavskii, V.A., 1976. *In: Photoperiodism in animals and plants.* Proc. Symp. Leningrad. USSR. pp. 212 (photoperiodic control).

9. Hormones

Early work

Early work on insect hormones, particularly with *P. brassicae*, relied very much on testing the effects of crude homogenates, or testing the effect of mammalian hormones on the insect. In 1927 Magaudda tested the effect of thyroid and testicular hormones from a mammal on larvae and pupae of *P. brassicae*, but did not notice any appreciable effects. Peredielsky (1940) and Florey (1951) prepared homogenates of fifth instar *P. brassicae* larvae close to ecdysis and found that these stimulated the pigment in the melanophores of the fins of the Minnow *Phoxinus phoxinus* (Linnaeus) (= *Phoxinus laevis*) (Cyprinidae), as well as having an effect on metamorphosis of tadpoles of the frog (*Rana temporaria* Linnaeus). Peredielsky (*loc. cit.*) also found that homogenates of young *P. brassicae* pupae had a significant effect on accelerating the growth of tadpoles.

Attention must be drawn to the important anatomical and developmental studies on the corpora allata and prothoracic glands of various insects including those of *Pieris* sp. made by Kaiser (1949). He published several plates depicting cells taken from both these glands and his major finding was that the size ratio of the prothoracic glands to the corpora allata increased by a factor of 29 from the first instar to pupation. He postulated that secretions from these glands contributed to the process of moulting, and found that the corpora allata becomes inactive in the imago. However, Novak (1975) commenting on Kaiser's work, said that the interaction of both the corpora allata and the prothoracic glands which he had described had not received much support.

Introduction

Today 30 behaviour-modifying chemicals, mostly hormones, have been found in 50 species of lepidoptera, and these are representative of nine

256

families out of 100 families analysed (Tamaki, 1977). A great deal of work has been devoted to the hormones of insects and in particular those of *P. brassicae* during the last 10 years (Wigglesworth, 1964, 1970; Srihari, 1974; Novak, 1975; Lafont, 1975; Lafont *et al.*, 1975, 1977; L'Hélias, 1975; Hepburn, 1976; Matsuda, 1976; Ushatinskaya, 1976; Chippendale, 1977; Highnam & Hill, 1977; Blais, 1978; Beydon *et al.*, 1979; Mauchamp *et al.*, 1979a,b; Guillet & Fourche, 1976).

In *P. brassicae* hormones have so far been demonstrated or suggested to play a part in hardening of the cuticle, pupation, oogenesis and yolk formation, diapause, pigmentation including the synthesis of and regulation of brown pigments, bile pigments, ommochromes and melanin formation, metabolism of food reserves during imaginal life and degeneration of the muscular system of the imago.

The only hormones so far identified in *P. brassicae* are α- and β-ecdysone (Lafont, 1975; Lafont *et al.*, 1975, 1977), juvenile hormone (Varjas *et al.*, 1976) and one of its forms juvenile hormone 1 (Mauchamp *et al.*, 1979a,b), and the pheromone tricosane (Bergström & Lundgren, 1973). The presence of another hormone, bursicon, a cuticle-hardening hormone (initially found and described from the housefly (*Musca domestica* L.) by Fraenkel & Hsiao, 1965) has been indicated by Post (1972), Post & de Jong (1973), Post *et al.* (1974) (Table 9.1). There are also accounts of hormones in *P. brassicae* which are unnamed and uncharacterised, which have been postulated to explain the workings of a particular system.

The mammalian "hormone" adrenalin and the neural transmitter, noradrenaline, have been found in *P. brassicae* larvae (von Euler, 1961) but their presence in insects has not been sufficiently explained. Adrenalin was present in negligible amounts (< 0.005 μgm/g), while noradrenaline was found in very much greater concentrations (0.33 μgm/g). *P. brassicae* is not apparently peculiar in having these hormones as they have a general distribution in other invertebrates and acraniates, viz: Mollusca, Annelida, Crustacea, Echinodermata, Urochordata and Acrania. However, von Euler (1961) did not elaborate on the functional significance of these hormones in insect tissues.

It appears from this collation of hormone data on *P. brassicae* that there is a great need for someone to study the morphology and anatomy of each endocrine gland of *P. brassicae*, as very little has been done on this species, particularly since Kaiser's (1949) study of the corpora allata and the prothoracic glands. It has been necessary therefore to rely on the generalised statements on the structure of endrocrine glands as published in Novak's (1975) *Insect Hormones*, and those of other workers to compensate for this deficiency.

257

Table 9.1 Site of secretion of hormones in *P. brassicae.* nsc = neurosecretory cells, ? = unnamed hormones.

Endocrine gland	Hormone	Authority
Principle sites		
nsc of brain	?	Rehm, 1955
	diuretic hormone	Nicolson, 1976c
corpora cardiaca	diuretic hormone (stored only)	Nicolson, 1976c
nsc of suboesophageal ganglia	yolk deposition hormone	Street, 1975
corpora allata	juvenile hormone	Kopec, 1922
	juvenile hormone (possibly)	Benz, 1970
prothoracic glands	"humoral factor", ecdysone	Ohtaki, 1960; Varjas *et al.*, 1976 Rothschild *et al.*, 1975
androconial scales	tricosane	Bergström & Lundgren, 1973
General areas		
head	hormone which regulates pupation. Has no effect on emergence of parasite *Apanteles glomeratus* larvae from host	Johansson, 1951
anterior part of the body	melanin stimulating hormone	Kayser-Wegmann, 1975
anterior part of the body	melanin inhibiting hormone	Kayser-Wegmann, 1975
anterior part of the body	"some hormone factor" responsible for ommochrome distribution	Rothschild *et al.*, 1975
cuticle	bursicon	Post, 1972; *et al.*, 1974; Post & de Jong, 1973
brain	hormone responsible for melanisation and bile pigment synthesis	Kayser & Angersbach, 1974
brain	hormone responsible for synthesis of protein F	Lamy, 1967

Sites of secretion

In insects generally, hormones may be secreted from nine possible sites situated in the anterior of the insect body (Novak, 1975, p. 4). In *P. brassicae* however, hormones are associated with five of these sites but do not appear to have been associated with the remainder, i.e. the pericardial cells, oenocytes, corpus luteum and the neurosecretory cells of the ventral nervous system (Table 9.1).

Basically, there are four major components of the typical insect endocrine system: the neurosecretory cells of the brain, the corpora cardiaca, the corpora allata and the thoracic glands or their equivalent. The most important of these organs is the corpora allata which stores and releases neurosecretory hormones from the brain (Highnam & Hill, 1977). The androconial wing scales may also be included here as they produce a pheromone which is discussed in chapter 12.

Neurosecretory cells of the brain

a) *Structure.* The neurosecretory cells of the brain are usually arranged in two groups, placed symmetrically on the upper surface of each hemisphere near the median furrow in the pars intercerebralis protocerebri (Novak, 1975). In insects generally there are 4 – 15, or more, neurosecretory cells in each group, often being very conspicuous in a living brain dissected in Ringer because of their milky-white or slightly bluish colour. The transparent nucleus often appears as a dark spot in the middle of each cell (Novak, 1975). Panov & Kind (1963) said that typically the neurosecretory cells of the insect brain are made up of six groups of cells, three median (M_1, M_2, M_3) and three lateral (L_1, L_2, L_3) in the protocerebrum, but in *P. brassicae* L_1 cells are absent and M_3 cells only occur in the pupa. These workers published diagrams of the arrangements of these cells in the larva, pupa and imago.

b) *Function.* The neurosecretory cells of the brain produce an "Activation Hormone" (Novak, 1975).

Rehm (1955) found that there was a great accumulation of stainable material indicating cellular activity in the cytoplasm of the neurosecretory cells of the brain after the liberation of the secretion in *P. brassicae*.

Nicolson (1976a,b) investigated the weight loss of freshly emerged *P. brassicae* imagines caused by the reduction in the haemolymph volume (*in vitro*). She found that diuresis in the Malpighian tubules was under control of a diuretic hormone which was produced in the brain and released from the corpora cardiaca where it was stored at eclosion. Diuretic hormones have also been found in other insect orders such as Orthoptera, Phasmidia and Hemiptera where they originate from the pars intercerebralis and thoracic glands (Novak, 1975, p. 378).

Neurosecretory cells of the suboesophageal ganglia

a) *Structure.* According to Novak (1975) several paired groups of neurosecretory cells have been described in the suboesophageal ganglia in various insect orders.

b) *Function.* Street (1975, thesis) showed that the suboesophageal ganglia were responsible for yolk deposition. In his work on hormonal control of growth and reproduction in *P. brassicae* he ligatured the heads of pupae and imagines but failed to prevent the reproductive development in either sex. However, yolk deposition in the imago was prevented thus suggesting that the suboesophageal ganglia were responsible for yolk deposition. He found that the application of juvenile hormone analogues always caused the death of larvae at the next ecdysis, but when applied at 0.5 μg/ova during embryogenesis they disrupted embryonic development in early larval life. At 10 μg/insect they disrupted larval and pupal development when applied after the last instar and before the pupal ecdysis. Novak (1975) stated that nothing was known of the way in which the neurosecretory substances of these cells reaches the target organs, e.g. the ovaries.

Prothoracic glands

a) *Structure.* According to Novak (1975) the prothoracic glands of insects are generally composed of two strips of glandular tissue, the basic components of which are glandular cells very similar to those of the corpora allata. Their cytoplasm is basophilic which stains deeply with those such as methylene blue and neutral red. Numerous deeply staining granules may be found in the cytoplasm.

b) *Development.* In *P. brassicae* the prothoracic glands increase in size during larval development up to a peak at formation of the pupa and decrease in size until they have disappeared in the two day old imagines (Kaiser, 1949; Karlinsky & Srihari, 1973; Novak, 1975: Figure 9.1). In the larva the prothoracic glands are long structures, with a swelling at the posterior end, which lie on either side of the salivary glands, but in the pupa they shorten about a quarter and then elongate as very thin lobed glands in the imago at eclosion.

c) *Function.* As early as 1922, Kopec believed that the prothoracic glands of insects produced a hormone which initiated pupation (Novak, 1975), but more recently Otaki (1960) suggested that a "humoral" factor was released from the prothoracic glands of *P. brassicae* and was controlled by the brain through the oesophageal commisures.

It is apparent however, that the prothoracic glands, on stimulation from a hormone released from the neurosecretory cells of the brain, secrete a

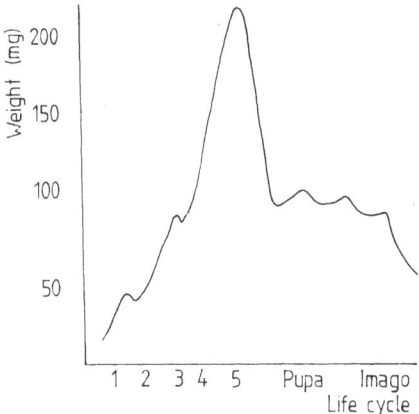

Fig. 9.1 Weight changes of prothoracic gland of *P. brassicae*
(after Karlinsky & Srihari, 1973)

hormone called ecdysone or moulting hormone (McDaniel & Berry, 1967).
α-ecdysone is secreted by the prothoracic glands and has a typical steroid
structure (Figure 9.3). Highnam & Hills (1977) gave a more detailed
account of the processes involved: a hormone from the neurosecretory
cells of the brain passes via the neurosecretory cell axons to the corpora
cardiaca and then via the haemolymph to the prothoracic glands. The
ecdysone which is then released from the prothoracic glands stimulates
cells in the process of moulting.

β-ecdysone

β-ecdysone was first identified in the haemolymph of pupae of *P. brassicae*
by Lafont *et al.* (1974), and was shown to reach a peak concentration of
just over 2 μg/ml which corresponded closely with the incorporation of
labelled uridine in RNA synthesis of wing buds at the same time (Figure
9.2a). Both α- and β-ecdysone were later identified in the pupae (Lafont
et al., 1975, 1976). In further work Lafont *et al.* (1975, 1977) demonstrated
that ecdysone levels increase when imaginal discs are formed and fat body
development takes place. However, Lafont *et al.* (1976) pointed out that
the determination of possible correlations between hormone levels and
developmental stages is difficult because of variable factors, such as, for
instance, the "competence" of the tissues for these hormones; some tissues
can convert α-ecdysone to β-ecdysone, some effects are cumulative, also
other factors may be present which elicit tissue responses.

Levels of circulation of α-ecdysone and β-ecdysone were also determined in diapausing and non-diapausing larvae of *P. brassicae* (Calvez,
1976). A peak of up to 600 pg/μl of α-ecdysone and β-ecdysone was found

a) Ecdysone in pupal haemolymph
    (after Lafont *et al.*, 1976)

b) Ecdysone in haemolymph
    (after Calvez, 1976)

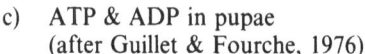

c) ATP & ADP in pupae
    (after Guillet & Fourche, 1976)

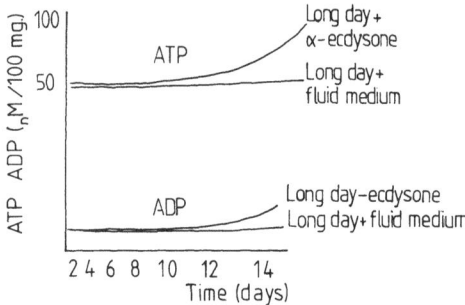

Fig. 9.2 Hormones ATP & ADP in *P. brassicae*

in fresh diapausing pupae, compared to about 530 pg/μl in non-diapausing larvae. In both cases the amount of ecdysone in the haemo-lymph falls off dramatically (Figure 9.2a,b). Injections of α-ecdysome into diapausing pupae were always found to break diapause, and at the same time they stimulated the production of large amounts of ATP and ADP, which is very different from that of non-diapausing pupae (Guillet & Fourche, 1976).

Lafont *et al.* (1977) further quantified the level of β-ecdysone in the haemolymph of fifth instar of *P. brassicae* and demonstrated two successive peaks of ecdysone production *in vivo*, at 44% and 70% of the larval duration between ecdysis (Lafont *et al.*, 1975: Figure 9.2a). These workers looked for correlations between levels of ecdysone and the genetically determined steps of imaginal wing disc development including RNA and DNA synthesis in the larva, but were unable to find any correlation between them. The high levels of ecdysone circulating in the haemolymph during the peaks do not appear to be necessary for development, although they can accelerate some developmental steps, as neck-ligatured *P. brassicae* at various intervals after the first critical period produced normal pupae.

Other effects

Moulting hormone may also have an effect on endoparasites of insects, indeed Maslennikova (1961) thought that the moulting hormone of *P. brassicae* possibly influenced the physiological state of the pupal parasite *Pteromalus puparum* (Linnaeus). Ten years earlier Johansson (1951) had experimented with applying ligatures at the pre-meso- and meta-thorax positions of the parasitised *P. brassicae* larvae but failed to induce the hymenopteron *Apanteles glomeratus* to emerge.

Hormones from the prothoracic glands also appear to be concerned with response of *P. brassicae* larvae to background, as Rothschild *et al.* (1975a,b) found that only those larvae reared on an artificial diet containing xanthophylls responded to background. They suggested that xanthophyll pigments were an essential link in the chain of events which resulted in the release of the "appropriate hormone" from the prothoracic glands, which in turn determines either homochromy or heterochromy in cryptic or aposematic species respectively.

L'Hélias (1959) found that diapausing pupae of *P. brassicae* reared in the absence of moulting hormone produced increased amounts of RNA and DNA and tumour-like growth in the presence of folic acid.

Corpora cardiaca

a) *Structure.* These consist of a pair of bodies situated immediately behind the brain, between the anterior end of the dorsal vein and the oesophagus, in front of the corpora allata with which they are connected by the nervi allati (Novak, 1975). They are sometimes fused with the corpora allata and are connected medially with the hypocerebral ganglion by a nerve bridge and with the suboesophageal ganglion by the "nerve allata". They send one or paired visceral nerves along the digestive tube to the visceral ganglia.

b) *Function.* It is currently thought that the corpora cardiaca relay hormones sent from the neurosecretory cells of the brain into the haemolymph whereupon they stimulate the thoracic glands (Highnam & Hills, 1977). Nicolson (1976a,b) found that the corpora allata acted as a storage region for diuretic hormone produced in the brain prior to its secretion at imaginal eclosion. It is interesting to note that evidence has suggested that the corpora cardiaca of *Schistocerca* sp. may actually produce diuretic hormone, which has been shown to stimulate Malpighian tubules in the laboratory.

Corpora allata

a) *Structure.* According to Novak (1975) the corpora allata are spherical, ovoid or pad-shaped bodies in most insects, connected with the corpora cardiaca by a nerve (nervi allati) which may become reduced so that the two kinds of glands are closely approximated or, in some cases, partially fused. They are enclosed in a mesenchymatous connective tissue membrane formed by plate-like cells with an oblong nucleus and homogenous surface layer. In the lepidoptera the corpora allata is usually made up of large cell types, often lobulae nuclei and numerous uniformly distributed chromatin granules.

b) *Development.* In insects generally the corpora allata are active during most of the larval life and become inactive only in the final instar (Highnam & Hills, 1977).

Juvenile hormone

The corpora allata produce juvenile hormone (JH) which is secreted throughout most of the larval life except the last larval instar when the glands become inactive (cf. Novak, 1975; Highnam & Hills, 1977). JH has a typical terpenoid structure and three forms have so far been recognised, JH_1, JH_2, JH_3 (Mauchamp *et al.*, 1979b: Figure 9.3). Novak (1975) pointed out that in 1970 it had been recognised that the corpora allata produced JH but this had not then been established in the lepidoptera.

Juvenile hormone was first bioassayed in fourth and fifth instar larvae of *P. brassicae* and in pupae by Varjas *et al.* (1976) at Wageningen, The Netherlands. They showed that during the fourth instar, several thousand *Galleria* units of JH were present, but that this level decreased rapidly at the beginning of the fifth instar and increased again in the pharate pupa to 60–120 GU/ml. Lavenseau & Surlève-Bazeille (1976) believed that the decrease in the amount of JH circulating in the haemolymph at the beginning of the pupal stage was probably responsible for cellular necrosis of the larval tissues.

More recent work by Mauchamp *et al.* (1979a) has shown the presence of JH_1 in the haemolymph of fifth instar larvae and prepupae of *P. brassicae*. Significant amounts of JH_2 and JH_3 were not found. Analysis of JH's by gas chromatography mass spectrometry methods failed also to detect the presence of JH_2 and JH_3. If they were present they must have

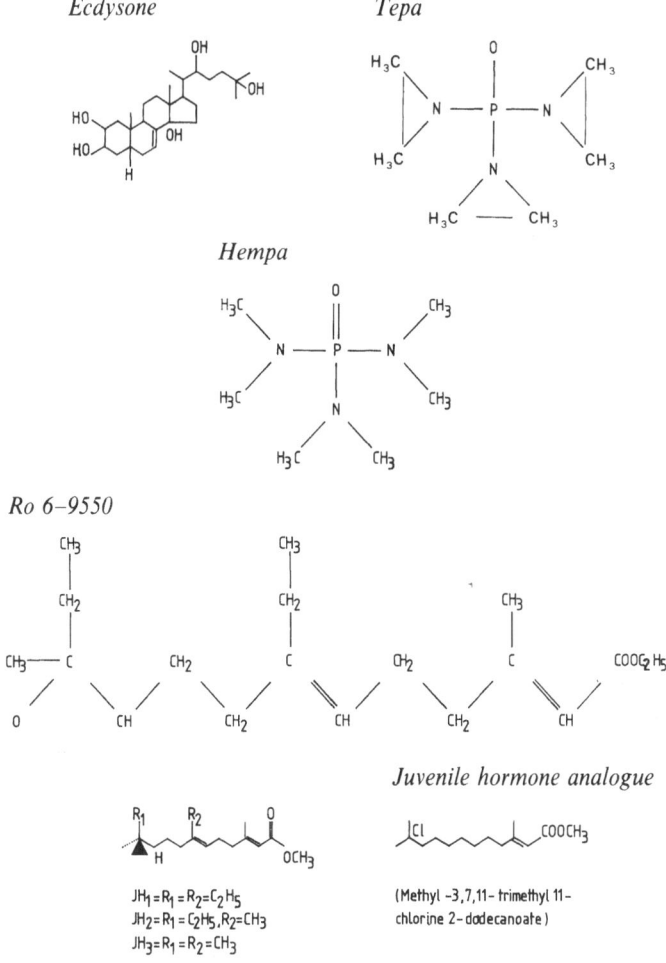

Fig. 9.3 Structure of ecdysone and analogues used on *P. brassicae*

been at concentrations less than 50 pg ml^{-1} (Mauchamp *et al.*, 1979b). The sequence of events during the last larval instar is described in this paper by Mauchamp's team. At the beginning of the fifth instar JH$_1$ values are high, but they fall off quickly to about 200 pg ml about 24 hours after ecdysis. JH$_1$ then increases slightly to 1 ng ml, then stays at very low levels until the end of the feeding period. In the prepupa, about 12 hours before pupal ecdysis, JH$_1$ then increases slightly to 1 ng ml, then stays rises to a noticeable level of 3.5 ng/ml and finally falls to a very low level in the young pupa.

   Some information on the function of juvenile hormones has come from the use of juvenile hormone analogues (JHA) and juvenile hormone derivatives (JHD) which are discussed later.

Juvenile hormone and muscle development

Auber-Thomay & Srihari (1973), who studied the development of inter-segmental muscles of fifth instar larvae and pupae of *P. brassicae*, found that application of JH resulted in "gigantic growth of larval muscles whose lengthening was attributed to an increase in the number of muscle cells" (cf. Novak, 1975). During metamorphosis the muscles of the larva are lost in the pupa and a novel set of imaginal muscles are produced. This process is thought to be controlled by hormones which originate from the head.

JHA applied topically to fifth instar larvae, however, was shown to induce degeneration of larval muscles (Karlinsky & Srihari, 1973) but this could be prevented by application of ligatures behind the thorax of the fifth instar larva (Srihari, 1974) or perpetuated by injection of β-ecdysone. Srihari (*loc. cit.*) believed that JHA and β-ecdysone controlled muscle degeneration, perhaps through the nervous system, indirectly. Some of his earlier work on ligaturing the head indicated that hormones from the major endocrine glands did not appear to influence muscle degeneration, although they appeared to influence metamorphic growth.

The experiments conducted by Street (*loc. cit.*) and Benz (*loc. cit.*) on the effects of JH analogues on larvae of *P. brassicae*, and the resultant extraordinary morphological features, has already been made in connection with corpora allata secretions.

Juvenile hormone and reproduction

Karlinsky, working at the University of Paris (VI), has made several studies of the hormonal control of ova development and vitellogenesis in *P. brassicae* (Karlinsky 1962, 1963, 1967a,b,c, 1970, 1971; Karlinsky & Poulaert, 1971).

Karlinsky (1962) showed that cutting out the corpora allata in fifth instar larvae just after the fourth moult resulted in a degeneration of the ovarian follicles. This was similarly demonstrated in freshly eclosed imagines which had their corpora allata-corpora cardiaca (ca-cc) complex removed. If the ca-cc complex is implanted into imagines which have had these glands removed, vitellogenesis is resumed only in the presence of JH (Karlinsky, 1967a). Even males, which still have their ca-cc complex intact, are able to stimulate vitellogenesis if an ovary is implanted into them (Karlinsky, 1967b). This work is in agreement with Lamy's (1967) suggestion that the ovary and the corpus allata complex are necessary for the release into the haemolymph of a "Female Protein" necessary for vitellogenesis to take place (Karlinsky, 1970; Lamy & Karlinsky, 1974; Lamy *et al.*, 1975). A clear sexual dimorphism of this protein was demonstrated, being present only in females (Karlinsky & Lamy, 1976).

Benz (1970) thought that certain factors such as water and sugar had an effect on the corpora allata which produced JH, which thus stimulated oogenesis. He also applied a juvenile hormone analogue named FAEE (Table 9.2) topically onto decapitated imaginal female *P. brassicae* and confirmed that oogenesis was stimulated. This substantiated Karlinsky's work and strongly suggested that the corpora allata stimulate oogenesis by releasing JH. Other JH derivatives (Table 9.2) were found by Benz (1973) to cause a reversal of spinning behaviour in fifth instar larvae. A juvenoid (RO 20–3600) applied to fifth instar larvae caused an extra-ordinary development of mandible-like processes on the galae which were themselves enlarged out of all proportion.

Table 9.2 Effects of juvenoids (juvenile hormone analogues, juvenile hormone mimetics) on *P. brassicae* physiology.

Key:	DFME	Dichlorofarnesenic acid methyl ester
	FAEE	Farnesenic acid ethyl ester (juvenile hormone derivative) marketed by Hoffmann La Roche & Co., Switzerland
	*HEMPA	A phosphoramide (hexamethyl-phosphoric triamide), mixture of the isomers of methyl-10,11-epoxy-7-ethyl-3,11-dimethyl-2,6-trideca-dienoate and 6,7-epoxy-3,7-dimethyl-1-(3,4-(methylenedioxy)-phenoxy)-2-nonene.
	JHA	Juvenile hormone analogue (methyl-3,7,11-trimethyl-11-chloro-2-docecanoate)
	Law's Mimic	made from the action of farnesoic acid plus ethanolic hydrogen chloride plus RO 6-9550 which gives methyl-10,11-tetramethyl-2-cis/trans-6-cis/trans dodecienoate
	*TEPA	An aziridine (tris(1-aziridinyl) phosphine oxide)

Hormone	Stage administered	Effect	Authority
Ecdysone (α- or β-)	L, injected	decreased level of diapause in *P. puparum* in *P. brassicae*	Maslennikova, 1961, 1968, 1970, 1971, 1972; Maslennikova & Chernysh, 1973
Ecdysone*	P, diapausing	diapause broken	Guillet & Fourche, 1976
Ecdysone	P, diapausing	development continued	Calvez, 1976
Ecdysone ^{3}H	P	—	Beydon *et al.*, 1979
JHA	L$_5$, P	disrupted larval and pupal development	Street, 1975

Table 9.2 continued

Hormone	Stage administered	Effect	Authority
JHA	L$_5$, topically	prolongation of larval life and increase in size of prothoracic gland	Karlinsky & Srihari, 1973
JHA*	L$_{1-5}$, injected	muscle degeneration	Srihari, 1974; Srihari & Gahukar, 1975
FAEE	I, virgin	suggested stimulation of oogenesis by CA	Benz, 1970
Roeller's compound DFME	I, virgin	no "sterilant effect"	Benz, 1970
FAEE	L$_5$	reversal of spinning behaviour	Benz, 1973
RO-20-3600	L$_5$	homeosis: enlarged galae with mandible-like structures distally	Benz, 1974
Law's mimic	I, female	promoted egg production	Street, 1975
TEPA*	P, PP, I (10–20 μg/insect)	sterilised both sexes	Street, 1975
HEMPA*	(100 μg/insect)	ineffective as chemo-sterilant in either sex	Street, 1975
H^4 folic acid	L (non-diapausing)	stimulated pteridine synthesis	L'Hélias, 1975
ethyl 3,7-dimethyl-9-cyclohexyl-2-4-nonadienoate	L	effective	Sehnal et al., 1976

*structure shown in Figure 9.3

Juvenile hormone and endocrine correlations

Recent work by Mauchamp et al. (1979a) on the epidermis and wing discs of *P. brassicae* did not suggest that the peak of JH$_1$ in the prepupa had any definite function, on account of two points; that neck-ligatured larvae before wandering can ecdyse to almost headless pupae (Lafont et al.,

1977), and that imaginal discs from fully-fed larvae exposed to ecdysone secrete first a pupal cuticle and then form scales and an imaginal cuticle (Blais, 1978). This is contrary to the thoughts of Varjas *et al.* (1976) and other workers that the rise in JH prevented precocious adult differentiation. Mauchamp *et al.* (1979a) conclude that the JH_1 peak could be involved in a regulatory process of ecdysone synthesis.

The rapid fall in JH_1 levels in feeding larvae is thought to be related possibly to the appearance of specific esterases able to metabolise the hormone even when bound to its lipoprotein carrier (see references cited in Mauchamp *et al.* (1979a)). 3-H-JH Acid and $3H-JH_1$ were identified, JH-esterase activity being near the end of the feeding period of a third instar larva and the other nine hours before larval–pupal ecdysis.

Other effects

It was suggested in an earlier paper by Chuang-Lung (1959) that the differentiation centre, which he found in the epidermis of the larval leg, is probably controlled by JH and has effect on the characteristics of the epicuticle. This centre also required a higher oxygen supply and probably absorbed more moulting hormone.

Juvenile hormone mimics and growth regulators

This is a field of control which is very much in its infancy (Table 9.2). Early in the 1970s, Benz (1971, 1973, 1974) induced reversal of spinning behaviour and homoetic transdetermination by introducing juvenoids in *P. brassicae*. Sehnal *et al.* (1976) found their juvenoid with a cyclohexone moiety in the molecule most effective on *P. brassicae* larvae. The growth regulating effects of azadirachtin on *P. brassicae* larvae were shown in photographs by Ruscoe (1972).

Scheurer *et al.* (1975) tested the effects of three insect growth regulators (CGA 13353, CGA 34301 and CGA 34312, which are derivatives of aryl-pentenoic acid and aryloxy-butenoic acid), on larvae of *P. brassicae* as well as the two parasites *A. glomeratus* and *P. puparum*.

On treatment with CGA 13353 *P. brassicae* larvae were unable to change into a supernumerary moult and died in the fifth instar. Pupae were often deformed but failed to eclose.

Application of these insect growth regulators did not have any significant effect on the mortality of *A. glomeratus* when the chemicals were applied to parasitised third and fifth instar *P. brassicae* larvae at a concentration of 0.05% a.i. Similar results were obtained with *P. puparum* although the numbers of these insects hatching was reduced at concentrations of 0.05% a.i. using CGA 13353 and CGA 34302.

Juvenoids and sterility

Certain juvenoids cause sterility in the female and act as control chemicals, whereas other juvenoids may act as possible target sites for insecticides and thus be useful (Watkinson & Clarke, 1973).

North (1975) gave a list of 14 lepidopterous species including *P. brassicae* in which induced sterility has been successfully achieved by using juvenoids.

Benz (1971) experimented with three compounds including FAEE (Table 9.2) which had been shown to have a "sterilant effect" on Hemiptera but none of these showed the same reaction on *P. brassicae*. Street (1975) however, demonstrated a high degree of sterility by using TEPA (Table 9.2) which he applied to the pupae of both sexes of *P. brassicae*. At concentrations of 20 μg/insect 100% sterility occurred. HEMPA, which is 10 times less active than TEPA, was not found to have any such effect.

References cited

Auber-Thomay, M. & Srihari, T., 1973. Evolution ultrastructurale des fibres musculaires intersegmentaires chez *Pieris brassicae* (L.) pendant le dernier stade larvaire et la nymphose. *J. Microsc.* 17: 27–36.

Benz, G., 1970. Stimulation of oogenesis in *Pieris brassicae* by the juvenile hormone derivative Farnesenic acid ethyl ether. *Experimentia* 26: 1012.

—— 1971. Failure to demonstrate sterilant effect of juvenile hormone mimetics in *Pieris brassicae* and *Galleria mellonella*. *Experimentia* 27: 581–582.

—— 1973. Reversal of spinning behaviour in last instar larvae of *Pieris brassicae* treated with juvenile hormone derivatives. *Experimentia* 29: 1437–1438.

—— 1974. Homeotic transdetermination caused by juvenoid on larvae of *Pieris brassicae* L. *Experimentia* 30: 1264–1265.

Bergström, G. & Lundgren, L., 1973. Androconial secretions of three species of butterfly of the genus *Pieris*. *Zoon. Suppl.* 1: 67–75.

Beydon, P., Sommé-Martin, G. & Lafont, R., 1979. Le métabolisme de l'ecdysone chez *Pieris brassicae* L.: synthèse, degradation. *5éme Coll. phys. Insect. Marseilles*. 27–29 Sept. 1979.

Blais, C., 1978. *Étude des effects des ecdysones sur le développement des disques imaginaux alaires de Pieris brassicae en culture in vitro*. Thèse 3éme cycle, Paris.

Calvez, B., 1976. Taux d'ecdysone circulante aux derniers âges larvaires et induction de la diapause nymphale chez *Pieris brassicae*. *C. r. Acad. Sci. Paris* D 282: 1367–1370.

Chippendale, G.N., 1977. Hormonal regulation of larval diapause. *A. Rev. Ent.* 22: 121–138.

Chuang-Lung, K., 1959. The differentiation centre inducing the development from larval to adult leg in *Pieris brassicae* (Lepidoptera). *J. embr. exp. Morph.* 7: 572–582.

Florey, (–), 1951. cited on p. 436 *in* Novak, V.J.A., 1975. *Insect Hormones*. John Wiley & Sons, New York. No reference given.

Fraenkel, G. & Hsiao, C., 1965. Bursicon, a hormone which mediates tanning of the cuticle in the adult fly and other insects. *J. Insect Physiol.* 11: 513–556.

Guillet, C. & Fourche, J., 1976. Les modifications de metabolisme énergétique lors de la levée de diapause par injection d'ecdysone chez *Pieris brassicae. Bull. biol. Fr. Belg.* 110: 31–44.

Hepburn, H.R., 1976. *The Insect Integument.* Elsevier Scientific Publishing Company. Amsterdam, Oxford, New York.

Highnam, K.C. & Hill, L., 1977. *The comparative endocrinology of the invertebrates.* Arnold.

Johansson, A.S., 1951. The food plant preference of the larvae of *Pieris brassicae* L. (Lepidoptera: Pieridae). *Norsk. ent. Tidsskr.* 8: 187–195.

Kaiser, P., 1949. Histogische Untersuchungen über die Corpora allata und Prothoraxdrüsen der Lepidopteren in Bezug auf ihre Funktion. *Arch. entMech. Org.* 144: 99–131.

Karlinsky, A., 1962. Effets de l'ablation des corpora allata larvaires sur le développement ovarien de *Pieris brassicae* L. (Lépidoptère). *C. r. Acad. Sci., Paris* 255: 191–193.

—— 1963. Effets de l'ablation des corpora allata imaginaux sur le développement ovarien de *Pieris brassicae* L. (Lépidoptère). *C. r. Acad. Sci., Paris* 256: 4101–4103.

—— 1967a. Influence des corpora allata sur le fonctionement ovarien en milieu mâle de *Pieris brassicae* L. (Lépidoptère). *C. r. Acad. Sci., Paris* 265: 2040–2042.

—— 1967b. Corpora allata et vitellogenèse chez les Lépidoptères. *Gen. comp. Endrocrinol.* 9: 511.

—— 1967c. Reprise de la vitellogenèse après implantation de corpora allata chez *Pieris brassicae* L. (Lepidoptere). *C. r. Acad. Sci., Paris* 264:1735–1738.

—— 1971. Mode d'action de l'hormone juvénile sur l'ovarie de *Pieris brassicae* L. (Lépidoptère). *Proc. XIII int. Congr. Entomol., Moscow* 2–9 Aug. 1968, 393–395.

Karlinsky, A. & Lamy, M., 1976. Hormone juvénile et vitellogenèse en milieu mâle chez *Pieris brassicae* L. (Lépidoptère, Pieridae). *Soc. fr. Endocrinol.,* Journées Provinciales Toulouse. 16–19 Sept. 1976.

Karlinsky, A. & Poulaert, J., 1971. Evolution du tissue adipeux au cours de la vie imaginale de *Pieris brassicae* L. (Lépidoptère). *Bull. Soc. zool. Fr.* 96: 453–466.

Karlinsky, A. & Srihari, T., 1973. La glande prothoracique au cours du développement post-embryonnaire chez *Pieris brassicae* L. (Lépidoptère). *Bull. Soc. zool. Fr.* 98: 243–262.

Kayser, H. & Angersbach, D., 1974. Action spectrum for light controlled pupal pigmentation in *Pieris brassicae* melanisation and level of bile pigment endrocrine control. *J. Insect Physiol.* 20: 2277–2286.

Kayser-Wegmann, I., 1975. Untersuchungen zur Photobiologie und Endokrinologie der Farbmodifikationen bei der Kohlweisslingspuppe *Pieris brassicae.* Zeitverlauf der sensiblen und kritischen Phasen. *J. Insect Physiol.* 21: 1065–1072.

Kopec, S., 1922. Studies on the necessity of the brain for the inception of insect metamorphosis. *Biol. Bull.* 42: 323–342.

Lafont, R., 1975. *Les aspects biochimiques et enzymologiques de la différentiation des ébauches alaires chez Pieris brassicae.* Thèse de Doctorat, Paris.

Lafont, R., Delbecque, J.P., Hys, L.D., Mauchamp, B. & Pennetier, J.L., 1974. Étude du taux de B-ecdysone dans l'hémolymphe de *Pieris brassicae* L. (Lépidoptère) au cours du stade nymphal *C. r. Acad. Sci., Paris* 279: 1911–1914.

Lafont, R., Mauchamp, B., Pennetier, J.L., Tarroux, P., Hys, L.de & Delbecque, J.P., 1975. α- and β-ecydsone levels in insect haemolymph: correlation with developmental events. *Experimentia* 31: 1241–1242.

Lafont, R., Mauchamp, B., Tarroux, P. & Blais, C., 1976. Biochemical parameters of imaginal wing disc development in *Pieris brassicae*. *Abstract from Insect Biochemistry Symposium, Hamburg, August, 1976*. p. 649.

Lafont, R., Mauchamp, B., Blais, C. & Pennetier, J.L., 1977. Ecdysones and imaginal disc development during the last larval instar of *Pieris brassicae*. *J. Insect Physiol.* 23: 277–283.

Lamy, M., 1967. Physiologie des insectes: une protéine vitellogène dans l'hémolymphe de l'imago femelle de la Piéride du Chou. *C. r. Acad. Sci., Paris* D. 265: 990–993.

Lamy, M. & Karlinsky, A., 1974. Vitellogenèse protéique en milieu mâle chez la Piéride du Chou, *Pieris brassicae* L. (Lépidoptères). *C. r. Acad. Sci., Paris* D. 278: 91–94.

Lamy, M., Karlinsky, A. & Julien-Laferreiere, N., 1975. Vitellogenèse protéique anormale en milieu mâle chez deux Lépidoptères: *Pieris brassicae* L. et *Bombyx mori* L. *Bull. Soc. zool. Fr.* 100: 254.

Lavenseau, L. & Surleve-Bazeille, J.E., 1976. Nécrose cellulaire et morphogenèse alaire chez quelques Lépidoptères. *Bull. Soc. zool. Fr.* 101: 69–74.

L'Hélias, C., 1959. Facteur inducteur de tumeur provoqué par l'acide folique chez *Pieris brassicae* en état de diapause. *Ann. Biol.* 63: 237–247.

—— 1975. Hormone juvenile de *Pieris brassicae* diapausante et mutations. *Ann. Épiphyt.* 36: 63–85.

Magaudda, P., 1927. Effect of administering of thyroid and testicular substance on development. Experiments on caterpillars and pupae of *Pieris brassicae*. *Boll. Soc. Ital. Biol. Sperim.* 2: 791–794.

Maslennikova, V.A., 1961. The effect of the host's hormones on the diapause in *Pteromalus puparum* L. *Akad. nauk. SSSR. Dok.* 139: 249–251.

—— 1968. The hormonal mechanism of regulation of pupal diapause in *Pieris brassicae* L. (In Russian.) *Entomol. Obozr.* 47: 429–439.

—— 1970a. Hormonal regulation of diapause in *Pieris brassicae*. (In Russian.) *Dokl. Akad. nauk. SSSR* 192: 942–945.

—— 1971. An investigation of the hormonal mechanism of diapause on *Pieris brassicae* by use of the parasite *Pteromalus puparum* as test subject. (In Russian.) *All Union ent. Soc. Leningrad* 1: 415.

—— 1972. Influence of hormonal balance of diapausing insects on their re-activation (*Pieris brassicae, Pteromalus puparum, Orgyia antigua*). *In Problemy Fotoperiodizma i Diapauzy Nasekomtkh.* p. 229–241.

Maslennikova, V.A. & Chernysh, S.I., 1973. Effect of ecdysone on determination of diapause of *Pteromalus puparum* L. (In Polish.) *Dokl. Akad. nauk. SSSR Ser. Biol.* 213: 480–482.

Matsuda, R., 1976. *Morphology and Evolution of the Insect Abdomen.* Pergamon Press, Oxford.

Mauchamp, B., Lafont, R. & Jourdain, D., 1979. Mass fragmentographic analysis of juvenile hormone 1 levels during the last larval instar of *Pieris brassicae*. *J. Insect Physiol.* 25: 545–550.

Mauchamp, B., Lafont, R., Hardy, M. & Jourdain, D., 1979b. Analysis of insect juvenile hormones by gas chromotography mass spectrometry; problems of sample preparation and choice of detection procedures. *Biochem. mass Spect.* 6: 276–281.

272

McDaniel, E.N. & Berry, S.I., 1967. Activation of the prothoracic glands of *Antheria polyphenus*. *Nature, Lond.* 214: 1032–1034.

Nicolson, S.W., 1976a. Diuresis in the cabbage white butterfly, *Pieris brassicae*: fluid secretion by the Malpighian tubules. *J. Insect Physiol.* 22: 1347–1356.

—— 1976b. Diuresis in the cabbage white butterfly, *Pieris brassicae*: water and ion regulation and the role in the hind-gut. *J. Insect Physiol.* 22: 1623–1630.

—— 1976c. The hormonal control of diuresis in the cabbage white butterfly, *Pieris brassicae*. *J. exp. Biol.* 65: 565–575.

North, D.T., 1975. Inherited sterility in Lepidoptera. *A. Rev. Ent.* 20: 167–182.

Novak, V.J.A., 1975. *Insect Hormones.* John Wiley & Sons, New York.

Ohtaki, T., 1960. Humoral control of pupal colouration in the cabbage white butterfly, *Pieris rapae crucivora*. *Annot. zool. Jap.* 33: 97–103.

Panov, A.A. & Kind, T.V., 1963. The histology of the neurosecretory cell system in the lepidopteran brain. (In Russian.) *Dokl. Akad. nauk. SSSR* 153: 1186–1189.

Peredielsky, (–), 1940. cited *in* Novak, V.J.A., (1975). *Insect Hormones.* John Wiley & Sons, New York. No reference given.

Post, L.C., 1972. Bursicon: its effect on tyrosine permeation into insects. *Biochem. Biophys. Acta* 290: 424–428.

Post, L.C. & de Jong, J., 1973. Bursicon and the metabolism of tyrosine in moulting cycle of *Pieris brassicae* larvae. *J. Insect Physiol.* 19: 1541–1546.

Post, L.C., de Jong, B.J. & Vincent, W.R., 1974. 1-(2,6-disubstituted benzoyl)-3-phenylurea insecticide: Inhibitors of chitin synthesis. *Pestic. Biochem. Physiol.* 4: 473–483.

Rehm, M., 1955. Morphologische und histologische Untersuchungen an neurosekretorischen Zellen von Schmetterlingen. *Z. Zellforsch.* 42: 19–58.

Rothschild, M., Gardiner, B.O.C., Valadon, L.R.G. & Mummery, R., 1975a. Lack of response to background colour in *Pieris brassicae* pupae reared on carotenoid-free diet. *Nature, Lond.* 254: 592–594.

—— 1975b. Exhibition of *Pieris brassicae*. *Proc. R. ent. Soc. Lond.* 39: 39.

Rothschild, M. & Schoonhoven, L.M., 1977. Assessment of egg load by *Pieris brassicae* (Lepidoptera: Pieridae). *Nature, Lond.* 266: 352–355.

Ruscoe, C.N.E., 1972. Growth disruption effects of an insect antifeedant. *Nature, New Biol.* 236: 159–160.

Scheurer, R. Fluck, V. & Ruzette, M.A., 1975. Experiments with insect growth regulators (IGRs) on lepidopterous pests and some of their parasites. *Mitt. Schweiz. Entomol. Ges.* 48: 315–321.

Sehnal, F., Romanuk, M. & Streinz, L., 1976. Potent juvenoids with cyclehexane moiety in the molecule. *Acta Entomol. Bohemsoslov.* 73: 1–12.

Srihari, T., 1974. Effects of insect hormones on the growth and degeneration of muscles in *Pieris brassicae* L. (Lepidoptera: Pieridae). *Bull. Soc. zool. Fr.* 99: 325–333.

Srihari, T. & Gahukar, R.T., 1975. The influence of juvenile hormone on food consumption and growth in the last larval instar of *Pieris brassicae* L. (Lepidoptera: Pieridae). *Bull. Soc. zool. Fr.* 100: 327–333.

Street, M.L., 1975. *The growth and development of the reproductive system of Pieris brassicae L. and their control by hormones.* Ph.D. thesis. University of Newcastle upon Tyne.

Tamaki, Y., 1977. Complexity, diversity and specificity of behaviour modifying chemicals in Lepidoptera and Diptera. *In: Chemical Control of Insect Behaviour.* Edited by H.H. Storey and J.J. McKelvey. J. Wiley & Sons. pp. 253–285.

Ushatinskaya, R.S., 1976. Insect dormancy and its classification. *Zool. Jb. Syst.* 103: 76–97.

Varjas, L., Paguia, P. & De Wilde, J., 1976. The juvenile hormone titre of the haemolymph in the caterpillars of *Pieris brassicae* L. and *Mamestra brassicae* L. (Lepidoptera) in the development phase before the larval pupal transformation. (In Hungarian.) *Allattani Kohl.* 63: 211–217.

von Euler, U.S., 1961. Occurrence of catecholamines in Acrania and Invertebrata. *Nature, Lond.* 190: 170–171.

Watkinson, I.A. & Clarke, B.S., 1973. The insect moulting hormone system as a possible target site for insecticidal action. *Pans* 19: 488–506.

Wigglesworth, V.B., 1964. Hormonal control of growth and reproduction in insects: Review. *Adv. Insect Physiol.* 2: 247–336.

—— 1970. *Insect Hormones.* Oliver & Boyd, Edinburgh.

10. Biochemistry

Introduction

General considerations

There is nothing which is unique in the biochemistry of *P. brassicae* which sets it apart from other phytophagous insects, and indeed the information recorded here can be compared quite readily with data from other lepidopterous species. There are, however, a number of extrinsic factors which vary the results of any biochemical analysis both quantitatively and qualitatively in all insects.

a) *Foodplants.* It is to be expected that the biochemistry of foodplants will vary interspecifically and with an oligophagous feeder such as *P. brassicae* the problem of understanding which compounds are ingested is made more difficult by the fact that it eats a wide variety of *Brassica* cultivars.

When studying particular groups of compounds such as the lipids and vitamins of insects it is important to define the nature and the composition of the food source accurately, since these may materially affect the compounds being studied. It has been shown for example that plants grown from the same seed packet may vary quantitatively in their pigment content (Feltwell & Valadon, 1972, 1974). Indeed leaves from different parts of the same plant may vary, and the concentration of the pigments can also be affected by the age of the plant, thus introducing subtle variations into the *P. brassicae* biochemistry. Attention has been drawn to the usefulness of using half leaves in feeding experiments, but here one reaches the limits of practical analysis, as it is impossible to analyse the biochemical nature of the food before it is eaten.

b) *Individual variation.* It must be pointed out that certain biochemical differences can occur in *P. brassicae* and between members of the same strain. Harmsen (1966a) noted that individuals of the Cambridge strain of

P. brassicae which had been inbred for 10 years varied in imaginal size and pteridine pigmentation, while Clements (1967) noted differences in the esterases in different races and strains of *P. brassicae*. It is perhaps worth mentioning that many of the experiments which have been conducted on *P. brassicae* in various institutions throughout western Europe, have used specimens initially supplied by David & Gardiner from Cambridge.

c) *Bacterial commensals.* In a number of biochemical investigations it is necessary to ascertain whether micro-organisms manufacture the compounds under investigation; however, this does not seem to have been widely done. This is particularly relevant as at least six species of bacteria have been found in the gut of *P. brassicae* and these can cause infection in the haemolymph after insecticidal infection (Chapter 15).

Clements (1967) found 13 esterases in the tissues of *P. brassicae* but he ruled out the possibility that bacteria synthesised any of them by quoting an observation made by J.R. Norris, who had found the bacterium *Streptococcus faecalis* Andrews & Horder confined to the villi of the midgut epithelia of the larva. However, Harmsen (1966a), working on pteridine pigments, stated that no bacterial commensals were to be found in either the gut or the fat body of *P. brassicae*. Carotenoid pigments are also found in *P. brassicae* and are also manufactured by bacteria, so it appears necessary that only aseptic material should be analysed. In a recent paper by Britton *et al.* (1977) it was conjectured that micro-organisms might play a part in vitamin A synthesis in the ladybird beetles (*Coccinella* sp.).

Carbohydrates

Introduction

Five carbohydrates have been identified in *P. brassicae*; the hexose glucose, the disaccharide lactose and trehalose, the polhydric alcohol sorbitol and the polysaccharide glycogen. It has been established that there is a particular cell on the maxillae of the larvae which is sensitive to the disaccharide sucrose which acts as a feeding stimulant. Wei Chun Ma (1972) working at Wageningen has screened several carbohydrates for their possible importance in chemoreception.

One of the prime functions of carbohydrates in *P. brassicae* appears to be in their involvement as solutes, in physiological adaptation for overwintering i.e. diapause, although the actual energetics of this have not yet been elucidated. Carbohydrates as well as fats are used as fuel in flight (Johnson, 1976) and probably they are also used as an energy source during ordinary metabolic processes in the body. Beenakers (1969) believed that prior to flight carbohydrates may well be metabolised in order to create the required temperature necessary for flight during the process of wing vibration.

Glycogen

Chippendale & Kilby (1969) recorded glycogen present in the fat body of *P. brassicae* larvae and showed how it increased in concentration in the early fifth instar to the prepupa from 0.3 to 47 mg per dry weight. During the formation of the pupa the amount of glycogen decreased to 3.9 mg per dry weight suggesting that some of it may be used for chitin or sorbitol synthesis (Sømme, 1967).

Irregular deposits of glycogen have also been found in the cylindrical cells of the mesenteron of *P. brassicae*, while glycoproteins were found in goblet cells (Plantevin & Nardon (1970)). Non-significant amounts of glycogen (0.2–9.49 mg) were found in non-diapausing and diapausing pupae of *P. brassicae* (Fourche *et al.*, 1977).

Glycogen is used in other insect species as a solute used in the protection against super-cooling but it was not recorded with this function in *P. brassicae* larvae by Sømme (*loc. cit.*). Moreau (1969) suspected that glucose is synthesised to trehalose and then to glycogen.

Sorbitol

There is a steady increase in the amount of sorbitol in the haemolymph of diapausing larvae up to 80 days after hatching from the ova (Figure 10.1). The concentration of sorbitol may vary geographically and seasonally in

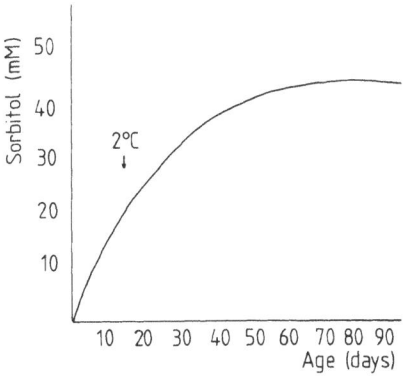

Fig. 10.1 Concentration of sorbitol in pupae of *P. brassicae* (after Sømme & Velle, 1968)

diapausing pupae (Sømme & Velle, 1968). These workers recorded a concentration of sorbitol in the pupae as 50 mM 12 weeks after pupation, representing about half the concentration reported by Sømme (1967) for Norwegian specimens. A high concentration of sorbitol provides a lower super-cooling point thus giving a better adaptation of *P. brassicae* to cold conditions.

Sucrose and glucose

Small amounts of glucose were demonstrated by chromatography in the haemolymph of fifth instar larvae by Howden & Kilby (1961). Schoonhoven (1969) showed that the second cell on the lateral maxilla sensilla styloconica was sensitive to sucrose. However, Ma & Schoonhoven (1973) were unable to demonstrate any similar cell response to sugars on B-type setae of the tarsi. Ma (1969) also showed that cells on the medial sensillum styloconicum were also responsive to sucrose; a maximal response being evoked by an 0.1 M solution of sucrose. When the mustard oil glycoside sinigrin was given to the larva together with sucrose, an even greater response was evoked than would have been predicted by each chemical separately. This suggested that sucrose acts as a feeding stimulant.

Ma (1972), in his electrophysiological study of the feeding responses of *P. brassicae* larvae, tested 18 different carbohydrates (Table 10.1) and found that only two sugars, sucrose and glucose induced any response which was significantly different from the controls. This tentatively demonstrated that *P. brassicae* larvae have a very high selective response for sucrose and glucose only. Ma designated a "sugar" cell on the styloconium based on the criteria that: a) "its impulse frequency is possibly correlated with changes in sucrose concentrations" and b) "different impulse frequencies are evoked when different kinds of sugar at equal concentrations serve as stimuli".

Table 10.1 Carbohydrates tested on the chemosensory mechanism of *P. brassicae* (after Ma, 1972).

Pentoses:	d-arabinose	Disaccharides:	lactose
	l-arabinose		d-maltose
	l-fucose		sucrose
	d-ribose		d-mannose
	d-xylose	Trisaccharides:	melezitose
Hexoses:	d-fructose		raffinose
	d-galactose	Polyhydric	
	d-glucose	alcohols:	inositol
	d-mannose		sorbitol
	l-sorbose		

Trehalose

Moreau's (1969) study on trehalose in the life of *P. brassicae* was the first of such studies in an insect (Nettles *et al.*, 1972). However, large amounts of trehalose had been identified in the haemolymph of fifth instar larvae by Howden & Kilby (1961). Moreau found that the amount of trehalose dropped off with imaginal life from 6% of dry body weight at the end of the fifth instar to 0.1% dry body weight at death of the imago. There is twice as much trehalose in diapausing as in non-diapausing pupae, when the larvae are kept at 23°C on synthetic diet. Diapause is broken when the trehalose is used up.

Other sugars

As well as glucose and trehalose, Sømme (1967) also recorded sorbitol and lactose in pupae of *P. brassicae* and showed how these increased in concentration with super-cooling. Cylindrical cells of the gut of *P. brassicae* were found by Plantevin & Nardon (1970) to contain acidic monosaccharides, while mucopolysaccharides were absent from goblet cells.

Lipids

Introduction

Seppo Turunen from the University of Helsinki has extensively studied the fatty acids (FA), lipids, phospholipids (PPL) and acetate metabolism in *P. brassicae* (Turunen, 1973a,b, 1974a,b,d, Turunen, 1975c, Turunen & Junnikkala, 1974) and has given special attention to the metabolism of linolenic acid (Turunen, 1974c), palmitate (Turunen, 1975a) and vitamin E (Turunen, 1976). He has also investigated the effects of the insecticide gamma-BHC on lipid metabolism. Most of the work has been performed on the larva of *P. brassicae*, but lipid metabolism in the whole of the life cycle has been dealt with in Turunen (1974d).

Two lipids are actively synthesised in *P. brassicae*, palmitoleate which is produced when larvae are reared on artificial diet (Turunen, 1973a), and diglycerides (DGL) which are released from the gut wall into the haemolymph and pass to the fat body (Turunen, 1975c; Turunen & Junnikkala, 1974). All the other work by Turunen is based on analysing the lipids present in the larva after incorporating certain known lipids into artificial diets or introducing labelled lipids.

Kayser & Angersbach (1975) work in part substantiates Turunen's results with fatty acids, and he shows that the carotenoid lutein forms monoesters and diesters with these fatty acids.

Early work

The lipids contained in 500 whole larvae (age not stated) and 308 diapausing pupae of *P. brassicae* were extracted and compared by Timon-David (1929–1932). He showed that about 5.3 parts of linolenic acids per 100 were present in larvae and that there were lesser amounts of lipids present in pupae which had overwintered and had been analysed in the middle of April. He concluded from this that pupae metabolise some of the non-saturated fatty acids which they store. Strogaya (1962) investigated fat metabolism in *P. brassicae* and *Aporia crataegi* (Linnaeus) and showed that endogenously produced fat is useful for providing energy for the first generation via the ova and for the following generations as fat stores in overwintering pupae.

Ovum

Turunen (1974d) demonstrated that if larvae were reared on lipid free diets their growth was adversely affected at the second instar, and that their growth would be restored if a lipid source was given at this stage. This therefore suggested that the ova must have their own lipid source. When labelled linolenic acid was fed to female imagines the label went into PPL (42.2%) and neutral lipids (NL) (57.8%) (Turunen, 1973a, 1974c).

Larva

It was shown by Turunen (1973a) that if larvae were fed on standard artificial diets three lipids were found in their tissues: first, oleate was preferentially accumulated in the fifth instar (43.8%); second, the amount of linolenate accumulated was found to be dependent on the concentration of linolenate and linoleate in the diet, and third, palmitoleate was found to be synthesised as previously stated.

Turunen then incorporated lipids into *P. brassicae* larvae and discovered some interesting results, among these being that lipids are vital for growth, and are important for flight and energy release and possibly ova metabolism.

In 1974 (d) Turunen investigated lipid metabolism in respect to the whole life cycle of *P. brassicae* and came to three conclusions about lipids and growth; first, that lipids are required for growth after the second instar; second, that the early instars are most susceptible to a lack of lipids, and third, that there are sexual differences in lipid metabolism (see later section). He incorporated cholesterol and trilolein or trilinolenin as the only lipid sources into the diet of the larva and found that trilinolenin was the most effective substance which stimulated growth, and that the amounts of palmitoleate and stearate were greatly enhanced (Turunen, 1974b). Other fatty acids synthesised on artificial diet were oleate, palmitrate and palmitoleate (Turunen, 1974a).

The salient features of the other labelling experiments carried out by Turunen (1973b) were that: a) labelled acetate was incorporated as PPL in the fat body; b) labelled PPL was incorporated as phosphatidylcholine (PTC) and phosphatidylethanolquine (PTE); c) labelled palmitate was incorporated into sphingomyelin and (possibly) phosphatidylsemin, and in Turunen (1975c) that; d) labelled oleate and oleic acid caused triolein to be hydrolysed to free fatty acids (FFA, DGL, monoglycerides (MGL). Some of the FFA was excreted. This suggested that the gut wall released PPL.

The fat body of the fifth instar was found to store lipids as triglycerides (TGL), and to release them as DGL, TGL and sterol esters (Turunen, 1975c). In experiments with incorporating vitamin E Turunen (1976) found that linoleate accumulated in the fat body as well as in the mid-gut and other fatty tissues.

The haemolymph of the fifth instar was found to be rich in palmitate and NL but lower in oleate than the fat body. The lipid content of the haemolymph was 0.54% w/v of which 29% was PPL. In larvae weighing an average of 451 mg there is approximately 0.8 mg of lipid in the haemolymph (Turunen & Junnikkala, 1974).

Kayser & Angersbach (1975) found that the carotenoid lutein forms mono- and diesters with the same fatty acids, palmitoleic, stearic, oleic, linoleic and linolenic acids, and that the fatty acids extracted by him agree exactly with those found by Turunen. More lutein was found to be esterified in the female than in the male. Clements (1967) demonstrated slight lipolytic activity from gut walls dissected from fifth instar larvae.

Action of insecticides on lipid metabolism

Turunen (1975b) studied the effect of a sub-lethal dose of gamma-BHC on the utilisation of tritiated triglyceride on *P. brassicae* larvae and found four effects: first, that there was a lower NL/PPL ratio, second, that there was more complete hydrolysis of the TGL in the gut resulting in more MGL, third, that there was a deviation from the normal of the synthesis of gut PPL; and fourth, that lipids were released from the fat body into the haemolymph more than normally.

Prepupa

Turunen (1974a) injected *P. brassicae* pupae with labelled acetate (Table 10.10) but failed to demonstrate incorporation into the sterol fraction. Incorporation of acetate has been noted in other insects and so Turunen's experiment seems to indicate that *P. brassicae* is unable to synthesise the sterol ring (Turunen, unpublished, personal results *in* Turunen, 1974a).

Pupa

Diapausing and non-diapausing pupae do not appear to have significant differences in amounts of lipids, these varying from 11–26 mg/pupa at moulting (Fourche *et al.*, 1977).

Fat as an energy source for flight

Energy in imaginal flight muscles was shown by Turunen (1974a) to come from oleate, palmitate and palmitoleate. His work suggested that relatively large amounts of FFA are present in flight muscles, and with increasing age the TGL remained fairly stable. TGL are present in flown muscle cells in trace amounts. Turunen (*loc. cit.*) put forward the view that DGL and FFA function in lipid transport in insects. Cardialipin was found to be the major PPL of flight muscles and PTE appeared to be more labile than PTE or cardiolipin. There appears to be a distinct sexual dimorphism in that males utilise more fat than females, presumably from the oxidation of oleate (Turunen, 1975a).

Turunen (1974c) found that lineolenic acid appeared to play an important part in flight muscle membranes and organelles. Thorax flight muscles of females reared on artificial diet were shown to increase their PPL content from 67.9 to 83.6% from Day 1 to Day 9 imagines. Thorax NL were also shown to increase in imagines, more so in the male.

Wax (Epicuticular lipids)

Martini (1975) showed that surface wax of pupae comprised mostly C_{29} and C_{31} alkanes (equal proportions of each) and fatty acids, while small amounts of esters, primary alcohols and unknowns were also present. Subsurface wax consisted of hydrocarbons and fatty acids. The non-lipid wax fraction obtained from pupal exuviae contained phospholipids including l-lethicin. The surface wax does not appear to be held very strongly on the cuticle as 90% of it can be removed after two washings in chloroform (Martini, *loc. cit.*).

Koizumi (1953) investigated the effect of changes of temperature, humidity and light and dark on the secretion of wax onto the epidermis. He found that the amount of wax secreted onto the epidermis almost doubled with an increase in the humidity (from 85.6 mg/g at 65–85% RH to 154.8 mg/g at 100% RH), and had a lower melting point (43.2 to 36.5°C). Larvae reared under a dark regime secreted more epicuticular lipids, which also had a lower melting point. He explained these results by saying that lipids were secreted onto the surface of the epidermis by the epidermal cells during the inter-moult periods (he proved this in another species, the Rice Corn Borer, *Chilo simplex*). Previously, wax was only thought to have been secreted onto the epidermis during ecdysis and it was believed that, if it was removed, no more could be produced. However, Koizumi (*loc. cit.*) showed that *more* wax can be produced. This ability of the insect to compensate for changes in the environment therefore serves to demonstrate a homeostatic mechanism in *P. brassicae*.

Proteins

Ovum

Michel Lamy (1967) found by cellulose acetate electrophoretic methods that there were three or four soluble protein bands (unidentified other than by Rf value) in the ova; and that one of these was a particular protein termed "Protein F" found also in the female imagines. Later Lamy & Karlinsky (1974) found six proteins in the ova.

Larva, prepupa and pupa

The proteins of the larvae of *P. brassicae* have been studied extensively. Van der Geest & Borgsteede (1969) pointed out that the concentration of protein fractions in *P. brassicae* varied so much intraspecifically that it is difficult to interpret results.

The total protein concentration of the haemolymph of the fifth instar larva increases seven to eight fold in just under six days (Van der Geest, 1968), so unless the actual age of the fifth instar larva is given the concentrations published by others are not very useful in comparison. However, Van der Geest (*loc. cit.*) quoted a figure of c. 6.62 g/100 ml for a fifth instar larva of 4.75 days, Munn & Greville (1969) quoted 0.62 g/ 100 ml for the larvae (sic) and Chippendale and Kilby (1969) quoted 22 mg/ml for a newly hatched fifth instar larva.

Stamm & Aguirre (1955) calculated the amount of total proteins, albumens and globulins in *P. brassicae* and showed how the protein total increased throughout larval, pupal and imaginal life, though some is lost in the exuviae of the larva and pupa (Figure 10.2a). Junnikkala (1966) demonstrated seven protein fractions in larval haemolymph by starch gel electrophoresis. Chippendale & Kilby (1967) using labelled proteins (see Table 10.10) demonstrated four larval proteins in the haemolymph of the fifth instar and two in the prepupal stage. These two last proteins disappeared in the pupa. Munn & Greville (1969) noted that none of the proteins found by them in haemolymph of larvae exceeded a molecular weight of 540,000. It has been shown that the haemocytes do not significantly contribute to the synthesis of these proteins in holo-metabolous larvae and pupae such as *P. brassicae* (Chippendale, 1977).

By far the most abundant protein extracted from the haemolymph of *P. brassicae* larvae was Peptide I (Junnikkala, 1969), shown by hydrolysis to be made up of possibly 10 amino acids (Junnikkala, 1968). The close similarity of its absorption spectrum (277 & 270 nm) to that of tyrosine, together with its high concentration prior to pupation, suggested that it was involved with the tanning process.

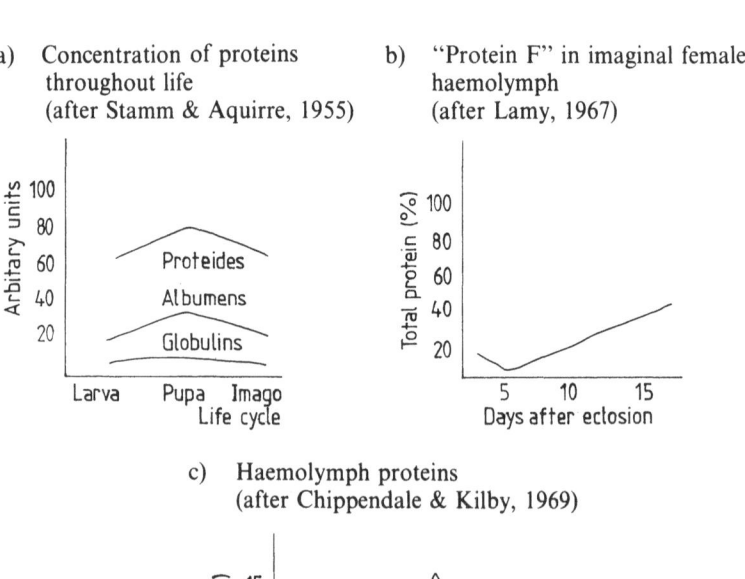

a) Concentration of proteins
    throughout life
    (after Stamm & Aquirre, 1955)

b) "Protein F" in imaginal female
    haemolymph
    (after Lamy, 1967)

c) Haemolymph proteins
    (after Chippendale & Kilby, 1969)

Fig. 10.2 Proteins in *P. brassicae*

Van der Geest & Borgesteede (1969) demonstrated by gel electrophoresis the presence of several glycoproteins and one lipoprotein in the haemolymph of *P. brassicae* larvae (fourth and fifth instar). Chromoproteins in the form of biliproteins and carotenoproteins were also present, with the bile pigment being attached to at least four protein fractions.

According to Lafont *et al.*, 1976a,b, the protein level of the pupae stays constant until two days before eclosion of the imagines. This is when pterins are synthesised. However, amounts of protein vary complimentarily in the fat body and wings which shows that the fat body acts as a source of materials for imaginal tissue development.

Incorporation of ^{14}C leucine showed two peaks which corresponded with protein synthesis in imaginal discs. During scale development protein synthesis is much reduced, later it increases as definite imaginal structures are formed. Eventually the protein level falls by 50% as numerous cell "degenerescences" occur.

Imago

Lamy (1967) found 10 protein bands in females, one of which he designated "Protein F" and which he also found in the ova. The concentration of "Protein F" in haemolymph increased up to 15 days after emergence, representing 35% of total proteins (Figure 10.2b). It was thought that control of this protein was governed by a hormone secreted from the head. It was shown subsequently that implantation of the ovaries into the abdomen of males induced atypical synthesis of "Protein F" using haemolymph carotenoproteins (Lamy & Karlinsky, 1974).

Nitrogen content

Junnikkala (1966) recorded the total nitrogen content of the larva as 600 to 1700 mg/100 ml which contrasts with a figure of 2.36 mg/ml of amino compounds in early fifth instar larvae (sic) (Van der Geest, *loc. cit.*). Junnikkala (1966) studied the amount of nitrogen taken in by larvae during the fourth and fifth instar period (Figure 10.3a,b). The graph shows two peaks of increase in nitrogen levels which are produced by active feeding. During the preparatory period low values are attained. Female larvae also have more total nitrogen than males during the fifth instar.

Although Van der Geest's results show that a peak of protein concentration in the fifth instar occurs from day four to 5.5 after ecdysis, Chippendale & Kilby (1969) recorded the peak in the prepupal stages (Figure 10.2c). There is a sexual dimorphism exhibited, as females were

a) Concentration of total nitrogen
   in fourth instar larval haemolymph
   (after Junnikkala, 1966)

b) Concentration of total nitrogen
   in larvae (fifth instar)
   (after Junnikkala, 1966)

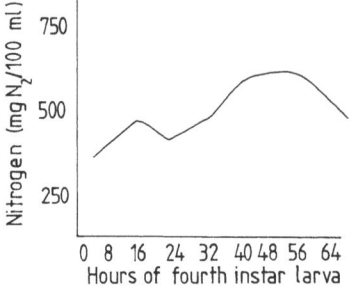 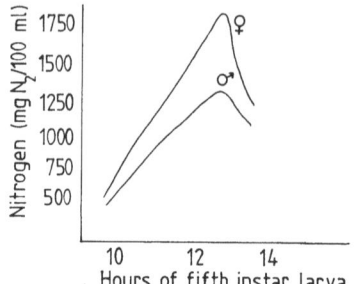

Fig. 10.3 Nitrogen in *P. brassicae*

found to have a higher concentration of protein in the second half of the fifth instar; this effect disappeared in the prepupal stage (Van der Geest, *loc. cit.*). Junnikkala (1966) also demonstrated that the protein concentration of the haemolymph does not alter significantly when larvae are reared on normal cabbage, artificial diet with or without cabbage powder, or on normal swede leaves. For nitrogen excretion in the meconium, wings and body, see Chapter 8.

Amino acids

The Spaniards Stamm & Aquirre (1955) made a special study of the amino acids of *P. brassicae* and recorded 12 from the larvae, pupae and imagines (Table 10.2). Junnikkala (1966, 1968, 1969) published several papers on the effect of braconid parasitisation and semi-synthetic diets on nitrogen metabolism of *P. brassicae* and recorded 23 amino acids by paper chromatography of the larval haemolymph. Four amino acids were found to be most concentrated: glutamine, proline, histidine and asparagine, in fact the total concentration of amino acids was made up by the first three compounds. Junnikkala (1966) also noted 17 amino acids in the defensive salivary secretion of *P. brassicae*, and arginine was conspicuous by its absence, as it was present in the swede foodplant. Larvae parasitised (by *A. glomeratus*) tended to have less nitrogen than non-parasitised larvae and the difference in nitrogen was presumed to have been used by the parasitic larvae.

Two dimensional separation of *P. brassicae* haemolymph amino acids has been published in separate research topics at Imperial College (Champ, 1958; Miah, 1976). Miah (*loc. cit.*) noted that asparagine and

Amino acids	Stamm & Aquirre, 1955			Champ, 1958	1966	Junnikkala		
	L	P	I			1966	1966	1969
Alanine (isomer not stated)	+	+	+					
α-Alanine				+++	+	+	+	+
β-Alanine				+++	++	++	++	++
γ-Aminobutyric acid				+++	−	+	+	+
Arginine				+				+
Asparagine				+				+
Aspartic acid		+		+++	+++	+++	+++	+++
Glutamic acid				+++	+++	+++	+++	+++
Glutamine		+	+?	++	++	++	++	++
Glycine	++	+	+	+++	+++	+++	+++	+++
Histidine	++			+				+
Hydroproline								
Leucine or Isoleucine				+	+	+	+	+
Isoleucine								
Lysine	+	+	+	+	−	+	+	+
Methionine/						meth. only		meth. only
Tryptophan/					+	+	+	+
Valine					all three		all three	
Methionine sulfoxide				+	+	+	+	+
Ornithine								
Phenylalanine	+++	+++	+++	+++	+++	+++	+++	++
Proline	+++	+++	+++	+	+++	+++	+++	+++
Serine	++	++	++		+++	+++	+++	++
Taurine								
Threonine				+				
Tryptophan	+++	+++	+++	++	++	++	+	++
Tryrosine	++	++	++		+	+	+	+++
Valine	++	++	++	++		+		++
Unspecified spot no. 19					+			

hydroxyproline were found only when large amounts of haemolymph were used (20–40 μl). Leucine, isoleucine and phenylalanine were only incompletely separated and proline was poorly defined. Miah also published graphs showing fluctuations in 15 amino acids in the pupal stage.

Stamm & Aquirre (*loc. cit.*) made quantitative measurements of tyrosine, tryptophan and phenylalanine at three stages in the life cycle of *P. brassicae*, and Figure 10.4 shows clearly that all three show maximum concentration during the pupal stage, tyrosine being the most abundant of the three. All three though are lost in varying amounts in the exuviae of the larvae and pupae.

Schoonhoven (1969) showed that the first cell on the lateral sensillum styloconicum of *P. brassicae* larvae was responsive only to amino acids (Table 10.3). Ma (1969) showed that 0.01 and 0.001 M concentrations of l-proline produced a feeding response only in the presence of sucrose; dl-methionine only weakly evoked a response, and cysteine even stimulated a deterrent effect.

Tryptophan and essential amino acids have been found in the gut villi (Plantevin & Nardon, 1970).

Enzymes and coenzymes

Introduction

At least forty enzymes have been recorded from *P. brassicae* (Table 10.4); nine enzymes were found associated with flight and 13 with digestion and other unknown functions.

P. brassicae enzymes have been found to be associated with most of the

a)   Concentration of three amino acids
     (after Stamm & Aquirre, 1955)

b)   Concentration of tyrosine
     (after Post & de Jong, 1973)

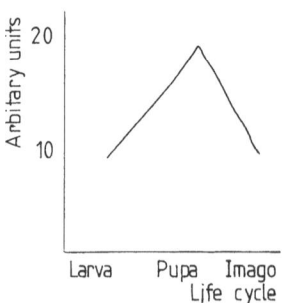
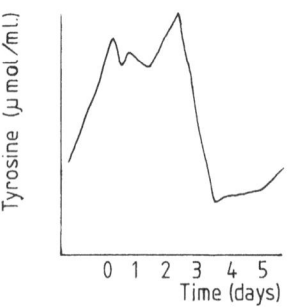

Fig. 10.4 Amino acids in *P. brassicae*

288

Table 10.3 Location of chemosensory cells on the maxillae of *P. brassicae* larvae (from Schoonhoven, 1969).

Styloconium		Chemosensory cell
Medial sensillum	Cell 1	salt
	Cell 2	salt
	Cell 3	some mustard oil glycosides
	Cell 4	undetermined
Lateral sensillum	Cell 1	amino acids
	Cell 2	sucrose
	Cell 3	mustard oil glycosides
	Cell 4	some anthocyanins

physiological aspects of metabolism that one might expect such as digestion, development of the fat body and ovaries, nitrogenous excretion including pteridine synthesis, catalysis of carbohydrates, conversion of tyrosine during cuticle synthesis, flight and energy relationships. There is, though, plenty of room for further research in this field, particularly in excretion and pigmentation.

Functional importance of enzymes

Central nervous system. No enzyme activity was reported for the intact central nerve cord of *P. brassicae* larvae by Wahla *et al.* (1974, 1976) who were working on the poisonous effects of diazinon, but this was qualified by saying that either the threshold levels of the enzyme were too low to detect or they were testing the wrong substrates.

In other work on the effect of an insecticide David (1959) using a paraoxon showed that when ova of *P. brassicae* were exposed to high concentrations of acetylcholine the ova did not hatch. He took this to suggest that there may be high concentrations of the enzyme cholinesterase in the ova, which is involved with neutral transmission and which would presumably be useful in the larva.

Cold hardiness. Sømme & Velle (1968) isolated a polyol dehydrogenase from the haemolymph of diapausing pupae of *P. brassicae*, which reduced dl-glyceraldehyde in the presence of NADPH. Their experiments indicated that this enzyme might be responsible for catalysing the reversible reactions of several sugars and polyols, and that the coenzymes NADP and NAD might also associate with this enzyme.

Cuticle synthesis. Post & de Jong (1973) both repeated the widely held view that the hormone bursicon renders the haemocyte membranes

Table 10.4 Enzymes recorded in *P. brassicae*. L=larva; L_{1-5}=larval instar;
* = thesis, not read.

Enzymes listed by function	Tissue or organ	Authority
Digestion		
Endopeptidases I & II	L, gut contents	Lecadet & Dedonder (1964)
Proteases	L, anterior gut	Lecadet & Chevrier, 1966
Proteases	L_5, whole gut	Clements, 1967
Lipases	L_5, gut wall	Clements, 1967
Leucinaminopeptidases (LAP)	$L_{4,5}$, haemolymph	Van der Geest & Borgsteede, 1969
Excretion		
Xanthin dehydrogenase (XDH)	L_5, fat body, gut	Lafont & Papillon, 1972
Uricase, Allantoinase	—	Razet (1961)*
Uricase	L_5, fat body, gut	Mauchamp & Lafont, 1975
Guanine dehydrogenase (GD)	L_5, fat body, gut	Mauchamp & Lafont, 1975
Flight		
Arginine kinase (AK)		
Glyceraldehydephosphate dehydrogenase (GAPDH)		
Citrate cynthase (CS)		
Glycerol-3-phosphate dehydrogenase (GDH)		
3-hydroxyacyl-CoA dehydrogenase (HOAD)	flight muscles	Beenakers, 1969
Carnitine acetyltransferase (CAT)		
Succinate dehydrogenase (SDH)		
Glycerol-3-phosphate oxidase (GP-ox)		
Lactate dehydrogenase (LDH)		
Cold Hardiness		
Polyol dehydrogenase	P, haemolymph	Sømme & Velle, 1968
Melanisation		
Tyrosinase	wing	Onslow, 1917
Cuticle formation		
Bursicon	haemocytes	Post, 1972
Central nervous system		
Cholinesterase (possibly)	ova	David, 1959

Table 10.4 continued

Enzymes listed by function	Tissue or organ	Authority
Esterases		
13 distinct bands on starch gel electrophoresis SGE using tris-citrate/ borate buffer Ef values: 5,6,8,10,15, 18,24,30,37,45,50,55,66	whole guts, gut wall, gut contents, regurgitated juice, frass, haemolymph, larval heads, flight muscles, of $L_{4,5}$, P, A	Clements, 1967
Non specific esterases	$L_{4,5}$, haemolymph	Van der Geest & Borgsteede, 1969
Phosphatases		
Alkaline phosphatases (ALP)	$L_{4,5}$ haemolymph	Van der Geest & Borgsteede, 1969
Acidic phosphatases (ACP)	$L_{4,5}$ haemolymph	Van der Geest & Borgsteede, 1969
Tissue respiration		
Cytochrome oxidase	larvae and pupae	Tikhonravova, 1973
Succinate dehydrogenase	larvae and pupae	Tikhonravova, 1973

permeable to tyrosine, the latter then reacting with the liberated "enzyme system" converting it to dopa and dopamine. This was supported by work done by Post (1972) and Post & de Jong (*loc. cit.*) which showed that tyrosine in the haemolymph of *P. brassicae* was converted by bursicon *in vitro*.

Digestion

Lecadet & Dedonder (1964) investigated the proteases of the cycle (gut contents) of *P. brassicae* larvae and isolated two endopeptidases (both with a molecular weight = 32,000); one of these was responsible for hydrolising peptide links at the carboxyl group of arginine, the other responsible for attaching the aromatic link (like chymotrypsin). It is interesting to note that Clements (1967) found eight esterase bands by SGE in regurgitated juice of fifth instar larvae, and he thought that those esterases isolated from the gut wall of the fifth instar larva were involved with digestion. Later Lecadet & Chevrier (1966) isolated zymogens corresponding to proteases in the anterior part of the gut of *P. brassicae*.

These could be activated *in vitro* at pH 9.5 and with trypsin at 8.0. Clements (1967) also found proteases present in whole gut extracts from fifth instar larvae and confirmed that they were esterases from their high sensitivity to organophosphorus compounds.

Excretion

Xanthin dehydrogenase (XDH) has been found in several organs of *P. brassicae* (fat body, epidermis, gut) (Mauchamp & Lafont, 1975) and is thought to play an essential part in ovary development and pteridine synthesis, especially of isoxanthopterin and leucopterin (Lafont & Papillon, 1972; Lafont & Tarroux, 1974; Table 10.4).

Lafont & Papillon (1972) investigated the activity of XDH in the pupa and imago of *P. brassicae* and showed that there were two distinct phases, a peak of activity shortly before eclosion and another in the female only (also reflected in the fat body and ovaries (Figure 10.5a,b), 2–8 days after eclosion. In the male, however, the activity of XDH is about twice as

a) Variation of XDH during development
     (after Lafont & Papillon, 1972)

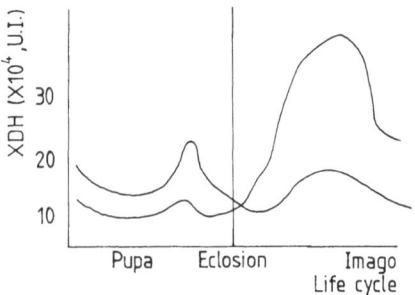

b) Variation of XDH in heads and ovaries
     (after Lafont & Papillon, 1972)

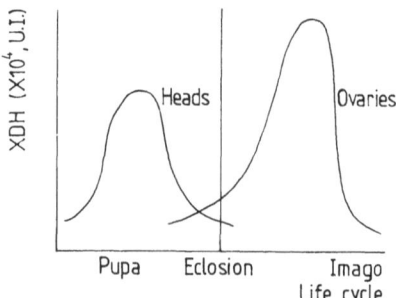

Fig. 10.5 Xanthine dehydrogenase (XDH) in *P. brassicae*

much at the end of the fifth instar as in the female. The concentration of XDH in pupal heads was found to increase before imaginal eclosion, while in the haemolymph, gut and testes little XDH activity was recorded.

Mauchamp & Lafont (1975) also found other enzymes in the fat body and gut of fifth instar larvae, namely uricase, guanine deaminase and allantoinase which are responsible for uric acid breakdown. These enzymes increase in concentration in the larvae (with age), then it rapidly decreases towards the end of the fifth instar when it increases again slightly. Little activity of the enzymes was found in the Malpighian tubules and gut wall, whereas much was found in the fat body and body wall. GD was found in low concentrations at the start and finish of larval life. Uricase, GD and XDH were also recorded from *P. brassicae* by Lafont & Pennetier (1975).

Lafont & Papillon (*loc. cit.*) pointed out that little is known about the physiology of pteridine pigments in *P. brassicae* and that few enzymes have been found, unlike other insects such as *Apis* sp., *Bombyx* sp. and *Drosophila* sp. in which important enzymes such as sepiapterine deshydrogenase, pterine desaminase, isoxanthopterine desaminase have all been identified. A great deal is known about the substrates in *P. brassicae* and the products of enzyme action but not a lot is known about the enzymes themselves. This therefore presents plenty of scope for research.

Some enzymes are voided in the frass of *P. brassicae*, five bands were identified by SGE by Clements (1967).

Flight

Beenakers (1969) recorded the activities of nine enzymes associated with flight muscles of *P. brassicae* (Table 10.4), and found that they varied in activity from above 10^4 moles per fresh weight per hour for arginine kinase to less than 10 moles per fresh weight per hour for lactate dehydrogenase (LDH). The other enzymes found by Beenakers are listed in Table 10.4 in order of decreasing activity.

The high activity of AK was thought by Beenakers (*loc. cit.*) to be comparable to creating kinase activity in vertebrate muscles. The level of glyceraldehydephosphate dehydrogenase (GAPDH) in comparison to other lepidoptera was high and was thought to reflect the flight abilities of the insect bearing in mind that *P. brassicae* can sustain prolonged flight (Chapter 11). In contrast to the lepidoptera which do not feed as an imago *P. brassicae* has a much higher GAPDH activity.

Melanisation

Onslow (1917) stated that when wing buds of *P. brassicae* pupae were dipped into tyrosinase black pigment appeared at the tips, but when they were dipped into tyrosine solution the wings blackened all over. This indicated that tyrosine was localised, whereas the enzyme tyrosinase was present all over the wing.

Other physiological functions

Esterases. Clements (1967) identified 13 esterases by electrophoresis from many tissues and organs of *P. brassicae* (Table 10.4), and all of them were thought by him to have been synthesised by the insect rather than by bacteria. Three enzymes were always found in particularly high concentrations (Ef. 30, 37, 45).

During metamorphosis certain qualitative changes take place; in the larval gut several new but minor esterases appear for the first time in the fifth instar, and then in the pupa two major esterases disappear. Esterase activity in the flight muscles was found to be 10 times that of the gut.

Inhibition studies revealed that the esterases were in fact carboxyl esterases and were not arylesterases or cholinesterases. Clements (*loc. cit.*) stated that "not a great deal is known about the functions of carboxyl esterases" and that presumably those in the gut have a digestive function; "whether these enzymes have a hydrolytic or synthetic function in other regions of the body remains to be shown".

Van der Geest & Borgsteede (1969) separated by gel electrophoresis (3 & 5% polacrylamide) esterases and one leucinaminopeptidase (LAP) from the haemolymph of fourth and fifth instar larvae. In *Drosophila* five LAPs have been recorded but this was from whole larvae; thus it is possible that more LAPs will be found in *P. brassicae* larvae. Van der Geest & Borgsteede (*loc. cit.*) also stated that LAP isozymes may also be found in other tissues of *P. brassicae*.

Phosphatases. Van der Geest & Borgsteede (*loc. cit.*) also recorded the presence of both alkaline and acidic phosphatases (ALP, ACP) in the haemolymph of fourth and fifth instar larvae.

Vitamins

Introduction

Very little work has been done on the identification of vitamins in *P. brassicae*, and in fact only vitamin A has been identified in its tissues; instead a great deal of work has been done on the effect of a lack of vitamins on the physiology of *P. brassicae* larvae by omitting vitamins from synthetic diets.

Fat soluble vitamins

Vitamin A. Vitamin A was identified quantitatively in a sample of 89 ova (11 International units (I.U.)) and a sample of 14 heads of *P. brassicae* (2 I.U.) (Feltwell, 1973). Four vitamin A potent carotenoids (ones which yield vitamin A under enzymatic action) were also found in the heads and may serve as an immediate source of vitamin A in these organs.

Vitamin E. Turunen (1976) experimented with the inclusion of vitamin E (alpha tocopherol) into an artificial diet of *P. brassicae*. Vitamin E is thought to protect the nature of the subcellular membranes and may also act as a dietary antioxidant. When vitamin E is absent from the diet there is a decrease in the accumulation of linoleate and linolenate (both polyunsaturated fatty acids, PUFAs) in the fat body and mid-gut. When PUFAs are present in the diet there is a decrease in the accumulation of monosaturated fatty acids, e.g. palmitate.

Water soluble vitamins

B Complex. It was mentioned in Chapter 5 (Table 5.2) that at least eight workers had experimented with the inclusion of 11 different members of the vitamin B complex into synthetic diets and had observed the effects on *P. brassicae*, although in a random manner.

Van der Geest & Sloog-Hoebe (1972) found that three of the nine B vitamins tested has an extremely detrimental effect on the larvae (i.e. 0% pupated after 20 days) when they were omitted from the diet. These were folic acid, nicotinic acid and calcium pantothenate. A lack of aneurine hydrochloride and B_{12} only caused a small percentage mortality. These workers suggested that higher concentrations of these vitamins should be included in synthetic diets, as they had maintained a culture of *P. brassicae* on synthetic diet for several years and their growth rate was well below what would normally be expected.

Harmsen (1966a) at Cambridge, England, failed to measure accurately folic acid and riboflavin in cabbage leaves, but quoted Long as saying that 50 μ of riboflavin was present in 100 g of fresh cabbage. 330 μg of folic acid were present in cauliflower. L'Hélias (1959b) found that folic acid promoted formation of melanin nodules (cf. 1959a; 1960, 1961, 1975).

B vitamins were shown to have an effect on more than one cell on the styloconica of the maxillae of the larvae but further work on this was not pursued (Schoonhoven, 1969).

Vitamin C. Ma (1972) tested the responsiveness of the maxillae of *P. brassicae* larvae to ascorbic acid together with sucrose and found that there was a positive relationship between the concentration of the vitamin and the response, in other words there was an increase in response up to a maximum of 0.002 M, after which sucrose acted as an inhibitor. The significance of this is that it adds to the sophisticated chemosensory mechanism of foodplant selection.

Pigments

Insect tissues contain a variety of pigments, some of which are derived exogenously from plant tissues on which the larvae feed and others which are synthesised *de novo* (endogenously) in the insect.

The pigments found predominantly in insects are pteridines, ommochromes, carotenoids, bile pigments, melanin (Figure 10.6), flavones, anthocyanins and quinines all of which, except quinines have been found in *P. brassicae*. In total *P. brassicae* has been accredited with nine pteridine pigments, four ommochromes, two bile pigments and 17 carotenoids in various parts of the body, but particularly in the haemolymph and integument.

The function of many of these pigments is in colouration (cf. Thompson, 1960, p. 21, Rothschild, 1975), particularly in the case of melanin, but other functions attributed to other pigments include photoreception (carotenoids) and stabilisation of proteins (carotenoids and bile

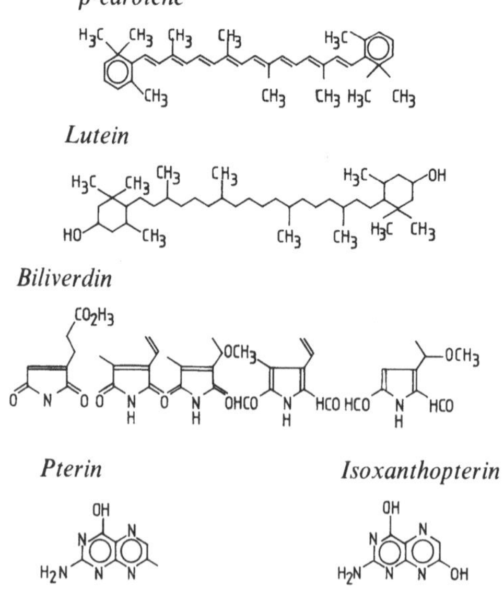

Fig. 10.6 Structure of main pigments in *P. brassicae*

pigments), while in themselves pteridine pigments are useful to the insect in colouration as white excretory products.

Pigments can be obtained readily from the insect material after preliminary extraction and then separation and purification by column chromatography, using aluminium oxide and celite (carotenoids), Sephadex (ommochromes), TLC (bile pigments) or by chromatograming on Whatman No. 1 paper (pteridines).

Anthocyanins

Anthocyanin pigments were recorded together with carotenoids in *P. brassicae* by Meyer (1930) (cf. Portier, 1949) and again as flavones in the haemolymph and epidermis of *P. brassicae* (Manunta, 1935). Apart from these two works very little appears to have been done on these pigments in *P. brassicae*.

It is established though, that the fourth cell on the lateral hair of the sensillum styloconicum on the maxillae of the larva, responds to anthocyanins. Those to which the cell is receptive are pelargonin (10^{-3} M), malvin (10^{-3} M), cyanidin mannoside (10^{-2} M) and two cyanidin compounds (10^{-2} M), both extracted from red cabbage. Their aglycones and delphinin (5×10^{-3} M) were found to be ineffective.

Bile pigments

Weiland & Tartter (1940) first investigated a blue chromoprotein which they extracted from about 1 million cabbage butterflies (cf. Thompson, 1960). The presence of this blue bilin (pterobilin) was confirmed by Hackman (1952).

In 1968, Ruediger and workers extracted this pigment from the integument of 200 *P. brassicae* larvae and confirmed the presence of pterobilin and biliverdin-IXy (cf. Choussy & Barbier, 1975).

Later, Ruediger *et al.* (1969) found that by injecting ^{14}C glycine (a precursor of bile pigments in vertebrates) into 220 fifth instar larvae biliverdin-IXy was biosynthesised from simple precursors in the larvae, rather than from compounds in their food. Vuillaume *et al.* (1970) found that pterobilin was widely distributed in the lepidoptera; they screened 80 species and noted that there was about 10γ of pterobilin per pupa of *P. brassicae*. Pterobilin was always found in the larvae and pupae but not in the wings of the imago. Vuillaume *et al.* (1971) further showed that the green tegmental pigment, pterobilin, was of fundamental importance in larval growth and as a photoreceptor in diapause. They used red light of 620–650 nm.

Barbier *et al.* (1970) recorded that there were smaller amounts of bile pigment present in diapausing pupae than in non-diapausing pupae and went on to re-affirm that there appeared to be a relationship between diapause and tetrapyrolic pigments. It would appear that photoperiod conditioning in the larvae of *P. brassicae* is under pterobilin control. Synthesis of pterobilin (with melanin) was noted by Kayser & Angersbach (1975) as a result of short wave light. Above 500 nm the action spectrum of bile pigment represents the mirror image of melanin synthesis.

Carotenoids

There are over 400 carotenoids in nature and 150 have been recorded from insects (Feltwell, 1973, 1978; Feltwell & Rothschild, 1974). Seventeen carotenoids have been recorded from *P. brassicae* and this vies with the 7-spot ladybird (*Coccinella septempunctata* Linnaeus) which has been recorded with 25 carotenoids, the highest number of any insect. The spectrum of yellow green pigment extracted by MacMunn (1883) from *P. brassicae* was believed to be very similar to chlorophyll which however, is probably always broken down in insects.

Carotenoids are C_{40} compounds which are of plant origin and are responsible for yellow, orange and red hues. In invertebrates as well as vertebrates, carotenoids serve as vitamin A precursors; it is thought that their molecules are split evenly by an enzyme called β-carotene, 15,15′, oxygenase, which however, has not yet been identified in insects.

All stages of *P. brassicae* have been recorded with carotenoids, higher concentrations being found in the ova, originating from the female which passes them on transovarially (Table 10.5). Carotenoids are also present in the wings, bodies and head of imagines as well as the haemolymph, fat body, gonads, brains, gut, integument and exuviae. Meyer (1930) reported that there were sexual differences in the amounts of carotenoids and that these gave a different hue to the haemolymph. Portier noted that the haemolymph of *P. brassicae* female larvae was golden yellow, while that of males is brilliant green.

Some carotenoids which are abundant in cabbage, such as β-carotene and the xanthophyll lutein are found in most stages of *P. brassicae*, while others, such as cryptoxanthin, which are found in the plant only in very small amounts are often difficult to identify in the insect unless large quantities are used. A study of the carotenoids in cabbage throughout the year showed that the type of carotenoid varied from month to month (Feltwell & Valadon, 1974), thus varying the amounts of carotenoids one would expect to find in the insect at any one time. Some modifications of the carotenoid molecules once in the insect is well documented now and this may account for the presence of ξ-carotene in the insect when it was not observed in the plant (Feltwell & Valadon, 1972).

Table 10.5 Carotenoids found in *P. brassicae* (after Feltwell, 1978).

Stage or part	Carotenoids present
Ovum	β-carotene, 5,6-monoepoxy-β-carotene, lutein
Larva (L_{1-5})	β-carotene, lutein, 5,6-monoepoxy-β-carotene
Larval haemolymph	α-carotene, β-carotene, violaxanthin, lutein
Prepupa	β-carotene, 5,6-monoepoxy-β-carotene, lutein, 5,6-mono-epoxylutein
Pupa	β-carotene, 5,6-monoepoxy-β-carotene, lutein, lutein esters
Imago	α-carotene, β-carotene, ξ-carotene, β-zeacanotene, 5,6-monoepoxy-β-carotene, mutatochrome, cryptoxanthin, 5,6-monoepoxy lutein, lutein, neoxanthin, violaxanthin, chrysanthemaxanthin, auroxanthin
Integument	α-carotene, taraxanthin
Epidermis	xanthophylls
Hypodermis	β-carotene, taraxanthin, phoenicoxanthin (adonirubin)
Imaginal haemolymph	carotenoids, xanthophylls
Fat body	β-carotene, lutein epoxide, lutein
Silk	α-carotene, taraxanthin
Gonads	carotenoids (possible)
Intestine	β-carotene
Imaginal heads	β-carotene, mutatochrome, violaxanthin
Imaginal brain	lutein, β-carotene
Larval exuviae (L_5)	β-carotene, lutein

Unusually large amounts of carotenoids are found in *Tropaeolum* (6,756 μg/g), a very readily accepted foodplant, and these are reflected in *P. brassicae* reared on this plant (Rothschild *et al.*, 1975a,b, 1977).

Carotenoids are also incorporated by the larvae of the endo-parasite *Apanteles glomeratus*, whose larvae spin yellow cocoons around the dying bodies of its host. Their larvae probably obtain their carotenoids from the non-essential body tissues, such as the fat body of the host. The yellowness of *P. brassicae* haemolymph probably serves as an aposematic warning colouration for predators, as these pigments are often an essential ingredient of defensive sprays in other insects.
sprays in other insects.

The role of vitamin A in *P. brassicae* as in other insects is not at all clear. Certainly *P. brassicae* can be reared on an artificial diet where cabbage powder is omitted and apparently healthy imagines result (David & Gardiner, 1966). In some insects, however, slight changes take place in the absence of vitamin A, such as lessened sensitivity of the visual receptors, but none which seem to be to the detriment of life, as in vertebrates. There is a possibility that carotenoids may be produced by symbiotic bacteria present in the insects, but this is only conjecture so far (cf. Britton *et al.*, 1977).

A particularly significant function of carotenoids in *P. brassicae* is that they are involved with photoreception in relation to diapause determination (see Chapter 8), acting either though the integument or through the yellow triangle in the head of the fifth instar larvae. Rothschild *et al.* (1975b) stated that large amounts of carotenoid were often found in diapausing pupae. The *pars intercerebralis* of the brain is bright yellow and contains β-carotene (Seuge & Veith, 1976).

Melanin

Introduction. With the exception of the ova, melanin pigment is present in all stages of the Large White Butterfly. It occurs as variable black flecks on the larva and pupa and as black scales on the wings of the imago. The haemolymph of the larva and imago also has the capacity to produce melanin by the action of the enzyme tyrosinase. Early workers investigated the chemistry and black patterns of the larva, pupa and imago without elucidating how melanisation occurs. Recently work has continued on melanin deposition in the cuticle and the effects of light and hormones on melanisation.

Classification of melanin patterns. The patterns of markings on the larval and pupal cuticles are fairly complex and in the pupa these patterns take about 48 hours to develop completely at normal temperatures (Gardiner, 1963). Many of the variations described for the imagines are variations of the degree of black and greyness of the wings (David & Gardiner, 1961; Gardiner, 1974). Classification of the degree of melanisation of the larva has therefore been necessary in order to compare the effects of different factors on melanisation.

Oltmer (1968) made five divisions on the basis of comparative intensities of colour and this method has been used successfully on *P. brassicae* by Johansson (1954), Angersbach (1975) and Kayser & Angersbach (1974), the last two having made half divisions of Oltmer's classification. Another mode of classification has been by spots (Bückmann, 1971).

Biochemistry: Early work. Coste (1890) chemically investigated the white, black, brown, red and cream pigmentation of at least 26 species of lepidoptera including *P. brassicae*, by immersing samples of scales from the wing in solutions of a variety of chemicals (24 types including 11 acids, 2 alkalis, inorganic salts, phenol, trinitrophenol and tannin). He noted that the black pigmentations of *P. brassicae*, like that of all other species, was "utterly unassailable" by any of his chemicals and he came to the conclusion that the black pigment was not pigmental. (He also noted the

300

same results with the white pigment and concluded that this was a physical colour.)

Coste (*loc. cit.*) also incorporated dyes into the wings of *P. brassicae* using methyl aniline green, violet and iodine, with a view to understanding colour changes and he concluded that air spaces must be present. The main theme of Coste's work was that insect colours were "phylogenetically progressive" and that "retrogressive modification" of colour was possible to achieve by applying chemicals (i.e. that black colours could be changed back to red and thus to yellow and white in a progressive sequence). However, Coste proved his own preconceived ideas were wrong.

The development of black markings on the wings of *P. brassicae* were described by Urech (1892) and Onslow (1916, 1917). Onslow described how the black markings on the developing wings of the pupa are first yellow and then black as air, which passes between the pupal case and the wing bud, oxidises the colourless chromagen present in the presence of tyrosinase which is present on the wing and haemolymph It is worth noting that Fraenkel & Rudall (1947) were unable to find sufficient tyrosin in fully grown fifth instar larvae.

Current knowledge on melanin. Although much about the chemical structure of melanin is still unknown (Kayser & Angersbach, 1974), a considerable amount of information is known about factors which affect the amount of melanin production, for instance it is known for certain that:

a) the direction of light is important (Angersbach, 1970, 1975).

b) the wavelengths of < 500 nm promote melanogenesis and > 500 nm inhibit melanogenesis (Kayser & Angersbach, 1974, 1975).

c) at wavelengths of < 500 nm the degree of melanin and the level of the bile pigment biliverdin are correlated positively while at < 500 nm they are negatively correlated (Kayser & Angersbach, 1975).

d) the critical stage for the reception of stimuli is in the fifth instar stage (Kayser-Wegmann, 1975, cf. Kayser & Angersbach, 1974, 1975).

e) the hormone factors responsible for melanogenesis originate from the first thoracic segment (Oltmer, 1968; Kayser & Angersbach, 1974; Kayser-Wegmann, 1975).

f) melanisation is dependent upon light received through two or more ocelli, the ventral one being the most important (Angersbach, 1975).

g) that blue light stimulates melanisation (Kayser-Wegmann, 1975).

Effect of light. Since 1938 it has been established that blue light, especially ultra-violet, stimulates melanogenesis and that yellow light has very little effect (Brecher, 1938). This has subsequently been confirmed by Kayser-Wegmann (1975) and Kayser Angersbach (1975) with their studies on melanin and bile pigment synthesis referred to in the previous section. They suggested that a central switch mechanism might explain this phenomenon whereby light of < 500 nm promoted melanogenesis and > 500 nm inhibited it. Melanogenesis was also found to be positively correlated with bile pigment synthesis at < 500 nm and negatively at > 500 nm, and that these two pigments served to protect the insect against harmful effects of light (Kayser & Angersbach, 1974).

Light is probably absorbed through the same structures for melanogenesis as for stimulation of green and brown pupae, i.e. through the triangle on the clypeus as well as through the dorsal patch on the head. Furthermore, it has been shown by Angersbach (1975) that the ocelli are important in light reception which governs the degree of melanogenesis in *P. brassicae*. Blinding the most ventrally placed ocellus (No. 1) produces maximal melanogenesis indicating that this is probably the most important ocellus for reception of light. Light reaching the larva's head from above penetrates the extra-ocular light receptor (the dorsal patch) and is able to differentiate changes in spectral ranges and brightness contrast.

Hormonal control. Melanin synthesis appears to be controlled by hormones released by the anterior part of the prepupa, as evidence for this has come from ligature experiments performed by Oltmer (1968) and Kayser-Wegmann (1975). It is now universally recognised that the titre of hormone in the haemolymph controls various colour changes. Oltmer (1968) demonstrated that when prepupae are ligatured between the second and third thoracic segment melanogenesis occurs anteriorly to this point, and that deposition of melanin into the cuticle was due to a "melanin inhibiting hormone". He also proposed that another hormone, the "pigment promoting hormone", was responsible for incorporating brown pigments, and that greatest melanogenesis occurred in bright illumination against a dark background. Kayser & Angersbach (1975) believed that one hormone may be responsible for the distribution of melanin and bile pigment in *P. brassicae*; Rothschild *et al.* (1975a) believed that melanisation of the cuticle and the distribution of ommochromes are controlled by different hormones.

Melanin deposition in the cuticle. Brunet (1965) posed the question: What is the difference between the white and black parts of the wing of *Pieris* ? – a question which previously puzzled Coste in 1890. A quotation from Brunet's paper is worthwhile considering, it contains many uncertainties: "The black and white colours might be a result of the state of oxidation of the tanned proteins of the cuticle: catechol-proteins are pale, quinone-proteins dark; but, although there is all too little concerning this aspect of the subject in the literature, it would appear that quinone proteins are only black if two conditions are fulfilled, namely when dopa is the chromogen, and when the protein has sulphydryl groups for attachment to the chromogen. ... We are left to suppose that where the cuticle is black this must be the result of incorporation of the oxidation products of dopa as a supplementary pigment, not directly attached to the protein, but in a polymerised form, either pure or co-polymerised with the excess tanning quinone."

Kayser-Wegmann (1976a,b) studied the ultrastructure of the cuticle of larvae and pupae of *P. brassicae* and showed by light microscopy that there was a dark homogenous layer limited to the epicuticle of the exocuticle of larvae. In the pupa, black pigment was homogenously deposited in the distal parts of the exocuticle, and there were brown areas at the base of bristles, the dark pigment being "thicker" than other dark spots. No distinct grana or homogenous layer was found with electron microscopy. Kayser-Wegmann's results indicate that quinoid sclerotization may occur in the black spots, whereas tanning by the side chain process may occur in the colourless parts of the cuticle.

Darkening of the haemolymph. An early reference which appears to refer to darkening of the haemolymph of *P. brassicae* is that of Schmalfuss (1926). He demonstrated the synthesis of "pigments" in a 40-hour experiment in which he incubated three "microdrops" of *P. brassicae* haemolymph with amino acids in a phosphate buffer. After 24 hours pigments were produced and after another 20 hours a grey colour developed.

The haemolymph of *P. brassicae* darkens in contact with air (Post, 1972) and precautions have to be taken to prevent this happening when samples are taken in the laboratory (see Chapter 7). Melanin is produced when the haemolymph darkens and this is brought about by the reaction of the enzyme tyrosinase with air (Onslow, 1916). Levels of tyrosin in the haemolymph of fourth and fifth instar larvae were calculated by Post & de Jong (1973) (Figure 10.7).

Ommochromes

Very little work has been carried out on these pigments in *P. brassicae* and Grassé (1975) recorded that the larval secretions of *P. brassicae* contain the ommochrome pigments rhodommatine and ommatine D. Apparently the red frass pellets produced by the fully fed fifth instar larva contain ommochromes together with uric acid, and are thus voided from the animal as unwanted material.

Stratakis (1976) studied the ommins and xanthommatins and the total ommochromes in the eyes of *P. brassicae* using Sephadex column chromatography and found that the ratio of ommin to xanthommatin was

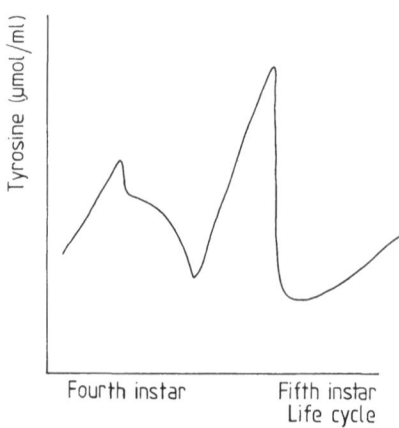

Fig. 10.7 Tyrosine levels in *P. brassicae* haemolymph
(after Post & de Jong, 1973)

higher in female eyes than male ones; however, the ratio of ommin and total ommochromes was higher in males than females.

Pteridines

Structure. Pteridines are C_6 compounds which have six atoms of nitrogen incorporated into their structure (Figure 10.6). As they are nitrogenous compounds, pteridines have often in the past been studied with purine metabolism. Pteridines are found both in plants and animals and exist in nature in three forms; 2-amino-4-hydroxy derivatives which are termed pterins, conjugated pterins which may be bound to folic acid, and unconjugated pterins. The 2-amino-4-hydroxy derivatives or pterins are those which are predominantly found in the class Insecta.

Function. Pteridines function as co-factors in hydroxylation reactions in the cell, and may function as neurohormones or as sex attractants in some insects (Ziegler & Harmsen, 1969). They also play a part as screening pigments and in the filtering of light in the near ultra-violet and the blue regions, thus complimenting ommochrome pigments which filter light in the visible region. In the Pieridae pterins are responsible for the white colour of the wings (the "ornament" function proposed by Sir Gowland Hopkins in 1889) and thus serve for specific and sexual recognition, and at the same time serve a dual function in being stored excretory products of nitrogen metabolism (Ziegler & Harmsen, 1969).

One particular pterin, sepiapterin, is responsible for the yellow colours of the aberration *jauni* (Harmsen, 1964); however, Harmsen (*loc. cit.*)

pointed out that little work had been done on the biological importance of pterins in insects, particularly on the degree of selection which had favoured pterins as excretory products or as useful colouration pigments (cf. Maddrell, 1971).

Early work. Sir Gowland Hopkins, once a Demonstrator of Physiology at Guy's Hospital, was interested in the part played by excretory products in the colouration of pierid butterflies including *P. brassicae*, and isolated a white pigment which he thought was uric acid (UA) and a yellow pigment (a derivative of UA) which he named "Lepidoporphyrin". About equal amounts of his white and yellow pigments were found in each imago. However, 31 years later Hopkins' UA was identified as being a pteridine called leucopterin which is in fact chemically similar to UA (Schoff & Wieland, 1926; Harmsen, 1966b).

Recent work. Harmsen (1966a,b,c) thoroughly investigated pteridine excretion in insects, concentrating mainly on *P. brassicae*, and found that the principal metabolic end products were purines and to a lesser extent pteridines. These were mainly stored in the body, few being excreted. On the other hand the main excretory products, kynurenine and urea, were greatly excreted in the frass.

Altogether Harmsen (*loc. cit.*) isolated nine fluorescing substances (pterins) and three ultra-violet absorbing pigments (purines) from various tissues and organs of *P. brassicae* (Table 10.6) and showed that the most abundant pterins in this species are isoxanthin and leucopterin, while UA and xanthine are the most abundant purines (cf. Harmsen, 1963).

With regard to their function Harmsen (*loc. cit.*) thought that sepiapterin was responsible for providing the yellow colour of the epidermis of the larva and that there was appreciably more of this pigment in the yellow variety of *P. brassicae* than in the grey variety.

Pteridines in the life stages

Larva. The concentration of leucopterin, isoxanthopterin, sepiapterin and UA was measured accurately at five periods during the life of the fifth instar larva by Harmsen (1966a), during which period the total pterin content increased from 69- to 134 µg per larva (Table 10.7, cf. Figure 10.8). Most of the pterins involved were assumed to be end products of folic acid or riboflavin metabolism, while some were probably synthesised by the insect.

A certain amount of pteridines are ingested with cabbage and this was estimated to be no more than 750 µg/100 g of fresh material (Harmsen,

Table 10.6 Pteridines and purines in *P. brassicae*. Data taken from Harmsen (1966a,b), but also see Hopkins (1895a,b, 1889) and Schoft & Wieland (1926). Watt & Bowden (1966) made quantitative measurements of those marked with an asterix.

	Pteridine or purine	Tissue or organ
a)	*Fluorescing substances*	
	Biopterin	all stages, only in living tissues in low concentration
	Erythropterin	
	(and breakdown products)	wing scales (small concentrations)
	*Isoxanthopterin (definite)	all stages, all tissues, wings
	Kynurenine (most likely)	imago, gut, meconial sac, meconium, frass
	*Leucopterin	all stages, all tissues (more concentrated than kynurenine)
	Pteridine-7-carbonic acid	traces in wings and scales
	*Sepiapterin	larval epidermis
	*Xanthopterin	traces in wings and scales
	One unidentified substance recorded	—
b)	*Absorbing substances*	
	Hypoxanthine (probably)	traces only
	Uric acid	all tissues (large amounts) in all stages, also frass
	Xanthine	imago, fat body

Table 10.7 Fluctuations in pteridines and purines during the larval and pupal stages of *P. brassicae*. (in μg/insect; + indicates presence, − indicates absence).

	Stage			
Pteridine or purine	Fifth instar (24 hrs)	Fifth instar (5 days: spun)	Prepupa	Pupa
Isoxanthopterin	12.1	26	24.7	82.4
Leucopterin	55	107	151.2	549.7
Sepiapterin	2.5	+		
Other pterins	+	+	+	+
Total pterins	69	134	180	650
Hypoxanthin			−	+
Kynurenine			−	46.9
Uric acid	442	1752	1621	1708
Xanthin			−	1200

1966a), and the average pterin intake per larva was worked out not to exceed 150 μg.

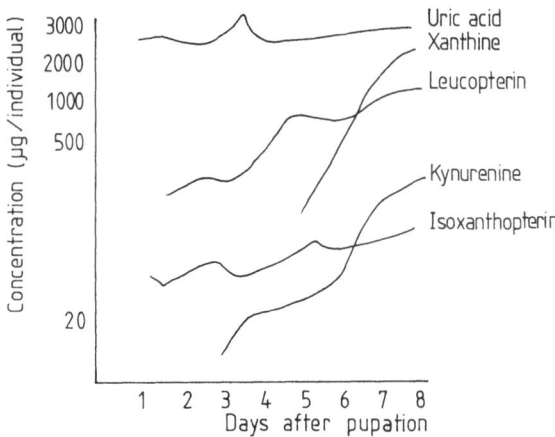

Fig. 10.8 Fluctuations of pteridines, purines and kynurenines in pupae.
(after Harmsen, 1966a)

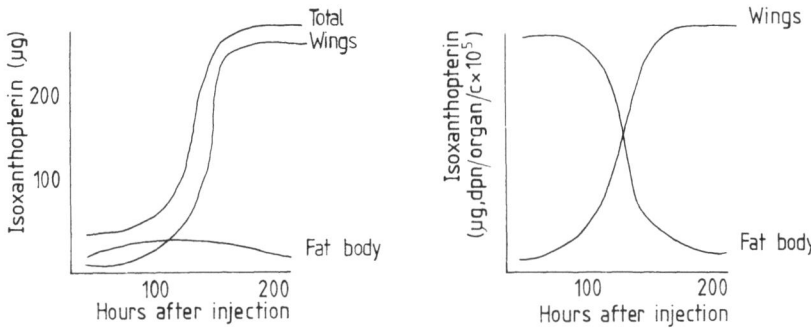

Fig. 10.9 Concentration of isoxanthopterin in pupal life
(after Lafont et al., 1971)

Pupa. Lafont *et al.* (1971) investigated the biosynthesis of pteridines
during development of the pupa of *P. brassicae* (Figure 10.9), and showed
that the concentration of isoxanthopterin rose 30 times in the wings
during the first 100 hours, but little was found in the fat body. By using
labelled isoxanthopterin they demonstrated that this was lost from the fat
body and gained by the wings.

In a further paper Lafont (1972a) investigated the total pteridines
in various organs of the pupa and showed how they increased more in the
male (30–220 μg) than in the female (30–120 μg). Isoxanthopterin present
in the head and integument increased in concentration following a typical
sigmoid curve as the pupa becomes older, and Lafont concluded that

isoxanthopterin was synthesised in the fat body and transported to the wings, head and scales.

Enzyme studies on xanthin-dehydrogenase showed that least activity occurred in the fat bodies of pupae during the period of active pigmentation, indicating that pteridine synthesis does not originate from these organs (Lafont & Papillon, 1972). In fact the highest activity was found in the wings, twice as much in the males as in the females.

Having shown that most uric acid and pteridines are stored in the wings Harmsen (1966a,b) tried wing amputation by ligaturing to see what effect this would have on purine deposition. The total content of pteridines went down and the concentration of pteridines in the meconium went up significantly. In the wings of the pupa he found the concentration of isoxanthopterin to be 44 μg/insect and leucopterin 239 μg/insect while in the bodies they were 17 and 69 μg respectively.

Imago

a) *Wings*. Pteridine pigments were some of the first pigments discovered in, the wings of butterflies and Hopkins (1885a,b; 1889) isolated xanthopterin, a yellow pigment, from the wing scale of *P. brassicae*. Yagi (1956) also investigated the pterin pigments found in the wings of pierids with respect to their bearing on systematics and noted that there were considerable sexual differences in the form of pterin "masses" found there. In *P. brassicae* the differences were not so conspicuous but pterin and isoxanthopterin were noted as present.

More recently much smaller oval granules termed "pterinosomes" have been reported in wing scales by electron microscopy (Descimon, 1969 – thesis). These granules are 5–10,000 Å by 1500–1800 Å and contain pigments. Apparently the pterinsosomes are derived from threads of vesicles in the endoplasmic reticulum.

The amount of leucopterin found in the wings of one individual has been calculated by Weygand *et al.* (1961) to be 500 μg and 350 μg by Harmsen (1966c). Watt & Bowden (1966), who were working principally on the pteridines of the *Pieris napi-bryoniae* Ochsenheimer complex, also analysed the wing pterins of *P. brassicae* quantitatively by way of comparison, and confirmed the presence of isoxanthopterin, leucopterin, xanthopterin, erythropterin and sepiapterin. Leucopterin was the most concentrated pigment in the wings.

b) *Ovaries and fat body*. The ovaries and fat body actively synthesise pteridine pigments and the presence of the enzyme XDH has been found there (Lafont & Papillon, 1972). When Lafont (1974) injected females with labelled isoxanthopterin, pteridines increased in the ovaries and decreased in the fat body.

c) *Aberrations.* Harmsen (1966a) studied the distribution of pteridine pigments in three aberrations, ab. *coerulea, jauni,* and *albinensis,* and found that xanthopterin and erythopterin were present in *jauni,* absent in *coerulea* and that those of *albinensis* did not vary significantly from those of *P. brassicae* wild type. In further work, Harmsen (1966b) noted that in the partly scaleless *coerulea* there was increased pteridine excretion coupled with a decrease in overall pteridine secretion.

Nucleotides

Ribonucleic acid (RNA)

Initial work by L'Hélias (1959) demonstrated that when folic acid was injected into larvae during inactivity of the corpora allata at a time when hormones are not being produced, an increase in the amounts of both RNA and DNA resulted.

Lafont (1969, 1970a) later showed that nucleic acid synthesis in the wing buds of fifth instar larvae followed very closely the rise in their dry weight (Figure 10.10d; cf. a,b,c) and that the synthesis could be stopped by using 5-fluorouracil (Lafont, 1970b). Adenosine was found to be involved with the fat body of the male (Lafont, 1974) and on injection of β-ecdysone into the pupa, a corresponding rise in RNA was produced (Lafont, 1974).

Tarroux (1975) demonstrated by injecting ^{3}H-adenosine into pupae that large amounts of the label became incorporated into ribosomal RNA from wing discs. Enzymatic breakdown of the Poly A fragments of the RNA by pancreatic and T_1 ribonucleases and purification by affinity chromatography showed that they were made up of about 150 nucleotide units. Evidence showed that Poly A segments shortened during maturation of RNA. The Poly A segments were closer in size to those from maturation cells than to eukaryotic organisms, a fact which is interesting as it is believed that Poly A segments increase in length with the evolution and age of the species. Tarroux (*loc. cit.*) did not show that the length of Poly A increased significantly over the period when most β-ecdysone is concentrated in the wing.

Lafont *et al.* (1976a,b) found that the concentration of RNA in whole fresh bodies of *P. brassicae* was 2–6 mg/g from autoradiographic data, and that the amount of RNA in males decreases during the development of the pupae. Three different phases of RNA synthesis were observed: a) an increase in the pharate pupa (Lafont *et al.*, 1975a) leading to histolysis of larval tissues, b) (another peak?) corresponding to growth of imaginal tissues, c) (another peak?) corresponding to cellular breakdown during

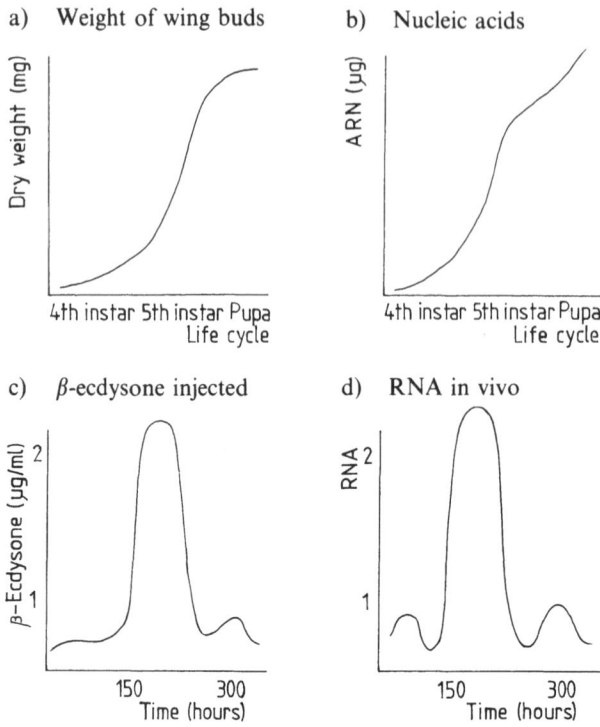

Fig. 10.10 Nucleic acids in *P. brassicae*
(from Lafont, 1970a, 1974)

imaginal differentiation. About two thirds of the RNA is lost at this stage.

Incorporating tritiated uridine into pupae gave one large peak during the maximum period of histogenesis. Synthesis of ribosomal RNA occurs in the wings and one of the peaks after incorporation of corresponded to scale development. In the fat body the concentration of RNA stayed constant but falls in keeping with DNA, protein and nitrogenous excretory products at the end of pupal development. A second peak of uridine incorporation corresponded with the renewal of enzymatic activity before eclosion (that is of xanthine dehydrogenase and guanine deaminase).

Deoxyribonucleic acid (DNA)

Very small amounts of DNA were found in imaginal discs, fat body cells and "carcass" (larval tissues and imaginal discs except wings) of fifth instar *P. brassicae* larvae (Lafont *et al.*, 1975a,b). The DNA in the imaginal discs increases by more than a 100-fold during development, unlike that of the fat body cells, while the ratio of RNA to DNA in the fat body was about 80, imaginal discs about 10 and larval disc between 30–40.

Lafont *et al.* (1976) made extensive investigations into RNA and DNA synthesis in *P. brassicae* pupae and imagines. They found that the DNA content during the pupal–imaginal development does not vary very much but that three phases were noticed: a) a slow decrease at the beginning of the pupal stage which is possibly due to histolysis of larval tissues (cf. Lafont *et al.*, 1975a), b) an increase in DNA synthesis by wings, legs and other organs, c) a decrease due to cell degeneration in the late pharate phase.

When these workers incorporated labelled thymidine, there were three peaks during pupal metamorphosis, their second one being higher in the female and probably related to their reproductive development; the other two were attributed to imaginal wing disc development. One of these latter peaks is related to the trichogen and tormogen cells, the other to "reprogression" of cells and synthesis of new enzymes.

In the wings DNA synthesis appears to be correlated with levels of β-ecdysone; when there are high levels of the hormone (> 0.4 g/nl), there are usually low levels of DNA synthesis.

Mustard oil glycosides

The chemical affinity of the mustard oil glycosides of the Cruciferae, Tropaeolaceae, Capparaceae and Resedaceae were established by Guignard (1890, 1893) *in* Gautier & Riel (1919), cf. Table 10.8.

Two mustard oil glycosides, allylisothiocyanate and sinigrin, together with sulphur, have been identified in both *P. brassicae* and its foodplants. These compounds were only identified qualitatively on account of the instability of these extracts. When reared on artificial diets lacking these substances pupae of *P. brassicae* were acceptable to birds and when extracts were interperitoneally injected into mice these still caused death. However, when reared on cabbage, pupae were rejected by birds and it was suggested that the presence of mustard oil glycosides would have contributed to their unacceptability.

The mustard oil glycosides found by Aplin *et al.* (1975) in *P. brassicae* were assumed to be sequestered from its foodplant during the larval stage, and those in the ova must therefore be transferred transovarially by the female. The role of mustard oil glycosides in foodplant selection and chemoreception is described in Chapter 4 (Foodplants) and Chapter 12 (Senses).

Table 10.8 Mustard oil glycosides in *Brassica* and *P. brassicae* and various food-plant families.

Group	Mustard oil glycoside	Authority
Brassica oleracea varieties		
B. oleracea var. *capitata*		David & Gardiner,
Linnaeus	sinigrin	1966
B. oleracea var. *gongyloides*	glucoerucin,	David & Gardiner,
Linnaeus	glucoraphinin	1966
B. oleracea var. *sabauda*		
Linnaeus		David & Gardiner,
(Savoy)	glucoiberin	1966
B. oleracea var. *capitata*		
Linnaeus	allylisothiocyanate,	
(Greyhound cultivar)	sinigrin	Aplin *et al.*, 1975
Pieris brassicae (Linnaeus)		
ova	allylisothiocyanate,	Aplin *et al.*, 1975
	sinigrin absent	cf. Marsh &
		Rothschild, 1974;
larvae	present	Aplin *et al.*, 1975
		cf. Marsh &
		Rothschild, 1974;
pupae	allyliosothiocyanate,	Aplin *et al.*, 1975
	sinigrin, another	cf. Marsh &
	unidentified	Rothschild, 1974;
imagines	allyisothiocyanate,	Aplin *et al.*, 1975
	sinigrin absent	cf. Marsh &
		Rothschild, 1974;
		Rothschild, 1974
Foodplant families		
Capparaceae	sinalbin	Gautier & Riel, 1919
Cruciferae	myrosin (+benzyl	
	isosulfocyanate)	Gautier & Riel, 1919
Resedaceae	glucotropaeolin	
	(+phenyl-	
	ethylisothiocyanate)	Gautier & Riel, 1919
Tropaeolaceae	sinigrin (+benzyl	
	isosulfocyantate	Gautier & Riel, 1919

David & Gardiner (1966) tested the effect of increasing concentration of nine mustard oil glycosides, four of which occurred in cabbage, as feeding stimulants to the fifth instar larva. When the concentrations of these was increased, it caused an "irregular" increase in feeding.

Biochemistry of silk

According to Wigglesworth (1967) silk is made up of a number of amino acids which vary according to family. Glycine, alanine and tyrosine are the main amino acids and a gelatinous protein called sericin is also present. As far as is known the composition of *P. brassicae* silk has not been examined. Carotenoid pigments are present in the silk of the silkworm *Bombyx mori* Linnaeus but they do not appear to be present in *P. brassicae* silk as the silk is colourless.

Fabre (1912) stated that substances contained in the ova case of the first instar larvae were essential nutrients of silk manufacture. The ova case is known to contain proteins and carotenoids.

Biochemistry of pupal exuviae

Martini (1975) made a thorough analysis of the organic components of pupal exuviae of *P. brassicae* and determined the amounts of eight fractions: surface wax, sub-surface wax, other lipids, free phenols, soluble protein, residual protein, condensed phenols and chitin residues (Table 10.9).

The dry matter of the pupal exuviae was 96%, comprising 70–80%

Table 10.9 Biochemical content of pupal exuviae of *P. brassicae* (figures taken from four replicates published by Martini (1975)).

Fraction	% Dry matter	Compounds present
Surface wax	3	Hydrocarbons: C_{29} & C_{31} alkanes Fatty acids, Esters, primary alcohols, unknowns
Subsurface wax	2–4	Hydrocarbons and fatty acids
Non-lipid wax fraction	3	Phospholipid: l-lecithin
Soluble protein and Residual protein	54–58	Water soluble protein: 18%
Phenolic compounds	3–7	75% pigmented material by methanolic extraction
Polyester compounds	3–7	Ester soluble acids: C_{16}, C_{18} saturated and C_{18} unsaturated acids
Chitin residue	34–37	69–85% chitin by spectrophotometric analysis, 71% by GLC
Sterols or Triglycerides	–	No significant proportion found

chitin. Various extracts of proteins, waxes and fats were found and these have been described in previous sections. It is important to point out that this appears to be the first appearance in the literature of the presence of alkanes, esters and primary alcohols in *P. brassicae*, although again this is not altogether surprising because of their widespread abundance. Cuticulin, a polymer similar to cutin or suberin in plants, was not found in pupal cases of *P. brassicae* (Martini, 1975).

Radionuclides used on *P. brassicae*

As can be seen from Table 10.10 quite a number of compounds have been incorporated into *P. brassicae* either in the larval or imaginal stage, the compounds differing according to the subject studied, i.e. amino acids in protein or pigment synthesis, carbohydrates and fats in pure biochemical investigations, excretory products with special regard to pteridine synthesis, hormones and nucleotides with growth and carotenoids with colouration. One of the points worth emphasising, as it applies to many such studies, is that often the radionuclides become incorporated into the fat body. In the experience of the author, the radionuclide is liable to be incorporated into many areas of the body, i.e. the haemolymph, fat body and gut, thus providing problems of interpretation. The Italians Cavallero & Cirio (1967) experimented with incorporating labelled strontium, caesium and iodine in a variety of insects from four orders. They measured the activity of the labels in eight different parts of the insect body but only found large amounts in three regions in *P. brassicae* (Table 10.10).

Miah (1976) in his doctoral thesis investigated the effects of administering 6–12 Krads of gamma radiation from Cobalt 60 on larvae and pupae of *P. brassicae*. He studied the development of the ovum up to the black head stage, the development and mortality, activity, feeding, excretion, growth and oxygen consumption of third instar larvae, as well as the amino acid content of the haemolymph, mating frequency, fecundity, number of spermatophores in the female and sterility of ova. The data from many of his controls have been incorporated into this book and the reader is referred to the relevant chapter for further information. Miah (*loc. cit.*) found that gamma radiation caused four effects: a higher mortality, a prolongation of larval life, reduced pupation and a higher incidence of crippledness.

Bauer (1967) x-rayed male imagines of *P. brassicae* with 6000 R units and crossed them with untreated females. This produced 19.9% lethal zygotes in the F_1. 69.9% of the resultant larvae had in their spermatocytes 1–4 heterozygous translation rings or chains consisting of 5–14 chromosomes.

Table 10.10 Radionuclides which have been incorporated into *P. brassicae*.

	Radionuclide	Stage administered	Effect/ incorporation (FB = fat body, H = haemolymph RNA = Ribo- nucleic acid)	Authority
a)	*Amino acids*			
	Folic acid [^{4}H]	L, diapausing	stimulated pteridines synthesis but not in diapausing larvae	L'Hélias, 1959a,b 1975
	Glycine [^{14}C]	L$_4$ injected	into quanine, hypoxanthin, triamino-6- pyrimidine	Weygand & Waldschmidt, 1955
	Glycine [^{3}H]	P	proteins	Fourche et al., 1981
	Glycine −2[^{14}C]	L$_{1.5}$	pigments	Mauchamp & Lafont, 1975; Simon et al., 1964
	Leucine [^{14}C]	L$_5$	FB	Chippendale & Kilby, 1969
	Proline [L^{3}H]	L$_5$	does not affect protein synthesis	Post & Vincent, 1973
	Proline [6^{14}C]	L$_5$	does not affect protein synthesis	Post et al., 1974
	Tyrosine [L^{-3}H]	L$_5$	does not affect protein synthesis	Post et al., 1974
	Thymidine [^{3}H]	L	FB	Chippendale & Kilby, 1969
	Thymidine [^{3}H]	P	Wings	Lafont, 1970a,b
	Glutamates [^{14}C]	L	Decrease in protein synthesis and increase in glutamates and α-ketoglutarate	Champ, 1958
b)	*Carbohydrates*			
	Glucose [$^{1-14}$C]	L,P injected	utilization of glycogen trehalose, pentoses at eclosion	Moreau, 1973; Moreau et al., 1974, 1977
	Glucose [$^{6-14}$C]	L injected	respiratory CO_2 labelled	Moreau et al., 1974, 1977

Table 10.10 continued

Radionuclide	Stage administered	Effect/ incorporation (FB = fat body, H = haemolymph RNA = Ribo- nucleic acid)	Authority
Carbohydrates — continued			
Glucose [^{14}C]	L$_5$	incorporated into precursor of chitin: uridine diphosphate N-acetyl glucosamine	Post *et al.*, 1974
Glucose [^{3}H]	L$_5$	abolition of endocuticle chitin	Post *et al.*, 1974
Glucose [$^{6-14}$C]	L$_5$	cuticular tissue (micro- autoradiography)	Post & Vincent, 1973
Glucose [$^{1-14}$C][$^{6-14}$C]	injected	$^{14}CO_2$	Moreau *et al.*, 1977
Inulin [^{14}C]	I ileum *in vitro*	used as a volume marker	Nicolson, 1976b
Ribose [$^{1-14}$C]			Simon *et al.*, 1963
c) *Excretory products and associated enzymes*			
Guanosine [^{14}C]	I (injected)	into nucleotides and wing pterins	Weygand & Waldschmidt, 1955
Guanine & Guanidine [^{14}C]	L$_5$ injected	FB	Lafont, 1975; Mauchamp & Lafont, 1975
Guanine [$^{2-14}$C] Guanine deaminase [^{14}C]	L	Imaginal discs	Lafont, 1975
Uric acid [^{14}C],[^{3}H]	L, I		Chippendale & Kilby, 1969
Uridine [^{3}H]	P	RNA	Lafont *et al.*, 1970a,b, 1974 Lafont & Tarroux, 1974
Uridine [^{3}H]	L	FB	Chippendale & Kilby, 1969
Uricase, Uric acid [$^{2-14}$C], guanine, guanosine	P (young)	Imaginal discs	Lafont, 1975

Table 10.10 continued

Radionuclide	Stage administered	Effect/ incorporation (FB = fat body, H = haemolymph, RNA = Ribo- nucleic acid)	Authority
d) *Fats*			
Glycerol tri [$1-^{14}$C] oleate	L	gut phospholipids	Turunen, 1975a,b
e) *Hormones*			
β-ecdysone [^{3}H]	P	haemolymph	Lafont *et al.,* 1974
f) *Nucleotides*			
Adenosine [^{3}H]	P (*in vivo*) P (*in vitro*)	RNA	Tarroux, 1975
Nucleotides [^{3}H]	P	nucleic acids	Lafont, 1969
5-fluorouracile C_i	P	wing nucleic acids	Lafont, 1970b
g) *Pigments*			
i) Carotenoids β-carotene [^{3}H]	L (orally)	FB, H	Feltwell, 1973
ii) Pterins Isoxanth- opterin [^{14}C]	L	FB	Chippendale & Kilby, 1969
Isoxanth- opterin [^{14}C]	I (injected)	H	Lafont & Pennetier, 1975
h) *Free radionuclides*			
P^{32}	—	used for tagging purposes, 1000–60,000 tpm, 50–400 tpm	Noordink, 1962, 1963
P^{32} phosphate	I	investigated several physiological criteria, see text	Miah, 1976 – thesis
C^{14}	sprayed	guanine [$2-^{14}$C], guanine [$2-^{14}$C], xanthopterin and leucopterin present in frass	Weygand & Waldschmidt, 1955
^{60}CO	O,L,P	investigated several physiological criteria, see text	Miah, 1976 – thesis

Table 10.10 continued

Radionuclide	Stage administered	Effect/ incorporation (FB = fat body, H = haemolymph, RNA = Ribo- nucleic acid)	Authority
Free radionuclides — continued			
P	L	H	Rogers *et al.*, 1966
Cs^{137} and I^{131}	Larva orally	most appeared in muscles, central nervous system, and fat body	Cavalloro & Cirio, 1967
Acetate $[^{1-14}C]$	First instar larva	oxidation of fats	Moreau, 1973
Acetate			Turunen, 1974a
Amino-5- imidazole- carboxamide	Pupa	nitrogenous compounds	Mauchamp & Lafont, 1975
4-amino- imidazole- carboxamide $[^{4-14}C]$	Pupa	—	Simon *et al.*, 1963
Glycine $[^{2-14}C]$	Pupa	—	
Austauch Barium carbonate and Potassium cyanide		—	
Barium carbonate $[^{14}C]$		—	Simon *et al.*, 1963
Natriumcyanid $[^{14}C]$		—	
2.4.5-triamino- 6-hydroxy- pyrimidine- $[^{2-14}C]$		—	
Bacillus thuringiensis $[^{3}H]$	Larva ingested	label appeared at base of gut epithelial cells 18 hours later	White, 1968
Nicotine ^{14}C	Larva *in vitro*	concentrated label in Malpighian tubules	Maddrell & Gardiner, 1976

Inorganic ions

Sodium, potassium, calcium, carbonate, magnesium, manganese, nitrogen, urate, iron and chloride ions have been identified in muscles, haemolymph, Malpighian tubules and gut of *P. brassicae* (Table 10.11).

Nitrogen is important in protein metabolism and amounts were shown to fluctuate from 600–1,700 mg/100 ml in the haemolymph of larvae (Junnikkala, 1966). During the fifth instar a peak of nitrogen circulating in the haemolymph was noted and this was probably associated with the peak of protein synthesis occurring at the same time. Junnikkala (*loc. cit.*) found significantly higher amounts of nitrogen in females than males.

Sodium and potassium concentration was measured in haemolymph, urine and meconium, Malpighian tubules and hind-gut of *P. brassicae* imagines (Nicolson, 1976b). Excised ileum absorbed potassium from the surrounding medium thus making the osmolarity of the urine low.

Table 10.11 Inorganic ions in *P. brassicae*.

Ions	Tissue/organ	Authority
Cl	haemolymph	Brecher, 1925, 1929
effects of several compounds on electro-physiological responses (see text)	—	Ma, 1972; Schoonhoven, 1977
N,K	Transverse tubular system of flight muscles (TTS)	Njio & Piek, 1977
Ca	flight muscles	Donnellan & Beechey, 1969
N	haemolymph	Junnikkala, 1966
N,K,Ca,Mg	Malpighian tubules	Nicolson, 1976a
N,K	urine, haemolymph, gut	Nicolson, 1976b
N, K, Ca, Na, Fe, Mn, PO_4 (trace), Cl. $Ca,Mg,P,PO_4,CO_4,$	Malpighian tubules larval hypodermis unpigmented imaginal cells	
potassium urate	mid-gut of pupa, early imago	Lhonoré, 1977
oxalates	Malpighian tubules of larva	
calcium phosphate potassium urate	Malpighian tubules of imago	
P. brassicae reared on diets deficient in K,N,P,Fe	—	Allen & Selman, 1957

Allen & Selman (1957) found that by omitting some or all of the elements nitrogen, potassium and iron from the diet of larvae, they developed a reduced larval weight, a reduced relative growth rate, an increased larval mortality and delayed mortality. Ma (1972) studied the effects of sodium chloride, sodium nitrate, sodium mono- and dihydrogenphosphate, and calcium chloride on gustatory responses in larvae. Sodium chloride was found to inhibit at 0.1 M and calcium depressed the reactivity of the sugar receptors. Unpredictable results were obtained with other inorganic compounds, ammonium chloride, potassium, lithium, magnesium chloride, barium chloride, sodium nitrate, sodium sulphate, sodium carbonate and sodium phosphate (Ma, 1972).

Sexual similarities and differences

Biochemical

a) *Haemolymph proteins.* Female imagines appear to have more proteins than males; Van der Geest (1968) found that the haemolymph of females in the second half of the fifth instar larvae possessed a higher protein concentration than males, and Lamy & Karlinsky (1974) recorded that a protein called "Protein F" was present only in female haemolymph. Feltwell (1973) also recorded that females have slightly more carotenoid pigment bound to proteins than males, thus suggesting that more proteins may be available for binding in females.

b) *Lipids.* No significant differences were found in the lipid content of the wandering stage of the larvae of each sex (Turunen & Junnikkala, 1974). However, Turunen (1974d) found that when larvae were reared on diets containing low amounts of lipids, male imagines were more prone to develop deformed wings. When Turunen (1975a) incorporated labelled palmitic acid into larvae he found that the females synthesised more palmitate, and that the males used more fat during flight, presumably by oxidation of oleate.

c) *Pigments.* Females also have a higher ratio of the ommochrome pigments ommin and xanthommatin in the imaginal eye than the males (Stratakis, 1976). However, more total ommochrome was found in male eyes than female eyes.

d) *Inorganics.* There is significantly more total nitrogen in female larvae than males (Junnikkala, 1966).

Physiological

a) In non-diapausing pupae at all temperatures, males have a lower respiratory rate than females (Fourche, 1977).
b) More glycogen is used up during the pupal stage in males than in females (Karlinsky & Poulaert, 1978).

Morphological, anatomical

a) There are more "B" type setae on the tarsi of females than males (Ma & Schoonhoven, 1973).
b) No significant difference was found in the haemolymph volume of either sex during the wandering stage (Turunen & Junnikkala, 1974).

References cited

Allen, M.D. & Selman, I.W., 1957. The response of larvae of the large white butterfly (*Pieris brassicae* L.) to diets of mineral deficient leaves. *Bull. ent. Res.* 48: 229–242.
Angersbach, D., 1970. *Die Bedeutung der Einfallsrichtung und der Wellenlänge des Lichtes bei der Melanisierung der Kohlweisslingspuppe Pieris brassicae* L. Dissertation. University of Giessen.
—— 1975. The direction of incident light and its perception in the control of pupal melanisation in *Pieris brassicae*. *J. Insect Physiol.* 21: 1691–1696.
Aplin, R.T., D'Arcy, Ward & Rothschild, M., 1975. Examination of the large white and small white butterflies (*Pieris* spp.) for the presence of mustard oils and mustard oil glycosides. *J. Ent.* A 50: 73–78.
Barbier, M., Bergerard, J., Hurpin, B. & Vuillaume, M., 1970. Conditionnement de la diapause et pigments tétrapyrroliques chez *Pieris brassicae*. *C. r. Acad. Sci., Paris* 271: 342–345.
Bauer, H., 1967. Die kinetische Organisation der Schmetterling-Chromosomen. *Chromosoma* 22: 101–125.
Beenakers, A.M.T., 1969. Carbohydrate and fat as a fuel for insect flight. A comparative study. *J. Insect Physiol.* 15: 353–361.
Brecher, L., 1925. Physiko-chemische und chemische Untersuchungen am Raupen- und Puppenblute (*Pieris brassicae, Vanessa urticae*). *Z. vergl. Physiol.* 2: 691–713.
—— 1929. Die anorganischen Bestandteile des Schmetterlingspuppenblutes (*S. pinastre, Pieris brassicae*). Veränderungen im Gehalt an anorganischen Bestandteilen beim Verpuppen. *Biochem. Z.* 211: 40–64.
—— 1938. Der Weg der Farbanpassung bei Schmetterlingspuppen vom Rezeptor bis zum Effektor. *Biol. Gen.* 14: 212–237.
Britton, G., Goodwin, T.W., Harriman, G.E. & Lockley, W.J.S., 1977. Carotenoids of the ladybird beetle, *Coccinella septempunctata*. *Insect Biochem.* 7: 337–345.
Brunet, P.C.J., 1965. The metabolism of aromatic compounds. *In: Aspects of Insect Biochemistry*. Biochemistry Society Symposium. No. 25. London. pp. 49–77.

Bückmann, D., 1971. Melanisierungsverlauf und Melanissierungshemmung bei der Kohlweisslingspuppe, *Pieris brassicae* L. *Wilhelm. Roux Arch. EntwMech, Org.* 166: 236–253.

Cavalloro, R. & Cirio, U., 1967. Distribution of radioactive Sr, Cs, I in the body of some insects. (In Italian.) *Redia* 50: 187–196.

Champ, B.R., 1958. *The effect of sub-lethal dosages of insecticides on parasitism in Pieris brassicae.* Ph.D. thesis. University of London (Imperial College).

Chippendale, G.M., 1977. Hormonal regulation of larval diapause. *A. Rev. Ent.* 22: 121–138.

Chippendale, G.M. & Kilby, B.A., 1969. Relationship between the proteins of the haemolymph and fat body during development of *Pieris brassicae. J. Insect Physiol.* 15: 905–926.

—— 1970. Protein biosynthesis on larvae of the large white butterfly, *Pieris brassicae. Comp. biochem. Physiol.* 34: 241–243.

Choussy, M. & Barbier, M., 1975. Pterobiline and neopterobilines, reactivity and structures. *Helv. chim. Acta.* 58: 2651–2661.

Clements, A.M., 1967. A study of soluble esterases in *Pieris brassicae* (Lepidoptera). *J. Insect Physiol.* 13: 1021–1030.

Coste, F.H.P., 1890. Contribution to the chemistry of insect colours. *Entomologist* 23: 123–132, 155–159, 181–187, 217–223, 247–252, 283–287, 309–314, 338–343, 370–374; 24: 9–15, 37–40, 53–60, 86–91, 114–119, 132–139, 163–170, 186–192, 207–211.

David, W.A.L., 1959. The systemic insecticidal action of paraoxon on the eggs of *Pieris brassicae* (L.). *J. Insect Physiol.* 3: 14–27.

David, W.A.L. & Gardiner, B.O.C., 1961. The mating behaviour of *Pieris brassicae* (L.) in a laboratory culture. *Bull. ent. Res.* 52: 263–280.

—— 1966. Mustard oil glycosides as feeding stimulants for *Pieris brassicae* larvae on a semi-synthetic diet. *Entomologia expl. appl.* 9: 247–255.

Descimon, H., 1966. Variations quantitatives des ptérines de *Colias croceus* (Fourcroy) et de son mutant *helice* (Hbn.) (Lepidoptera: Pieridae) et leur signification · dans la biosynthèse des ptérines. *C. r. Acad. Sci. Paris* 264: 390–393.

—— 1967. La dihydroxanthoptérine, un nouveau pigment naturel des lépidop-tères. *Bull. Soc. chim. biol.* 49: 1164–1166.

—— 1969. *Contribution à l'étude de la pigmentation ptérinique des Pieridae.* La *variation phénotypique et taxonomiques de la pigmentation ptérinique chez les Pieridae et quelques autres familles de Lépidoptères.* Thèse Doctorat d'état. Chapter III pp. 135–152.

—— 1971a. Métabolisme et excrétion de la guanine chez *Colias croceus. J. Insect Physiol.* 17: 1517–1531.

—— 1971b. Les ptérines des Pieridae (Lepidoptera) et leur biosynthèse. I. Identification des principales ptérines de *Croceas croceus* (Fourcroy) et de quelques autres espèces de Pieridae. *Biochimie* 53: 407–418.

—— 1976a. *Biology of pigmentation in Pieridae butterflies. In:* Chemistry and Biology of Pteridines. Walter de Gruyter, Berlin & New York.

—— 1976b. The pterins of the Pieridae and their biosynthesis: metabolism of D-erythro-Neopterin and its 7,8-Dihydro derivative in *Colias croceus. In: Chemistry and Biology of Pteridines.* Walter de Gruyter. Berlin & New York.

Descimon, H. & Lafont, R. et F., 1971. Localisation de la biosynthèse des pigments ptériniques chez *Pieris brassicae* (Lepidoptera: Pieridae) au cours de la vie nymphale. *C. r. Acad. Sci., Paris* 273: 1484–1487.

Donnellan, J.F. & Beechey, R.B., 1969. Factors affecting the oxidation of glycerol-1-phosphate by insect flight muscle mitochondria. *J. Insect Physiol.* 15: 367–372.

Fabre, J.H., 1912. *The life of the caterpillar.* Translated by A.T. de Mattos. Hodder & Stoughton.

Feltwell, J.S.E., 1973. *The metabolism of carotenoids in Pieris brassicae (L.) and in its foodplant, Brassicae oleracea var. capitata L. (Cabbage).* Ph.D. thesis. University of London (Royal Holloway College).

—— 1978. Carotenoids in the class insecta. *In: Plant and Animal Co-evolution.* Edited by G. Harborne. Academic Press, London & New York. pp. 277–307.

Feltwell, J.S.E. & Rothschild, M., 1974. The carotenoids of 38 species of lepidoptera. *J. Zool. Lond.* 114: 441–465.

Feltwell, J.S.E. & Valadon, L.R.G., 1972. Carotenoids of *Pieris brassicae* and of its food plant. *J. Insect Physiol.* 18: 2203–2215.

—— 1974. Carotenoid changes in *Brassicae oleracea var. capitata* L. with age in relation to the large white butterfly, *Pieris brassicae* L. *J. agric. Sci. Camb.* 83: 19–26.

Fourche, J., Bosquet, G., Guillet, C. & Calvez, B., 1981. Cold acclimation during the wintering of diapausing pupae of *Pieris brassicae* (Lepidoptera). *Comp. biochem. physiol.* (in press).

Fourche, J., Guillet, C., Calvez, B. & Bosquet, G., 1977. Le métabolisme énergétique des nymphs de *Pieris brassicae* (Lépidoptère) au cours de la métamorphose et de la diapause. Essai d'établissement d'un bilan. *Ann. zool. ecol. Anim.* 9: 51–61.

Fraenkel, G. & Rudall, K.M., 1947. The structure of insect cuticle. *Proc. R. Soc.* B. 134: 111–147.

Gardiner, B.O.C., 1963. Genetic and environmental variation in *Pieris brassicae*. *J. Res. Lepid.* 2: 127–136.

—— 1974. Observations on green pupae in *Papilio machaon* L. and *Pieris brassicae* L. *Wilhelm. Roux. Arch. EntwMech. Org.* 176: 13–22.

Gautier, C. & Riel, P., 1919. Sur l'alimentation des chenilles des genres *Pieris* et *Euchloe*. *C. r. Séanc. Soc. Biol.* 82: 1371–1374.

Grassé, P.P., 1975. *Traité de Zoologie.* Masson, Paris. Vol. 8.

Guignard, E., 1890, 1893 in Gautier & Riel, 1919.

Hackman, R.H., 1952. Green pigments of the haemolymph of insects. *Archs. Biochem. Biophys.* 4: 166–174.

Harmsen, R., 1963. *The storage and excretion of pteridines in Pieris brassicae L. and some other insects.* Ph.D. Thesis. University of Cambridge.

—— 1964. Genetically controlled variation in pteridine content of *Pieris brassicae* L. *Nature, Lond.* 204: 1111.

—— 1966a. A quantitative study of the pteridines in *Pieris brassicae* L. during post-embryonic development. *J. Insect Physiol.* 12: 9–22.

—— 1966b. Identification of fluorescing and ultra-violet absorbing substances in *Pieris brassicae* L. *J. Insect Physiol.* 12: 23–30.

—— 1966c. The excretory role of pteridines in insects. *J. exp. Biol.* 45: 1–13.

—— 1969. The biology of pteridines in insects. *Adv. Insect Physiol.* 6: 139–203.

Hopkins, F.G., 1889. Note on a yellow pigment in butterflies. *Proc. chem. Soc.* 5: 117–118.

—— 1895a. The pigments of the Pieridae: a contribution to the study of excretory substances which function in ornament. *Phil. Trans.* 186: 661–682.

—— 1895b. The pigments of the Pieridae: a contribution to the study of excretory substances which function in ornament. *Entomologist* 28: 1–2.

Howden, G.F. & Kilby, B.A., 1961. Biochemical studies on insect behaviour – II. The nature of the reducing material present. *J. Insect Physiol.* 6: 85–95.

Johansson, A.S., 1954. Diapause and pupal morphology and colour in *Pieris brassicae* L. *Norsk. ent. Tidsskr.* 9: 79–85.

Johnson, C.G., 1976. Lability of the flight system: a context for functional adaptation. *In: Insect Flight.* Edited by R.C. Rainey. Symposium of the Royal Entomological Society of London. Blackwell, Oxford. pp. 217–234.

Junnikkala, E., 1966. Effect of braconid parasitization on the nitrogen metabolism of *Pieris brassicae* L. A study of the nitrogenous compounds in the hemolymph of larvae at different ages reared in controlled laboratory conditions. *Annl. acad. sci. Fenn.* 100: 1–83.

—— 1968. A preliminary study of the character of peptide I (PI) in the haemolymph of *Pieris brassicae* L. *Annl. acad. sci. Fenn.* A 132: 1–10.

—— 1969. Effect of a semi-synthetic diet on the level of the main nitrogenous compounds in the haemolymph of larvae of *Pieris brassicae* L. *Annl. acad. sci. Fenn.* A 155: 1–10.

—— 1976. A possible storage form of tyrosine in the haemolymph of *Pieris brassicae* L. (Lepidoptera, Pieridae) and four other Lepidoptera. *J. Insect Physiol.* 22: 95–100.

Karlinsky, A. & Poulaert, J., 1971. The change in fat body during the imaginal life of *Pieris brassicae*. *Bull. Soc. zool. Fr.* 96: 453–466.

Kayser, H. & Angersbach, D., 1974. Action spectrum for light controlled pupal pigmentation in *Pieris brassicae* melanisation and level of bile pigment endrocrine control. *J. Insect Physiol.* 20: 2277–2286.

—— 1975. Dose effects in light controlled pupal melanisation in *Pieris brassicae*: specificities to spectral ranges. *J. Insect Physiol.* 21: 589–594.

Kayser-Wegmann, I., 1975. Untersuchungen zur Photobiologie und Endokrinologie der Farbmodifikationen bei der Kohlweisslingspuppe *Pieris brassicae*. Zeitverlauf der sensiblen und kritischen Phasen. *J. Insect Physiol.* 21: 1065–1072.

—— 1976a. Differences in black pigmentation in lepidopteran cuticles as revealed by light and electron microscopy. *Céll. Tiss. Res.* 171: 513–521.

—— 1976b. Ultrastructural differences between larval and pupal cuticles of *Pieris brassicae* (Lepidoptera). *Protoplasma* 90: 319–331.

Koizumi, K., 1953. Effects of environmental factors upon the interrelations between the amount and the melting point of the epicuticular lipid in some insects. *Ann. zool. Japan* 26: 168–175.

Lafont, R., 1969. Sur le dynamisme des acides nucléiques au cours de la différenciatation des ébauches alaires chez *Pieris brassicae*. *C. r. Acad. Sci. Paris* 267: 1639–1642. 268: 2133–2137.

—— 1970a. L'évolution des acides nucléiques dans les disques imaginaux alaires de *Pieris brassicae* au cours de cinquième stade larvaire. *C. r. Acad. Sci., Paris* 270: 1599–1602.

—— 1970b. Étude du développement des ébauches alaires de la chrysalide de *Pieris brassicae* en présence de 5-fluorouracile. *C. r. Acad. Sci., Paris* 271: 2186–2189.

—— 1972a. Modalités de l'excrétion azotée (pterines et acide urique) chez les lépidoptères Pieridae au cours de la vie nymphale. *Bull. Soc. zool. Fr.* 97: 401–411.

—— 1972b. Les ptérines des Pieridae (Lepidoptera) et leur biosynthèse. II. Synthèse et transport de l'isoxanthoptérine au cours de la vie nymphale et adulte chez *Pieris brassicae*. *Biochim.* 54: 73–81.

—— 1974. Adenosine, a sex-linked excretory product. *Experimentia* 30: 998.

—— 1975. *Aspects biochimique de la différenciation des disques imaginaux alaires des lépidoptères.* Thèse Doctorat. d'état. Ecole Normale Supérieur, Paris.

Lafont, R. & Blandin, P., 1968. Phases of differentiation of wing buds in *Pieris brassicae*: chronological and quantitative study of nucleic acids and isoxanthopterin. *C. r. Acad. Sci., Paris* D 267: 1639–1641.

Lafont, R., Delbecque, J.-P., De Hys, L., Mauchamp, B. & Pennetier, J.-L., 1974. Etude du taux de β-ecdysone dans l'hémolymphe de *Pieris brassicae* L. (Lepidoptera) au cours du stade nymphal. *C. r. Acad. Sci., Paris* 279: 1911–1914.

Lafont, R., Lafont, F. & Descimon, H., 1971. Localisation de la biosynthèse des pigments ptériniques chez *Pieris brassicae* (Lepidoptera, Pieridae) au cours de la vie nymphale. *C. r. Acad. Sci., Paris* 273: 1484–1487.

Lafont, R., Mauchamp, B., Blais, C. & Pennetier, J.-L., 1977. Ecdysones and imaginal disc development during the last larval instar of *Pieris brassicae*. *J. Insect Physiol.* 23: 277–283.

Lafont, R., Mauchamp, B., Boulay, G. & Tarroux, P., 1975a. Developmental studies in *Pieris brassicae* (Lepidoptera Pieridae). I. Growth of various tissues during the last larval instar. *Comp. biochim. physiol.* 51B: 439–444.

Lafont, R., Mauchamp, B., Pennetier, J.-L., Tarroux, P., De Hys, L. & Delbecque, J.P., 1975b. α, & β-ecdysone levels in insect haemolymph: correlation with developmental events. *Experimentia* 31: 1241–1241.

Lafont, R., Mauchamp, B., Pennetier, J.-L., Tarroux, R. & Blais, C., 1976a. Biochemical correlation during metamorphosis in *Pieris brassicae*. *Insect Biochem.* 6: 97–103.

Lafont, R., Mauchamp, B., Tarroux, P. & Blais, C., 1976b. Biochemical parameters of imaginal wing disc development in *Pieris brassicae*. Abstract from Symposium on Insect Biochemistry, Hamburg, August 1976. p. 649.

Lafont, R. & Papillon, J., 1972. Les ptérines des Pieridae (Lepidoptera) et leur bisynthèse. III. Etude de l'activité xanthine-déshydrogénase (E.C.1.2.3.2.) au cours du développement. *Biochim.* 54: 365–370.

Lafont, R. & Pennetier, J.-L., 1975. Uric acid metabolism during pupal–adult development of *Pieris brassicae* L. *J. Insect Physiol.* 21: 1323–1336.

Lafont, R. & Tarroux, P., 1974. Données biochimiques préliminaires sur le développement des disques imaginaux alaires de *Pieris brassicae* L. (Lepidoptera). *Ann. zool. ecol. Anim.* 6: 205–207.

Lamy, M., 1965. Electrophorèse des protéines de l'hémolymphe de la piéride du chou au cours de son cycle biologique et de sa diapause nymphale. *P.-v. Soc. sci. phys-nat. Bordeaux* (1963–1964): 241–246.

—— 1967. Physiologie des insectes. Une protéine vitellogène dans l'hémolymphe de l'imago femelle de la Piéride du Chou. *C. r. Acad. Sci., Paris* D 265: 990–993.

—— 1969. *Etude electrophoretique des protéines de l'hemolymphe chez les lépidoptères.* Thèse Doctorat Sciences naturelles. Université de Bordeaux.

—— 1970. Les protéines impliquées dans la vitellogenèse des lépidoptères. *Ann. Endrocr.* 31: 485–503.

Lamy, M. & Karlinsky, A., 1974. Vitellogenèse protéique en milieu mâle chez la Piéride du chou, *Pieris brassicae* L. (Lépidoptères). *C. r. Acad. Sci., Paris* Ser. D 278: 91–94.

Lamy, M., Karlinsky, A. & Julien-Laferriere, N., 1975. Vitellogenèse protéique anormale en milieu mâle chez deux Lépidoptères: *Pieris brassicae* L. et *Bombyx mori* L. *Bull. Soc. zool. Fr.* 100: 254.

Lecadet, M. & Dedonder, R., 1964. Différenciation des protéases du chyle de *Pieris brassicae* par l'étude de leur spécificité. *C. r. Acad. Sci., Paris* 258: 3117–3120, 3380–3383.

—— 1966. Les protéases de *Pieris brassicae*. Spécificité. *Bull. Soc. chim. biol.* 48: 661–691.

Lecadet, M.M. & Chevrier, G., 1966. The proteases of *Pieris brassicae*, demonstration of zymogens. *C. r. Acad. Sci., Paris* 263: 944–947.

L'Hélias, C.L., 1959a. Facteur inducteur de tumeur provoqué par l'acide folique chez *Pieris brassicae* en état de diapause. *Année Biol.* 63: 237–247.

—— 1959b. Purification partielle du facteur viral induisant les tumeur provoquées artificiellement chez *Pieris brassicae* (Insecta, Lepidoptera). *C. r. Acad. Sci., Paris* 240: 3646–3648.

—— 1960. Tumour induction factor provoked by folic acid in *Pieris brassicae* during the diapause. *Folia Biol. Praha* 6: 310–318.

—— 1961. Transmission génétique du facteur inducteur de tumeur de *Pieris brassicae* à la drosophile. *C. r. Acad. Sci., Paris* 252: 2015–2016.

—— 1975. Hormone juvenile de *Pieris brassicae* diapausante et mutations. *Ann. Épiphyt.* 36: 63–85.

L'Honoré, J., 1977. Données histophysiologiques sur les accumulations minérales et puriques. *La Cellule* 72: 1–54.

Ma, W.C., 1969. Some properties of gustation in the larva of *Pieris brassicae*. *Entomologia exp. appl.* 12: 584–590.

—— 1972. *Dynamics of feeding responses in Pieris brassicae L. as a function of chemosensory input: a behavioural, ultrastructural and electrophysiological study.* Doctoral Thesis. Agricultural University, Wageningen, Netherlands.

Ma, W.C. & Schoonhoven, L.M., 1973. Tarsal contact chemosensory hairs of the large white butterfly, *Pieris brassicae* and their possible role in oviposition behaviour. *Entomologia expl. appl.* 16: 343–357.

Macmunn, (–), 1883. *In:* Portier, 1949. no reference given.

Maddrell, S.H.P., 1971. Mechanisms of insect excretory systems. *Adv. Ins. Physiol.* 8: 199–331.

Maddrell, S.H.P. & Gardiner, B.O.C., 1976. Excretion of alkaloids by Malpighian tubules of insects. *J. exp. Biol.* 64: 267–281.

Manunta, C., 1935. Richerche biochimiche su due lepidotteri: la tignuola degli alveari *Galleria mellonella* e il bombice del gelso *Bombyx mori*. *R.C. Acad. Lincei.* 6: 75–161.

Marsh, N. & Rothschild, M., 1974. Aposematic and cryptic lepidoptera tested on the mouse. *J. Zool. Lond.* 174: 89–122.

Martini, J.T., 1975. Composition of egg-shell and cast larval skin of *Rhodnius prolixus* and pupal case of *Pieris brassicae*. *Insect Biochem.* 5: 275–287.

Mauchamp, B. & Lafont, R., 1975. Developmental studies in *Pieris brassicae* (Lepidoptera, Pieridae). II. A study of nitrogenous excretion during the last larval instar. *Comp. Biochem. Physiol.* 51B: 445–449.

Mayer, A.G., 1896. The development of the wing scales and their pigment in butterflies and moths. *Bull. Mus. comp. zool. Harv.* 29: 209–236.

—— 1897. On the color and color-patterns of moths and butterflies. *Bull. Mus. comp. zool. Harv.* 30: 169–256.

Meyer, P.-F., 1930. Untersuchungen über die Aufnahme pflanzlicher Farbstoffe in den Körper von Lepidopterenlarven. *Z. vergl. Physiol.* 11: 173–209.

Miah, M.A.H., 1976. *Some effects of gamma radiation on the different stadia of Pieris brassicae L.* Ph.D. thesis. University of London (Imperial College). pp. 176.

Moreau, R., 1969. Observations on the variations of trehalose levels in the course of direct development and of diapause in *Pieris brassicae* L. *C. r. hebd. Séanc. Acad.* 268: 1441–1445.

Moreau, R., 1973. *Recherches sur quelques aspects des phenomenes physiques metaboliques et physiologique qui accompagnent a conditionnent l'expansion des ailes de Lepidopteres.* These Doctorale, Universite de Bordeaux 1.

Moreau, R., Dutrieu, J. & Olivier, D., 1974. Dégradation du glucose par le cycle des pentoses au cours de la fin de l'ontogenèse chez *Pieris brassicae* normal et aptère. *C. r. Séanc. Soc. Biol.* 168: 1285–1288.

Moreau, R., Gourdoux, L., Dutrieu, J., 1977. Utilisation comparée du cycle des pentoses en fonction des variations thermiques chez deux lépidoptères, *Bombyx mori* L. et *Pieris brassicae* (L.). *Comp. biochem. physiol.* B. 56: 175–180.

Munn, E.A. & Greville, G.D., 1969. The soluble proteins of developing *Calliphora erthyrocephala* particularly calliphorin, and similar proteins in other insects. *J. Insect Physiol.* 15: 1935–1950.

Nagel, H., 1970. *Die grünen Farbstoffe der Larven von Cerula vinula L. (Notodontidae) und einiger anderer Lepidopteren.* Dissertation. University of Giessen.

Nettles, W.C., Parro, B., Sharbaugh, C. & Mangum, C.L., 1972. Trehalose and other carbohydrates in diapausing and starving boll weevils. *Annl. ent. Soc. Am.* 65: 554–558.

Nicolson, S.W., 1975. *Osmoregulation, metamorphosis and the diuretic hormone of the cabbage white butterfly, Pieris brassicae.* Ph.D. thesis. University of Cambridge.

—— 1976a. Diuresis in the cabbage white butterfly, *Pieris brassicae*: fluid secretion by the Malpighian tubules. *J. Insect Physiol.* 22: 1347–1356.

—— 1976b. Diuresis in the cabbage white butterfly, *Pieris brassicae*: water and ion regulation and the role in the hind gut. *J. Insect Physiol.* 22: 1623–1630.

—— 1976c. The hormonal control of diuresis in the cabbage white butterfly, *Pieris brassicae*. *J. exp. Biol.* 65: 565–575.

Njio, K.D. & Piek, T., 1977. Localisation of sodium and potassium ions in a flight muscle of *Pieris brassicae*. *J. Insect Physiol.* 23: 919–929.

Noordink, J.P.W., 1962. Onderzoek met behulp van radioactieve isotopen (project 9-1-3). *Inst. Plantenz. Onderz.* 1962: 147–152.

—— 1963. Onderzoek met behulp van radioactieve isotopen (project 9-1-3). *Inst. Plantenz Onderz.* 1963: 129.

Noordink, J.P.W., 1965. Some applications of radioactive isotopes in ecological investigations. *Ent. Ber.* 25: 130.

Oltmer, A., 1968. Die Steuerung des Melanineinbaus in das Farbmuster der Kohlweisslingspuppe *Pieris brassicae* L. *Wilhelm. Roux. Arch. EntwMech. Org.* 160: 401–427.

Onslow, H., 1916. On the development of the black markings on the wings of *Pieris brassicae*. *Biochem. J.* 10: 26–30.

—— 1917. A contribution to our knowledge of the chemistry of coat-colour in animals and of dominant and recessive whiteness. *Proc. R. Soc.* B. 89: 36.

Paillot, A. & Noel, R., 1928a. Recherches histophysiologiques sur les tissus adipeux des larves d'insectes (*Bombyx mori* et *Pieris brassicae*). *Bull. hist. appl.* 5: 56–78.

—— 1928b. Recherches histophysiologiques sur les cellules péricardiales et les éléments du sang des larves d'insectes (*Bombyx mori* et *Pieris brassicae*). *Bull. hist. appl.* 5: 105–128.

327

Plantevin, G. & Nardon, P., 1970. Histologie et activité sécrétoire de l'intestin moyen des larves de *Pieris brassicae* et *Galleria mellonella*. Evolution au cours de la mue larvaires et de la nymphose chez *Galleria mellonella*. *Ann. zool. ecol. anim.* 2: 25–50.

Portier, P., 1949. *Biologie des Lépidoptères*. Lechavalier, Paris.

Post, L.C., 1972. Bursicon: its effect on tyrosine permeation into insects. *Biochim. Biophys. Acta.* 290: 424–428.

—— 1973. Bursicon and metabolism of tyrosine in moulting cycle of *Pieris brassicae* larvae. *J. Insect Physiol.* 19: 1541–1546.

Post, L.C., Jong, B.J. de., & Vincent, W.R., 1974. I-(2,6-Disubstituted benzoyl)-3-Phenylurea insecticides. *Pest Biochem. Physiol.* 4: 473–483.

Post, L.C. & Vincent, W.R., 1973. A new insecticide inhibits chitin synthesis. *Naturwissenschaften* 60: 431–432.

Razet, P., 1961. *Recherches sur l'uricolyse chez les insectes*. Thèse Doctorat. de Science Naturelles. Imprimerie Bretonne, Rennes.

Rogers, A.H., White, A.G., Wolf, J. & Gibson, N.H.E., 1966. The quantitative administration of bacterial pathogens to insect larvae. *Lab. Pract.* 15: 427–430.

Rothschild, M., 1975. Remarks on carotenoids in the evolution of signals. *In: Co-evolution of animals and plants*. Edited by L.E. Gilbert & P.H. Raven. University of Texas Press. pp. 20–51.

Rothschild, M., Gardiner, B.O.C., Valadon, L.R.G. & Mummery, R., 1975a. Lack of response to background colour in *Pieris brassicae* pupae reared on carotenoid-free diet. *Nature, Lond.* 254: 592–594.

Rothschild, M. & Schoonhoven, L.M., 1977. Assessment of egg load by *Pieris brassicae* (Lepidoptera: Pieridae). *Nature, Lond.* 266: 352–355.

Rothschild, M., Valadon, L.R.G. & Mummery, R., 1977. Carotenoids of the pupae of the Large White Butterfly (*Pieris brassicae*) and the Small White (*Pieris rapae*). *J. Zool. Lond.* 181: 323–339.

Rothschild, M., Valadon, L.R.G., Mummery, R. & Gardiner, B.O.C., 1975b. Storage of carotenoids in diapausing pupae of *Pieris brassicae* L. and *Pieris napi* L. and their influence on larval colouration (Lepidoptera: Pieridae). *Nature, Lond.* 254: 592.

Ruediger, W., Klose, W., Vuillaume, M. & Barbier, M., 1968. On the structure of pterobilin, the blue pigment of *Pieris brassicae*. *Experimentia* 24: 1000.

—— 1969. Biosynthesis of biliverdin -1Xy in *Pieris brassicae*. *Experimentia*. 25: 487–488.

Schmalfuss, H., 1926. Eine Betrachtung über Zellvorgänge. *Biochem. Z.* 178: 227–228.

Schopf, C. & Wieland, H., 1926. Über das Leukopterin, das weisse Flügelpigment der Kohlweissling (*Pieris brassicae* und *Pieris napi*). *Ber. dt. chem. Ges.* 59: 2067–2072.

Seuge, J. & Veith, K., 1976. Diapause de *Pieris brassicae*: rôle des photorécepteurs, étude des carotenoides cerebraux. *J. Insect Physiol.* 22: 1229–1235.

Simon, H., Wacker, H. & Walter, J., 1964. Further studies on the biogenesis of leucopterin. *In: Pteridine chemistry*. Proc. 3rd. International symposium. Stuttgart. Macmillan. 1964.

Simon, H., Weygand, F., Walter, J., Wacker, H. & Schmidt, K., 1963. Zusammenhange zwischen Purin und Leucopterin- Biogenese in *Pieris brassicae* L. *Z. Naturf.* 18B 757–764.

Sømme, L., 1967. The effect of temperature and anoxia on haemolymph composition and supercooling in the overwintering insects. *J. Insect Physiol.* 13: 805–814.

Sømme, L. & Velle, W., 1968. Polyol dehydrogenase in diapause pupae of *Pieris brassicae*. *J. Insect Physiol.* 14: 135–143.

Stamm, M.D. & Aquirre, L., 1955. Estudios sobre bioqúimica de insectos. II. Aminoácidos aromáticos y triptofána en la metamorfosis de *Pieris brassicae* y ocnogyna báctica. *R. esp. Fisiol.* 1: 69–74.

Steche, O., 1912. Beobachtungen über Geschlechtsunterschiede der Haemolymphe von Insecktenlarven. *Verh. dt. zool. Ges.* 22: 272–281.

Stratakis, E., 1976. Quantitative Analyse des Ommochrompigments der Augen von Männchen und Weibchen von *Pieris brassicae*. *Insect Biochem.* 6: 29–34.

Strogaya, G.M., 1961. Peculiar features of fat and water balance in the individual development of *Aporia crataegi* and *Pieris brassicae*, as a form of adaptation to the surrounding medium. *Akad. nauk. SSSR. Dok.* 139: 474–477.

Strogaya, G.M., 1962. Variation de la quantité d'eau et de la quantité des corps gras contenus dans l'organisme des papillons de la piéride de l'aubergine et de la piéride du chou au cours de leur développement individuel. (d'après les résultats de l'analyse biochimique). *J. Zool. Acad. Sci. USSR* 41: 92–100.

Tarroux, Ph., 1975. Characterization and purification of messenger RNA's containing poly A in insect imaginal discs. *Biochimie* 57: 757–763.

Thompson, R.H., 1960. Insect pigments. *In: XI int. Congr. Ent. Vienna.* Symp. 3. Insect Chemistry. pp. 21–43.

Tikhonravova, N.M., 1973. Respiratory characteristics of the cabbage butterfly during different stages of larval development. *Sov. J. Ecol.* 3: 320–323.

Timon-David, J., 1929–1932. Recherches sur les matières grasses des insectes. *Ann. Fac. sci. Marseilles* 4: 29–207.

Turunen, S., 1973a. Utilisation of fatty acids by *Pieris brassicae* reared on artificial and natural diets. *J. Insect Physiol.* 19: 1999–2009.

—— 1973b. Role of labelled dietary fatty acids and acetate in phospholipids during metamorphosis of *Pieris brassicae*. *J. Insect Physiol.* 19: 2327–2340.

—— 1974a. Notes on lipid requirement of a phytophagous lepidopteran. *Ann. ent. fennici* 40: 151–155.

—— 1974b. Lipid utilization in adult *Pieris brassicae* with special reference to the role of linolenic acid. *J. Insect Physiol.* 20: 1257–1269.

—— 1974a. Metabolism and function of fatty acids in a phytophagous lepidopteran. *Ann. ent. fennici.* 11: 170–184.

—— 1974d. Polyunsaturated fatty acids in the nutrition of *Pieris brassicae* (Lepidoptera). *Ann. ent. fennici* 11: 300–303.

—— 1975a. Effect of gamma-BHC on lipid digestion and utilisation in *Pieris brassicae* (Lepidoptera: Pieridae) reared on an artificial diet. *Ann. ent. fennici* 12: 275–279.

—— 1975b. Absorption and transport of dietary lipid in *Pieris brassicae*. *J. Insect Physiol.* 21: 1521–1529.

—— 1975c. Metabolism of palmitrate in the adult *Pieris brassicae* (Lepidoptera: Pieridae). *Insect Biochem.* 5: 135–140.

—— 1976. Vitamin E: effect on lipid synthesis and accumulation of linolenate in *Pieris brassicae*. *Ann. zool. fennici* 13: 148–152.

Turunen, S. & Junnikkala, E., 1974. Haemolymph fatty acids and lipid content in larval *Pieris brassicae* (Lepidoptera: Pieridae). *Ann. ent. fennici* 40: 145–149.

Urech, F., 1892. Über einen grünen Farbstoff in den Flügelchen (nicht in den Schuppen) der Chrysalide von *Pieris brassicae*. *Zool. Anz.* 15: 281–283.

Van der Geest, L.P.S., 1968. Effect of diets of the haemolymph proteins of larvae of *Pieris brassicae*. *J. Insect Physiol.* 14: 537–542.

Van der Geest, L.P.S. & Borgsteede, F.H.M., 1969. Protein changes in the haemolymph of *Pieris brassicae* during the last larval instars and the beginning of the pupal stage. *J. Insect Physiol.* 15: 1687–1693.

Van der Geest, L.P.S. & Sloog-Hoebe, A.A.M., 1972. Effects of various vitamins on the development of *Pieris brassicae* larvae. *Medd. Faculteit. Landb. Gent.* 37: 713–715.

Vuillaume, M., Choussy, M. & Barbier, M., 1970. Pigments tetrapyrroliques verts et bleus des ailes de lépidoptères pterobilin et neopterobilin. *Bull. Soc. zool. Fr.* 95: 19–28.

Vuillaume, M., Seuge, J. & Bergerard, J., 1971. Photopériode et pigment tégumentaire vert des chenilles de *Pieris brassicae*: conditionnement de la diapause. *C. r. Acad. Sci. Paris* 273: 1608–1610.

Wahla, M.A., Ford, J.B. & Gibbs, R.G., 1974. The histochemical study of cholinesterase activity in the 24 hour old *Pieris brassicae* (L.) larvae. (Lepidoptera: Pieridae). *Pakistan J. Zool.* 6: 9–11.

Wahla, M.A., Gibbs, R., Gibbs, G. & Ford, J.B., 1976. Diazinon poisoning in the large white butterfly larvae and the influence of sesamex and piperonyl butoxide. *Pestic. Sci.* 7: 367–376.

Watt, W.B. & Bowden, S.R., 1966. Chemical phenotypes of pteridine colour forms in *Pieris* butterflies. *Nature, Lond.* 210: 304–306.

Weiland, H. & Tartter, A., 1940. Über die Flügelpigmente der Schmetterlinge. *Annl. Chem.* 545: 197–209.

Weygand, F., Simon, H., Schliep, H.J. & Dahms, G., 1959. Weitere Studien über die Biogenese des Leucopterin. *Angew. Chem.* 71: 522.

Weygand, F., Simon, H., Dahms, G., Waldschmidt, M., Schliep, H.J. & Wacker, H., 1961. Über die Biogenese des Leucopterin. *Angew. Chem.* 73: 402–407.

Weygand, F., 1959. Über die Biogenese des Leucopterin. *Angew. Chem.* 71: 746.

Weygand, F. & Waldschmidt, M., 1955. Über die Biogenese des Leucopterin. *Angew. Chem.* 67: 328.

White, A.G., 1968. *The histopathological effects and the site of action of Bacillus thuringiensis in the larvae of Pieris brassicae and its bioassay against the larvae of Galleria mellonella.* Ph.D. thesis. University of Leeds. pp. 110.

Wigglesworth, V.B., 1967. *The principles of insect physiology.* Methuen & Co. Ltd., London.

Yagi, N., 1956. Electron microscope studies on the form of pterin pigments in the scale of Pieridae with reference to their bearing on systematics. *New Entomol.* 4: 1–20.

Ziegler, I. & Harmsen, R., 1969. The biology of pteridines in insects. *Adv. Insect Physiol.* 6: 139–203.

Further references

Becker, E., 1937. *Z. physiol. Chem.* 246: 177–180 (pteridines).

Berthold, G. & Henze, M., 1971. *J. Insect Physiol.* 17: 2375–2381 (role of pteridines in colouration).

Brecher, L., 1934. *Akad. Anz. Wien* 4: 26 (formation of tyrosine).

Busnel, R.G. & Drilhon, A., 1949. *Bull. Soc. Loal. Fr.* 76: 21–23 (pteridines in the pupa).

Candy, D.J. & Kilby, B.A., 1975. *Insect Biochemistry and Function.* Chapman and Hall, London (biochemistry of insect flight).

Claret, J., 1973. *Arch. zool. exp. gén.* 114: 271–275 (melanisation).

Cromatie, R.I.T., 1959. *A. Rev. Ent.* 4: 59–76 (general interest in pigments).

Ford, E.B., 1941. *Proc. R. ent. Soc. Lond.* A 16: 65–90 (anthocyanins in lepidoptera).

Frey-Wybling, A. & Blank, F., 1963. *Ber. schweiz. bot. Ges.* 53A: 550–578 (anthocyanins).

Gerould, J.H., 1927. *Q. Rev. Biol.* 3: 58–78 (butterfly pigments).

Geyer, K., 1913. *Z. wiss. Zool.* 105: 349–499 (chemistry of haemolymph).

Gilbert, L.I., 1967. *J. Insect Physiol.* 4: 69–211 (lipid metabolism).

Hardonk, M., 1950. *Tijdschr. Ent.* 96: 11–15 (pterobilin).

Hopkins, F.G., 1894. *Proc. R. Soc. Lond.* 57: 5,6 (pteridines).

Kabos, W.J., 1962. *Ent. Ber.* 22: 242–245 (pteridines).

Kettlewell, B., 1973. *The Evolution of Melanism.* Clarendon, Oxford.

Kozhantchikov, I.W., 1938. *Bull. Ent. Res.* 39: 103–114 (carbohydrate and fat metabolism).

Kozhanchikov, I.V., Kikhailova, R.I. & Voludina, E., 1936. *Summ. Sci. Res. Wk. Inst. Plant Prot. Leningrad* 1936: 51–52 (metabolism of hydrocarbons during diapause).

Needham, A.E., 1974. *The Significance of Zoochromes.* Springer-Verlag, Berlin.

Ohtaki, T. & Oltmer, A., 1968. *Wilhelm. Roux. Arch. EntwMech. Org.* 160: 401–427 (melanisation).

Schmidt, K., 1965. *Beiträge zur Biogenese des Leukopterins. Biogenetischer Zusammenhang zwischen Purinnucleotiden und Pterinen beim grossen Kohlweissling.* Thesis, Ludwig-Max. University of Munich.

Tartter, A., 1940. *Z. Physiol. Chem.* 266: 130–134 (uric acid in wings).

Thomson, J.A., 1975. *Adv. Insect Physiol.* 11: 321–398 (gene activity).

Tuzet, O. & Manier, J.F., 1957. *Biol. Fr. Belg.* 9: 264–270 (formation of tetractyes).

Veith, K., Seuge, J. & Vuillaume, M., 1974. *C. r. Acad. Sci. Paris* 278: 637–638 (absorption of pterobilin).

Vuillaume, M., 1969. *Pigments des Invèrtebres.* Masson, Paris.

Vuillaume, M. & Barbier, M., 1969a. *C. r. Seanc. Soc. Biol. Orsay* 163: 591–593 (pterobilins).

—— 1969b. *C. r. Acad. Sci. Paris* 268: 2286–2289 (pterobilins).

Weiland, H., Metzger, H., Schoft, C. & Bülow, M., 1933. *Justus Leibigis Annl. Chem.* 507: 226–265 (leucopterin).

Wyatt, G.R., 1961. *A. Rev. Ent.* 6: 75–102 (biochemistry of insect haemolymph).

11. Migration

'Flying Crooked'

The butterfly, cabbage-white,
(His honest idiocy of flight)
Will never now, it is too late,
Master the art of flying straight,
Yet has–who knows so well as I?–
A just sense of how not to fly:
He lurches here and here by guess
And God and hope and hopelessness.
Even the aerobatic swift
Has not his flying-crooked gift.

Robert Graves

Introduction

Williams (1936a,b) classified *Pieris brassicae* as a Class III immigrant, that is it breeds in this country and its numbers are reinforced each year. This is in contrast to those species which never breed in Britain (Class I) and those which occasionally breed in the summer (Class II). Williams *et al.* (1942) believed that the three pierid species, *P. brassicae*, *Pieris rapae* and *Pieris napi*, were usually and significantly found abundant together on migration. He referred to these three species as "hardy annuals" (Williams, 1935a). Caution must be excercised when considering migration patterns in *P. brassicae* as Williams *et al.* (1942) pointed out that it tends to be reported only in years of absence or special abundance. In some years there is a real abundance of all three species, in other years there is a paucity of all three (cf. Chapter 3 on abundance). It has often been said that *P. brassicae* would not be as common in the British Isles if it were not for the reinforcements from the continent which occur each year (Williams 1935a,b, 1936a,b; Williams *et al.*, 1942; Ford, 1976). Recently

Parker (1978) stated that the migratory behaviour of *P. brassicae* can be interpreted in the light of whichever prospect (higher fecundity or faster development offers the highest pay off at any time in the season.

Although *P. brassicae* is strongly migratory it is atypical in its migratory habits, and is unlike most other migrant insects in the temperate zone which move towards the Equator in the autumn and away in the spring (Williams, 1949a). Instead there is more evidence to show that *P. brassicae* moves south, eastwards and westwards in the spring (Williams, 1930). Evidence of a "return" migration is very scant in the eastern Pyrenees during September and October. A return migration off the foothills of the Himalayas has also been described.

There are two principle books on butterfly migration which include much information on *P. brassicae*; Williams (1930) *Insect Migration* and Johnson (1965, 1969) *Migration and Dispersal of Insects by Flight*. A new book by Baker (1978) *The Evolutionary Ecology of Animal Migration* gives many details on migration of pierids.

For factual information on migration records in Great Britain and Ireland from 1928 to 1956, a great deal can be found in the numerous papers of Dannreuther and Williams, while migration data pertaining to the Netherlands can be found in Lempke (1941).

Origin

The southern part of Scandinavia and the Baltic islands appear to be the main areas from which migrations of *P. brassicae* originate in northern Europe (Nordman, 1954, Williams, 1930, Williams *et al.*, 1942, Anglade *et al.*, 1963), although Williams (1939b) stated "somewhere in the north" and Roer (1959) suggested central Europe. Roer's theory proposed that *Pieris* spread out with man and his cruciferous crops during prehistoric times. According to Huxley (1942, p. 88) real agriculture was practiced in the Near East somewhere before 5000 B.C. Bailey (1927) stated that cabbage and its cultivars had been in existence since Pliny's time (about 2000 years ago), so this lends some support for Roer's theory. After all, the present day cabbage is believed to be derived from a cliff-dwelling wild cabbage ancestor; the present day example of which is indigenous to the British Isles.

There are 40 species of brassicas known in the Mediterranean region and there exists "a multitude of closely related races" (Clapham *et al.*, 1962). It is beyond doubt that *P. brassicae* has exploited many of the wild and cultivated brassicas and their cultivars (cf. Chapter 4 Foodplants).

More specifically, it is probable that the islands and coasts around southern Sweden, Denmark and the protruding tip of western Germany, Schleswig-Holstein, all above 54° latitude, are the regions where thousands of *Pieris* grow up each year, resulting from imagines which have presumably emerged from overwintering pupae. Circumstantial evidence for this is set out as follows:

a) *Earlier work*. Professor Blunck, who worked in the northern German town of Kiel, showed that the start of the migration of *P. brassicae* commenced shortly after the beginning of the third 10-day period of April (21st) which coincided with the finishing of *Syringa vulgaris* Linnaeus (Lilac) flowering and the opening of *Aesculus hippocastrum* Linnaeus (Horse Chestnut) blossom. After mid-May, *P. brassicae* imagines became scarce (Blunck, 1950).

b) *Foodplant availability*. i) Around the Kiel canal area, much cabbage was grown earlier this century, when Blunck studied migrations. Today the same areas still grow cabbage but on a much smaller scale (Feltwell, 1977a). ii) Anglade *et al.* (1963) suggested that the foodplant *Lepidium latifolium* Linnaeus (Dittander) was extremely abundant in this region and may be used as a foodplant. iii) Speyer (1948) suggested that the first generation of *P. brassicae* feed on the cruciferous plants which grow abundantly on the North Sea island of Amrum. He also published a photograph of *P. brassicae* larvae which had completely eaten a plant of *Cakile maritima* Scopoli (Sea Rocket) (Speyer, 1956). Ghikr & Loibl (1925) also found larvae of *C. maritima* (Sea Rocket) on the Baltic Island of Sylt.

c) *Lack of parasites*. It has been tentatively suggested that the predominantly westerly winds on the west coast of Denmark might tend to drive away the parasites and thus populations of *P. brassicae* would have a greater chance of reaching the imaginal stage (Roer, 1975, pers. comm.). Roer observed high population density of *Pieris* larvae in North Issi when he visited the region. It is interesting to point out that Campbell (1970) also noted the absence of parasites on the Inner Hebridean Island of Canna off the exposed western coast of Scotland.

Direction and timing

Williams (1939a,b) originally suggested that in north Europe migrating *Pieris* flew southwards in spring and passed through central Europe and ultimately to the Alps, but this is now thought to be carried out in two stages, with one, two and possibly three generations. Roer (1959) suggested that the spring brood went off in a south-west direction these movements being genetically fixed. On the contrary, Vepsäläinen (1968) gave conclusive evidence that the migration in May 1966 passed northwards through southern Finland and eastern USSR. This is substantiated by the belief that *Pieris* could not possibly overwinter in Finland and that populations in that country are totally dependant upon re-inforcements over the Gulf of Finland from the south (Utrio, 1975, pers. comm.) Warnecke (1955) also noted that millions of immigrants arriving on the island of Heligoland came on an east wind.

Perhaps the best descriptions of the directional flights of *P. brassicae* are given by Williams *et al.* (1942), Williams (1949a,b) (cf. Anglade *et al.*, 1963), who said that in the early spring the larvae develop in large numbers in southern Scandinavia and the Baltic Islands, and then at the end of July, and early August the resultant imagines pass southwards to central Germany, Austria and Switzerland. Some insects veer west and pass into the Netherlands, Belgium and ultimately across the North Sea. Hence immigrants arrive on the British coasts in Norfolk, Suffolk, Essex, Kent and Sussex. However, in the British Isles there appears to be a northerly migration in the spring and summer and in the autumn there is a southerly movement of insects (Williams, 1951a).

The difference in directional movements of different generations of *P. brassicae* is shown by two more examples. Blunck (1954) found that there was a slight difference in the directional movement of both generations, the spring migration moving northwards and the summer generation in Schleswig-Holstein always in the southwesterly to southerly direction. He was unable to find a "motivation" other than instinct to account for the movements. Starega (1976, pers. comm.) stated that in Poland, *P. brassicae* migrate northwards in spring and always to the south in summer, sometimes in masses.

Peripheral range

It appears that during the summer months migrations of *P. brassicae* reach into regions on the periphery of its range, which do not otherwise support populations all the year round because of limiting environmental factors. Such regions, which only support summer populations are northwest USSR (Estonia) (Kusnezov, 1930) and southern Finland (Utrio, 1975, pers. comm.), while in the Inner and Outer Hebrides off the west coast of Scotland this may be the case only in years which are unfavourable for overwintering pupae or lack of immigrants the previous summer (cf. Chapter 3).

The presence of *P. brassicae* in many of the above mentioned areas is thought to be due to immigrants from the south. In these northern areas, no evidence of permanent populations being established has been found and no pupae have been seen overwintering successfully. Once the insects have arrived in such places, breeding may occur only if the summer temperature is sufficiently high and of sufficient duration (cf. Chapter 3 Generations). No examples appear to exist of *P. brassicae* migrating to peripheral areas in the south of its range (i.e. north Africa).

An advantage of this dispersal is that populations already present in areas over which *P. brassicae* fly are reinforced, providing there is enough food, and such is thought to be the case in southern Scandinavia (Haugum, 1976, pers. comm.), in Estonia, USSR (Hansen & Merivee, 1971) and also in the British Isles.

Return migration

Mention has been made in the literature of return migration of *P. brassicae* but clarification of the meaning of "return" must be made first. Return migration is not, of necessity, a reversal of flight direction, but does imply the going back to the same source. However, the distances between the areas over which migrations of *P. brassicae* have been documented are so great (more than 800 kilometers in some cases, e.g. south of France to north Germany) that individual butterflies are presumably not involved in both the emigration and return migration to the source. In any case, there are no species in Europe which have a complete return migration. If a "return" migration occurs it is always very weak, or even absent in some years. Thus individuals of *P. brassicae* which participate in "return" migration must be members of the second, third or very rarely the fourth generations.

Most accounts of return migration refer to migrants originating from north Europe and passing southwards and westwards; therefore spring/early summer migration through the Pyrenees would be expected to go southwards. However, if migrants arrive in southern Europe from across the Mediterranean then this early migration through the same region might be expected to be northwards. Thus any "return" migration could of course be either northwards or southwards according to which theory is held. Those who have searched for a return migration in the Pyrenees have however, looked for insects passing south in late summer and autumn. Roer (1974) mentioned that *P. brassicae* is one species which has a peak of migration in Europe, northwards, early in the year and a return migration in the summer or early autumn. Williams (1951a) noted in his records up to 1949 that there was no evidence of a return migration in Britain.

All the data on return migration comes from either the Pyrenees or Himalayas. The Pyrenees make an excellent area to study migration as the mountains form a natural geographical funnel towards the eastern side. The coastline from Montpellier, France to the Spanish border acts as a natural line for insects when flying north or south.

Gray *et al.* (1953) and Williams *et al.* (1956) recorded that *P. brassicae* passes south, admittedly in small numbers, during September and October in the coastal regions and mountains of the Pyrenees. They were observed from September 18 until October 10 by Williams *et al.*, (*loc. cit.*) and were in company with other Lepidoptera, Diptera and Hymenoptera. Gray (*loc. cit.*) observed *P. brassicae* from September 8 to October 3 going south on each day at a height of 8,000 feet (2,439 m) in the Pyrenees at Port de Vebasque. No northerly migrations of *P. brassicae* have been reported through the Pyrenees as far as the author is aware. The duration of the southerly migration has not been truly established, but it is thought to be at least the end of August and into October (Williams *et al.* (1956)) although Gray (*loc. cit.*) found that it began early in September and reached a climax in October.

Evidence about a "return" migration in the Himalayas is limited to a statement by Williams (1958) that a return migration off the foothills possibly occurs.

In the Alps (Tyrol) Williams (1939b) recorded a diurnal return movement of *P. brassicae* up a valley in the morning and down the valley in the afternoon.

Speed

This has been variously estimated by different authors to be between 1–12 metres per second (Table 11.1) and it is clear that speed is influenced by external factors, such as wind. The wide differences in the speed rates is perhaps indicative of the different types of flight that *P. brassicae* can produce; many people must have witnessed how an insect can suddenly

Table 11.1 Flight speeds of *P. brassicae*.

Speed	Authority
(metres per second)	
9	Demoll, 1918 and Müllenhoff *in* Demoll, 1918
1.66	Magnan & Planiol, 1933
12	Goulliart *in* Magnan, 1934
2–4 (normally)	Blunck, 1954
10 (when helped by the wind)	Blunck, 1954
3.6–4.4	Rothschild, 1956
"speed of a running man"	Anglade *et al.*, 1963
6.2–7.44	Johnson *in* Baker, 1978
0.6–1.5 (feeding)	
1.1–2.2 (foraging)	Nikolaus, 1974
1.5–2.5 (migration)	
2.5 (average speed of sustained flight)	Nachtigall, 1978

increase its speed and change its pattern of flight when being chased by an ardent collector. Many times the insect will make off faster after being disturbed than the collector can follow. In fact Nikolaus (1974) drew attention to the different flight patterns and speeds of flight of various butterflies in different habitats around Moscow. He recognised four different flight patterns of *P. brassicae*, each of which had characteristic different speeds (Table 11.1). These figures he thought were higher than had previously been reported for other lepidoptera but Table 11.1 shows that actually they are some of the lowest for *Pieris brassicae*.

Some idea of the average speed of travel in the wild can be taken from the experiments of Roer, who marked and released *P. brassicae* in Bonn, Germany. After four days one of his insects reached Dusseldorf, a distance of 95 kilometres (Roer, 1961). If this is worked out on the basis of five hours flying time each day (the "normal" flight time is probably between 1100 hrs and 1600 hrs in the temperate regions) the average speed of flight is 13 m/sec. Roer (cf. Baker, 1971) recorded that the spring generation moved northwards at the rate of 6 km/day while the autumn generation were slower at 2 km/day.

Distance

It was stated by Blunck (1954) that migrating masses of *P. brassicae* could cover over 400 kilometres in several days, and that during one day they would fly somewhere between 70–130 kilometres. This was based on his observations made over several years in the Kiel region of northern Germany. Schütte, who now works in the same institute as did Blunck, also studied the rate at which *P. brassicae* fly and found that when individuals were tracked by car they covered distances of "up to 5 kilometres (beeline)" in one day (Schütte, 1966). He stated in his paper that the distance of 400 kilometres quoted by Blunck (1954) seemed to him to be rather too far for one generation of insects to fly. The insects he studied were also migratory. Nielson (1967) stated that *P. brassicae* was capable of prolonged flight.

During migration across water it may be an advantage for insects to alight on the surface and indeed, instances of various lepidoptera alighting on water have been documented. Although *P. brassicae* has been recorded alighting on the water, such as off the coast of Somerset in 1877 (Ormerod, 1878, cf. Williams *et al.* 1942), they have never actually been recorded as flying off from the water, although this may indeed be possible.

Wind

It was mentioned previously that wind has a profound effect on the speed of migrating insects, increasing it from 2–4 metres per second to 10 metres per second. In such cases the insect is taking advantage of the wind and flying with it. However, this is not always the case. There are a lot of accounts of *P. brassicae* flying into the wind especially if it is a gentle breeze. If the wind speed is too great the insect will stop flying and rest. William's (1958) extensive work on the direction of flight of migrant *P. brassicae* at Rothamsted Experimental Station, Harpenden, showed conclusively that they pursued the same direction all day although the wind veered through all points of the compass. Thus the insects are not necessarily influenced by the direction of the wind in the direction they choose at these wind speeds and under these particular conditions.

It is likely that *P. brassicae* uses the wind for covering distance and crossing unsuitable feeding areas, although little evidence is available (cf. Brecher, 1925; Warnecke, 1955). Schütte (1975, pers. comm.) believed that *P. brassicae* may well "ride" with the wind in order to cross ground quickly and thus put them in a position to find oviposition sites more quickly. Should a female encounter the odour of a cabbage field downwind then it will turn back into the wind and steadily follow the concentration gradient of cabbage odours back to the field. The author has also seen this take place just outside Kiel, northern Germany, and the insects fly into the wind keeping close to the ground.

Altitude

Table 11.2 shows that *P. brassicae* has been recorded flying on some of the highest mountains in the Himalayas, the French, Swiss and Austrian Alps, the Pyrenees and Finland and Norway.

The highest that *P. brassicae* has been recorded flying is over 6,000 metres in northwest India. Members of a British Expedition in the Himalayas saw specimens at 4000 metres in Nepal, not an uncommon sight, and at the same elevation in the Swiss Alps, frozen imagines have been found in the ice; only to revive and fly away from the warmth of the hand. There is some opinion that the insects in the Himalayan foothills are of a separate subspecies, and may therefore be more physiologically adapted to the environment. Even more surprising is the evidence of a glider pilot over northern Germany on July 29th, 1959 who saw *P. brassicae* migrating at 1,200–1,700 metres, 10 kilometres SW of Weissenburg (Roer, 1968).

Insects flying high in the air is not an uncommon occurrence. An indication is given in the account of the migration of "whites" at Calais in

Table 11.2 Altitudes recorded for *P. brassicae*.

Altitude (m)	Locality	Authority	Comments
4–6,300	North-west India, Amur	South, 1891	smaller size
4,440	Nepal	Westmacott & Williams, 1954	three females seen during British Expedition
3,996	Austria	Williams, 1939a,b	many imagines dead
3,700–4,440	Himalayas	Harcourt-Bath, 1897	geographical race?
uncertain	Himalayas	Moulton, 1915	few seen
3,700–4,100	Switzerland, Valais	Mellows, 1924	frozen in ice, with a great tenacity for life
3,460	Switzerland, Jungfrau	Pictet, 1918b	—
2,220–3,700	Himalayas, Dharmsala	Hingston *in* Williams, 1928	hundreds seen ascending every minute
2,323	Central Pyrenees	Muspratt, 1951	many sitings
(2,076	Alps, Savoy	Bretherton, 1939a	not present)
1,850	Austria, Oetzal	Cooke, 1937	present
1,728	Alps, Savoy	Jackson, 1937	seen
up to snowline	Alps, Tyrol	Gurney, 1927	present
1,110	Norway	Higgins, 1935	very common
1,110	Austria	Daniel, 1957	frozen in ice
1,036	Breconshire	Sanderman, 1946	several movements, insects basking
900–1000	Fennoscandia	Haugum (pers. comm.)	recorded
c. 900	Czechoslovakia	Spitzer (pers. comm.)	recorded
740	Cumberland, Alston	Dannreuther, 1933	seen
740	Pyrenees (Saurat) and south France (Ariege)		
481	Wales, Abergavenny	Nabokoff, 1931	present
		Tullech, 1941	seen

1540 by Nichols (1846) who quoted the *Chronicle of Calais* as saying that the insects flew "so highe". *P. brassicae* has also been seen flying very high in the sky in western Germany by Döring-Ilmenau, (1931).

Front

The width of a migrating front of *P. brassicae* has been estimated at distances up to four kilometres; these include 100 m (Hesse, 1918), 100–200 m in Schleswig-Holstein (Williams, 1936a,b), 183–274 m in Tyrol (Williams, 1939a,b) 800 m in Germany (Kruger, 1900) and 4 km in Germany where it has been estimated that two million butterflies flew past a 100 m transect each hour, equivalent to 300–400 million individuals for the duration of the flight.

Flight pattern

The reader is reminded of Robert Graves' poem (Graves, 1975) on the Large White *"Flying Crooked"*, which indeed characterises its typical erratic flight pattern. When migrating, *P. brassicae* tend to be more "purposeful" in their movements and exhibit a different flight pattern.

Pictet (1918a) noted that *P. brassicae* tends to fly round obstacles rather than go over them, and were also found during migrations either as single individuals, in groups of four or five or in "flotillas" of 10–40. Blunck (1954b) stated that they tended to follow each other *en masse*, and he rarely saw singles. Anglade *et al.* (1963) noted that in France, migrant *P. brassicae* kept close to the ground at about 50 cm to 1 m above and that they exhibited the following flight characteristics: a) "they fly truly free of all natural obstacles (i.e. they go round trees and mountains") b) they fly in a constant direction which is linked to the terrain, c) that the displacement of their general direction is probably constant, d) that they follow tracks and borders like migrating birds (Schütte, 1966). This last point is also true for other migrant lepidoptera. *Hipparchia semele* (Linnaeus) (Grayling) (Feltwell, 1977b).

Recent work involving the cine-photography of *P. brassicae* in flight (8 mm, 16 frames per second) confirmed the presence of four characteristic flight patterns, feeding, foraging, migration and alarm, all of which had different flight speeds (Nikolaus, 1974). Flight paths of *P. brassicae* were published by Nikolaus and from these it is easy to discern change in pattern and height.

Feeding flight is characterised by no abrupt spurts forward or to the side, but flight at 10–15 cm above the ground pausing to visit flowers. Foraging flight is more rapid and the insects fly for longer periods during which they do not alight. They fly distances of 5–20 m at about 40 cm above the ground and make frequent changes of direction. On migration *P. brassicae* fly at 1.5–2.5 m for prolonged periods and do not change direction or alight. Alarm flight is produced when *P. brassicae* is either disturbed or released from the net. Nikolaus (1974) noted that *P. brassicae* differs here from *P. rapae* and *P. napi* in flying off at about 3 m above the ground instead of about 15 cm. Another interesting point for comparison is that *P. napi*, which is smaller than *P. brassicae* was recorded as having higher figures for the four different flight patterns.

Orientation

It has been established that *P. brassicae*, in common with other butterflies such as *P. napi*, *Maniola jurtina* (Linnaeus), *Aglais urticae* (Linnaeus) and *Inachis io* (Linnaeus), orientates by using the sun, but does not compensate for its movement during the day (Baker, 1968a,b). In the past it was believed that butterflies used the wind (Williams, 1958) and in some quarters this is still thought to be the case.

Baker's initial technique for estimating flight direction (Baker, 1968a) consisted of using a wrist watch and compensating for the movement of the sun every hour. In his more refined method (Baker, 1968b) he calculated the peak flight direction mathematically (measured as an angle clockwise from the south) by finding the movement about north–south and east–west. This method calculated the angle of the peak flight direction to an accuracy of ± 22.5. However, in the Addenda of Johnson's (1969) book on *Migration and Dispersal of Insects by Flight*, the author claims that Baker's views are of doubtful validity and in some cases his data were not adequately presented.

Other information

Blunck's extensive study of the 1936 migration in northern Germany produced a lot of other interesting information. *P. brassicae* flies best on sunny days, more often in the morning until noon when the sun is at its zenith, then they rest. In the afternoon they fly until sundown and then choose resting sites protected by the wind. The sex ratio in most migratory flights is 1:1 during the first stages of the migration. When a migratory stream encounters a lake or sea it will usually follow the shore or coast. However, this may not always be possible as there are numerous accounts of *P. brassicae* found well out to sea.

Interspecific relationships

P. brassicae has been found migrating frequently with *P. rapae* and *P. napi* with which it is always significantly abundant (Williams *et al.*,

1942). *P. brassicae* has also been found flying with *Aglais urticae* (Linnaeus) (Small Tortoiseshell), *Autographa (Plusia) gamma* (Linnaeus) (Silver Y) and *Xanthorhoe fluctuata* (Linnaeus) (Garden Carpet) (Williams *et al.*, 1942).

From the Index Cards of C.B. Williams (unpublished) the following insects have been recorded some time or other in the company of *P. brassicae*: *V. atalanta* (Linnaeus), *C. cardui*, *Colias* sp., *Musca* sp., *Sphex* sp., *Aeschna* sp. and *Euclois* sp.

Examples of migration

At sea

There are numerous accounts of *P. brassicae* having been seen "arriving on the coast", or "from the sea", or even "going out to sea", but there are few accounts of large migrations found well out to sea.

The most famous and most quoted example of what was thought by Williams (1958) to be of *P. brassicae* was quoted by Nichols in 1846 to the Camden Society. The passage reads:

> "... the 9. of July, beinge relyke sonday, there was a sene at Calleys an innumerable swarme of whit buttarflye e somine out of the north-este and flyinge south-estewarde, so thicke as flakes of snowe, that men beinge a shutynge in Saint Petar's filde without the towne of Calleys cowld not se the towne of the cloke in the afternone, they flewe so highe and so thicke."

This migration at Calais in 1540 was witnessed and recorded in the *Chronicle of Calais*, the manuscript of which is held at the British Museum (Bloomsbury).

Kerry (1892) quoted an "immense immigration" which started on August 11th, 1892 and continued for several days, sighted 3.72 km off Harwich by lobster catchers; Newman (1931) recorded "droves" of Large Whites at Sark coming from the French coast riding a cold east wind. Greater numbers came when the temperatures increased. Effraimaison (1934) recorded an incredible story about a great swarm which flew above the funnel of a steamer for more than 70 km out into the Baltic off Copenhagen. Williams *et al.* (1942, p. 138) quoted an instance when *P. brassicae* imagines were seen at sea "on water" during a migration in 1936, and "whites", possibly *P. brassicae*, being seen six hours out from Copenhagen also in 1936 (Williams Index Cards, unpublished).

Muspratt (1951) also recorded a large migration of *P. brassicae* at sea off the French coast (1 km from the Isle of Chausey, SE of Jersey) between 1800–1900 hrs, which was heading north east and was two miles (3.22 km) long. This "snowstorm" must have been similar in intensity to the one recounted by Nichols. Ford (1945) also likened migrating *P. brassicae* to snowstorms and stated that they "have been encountered mid-Channel making towards England, and have been observed from the shore coming in across the sea in myriads".

Williams *et al.* (1942) recorded 17 accounts of *P. brassicae* being observed either abundantly or on more than four occasions at eight separate lighthouses in the North Sea or in the English Channel up to 1942. Eight of these accounts involved large numbers, five moderate numbers. All the sightings were in May, June, July and August.

Western Europe

Austria. Kusdas & Reichl (1973) gave the years of great abundance and migration as 1917, 1923, 1924, 1926, 1927, 1935, 1937, 1939, 1941, 1948, 1953 and 1955. The third generation of *P. brassicae* tended to occur much higher up in the valleys. Williams (1939b) recorded "1000's" passing into a slight south west wind on Grossglockner mountain (2630 m) in north west Carrinthia in the Austrian Tyrol on July 24th, 1937. Occasionally, migrating *P. brassicae* may be caught in the summer storms and Williams (1958) has illustrated drowned *P. brassicae* washed up on the shores of an Austrian lake. It was reported in 1957 that *P. brassicae* was migrating through the Kararanke mountains (1,370 m) on the Austrian/Yugo-slavian border (Lon. Nat. Hist. Soc., 1957).

France. One of the earliest references to migrations in France is by De Serres (1842) who stated that the annual flight of *P. brassicae* across the Mediterranean into the south of France in April and May was regarded in the Paris region as a forerunner of the return of the quail (Williams Index Cards).

Offshoots of the migratory stream of butterflies from southern Scandinavia reach France during the summer months and in some years considerable numbers have been recorded. Anglade *et al.* (1963) sum-marised accounts given by Williams (1930, 1939a, *et al.* 1942) and his own observations of the main years in which *P. brassicae* invaded the country, *viz.* 1917, 1931, 1932, 1936, 1937, 1939, 1953 and 1963. The last was a year particularly noted for the immigration of large numbers of the insect south west, the course of which has been carefully traced by Anglade *et al.* (*loc. cit.*).

In a six year study of the migrant lepidoptera on the north west coast of France, Wiltshire (1976) concluded that the Large White was a marginal immigrant although it probably had a resident population there.

344

Germany (BRD, DDR). One of the earliest accounts is by von Fischer (1807) (Williams Index Cards), of migrating *P. brassicae* moving like a snowstorm on August 22nd from Radibor towards Budissim, north of Dresden.

Many accounts of migration in Germany up to 1930 have been summarised by Williams in his book *"Insect Migration"* (William, 1930; cf. Kruger, 1900; Grund, 1900).

Accounts of migration in northern Germany used to occur with reliable regularity every few years, particularly in Schleswig-Holstein. In 1933 there was a migration which lasted several days during the period August 2nd–22nd and which went from north-west to south-west in Schleswig-Holstein (Williams, 1936a,b), and in the area around Jena and Arnstadt in the previous month (July 28th) there was a migration which went from north to south (Bergmann, 1953). Ebert (1933) noted that *P. brassicae* was a migrant at Obersdorf (Allgau) in Western Germany.

In 1939, Blunck published extensive observations on a migration at Kiel during August 1936 in which there were at least 3–10 million Large White migrating. This was also noted by Warnecke (1937) who witnessed movements in May, mid-July and August. At Flensburg he recorded 500 min and on another occasion he recorded the immigration to be 4 km long. Williams *et al.* (1942) quoted six cases of migrations in northern Germany during 1937–39 and in 1942–43 there was a migration at Butjadingen in Schleswig-Holstein (Speyer, 1948; Blunck, 1954).

The year of 1955 appears to have been a good year for migrations in northern Germany. Warnecke (1955) described how there was an outbreak on the island of Rügen. The following year there was an outbreak of larvae and imagines of the first summer generation at Brodersdorf in July 1956. Finally, mention has already been made of the migration observed from a glider over northern Germany in 1959 (Roer, 1968).

Great Britain and Ireland. The earliest documented immigrations of *P. brassicae* into Great Britain appear to be in 1842 and 1846 at Dover (Coleman, 1860; Spence, 1847 respectively). Tutt (1902) and Williams (1942) recorded subsequent immigration years as 1880–87 and 1889–1899. A migration was noticed by Poulton (1928) which occurred in September 1928 in Dorset and which went to the north-east.

Up to 1937, there were 11 accounts of immigrations in Great Britain, and from 1931–1939, there were another 11 major immigrations and some 54 examples of directional flight (Williams *et al.*, 1942). However, details of these later accounts are not included here. During the period 1930–1957, there were eight years of major immigrations in Kent (Chalmers-Hunt, 1960).

Systematic records on immigrations of *P. brassicae* before 1930 do not exist, but since that date, Rothamsted Experimental Station, Harpenden have recorded the following years of major immigrations; 1933, 1937, 1940, 1945, 1947 and 1955 (Skelton, 1974, pers. comm.).

The migrations of 1939 and 1940 were both given special attention by Williams *et al.* (*loc. cit.*). In 1939 there was an extensive immigration along the south and east coasts, viz. Norfolk, Kent, Sussex, Devon and 23 examples were given. It is interesting to note that in this year was the first known observation of a northerly migration of *P. brassicae* through Ireland. *P. brassicae* was also recorded in parts of Perthshire and Northumberland.

In 1940 there was a large immigration of *P. brassicae* in Britain which was attributed to a) 1939 invasion, b) lack of parasites, and c) (possibly) the severe winter of 1939–40 (cf. Williams *et al.*, 1942). The migration started in the middle of July and lasted until the early part of August and was predominantly travelling towards the south at Harpenden.

In the 1950's, Worms (1962) noted that a huge invasion from the continent in 1955 supplemented the population of *P. brassicae* in Wiltshire and Wooff (1958) recorded that occasional immigrants reached the Farne Islands, Northumberland in May–June and August–September. At the end of July 1968 a "definite migration" of *P. brassicae* was noticed on the Dingle peninsula in Co. Kerry, Ireland which involved some 50 specimens flying at the rate of two specimens per minute.

Accounts of migrations into Britain in the 1970s are few; in 1970 a "trickle" of *P. brassicae* moved from the north-west to the south-east near Llyn Cwmynach, Merioneth, Wales (Young, 1974), in the summer of 1974 in Dorset, fresh *P. brassicae* were recorded coming off the sea (Luckens, 1975), and in September 1977 large numbers of *P. brassicae* were recorded at Dungeness, Kent (Baker, 1978).

Large Whites come to the British Isles in two waves, first at the end of May and early June and second, at the end of July and beginning of August (Williams *et al.*, 1942, Williams, 1951a,b). The resident population receives reinforcements from the continent, but these immigrants do not reach the Shetland Isles, are found "sometimes in Orkney" and are "not universally common" in the Western isles and north Scotland. (Ford, 1945). Ireland also receives waves of *P. brassicae*. In 1913 thousands were observed in a field near Loch Dan in Co. Wicklow and these eventually went off south (Williams, 1936a,b). Mention has already been made of a migration in Ireland in 1939. When migratory waves of insects arrive in the British Isles the insects quite quickly disperse and move across the country. One such one arrived in Yarmouth, Norfolk on the 30th May and was observed in Brighton, Sussex on 5th June.

At Monk's Wood Experimental Station, Abbot's Ripton, Skelton investigated whether there was any correlation between weather and years of good immigration. He found that during years of good immigration the summer weather conditions were either "good" or "above average". The data for each summer were calculated from the sum of the mean temperatures for the three months, June, July and August. The summer of 1937 was "good", 1939 "close to average", 1940 was "above average", 1947 was a warm summer and 1970 was also good. The curious point about the summer of 1945 was that it was the 14th successive summer which was above average since 1801. None of the summers this century have been as bad as summers recorded last century viz. 1816, 1823, 1830, 1841, 1862, 1879 and 1888. Regrettably too few data of migrant insects exist before 1930 for any correlation to be made.

Switzerland. Pictet (1918a,b,c) described in detail several waves of migrating *P. brassicae* coming into the country during the summer of 1917. These were during the periods July 18th–24th, July 28th to August 3rd, and in September and October. The probable origin of the migrations were thought to be Alsace in Germany.

Fennoscandia

Denmark. A large migration was noticed in islands off Denmark in 1944 (Fabricius, 1945).

Finland. There are only two reports, both of spring migration, those of Ulvinen (1920) and Vepsäläinen (1968). Vepsäläinen estimated that more than 840,000 *P. brassicae* migrated into Finland in 1966 during a two day period from May 20th. The migration was very diffuse but over a large front and at the best time was progressing at a rate of 10–20 individuals past a transect of 50 m every 10 minutes. He substantiated his overall reports of the migration from correspondence with entomologists in 10 countries.

Sweden. Two accounts of immigrations in 1938 quoted by Williams *et al.* (1942) extended the then known migratory path of *P. brassicae* to its furthest point northwards.

Comments on the accounts of migration from different countries

Table 11.3 indicates those years in which there have been significant migrations in the various countries involved. Combined with this are years of serious damage inflicted on crops by larvae of *P. brassicae*. An abundance of larvae is often indicative of an influx of migrant insects the previous season, or from a previous generation. Large numbers of larvae

Table 11.3 Years of migrations and accounts of ravages in various countries. Ravages marked * (cf. Chapter 13: Economic importance). Migration and ravage in same year marked as 1.

Country	Years of migration and ravages
Austria	1917, 1923, 1926, 1927, 1935, 1937, 1939, 1941, 1947, 1948, 1953, 1955
Finland	1966
France	1917, 1931, 1932, 1936, 1937, 1939, 1953, 1963, 1970, 1971
Germany (BRD, DDR)	1933, 1936, 1937–39, 1942–43, 1949*, 1955, 1956, 1959
Great Britain	1846, 1855, 1880–87, 1892, 1898–99, 1909, 1913, 1915–17, 1930, 1933, 1937, 1939, 1940, 1945, 1947, 1955
Poland	1923
Rumania	1948–49*, 1955–56*, 1962–63*, 1970–71*
Sweden	1938
Switzerland	1917[1]
USSR	1894*, 1913*

can also be produced late in the year from an immigration of insects earlier in the year.

It is clear from Table 11.3 that most accounts are from Austria, Britain, France and Germany, but this is probably not indicative of distribution of depredations. The reason for this apparent lack of information from other European countries and the USSR could be accounted for by either a lack of published material or publication in obscure local journals.

Another point which is immediately evident from Table 11.3 is that most of the accounts of either migrations or ravages are for the period prior to the 1960s. The decline in the incidence of migration and major ravages is not clearly established. Some authorities blame the increasing use of insecticides and other agricultural chemicals, while others blame the increase in the important naturally occurring biological control agent, the granulosis virus, which is believed to have decimated the British population of *P. brassicae* quite considerably in 1955 (see Chapter 15).

Mark and recapture

Mark and recapture experiments on *P. brassicae* have been carried out principally by the Germans, Brecher (1925), Meder (1926), Blunck (1935), Roer (1961), Fahrenberg (1959), Reiser (1966), Eitschberger & Steiniger (1973a,b) but also in Great Britain by Campbell (1951), Birukow, (1966) and Owen (1978) (Figure 11.1).

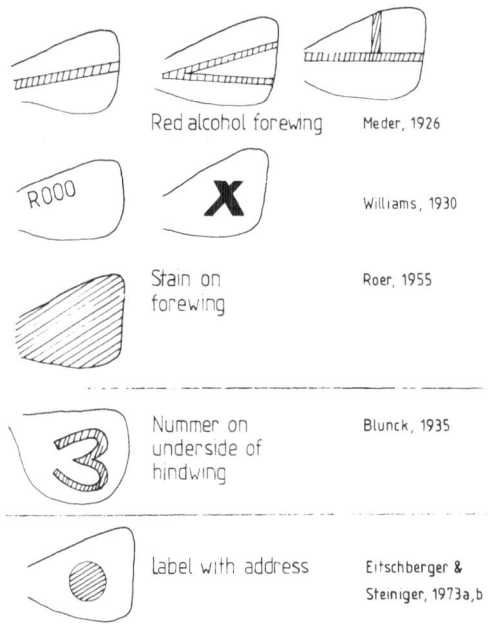

	Red alcohol forewing	Meder, 1926
R000	X	Williams, 1930
	Stain on forewing	Roer, 1955
3	Nummer on underside of hindwing	Blunck, 1935
	Label with address	Eitschberger & Steiniger, 1973a,b

Fig. 11.1 Designs for marking wings of *P. brassicae*

Brecher (1925) marked *P. brassicae* on Sylt Island in the North Sea off Schleswig-Holstein but they went off into the North Sea carried by the wind. Meder (1926) made quite extensive trials with *P. brassicae*, *P. rapae* and *P. napi*. He marked 4–5,000 specimens (numbers of each species not given) and released them from 12 locations in Schleswig-Holstein. Because of effective publicity and much help, Meder recaptured 850 specimens which represented a very high recapture percentage (recaptures of passerine birds usually average less than 5%). He stated that if much more help was available a much higher percentage would have been achieved. One of his workers who worked for four days managed to catch 600 of the 850 specimens caught at Lubeck (50 km away).

It was mentioned earlier that Roer (1961), who tagged *P. brassicae* in Bonn, recaptured one specimen to the north-west in Dusseldorf 95 km away. Reiser (1966) reared and marked 200 Large Whites (they were painted with black stripes on the forewing) and released them in Landshut, Bavaria. A college to the east was very helpful in returning some marked insects. However, no other details were given.

Several different methods of marking have been tried:

a) *Painting*
1) Cellulose paint: Campbell (1951) perfected a technique for *P. brassicae* using light blue paint which was painted on the wings. However, the

weather in Scotland that year was not good enough for sufficient *P. brassicae* to be caught; instead he used his technique on *Vanessa cardui* (Painted lady) and *Vanessa atalanta* (Linnaeus) (Red Admiral) quite successfully. Campbell combined his experimental work with an advertising campaign throughout Scotland which had good effect.

2) Lacquer: Blunck reared specimens for recapture experiments and marked them with a red model lacquer with alcohol. He gave each a large number covering the entire underside of the hindwing (see photo in Blunck (1935) p. 85), the number signifying the culture number.

3) Alcohol dye: Meder (1926) used this method by staining wings with a solution of a red alcoholic dye with the addition of a little shellac.

4) Aniline dyes: Roer (1961) stained the wings with this dye and also fed larvae food which contained the dye. The imagines came out pink.

b) *Tagging*. Attaching small adhesive information labels to the wings of insects is now a popular means of marking (see photo in Roer (1968) p. 305). The weight of the label does not appear to significantly affect the capacity to fly. Essential information, such as the name of the prominent museum e.g. "Museum Koenig" in Roer's case, implies that the recovered specimens be sent to that institution. Dispersal of information prior to mark and release experiments, to schools, colleges and the like has also been successfully employed by Roer.

c) *Tagging and dying*. Roer (1953) combined the two methods for more effective return. On this occasion he marked 16,980 imagines.

References cited

Anglade, P., Robin, J.C. & Roehrich, R., 1963. Observations de 1963 sur les migrations de la piéride du chou (*Pieris brassicae* L.) Lepidoptera: Pieridae, dans le sud ouest. *Rev. zool. Agric.* 10–12: 98–102.

Bailey, L.H., 1927. *The Standard Cyclopedia of Horticulture*. I. Macmillan, London.

Baker, P.J., 1978. Dungeness, Kent, 10th/11th September, 1977. *Proc. Brit. Ent. nat. Hist. Soc.* 11: 130–131.

Baker, R.R., 1968a. Possible method of evolution of the migratory habit in butterflies. *Phil. Trans. R. Soc. Lond.* B. 253: 309–341.

—— 1968b. Sun orientation during migration in some butterflies. *Proc. R. ent. Soc. Lond.* A. 43: 89–95.

—— 1969a. Evolution of migratory habit in British butterflies. *J. Anim. Ecol.* 38: 703–746.

—— 1969b. Die Entwicklung des Wanderverhaltens bei Schmetterlingen. *Umsch. Wiss. Techn.* 69: 626–627.

—— 1971. The evolution of the migratory habit in butterflies. *Proc. R. ent. Soc. Lond.* 36: 33–36.

—— 1978. *The Evolutionary Ecology of Animal Migration*. Hodder & Stoughton, Sevenoaks. pp. 1011.

Bergmann, A., 1933. Entomologische Beobachtungen und Versuche einer Expedition in Thüringer 1931. *Int. ent. Z.* 26: 415–422.

Birukow, G., 1966. Orientation behaviour in insects and factors which influence it. *In: Insect Behaviour. Symp. R. ent. Soc. Lond.* 2–12.

Blunck, H., 1935. Methodisches zur Zucht von *Pieris brassicae. Arb. Physiol. angew. ent. Berlin* 2: 78–87.

—— 1950. Zur Kenntnis des Massenwechsels von *Pieris brassicae* L. mit besonderer Berücksichtigung des Dürrejahres 1947. *Z. angew. Ent.* 32: 141–171.

—— 1954. Beobachtungen über Wanderflüge von *Pieris brassicae* L. *Beitr. Z. ent.* 4: 485–528.

Brecher, L., 1925. Physiko-chemische und chemische Untersuchungen am Raupen- und Puppenblute (*Pieris brassicae; Vanessa urticae). Z. vergl. Physiol.* 2: 691–713.

Bretherton, R.F., 1939a. Butterflies in the Alps, 1937 and 1939. *Entomologist* 72: 4–8.

—— 1939b. Migrating lepidoptera in the Oxford district. *Entomologist's mon. Mag.* 75: 250–251.

Campbell, J.L., 1951. An experiment in marking migratory butterflies. *Entomologist* 84: 1–6.

—— 1970. Macro-lepidoptera Cannae. Butterflies and moths of the Isle of Canna, Inner Hebrides. *Entomologist's Rec. J. Var.* 82: 1–27.

Chalmers-Hunt, J.M., 1960. Lepidoptera of Kent. *Entomologist's Rec. J. Var.* 72: 18–19.

Clapham, A.R., Tutin, T.G. & Warburg, E.F., 1962. *Flora of the British Isles.* Cambridge University Press, Cambridge.

Coleman, W.S., 1860. *British Butterflies, figures and descriptions of every native species.* Routledge, Warne & Routledge, London.

Cooke, B.H., 1937. Lepidoptera of the Oetztal in August. *Entomologist* 70: 121–127.

Daniel, F., 1957. Massenauftreten von *Pieris brassicae. Nachr. Bayer. Ent.* 6: 7–8.

Dannreuther, T., 1933. Immigration records. *Entomologist* 66: 17–19, 186–190, 209–212, 268–275.

—— 1934. Migration records. *Entomologist* 67: 10–14, 211–215, 254–256.

—— 1935a. Manx rhopalocera and migrant lepidoptera. *Entomologist* 68: 18.

—— 1935b. Migration records. *Entomologist* 68: 5–9, 185–188, 252–259.

—— 1936. Migration records. *Entomologist* 69: 1–6, 154–156, 182–186.

—— 1937. Migration records. *Entomologist* 70: 108–111, 165–166, 176–180, 200–202.

—— 1938. Migration records. *Entomologist* 71: 60–66, 176–178.

—— 1938. Migration records. *Entomologist* 72: 9–15, 273–283.

—— 1940. Migration records. *Entomologist* 73: 29–33, 62–63.

—— 1941. Migration records. *Entomologist* 74: 54–62.

—— 1943. Migration records. *Entomologist* 76: 73–80.

—— 1944. Migration records. *Entomologist* 77: 55–60.

—— 1945. Migration records. *Entomologist* 78: 49–56.

—— 1946. Migration records. *Entomologist* 79: 97–110.

—— 1947. Migration records. *Entomologist* 80: 107–112, 137–144.

—— 1948. Migration records. *Entomologist* 81: 73–83, 110–117.

—— 1949. Migration records. *Entomologist* 82: 73–78, 105–110; 83: 109–114.

—— 1951. Migration records. *Entomologist* 84: 85–90, 102–106.

—— 1952. Migration records. *Entomologist* 85: 125–131.

Demoll, R., 1918. *Der Flug der Insekten und der Vögel*. Fischer, Jena.

De Serres, M., 1845. *Des causes des migrations de divers animaux et particulaire des oiseaux et des poissons*. 2nd Edition. Paris.

Döring-Ilmenau, E., 1931. Sammelergebnisse im Jahre 1931 rings um den Kickelhahn. *Int. ent. Z.* 25: 289–294.

Ebert, H. von., 1933. Zur Rholaloceren Fauna der Umgebung von Oberstdorf (Allgäu). *Int. ent. Z.* 27: 129–133.

Effraimaison, R., 1934. Kaliperhosen Lentokestävyydesta ja-na peudesta. *Luonnon Ystärä* 38: 142.

Eitschberger, U., 1968. Jahresbericht 1967 der Deutschen Forschungszentrale für Schmetterlingswanderungen, Pieridae. *Atalanta* 2: 168–175.

—— 1969a. Die Unterscheidungsmerkmale der europäischen Arten der Gattung *Pieris* Schrank. *Atalanta* 2: 211–223.

—— 1969b. Jahresbericht 1968 der Deutschen Forschungszentrale für Schmetterlingswanderungen, Pieridae. *Atalanta* 2: 279–286.

—— 1970. Wanderfalterbeobachtungen im Mai 1969 in Spanien. *Atalanta* 3: 17–42.

—— 1970. Pieridae, Weiblinge. *Atalanta* 3: 62–75.

—— 1971. Pieridae, Weiblinge. *Atalanta* 3: 272–280.

—— 1972. Wanderfalterbeobachtungen im Juni 1970 in Spanien. *Atalanta* 4: 21–22.

—— 1973. Pieridae. *Atalanta* 4: 264–272.

Eitschberger, U. & Steiniger, H., 1973a. Appeal for international co-operation in the research of the migration of insects. *Atalanta* 4: 134–192.

—— 1973b. Wanderfalterbeobachtungen im Frühjahr 1972 auf der Iberischen Halbinsel. *Atalanta* 4: 309–321.

Fabricius, E., 1945. Mass migration of *Pieris brassicae* in the archipelago. *Notulae ent.* 25: 108–110.

Fahrenberg, U., 1959. Schmetterlinge mit Flügeletiketten. *Bildpost* 9: 8–9.

Feltwell, J.S.E., 1977a. Migration of whites. *Entomologist's mon. Mag.* 112: 88.

—— 1977b. Migration of *Hipparchia semele* L. *J. Res. Lepid.* 15: 83–91.

Ford, E.B., 1945. *Butterflies*. Collins, London.

Ford, R.L.E., 1976. The influence of the Microgasterini on the populations of British Rhopalocera (Hym: Braconidae). *Ent. Gaz.* 27: 205–210.

Ghika, G. & Loibl, H., 1925. Four days collecting in the Baltic in October 1924. *Int. ent. Z.* 19: 243–247.

Graves, R., 1975. *Collected Poems*. Cassell & Co., London.

Gray, J.H., Locke, M. & Putnam, C.D., 1953. Insect migration in the Pyrenees. *Entomologist* 86: 68–75.

Grund, F., 1900. Wanderung von *Pieris brassicae* L. *Illte. Z. ent.* 5: 352.

Gurney, G.H., 1927. Three weeks butterfly collecting in the Tyrol. *Entomologist* 60: 3–6.

Hansen, T. & Merivee, E., 1971. Cold-hardiness of the butterflies *Pieris brassicae* (L.) and *Pieris rapae* (L.). *Eesti Nsv. Tead. Akad. Toim.* 20: 298–303.

Harcourt-Bath, W., 1897. The probable causes of the decadence of British Rhopalocera. *Entomologist* 30: 55–58.

Hess, A., 1918. Der Kohlweissling. Ein Schädling in Schweizim im Sommer 1917. *Z. angew. Ent.* 4: 332–334.

Higgins, L.G., 1935. Butterfly collecting in Norway. *Entomologist* 68: 56–61.

Hingston, R.W.G., 1928. *Problems of instinct and intelligence*. London.

Huxley, J., 1942. *The Uniqueness of Man*. Book Club, London.

Jackson, F.W.J., 1937. A fortnight's butterfly hunting in Savoy in July, 1935. *Entomologist* 70: 145–146.

Johnson, C.G., 1965. Migration. *In: Physiology of the Insecta*, Ed. by M. Rockstein, Academic Press, London. pp. 187–226.

—— 1969. *Migration and Dispersal of Insects by Flight*. Methuen, London.

Kerry, F., 1892. Migration of the *Pieris brassicae* at Harwich. *Entomologist* 25: 321.

Kruger, D., 1900. Wanderung von *Pieris brassicae*. *Int. Z. ent.* 5: 299.

Kusdas, K. & Reichl, E.R., 1973. *The butterflies of Upper Austria. Auf. ent. Arb. Landmus. Linz* pp. 266.

Kusnezov, N.J., 1930. The dependence of the geographical distribution of the Pieridae on the distribution and chemical composition of their food plants. *Z. morph. Ökol. Tiere* 17: 778–793.

Lempke, B.J., 1942. Trevlinders in 1941. *Ent. Ber.* 11: 18–21. (see notes on migrations of insects in the Netherlands every two or three years up to the present published in the same journal).

London Natural History Society, 1957. Meeting of September 25th, 1956. *Entomologist's mon. Mag.* 93: 48.

Luckens, C.J., 1975. Notes on British butterflies, summer and autumn, 1974. *Entomologist's Rec. J. Var.* 87: 202–204.

Magnan, A., 1934. *Le vol des Insectes*. Hermann et Cie, Paris.

Magnan, A. & Planiol, A., 1933. *Sur l'excedent de puissance des insectes*. Hermann et Cie, Paris.

Meder, O., 1926. Über die Kennzeichnung von Weisslingen zwecks Erfassung ihrer Wanderungen. *Int. ent. Z.* 19: 325–330.

Mellows, C., 1924. Altitudes at which lepidoptera occur. *Entomologist* 57: 90–91.

Moulton, J.C., 1915. A note on collecting in the Himalayas where the east and the west meet. *Entomologist* 48: 230–235.

Muspratt, V.M., 1951. Observations sur quatre migrations de lépidoptères. *Rev. Fr. lépidopt.* 13: 125–127.

Nabokoff, V.V., 1931. Notes on the lepidoptera of the Pyrenées orientales and the Ariege. *Entomologist* 64: 268–271.

Nachtigall, W., 1978. *Insects in Flight*. George Allen & Unwin, Ltd., London. pp. 150.

Newman, L.H., 1931. Sark lepidoptera. *Entomologist's Rec. J. Var.* 43: 184–186.

Nichols, J.G., 1846. *The chronicle of Calais in the reigns of Henry VII and Henry VIII to the year 1540*. From MSS in British Museum (Bloomsbury). Printed for Camden Society. p. 7.

Nielsen, E.T., 1967. *Insekten auf Reisen*. Springer, Berlin.

Nikolaus, N.A., 1974. Flight speeds and types for certain Pieridae and Nymphalidae. Lepidoptera. *Mosc. Univ. Biol. Sci. Bull.* 59–62.

Nordman, A.E., 1954. The geographical distribution in Europe of the migrations of the Large White *Pieris brassicae*, and some remarks on their climatological causes. *Notul. Entomol.* 34: 99–106.

Ormerod, E.A., 1878. *Notes of observations on injurious insects. Report 1878*. West Newman & Co., London. Vol. 9.

Owen, D., 1978. Painted ladies in the schools. *Nat. Sci. in Schools* 16: 3–5.

Parker, G.A., 1978. Evolution of competitive mate searching. *A. Rev. Ent.* 23: 173–196.

Pictet, A., 1918a. Les migrations de la Piéride du chou en 1917 (*Pieris brassicae* L.) et leurs conséquences. *Arch. Sci. phys. natur.* 45:

—— 1918b. Les migrations de *Pieris brassicae* en Suisse en 1917. *Verh. schweiz. naturf. Ges.* 1917: 277–278.

—— 1918c. Observations biologiques sur *Pieris brassicae* en 1917. *Bull. Soc. lépidopt. Genève* 4: 53–66.

Poulton, E.B., 1928. Notes on the flight of *Colias croceus* and *Pieris brassicae*. near Bere Regis, Dorset. *Proc. ent. Soc.* 3: 78–79.

Reiser, M., 1966. Markierung von *Pieris brassicae* (L). bei Landshut (Südbayern) *Nachr. Bl. bayer Ent.* 15: 39.

Roer, H., 1953. Aufruf. Leaflet on red and blue markings of *Pieris brassicae*.

—— 1955. *Über Flug- und Wandergewohnheiten von Pieris brassicae* L. Doktorgrade, University of Bonn.

—— 1956. Rätsel um den grossen Kohlweissling (*Pieris brassicae*). *Orion* 11: 639–642.

—— 1957. Tagschmetterlinge als Vorzugsnahrung einiger Singvögel. *J. Orn.* 98: 416–420.

—— 1958. Über den Einfluss des Flugverhaltens auf den Massenwechsel des grossen Kohlweisslings (*Pieris brassicae* L.). *Bonn. zool. Beitr.* 9: 95–101.

—— 1959. Über Flug- und Wandergewohnheiten von *Pieris brassicae* L. *Z. angew Ent.* 44: 272–308.

—— 1960. Etikettierte Schmetterling auf Wanderung. *Orion Z. natur. Tech.* 8: 650–652.

—— 1961. Wanderfalter Forschung in Mitteleuropa. *In: Die Wanderflüge der Insekten.* Ed. by Williams, C.B., Hamburg & Berlin.

—— 1961. Ergebnisse mehrjahriger Markierungsversuche zur Erforschung der Flug- und Wandergewohnheiten europäischer Schmetterlinge. *Sond. zool. Anz.* 167: 456–463.

—— 1967. Wanderflüge der Insekten. *In: Die Strassen der Tiere.* Ed. by Hediger, H. Vieweg, Braunschweig. pp. 186–206.

—— 1968. Zur frage der Wanderschwarmbildung bei Tagfaltern als subsoziales phänomen. *Insectes Soc. Bull. Union Intn.* 15: 299–307.

—— 1973. Insektenmigrationen. *Nachr. Wanderinsek. Forsch.* 1: 6–19.

—— 1974. Wanderinsektenforschung in Europa. *Folia. ent. Hung.* 27: 49–70.

Rothschild, M., 1956. Note on insect migration on the north coast of France. *Entomologist's mon. Mag.* 92: 375–376.

—— 1962. Several large white moths commonly called butterflies seen for the first time this year. *Entomologist's mon. Mag.* 98: 167–170.

—— 1963. A further note on insect migration on the north coast of France. *Entomologist's mon. Mag.* 99: 163–164.

Sandeman, R.G., 1946. Autumn butterflies in Breconshire. *Entomologist* 79: 116–117.

Schmidt-Koenig, K., 1975. *Migration and Homing in Animals.* Springer-Verlag, Berlin, Heidelberg, New York.

Schütte, F., 1966. Observations on the flight of butterflies of the genus *Pieris*. *Z. angew. Ent.* 58: 107–193.

South, R., 1891. On the distribution in Eastern Asia of certain species of Lepidoptera occurring in Britain. *Entomologist* 24: 81–86.

Spence, W., 1847. Einige Fragen. *Stettin ent. Ztg.* 8: 376–381.

Speyer, W., 1948. Die Wandergewohnheiten und der Flug der *Pieris brassicae*. *Z. PflKrankh. PflPath. PflSchutz.* 55: 335–341.

—— 1956. *Pieris brassicae* L. in den Dünen der Nordseeinsel Amrum. Z. *PflKrankh. PflPath, PflSchutz.* 63: 12–14.

Tulloch, J.B.H., 1941. Mass movements of *Pieris brassicae* and *Pieris rapae. Entomologist* 74: 32–35.

Tutt, J.W., 1902. *Migration and dispersal of insects.* London.

Ulvinen, A., 1920. Kaaliperhosten Joukkoesiintymistä. *Luoonnen Ystävä* 24: 112.

Vepsäläinen, K., 1968. Immigration of *Pieris brassicae* L. into Finland in 1966, with general discussion on insect migration. *Ann. ent. fennici* 34: 223–243.

Warnecke, G., 1937. Zum Massenauftreten des grossen Kohlweisslings (*Pieris brassicae* L.). *Ent. Z.* 51: 220.

—— 1955. Die Grosschmetterlinge des Niederelbgebietes und Schleswig-Holsteins. *Sond. verh. nat. heim. Hamburg* 32: 24–66.

Westmacott, M.H. & Williams, C.B., 1954. A migration of lepidoptera and diptera in Nepal. *Entomologist* 87: 232–234.

Williams, C.B., 1928. Collected records relating to insect migration. *Trans. ent. Soc. Lond.* 1928: 79–91.

—— 1930. *The migration of Butterflies.* Oliver & Boyd, Edinburgh & London.

—— 1935a. Butterfly immigration in Britain. *Discovery.* February 36–39.

—— 1935b. British immigrant butterflies and moths. *British Museum (Natural History).* Pamphlet E 57.

—— 1936a. Our butterfly visitors from abroad. *Country-side.* Spring, No. 2.

—— 1936b. Collecting records relating to insect migration. 3rd. series. *Proc. R. ent. Soc. Lond.* A 11: 6–10.

—— 1939a. The migration of the cabbage white butterfly (*Pieris brassicae*). *7th. Int. Congr. Ent. Berlin,* 1938. 1: 482–493.

—— 1939b. Some butterfly migrations in Europe, Asia and Australia. *Proc. R. ent. Soc. Lond.* A. 14: 131–137.

—— 1949a. Migration in lepidoptera and the problem of orientation. *Proc. R. ent. Soc. Lond.* 13: 70–84.

—— 1949b. Rothamsted Experimental Station, Harpenden, Report for 1948. 66–69.

—— 1951a. Seasonal changes in flight direction of migrant butterflies in the British Isles. *J. Anim. Ecol.* 20: 180–190.

—— 1951b. Rothamsted Experimental Station, Harpenden, Report for 1950. 90–95.

—— 1957. Insect migration. *A. Rev. Ent.* 2: 163–180.

—— 1958. *Insect Migration.* Collins, New Naturalist Series, London.

—— 1961. *Die Wanderflüge der Insekten.* Paul Parey, Hamburg.

Williams, C.B., Cockbill, G.F., Gibbs, M.E. & Downes, J.A., 1942. Studies in the migration of lepidoptera. *Trans. R. ent. Soc. Lond.* 92: 101–280.

Williams, C.B., Common, I.F., French, R.A., Muspratt, V. & Williams, C.B., 1956. Observations on the migration of insects in the Pyrenees in the autumn of 1953. *Trans. R. ent. Soc.* 108: 385–407.

Wiltshire, E.P., 1976. Six year's notes on migrant lepidoptera in and near the coast of N.W. France. *Entomologist's Rec. J. Var.* 88: 165–175.

Wooff, W.R., 1958. *An ecological survey of the insects of the Farne Islands.* Ph.D. thesis. University of Durham.

Worms, N. de., 1962. *The Macrolepidoptera of Wiltshire. Wilts. arch. & nat. Hist. Soc.* pp. 177.

Young, M.R., 1974. An account of some of the lepidoptera of the Moorlands near Llyn Cwmynach, Merioneth. *Entomologist's Rec. J. Vr.* 86: 10–14.

Further references

NB. As so many of these references concern migrations in various localities, place names and countries are given where known. For Great Britain town and county are given.

Adamczewski, S., 1960. *Fragm. ent.* 11: 319–374 (Poland).
Ahlqvist, H., 1938. *Notul. Ent.* 28: 140 (Gulf of Finland).
Andrewes, H.L., 1928. *Proc. Ent. Soc. Lond.* 3: 78–79 (Bere Regis, Dorset).
Balland, R., 1947. *Feuille Nat.* 2: 106 (France).
Bandermann, F., 1932. *Int. Ent. Z.* 25: 311 (Germany).
Bandermann, A., 1931. *Int. Ent. Z.* 25: 209–213 (Germany).
Birkett, N.L., 1941. *Entomologist* 74: 64 (Watford).
Blunck, H., 1920. *Mitt. biol. Reichsanstalt* 21: 182–184 (Hamburg, Germany).
—— 1951. *Z. angew. Ent.* 32: 141–171 (Germany).
Boedijn, K., 1915. *De Levende Natuur* 19: 215–216 (Netherlands).
Bonn, V.H.R., 1959. *Sond. Z. aug. Ent.* 44: 272–309 (Germany).
Bowden, S.R., 1956. *Entomologist* 89: 131 (at sea, England).
Brockhuysen, G.J., 1953. *Opusc. ent.* 18: 244 (Ottenby).
Bryk, F., 1950. *Ent. Tidskr.* 71: 177–178 (Stockholm, Sweden).
Buckstone, A.A.W., 1926. *Entomologist* 59: 5–8 (south coast).
—— 1945. *Entomologist* 78: 157 (Ewell, Surrey).
Daniel, F., 1957. *Nachrbl. Bayer ent.* 6: 7–8 (Germany).
Dannreuther, T., 1935. *Entomologist* 68: 252–259 (off Lowestoft, Norfolk).
—— 1936a. *Entomologist* 69: 1–6 (Skegness, Lincolnshire).
—— 1936b. *Entomologist* 69: 154–156 (St. Malo, France).
—— 1938. *Entomologist* 71: 60–66 (Bamburgh Castle, Northumberland).
Duteurtre, M., 1957. *Rev. Soc. sav. Htes. Normandie* (5): 69–71 (Le Havre, France).
Eastern Evening News, Norwich, 1933. *Entomologist's mon. Mag.* 69: 163–164 (off Norfolk).
Eitschberger, U., 1970. *Atalanta* 3: 17–42, 62–75, 272–280 (several sightings).
—— 1972–1973. *Atalanta* 4: 21–42, 264–272 (several localities in Spain).
Eitschberger, U. & Steiniger, H., 1973. *Atalanta* 4: 134–192, 309–321 (several localities in Spain).
Ekholm, S., 1962. *Proc. XI Int. Cong. ent. Vienna* 1960: 24–26 (Finnish coast).
Eliot, N., 1943. *Entomologist* 76: 193–198 (discussion on migration and hibernation).
—— 1949. *Entomologist* 82: 245–250 (Riviera, France).
Fletcher, T.B., 1925. *Bull. ent. Res.* 16: 177–187 (India).
—— 1929. *Proc. ent. Soc. Lond.* 4: 103–104 (notes on migration).
Fraenkel, G.S., 1932. *Ergebn. Biol.* 9: 1–238 (Germany).
Garnett, D.G., 1947. *Entomologist* 80: 243 (Suffolk).
Gembloux, Belgium. Record cards (unpublished). (seen at Celles, Flandre, Ayuvaille, East of Belgium).
Goater, B., 1977. *Ent. Gaz.* 28: 81–84 (Cornwall).
Goulliart, M., 1939. *Bull. mens. Soc. ent. Nord. Fr.* (1) 2 (discussion).
Hagen, H.A., 1861. *Stett. ent. Z.* 22: 73–83 (30 accounts of lepidoptera migration).
Harford, H.C., 1927. *Entomologist* 60: 161–162 (Malta).
Harz, K., 1965. *Atalanta* 1: 21–31, 89–114, 136–145 (several accounts in C. Europe).

Heerdt, P.F., Bruyns, M., 1960. *Tijdschr. Ent.* 103: 225–275 (Ferschelling).

Kinzelbach, R., 1969. *Atalanta* 2: 245–248 (Germany).

Koch, M., 1961. *Sond. Ent. Z.*, 71: 1–7 (Germany).

—— 1961b. *Sond Mitt. Insekten. Jahr.* 5: 26–29, 59–62, 116–118; 6: 157–166 (Germany).

Kohler, K., 1938. *Mitt. natur. Vereines Troppau C.S.R.* 43: 18–20 (Schlesien).

Kohler, H.J., 1973. *Atalanta* 4: 199–201 (Katalonica).

Köllner, V., 1968. *Atalanta* 2: 132–135 (Germany).

Loran, A., 1902. *De Levende Natuur* 6: 235 (Netherlands).

Lycklama, H.J., 1914. *Ent. Ber.* 4: 109–113 (Netherlands).

Mazzucco, K., 1958. *Wien. ent. Ges.* 43: 4–12, 25–29, 36–43 (Germany).

Mikkola, K., 1967. *Ann. ent. fennici* 33: 65–99 (Finland).

Nordman, A., 1942. *Memor. Soc. F. Fl. Fenn.* 18: 127–184 (Finland).

Owen, D.F., 1954. *Entomologist's mon. Mag.* 90: 112 (SW France).

Peile, H.D., 1924. *Entomologist* 57: 234 (Bexhill, Sussex).

Ploeger, P.L., 1951. *Ent. Ber.* 13: 254 (Texel, Netherlands).

Prideaux, R.M., 1934. *Entomologist* 67: 186–191 (Feltwell, Norfolk).

Radovanovic, S., 1969. *Atalanta* 2: 295–300 (Yugoslavia).

—— 1970. *Atalanta* 3: 5–11 (Yugoslavia).

—— 1971. *Atalanta* 3: 300–309 (Yugoslavia).

Richter, V., 1957. *Nachrbl. Bayer Ent.* 6: 6–7 (Germany).

Row, A.W.H., 1947. *Entomologist* 80: 289–290 (Isle of Portland).

Sarel Whitfield, F.G., 1939. *Bull. Ent. Res.* 30: 365–442 (air transport).

Schütte, F., 1966. *Z. angew. Ent.* 58: 131–138 (N. Germany).

Selzer, A., 1918. *Int. Ent. Z.* 11: 226–228 (N. Germany).

—— 1919. *Int. Ent. Z.*, 13: 62–64 (Germany).

Shapiro, A.M., 1970. *Entomologist's Rec. J. Var.* 81: 85–86 (NE Scotland, Orkney).

Siviter, P., 1947. *Entomologist* 80: 244–245 (Birmingham).

Speyer, W., 1948. *Z. PflKrankh. PflPath, PflSchutz.* 55: 335–341 (migration habits).

—— 1956. *Dt. ent. Ges. Mitt.* 15: 57–59 (outbreak of larvae).

Sutton, S.L., 1959. *Entomologist* 92: 257–263 (Portland, Dorset).

Thijsse, J.P., 1902. *De Levende Natuur* 6: 210–211 (Netherlands).

—— 1938. *De Levende Natuur* 42: 191 (Norfolk to Netherlands).

Times, The, 1968. 9th. April, Review of Baker. *In: Phil. Trans. R. Soc. B.* 253: 309.

Tutt, J.W., 1902. *Migration and Dispersal of Insects.* London.

Valletta, A., 1953. *Entomologist* 86: 57 (Malta).

Van Rossem, G., *Tijdschr. Ent.* 99: 64–75 (Netherlands).

Walker, J.J., 1921. *Proc. ent. Soc. Lond.* 1921: 20–35 (summary of migrations).

—— 1931. *Entomologist's mon. Mag.* 67: 254–268 (insects at sea).

Warnecke, G., 1949. *Mitt. Faun. arb. Schleswig-Holstein Hamburg & Lubeck* 11: 32–33 (N. Germany).

Wereburg, A., 1874. *Der Schmetterling und sein Leben.* Berlin.

Williams, C.B., 1951. *J. anim. Ecol.* 20: 180–190 (seasonal changes).

Wittstadt, H., 1956. *Nachr. Bl. bayer. Ent.* 5: 12–15 (Germany).

—— 1959. *Pflanzenschutz* 11: 142–143 (Germany).

Wittstadt, H. & Mazzucco, K., 1957. *Pflanzenschutz* 9: 91–96 (Germany).

12. Senses

Introduction

The senses most frequently described in the Large White are visual, olfactory and tactile; these include chemosensory systems such as gustation in the larva and tarsal responses in the imago. Information about the gross morphology of ocelli and ommatidia has been known for a long time, but recent studies have revealed some more intricate details of these organs. Electrophysiological studies on insects and particularly on *P. brassicae* have only been pursued over the past 10 years or so. Larvae have also been credited with the sense of hearing and imagines with being able to detect changes in barometric pressure, but little work has been carried out in either of these two fields.

The senses of *P. brassicae* in both the larva and the imago are dealt with here under each particular sense. It is perhaps worth mentioning that little appears to have been investigated on the senses of the pupa. At the end of each chapter details of active and passive defense mechanisms have been included.

Visual senses

Larva

Larvae have 12 ocelli on the head, which are directed forwards, sideways and downwards (see Chapter 7: Figure 12.1). Götz (1936) found that *Pieris* larvae perceive colour and that they have a distinct preference for green (their foodplant colour) regardless of background. Prior to pupation this preference changes to brown and black, which are the colours of their pupation surfaces.

The structure of the lateral ocellus of *P. brassicae* larvae has been studied by Barrer (1969 – thesis). He also investigated the histology of the

Table 12.1 Sense organs and cells of *P. brassicae* (numbers of cells in brackets).

Sensory faculty	Larva	Imago
Chemosensory	*Labrum* campaniform sensilla epipharyngeal papilla-like sensilla tactile sensillae (3) trichoid sensilla *Lateral sensilla styloconia* amino acid sensitive cell anthocyanin sensitive cell mustard oil glycosides sensitive cell sucrose and glucose sensitive cell *Maxillary palp* sensillae at tip (8) *Median sensilla styloconia* alkaloids sensitive cell deterrents sensitive cell mustard oil glycosides sensitive cell salt sensitive cell 1 salt sensitive cell 2	*Proboscis* external spines modified trichoid sensillae (4), trichoid sensilla *Tarsal setae* B-type cells with water, salt and mustard oil glycoside sensitive cells
Olfactory	not known	androconia antennae olfactory pits at base of labial palp
Tactile	setae tactile hairs	sensitive cell associated with mechanoreceptor on tarsus tactile hairs: A-type cells
Visual	corneagen cells crystalline lens secreting cells extraocular receptor cells ocelli retinal cells	compound eye with ommatidia

corneal lens, corneagen cells, crystalline lens and crystalline lens secreting cells, retinula cells, and data on the refractive indices, radii of curvatures, fields of view of the ocelli and electrophysiology of the lateral ocellus.

In the study of the way in which *P. brassicae* larvae move around and survey their immediate vicinity, El-Dakroury (1972 – thesis) noted that they carry out two distinct body movements, a swaying of the head from side to side and an upward movement of the body involving the raising of several anterior segments.

Larvae blinded unilaterally exhibit typical photopositive response in

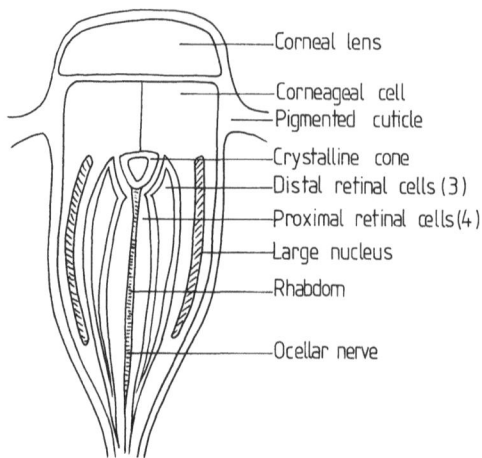

Fig. 12.1 Longitudinal section through turretted ocellus of fifth instar larva of *P. brassicae*
(modified from El-Dakroury, 1972) size not given

going towards the unblinded side if dark-adapted, and photonegative if light-adapted (El-Dakroury, *loc. cit.*). If lights of equal intensity are present larvae move midway between them. This suggested that their behaviour is normally trophotactic, i.e. based upon balancing the intensity of stimulation received via the two sets of receptors, rather than telotactic. Larvae were shown to be attracted to vertical stripes set against a light background, but only darker stripes set against a light background. White or grey stripes against a black background had little or no effect (El-Dakroury, *loc. cit.*).

Collenette (1945) studied the effect of various factors on the orientation of larvae after they had been taken off their foodplants. He found that on a sunny day most of the larvae regained the foodplant, while on a rainy day the larvae became lost and were unable to find their foodplant. This suggested that there was a visual stimulus. His overall impression was that the larvae went for the nearest object, purely visually, but that scent, sense of direction and wind direction played little part.

Functions of the ocelli

The ocelli have been found to be involved in the following: a) phototaxis (Bünning & Joerrens, 1960); b) induction and inhibition of diapause (Bünning & Joerrens, *loc. cit.*); c) perception of wavelength of light and direction of incident light (Angersbach, 1975); and d) light perception by lateral ocelli controlling pupal pigmentation (Seugé, 1973).

Angersbach (*loc. cit.*) found that No. 1 ocellus, which is the most ventrally situated, was the most receptive of light. Response to light had to be effected by the stimulus of two or more ocelli. In his experiments, Angersbach blinded larvae by painting over the ocelli with black, yellow or blue pigments, but he found that the ventral ocellus was still more sensitive, regardless of the spectral range; monochromatic light however, was more effective than the other colours used. Blinding larvae was the technique also used by Brecher (1924), who found that dark blue paint over the ocelli produced dark blue pupae, while yellow paint produced light pupae.

Seugé (1973) showed that when lateral ocelli of fourth and fifth instar larvae were cauterised, the photoperiodic control of diapause was not affected.

Extraocular receptor

The head capsule contains areas of cuticle which are transparent and these have been suggested to function as gateways for light passing inside the head (cf. Chapters 7, 8), or in the process of pigmentation determination (Oltmer, 1968). The exact situation of the cells of the extraocular receptors is not known, but it may be either in the integument, or in the central nervous system (Angersbach, 1975).

The extraocular receptors can detect light of different wavelengths but the spectral range of sensitivity has not been established. Angersbach (*loc. cit.*) painted over these regions with blue or yellow varnish and found that pupae were subsequently weakly or more strongly melanised respectively.

Imago

Compound eyes
In a recent study of the compound eyes of 15 insect species, Frantsevitch & Pichka (1976) found that in *P. brassicae*, as in other anthrophiles, the binocular zone enclosed 20–25% of the facets in each eye. "Middle sized insects had from 2000–9000 ommatidia". They carried out their work using ophthalmological methods, such as observation of the pseudopupils and the glow from the ommatidia, illuminated from the inside.

Ommatidia
Very little work appears to have been carried out on the ommatidia of *P. brassicae* apart from Ali's (1974) incidental treatment of them while studying the central nervous system (Figure 12.2).

The ommatidia are very similar to those of other insects in that they contain a lens and crystalline cone, a long axial rhabdom and surrounding

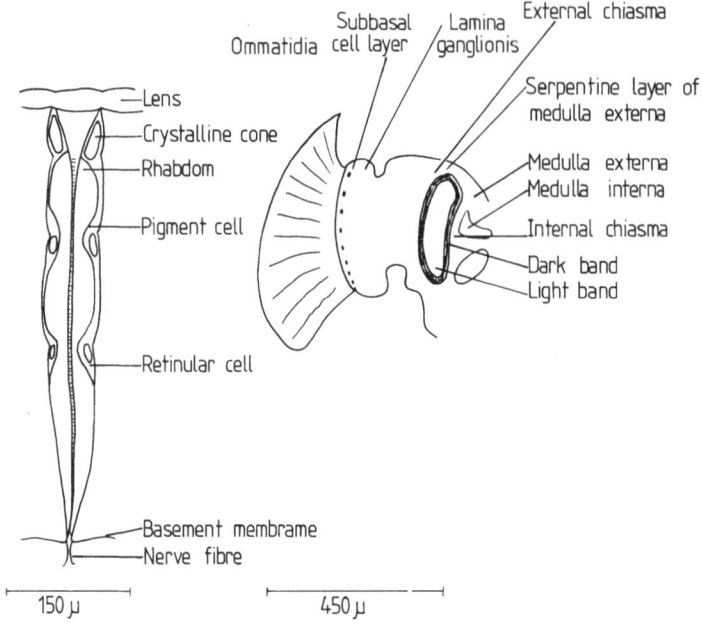

Fig. 12.2 Longitudinal section through ommatidium
(after Ali, 1974)

pigment cells. Nerve fibres leave the proximal end of the ommatidium as retinal fibres of the optic nerve. Below the basement membrane a layer of tracheae is present and some of these pass upwards supplying the retinular cells. The inside surface of the ommatidia is smooth.

The ommatidia arise from the epidermis and have no connection with the ocelli. Differentiation of the ommatidia starts about 72 hours after pupa formation and by the 120 hour pupal stage they are well developed, although they are telescoped together (see Ali, 1974, p. 401, 405; longitudinal section of ommatidium in a 120 hour old pupa).

Colour preferences

Ilse (1928, 1937) showed that as female imagines get older, their colour preferences change from selecting yellow-blue flowers for oviposition sites, to red-blue flowers for feeding (Figure 12.3). Schlieper (1928) also found that *P. brassicae* was sensitive to red but had a greater spectral range than other species of lepidoptera investigated. Eltringham (1933) pointed out that visits to red flowers by *P. brassicae* were infrequent. Eliot (1948), corroborating this view, recorded that *P. brassicae* (age not mentioned) had a preference for mauve flowers of *Cistus* sp. and blue

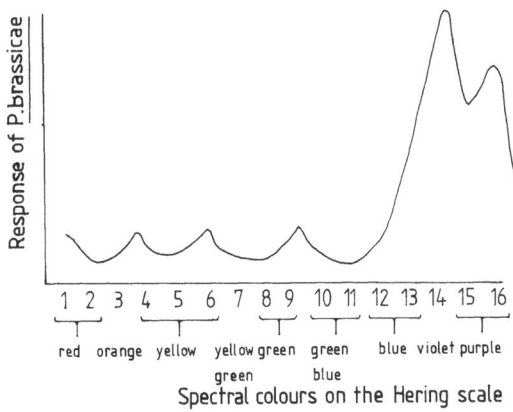

Spectral colours on the Hering scale

Fig. 12.3 Response to imago *P. brassicae* to colour
(after Isle *in* Portier, 1949)

Table 12.2 Flowers visited by imago of *P. brassicae*.

Species	Colour of flower	Authority
	W-white, P-purple, R-red, V-violet, Y-yellow, B-blue, M-mauve, O-orange	
Convallaria majalis L. (on a Lady's hat in Regent's St., London)	W	Bedford, 1897
Knautia sp.	P/R	Littlewood, 1912
Buddleja sp.	B/V	Scott, 1921
Valeriana sp., *Hedera* sp.	P,Y	Tulloch, 1928
Lavandula vera De Candalle, *Vicia* sp.	B/V	Scott, 1933
Trifolium sp., *Medicago sativa* L.	R/B	Frohawk, 1936
Lathyrus odoratus Linnaeus (in buttonhole of gent in city)	P	Bedford, 1940
Veronica sp.	B	Adkin, 1942
Valerian sp.	P	Dannreuther, 1946
Cistus sp., *Vinca* sp., *Oxalis* sp.	M,B,Y	Eliot, 1948
Aster sp.	P	Hamm, 1948
Inula viscosa Linnaeus	Y	Valletta, 1954
Medicago sativa Linnaeus	R/B	Heslop, 1956
Valerian sp., *Buddleja* sp.	R/B,V	Miles, 1957
Nasturtium sp.	O/R	Baynes, 1961
Viola (3 spp.)	V	Beattie, 1972
Cirsium sp., *Echium vulgare* Linnaeus	B/P,P	Feltwell, 1974 (unpublished)

Vinca sp., and that they shared the preference with *Gonepteryx rhamni* (Linnaeus) for yellow *Oxalis* sp. (Table 12.2). Isle & Vaidja (1956) stated that *Pieris*, like various other species, have a colour preference for yellow when they are feeding.

The examples of flowers visited quoted in Table 12.2 were taken after a thorough search of the literature, and show that all colours from red to blue are visited. Not surprisingly, white colours, such as white *Convallaria majalis* Linnaeus (Lily of the Valley), have been found to be attractive; this may explain why imagines were observed by Hertz (1927) to pursue airborne seedballs of *Taraxacum* sp. (Dandelion).

In experiments to test the sensitivity of *P. brassicae* to brightness of various colours, Portier (1949) recorded that both *P. brassicae* and *P. rapae* responded most to yellow No. 4 the brightest (on the Hering scale), while eight other species responded to green 7.

Speyer (1955) published an "intensity table" of the flowers visited by *P. brassicae*, *P. rapae* and *P. napi* during one day in July, 1943, in the Botanical Gardens of Berlin-Dahlem. For *P. brassicae*, nine species were visited frequently, 12 visited infrequently, 15 rarely and five not at all. Many of the flowers visited frequently were coloured purple to red.

Olfactory senses

Larva

The fact that *P. brassicae* larvae are gregarious and produce "an evil smell" (South, 1936) has raised the question of whether larvae use olfactory powers to keep together. Wojtusiac (1930) believed that *P. brassicae* larvae leave a detectable scent trail which is followed by other larvae when they cross it, in the same way that silk paths are supposed to function. Dethier (1947) believed that gregarious *P. brassicae* larvae could detect the odour from other larvae at a distance of 5 cm. On the contrary, Collenette (1945), who studied the orientation of larvae of *P. brassicae*, found that scent did not appear to influence their behaviour.

A choice chamber, designed to detect and test any olfactory powers of *P. brassicae* larvae (and other small invertebrates), was published by Feltwell (1977a). Results showed that groups of 25 larvae take at least two hours to locate cabbage when they cannot see it (unpublished).

Although larvae of *P. brassicae* prefer cut surfaces of leaves to intact ones, Sudah (1970) was unable to detect any pheromones which may have emanated from the saliva of larvae.

There is some evidence which suggests that *P. brassicae* imagines have a sense of smell, but lack of any concrete evidence is probably due to the inadequacy of devising techniques sophisticated enough to sample very small amounts of odours. The first piece of circumstantial evidence is that pheromones have been identified from androconia, which implies that males can smell. Second, there is some evidence that imagines can smell tetrachlorethane, a sweet-smelling fluid used as a killing agent in moth traps, as on nine separate occasions, a total of ten *P. brassicae* imagines have been caught in one locality in Wales (Nicklen, 1975, pers. comm.; details on Rothamsted Record Cards). A clear Tungsten 200 Watt bulb was used on each occasion.

Third, from observations made in the field in northern Germany, it appeared that imagines of each sex flew upwind to a cabbage field as if they were following a vapour trail (of volatile sulphur-containing molecules) (Feltwell, 1977b).

Other evidence about smell in imagines comes only from the comments of various people; Müller (1925) said that imagines were attracted to the scent of *Buddleja* sp., but this may also be due to visual effects. If imagines can smell, they must be selective or sensitive to only certain groups of volatile compounds, as they do not approach all scent-producing flowers (e.g. *Rosa* sp.).

Smell was also reputed to play a part in courtship and attraction of imagines at close quarters after visual cues have brought the insects together (Hertz, 1927). Below 20 cm olfaction becomes the dominant sense. Obara & Hidaka (1964, 1968) ruled out olfactory cues in recognition of partner and courtship behaviour in *P. rapae*, as males can be attracted to females enclosed in petri-dishes. Dora Isle (1928) believed that colours and odours were closely related sensory cues in recognition in insects.

On the smell of imagines to man, Longstaff (1912) (who "smelt" 19 spp. of Lepidoptera), noted that *P. brassicae* smelt like *Geranium* sp., but occasionally like *Viola* sp., or *Iris* sp. Later, Dixey & Longstaff (*in* Ford, 1945) stated that the male imago has a very slight smell of orris root (*Iris florentia* Linnaeus).

Experiments on the reaction of ovipositing imagines to squashed ova and to secretions of squashed female abdomens applied to leaves, seemed to suggest that imagines might respond to the smells which emanated from these (Rothschild *et al.*, 1975). Indeed, later experiments indicated that ova may well produce a scent that dissuades the female from ovipositing, and that the female may leave a pheromone produced from the tip of her abdomen on the leaves as a marker (Rothschild & Schoonhoven, 1977).

Antennae are used in insects for odour detection but in *P. brassicae* experiments do not indicate that all the olfactory powers lie in these organs. Wigglesworth (1967) stated that pits at the base of the antennae and palps were also concerned with olfaction. Eltringham (1933) stated that there is a large terminal cavity lined with pegs on the labial palps of *P. brassicae* and he thought that these too were possibly used in olfaction. Amputation or inactivation of the antennae of *P. brassicae* does not reduce the sense of smell by more than half (Minnich, 1924), or prevent oviposition (Ma & Schoonhoven, 1973). Work carried out at the Max Planck Institute near Munich demonstrated that *P. brassicae* antennae give no positive response to common aromatic compounds (Schneider, 1975, pers. comm.). However, this does not rule out the possibility that the antennae are sensitive to some other compounds hitherto untested.

Electromicrographs of antennae of the imagines of both sexes were prepared by Schoonhoven at Wageningen (Netherlands) and both were shown to have hair-like odour receptors at the tip, together with trichoid sensillae, and microtrichia. The trichoid sensillae are involved in the perception of volatiles from flowers, plant leaves and an oviposition deterring pheromone associated with conspecific ova (Plates VIII, IX). The latter substance is produced in the female's accessory gland (cf. Behan & Schoonhoven, 1978).

Androconia

Androconia of *P. brassicae* have been recognised for a long time and have been frequently figured (Westwood, 1854; SLENHS, 1914; Dixey, 1932; Barth, 1939; Warren, 1961; Kudrna, 1972, 1973; Bergström & Lundgren, 1973).

The structure of an androconium of *P. brassicae* has been published by Bergström & Lundgren (1973) in their scanning electromicrograph studies and it can be seen that it has a characteristic bladder which comes off from the base of the scale via a thin tube. These authors also summarised extensive studies of Barth (*loc. cit.*, in Portugese) on the morphology of androconia.

The bladder of the androconium, which contains the scent secretions, is situated in a cavity in the wing structure beneath the scale. When the male flutters around the female, the scales are erected from the surface of the wing and secretion is poured out into the lumen of the scale where it escapes through the pores of the androconium.

Androconia were also found in *P. rapae* and *P. napi* by Bergström & Lundgren (*loc. cit.*), each pierid species having characteristic pheromones.

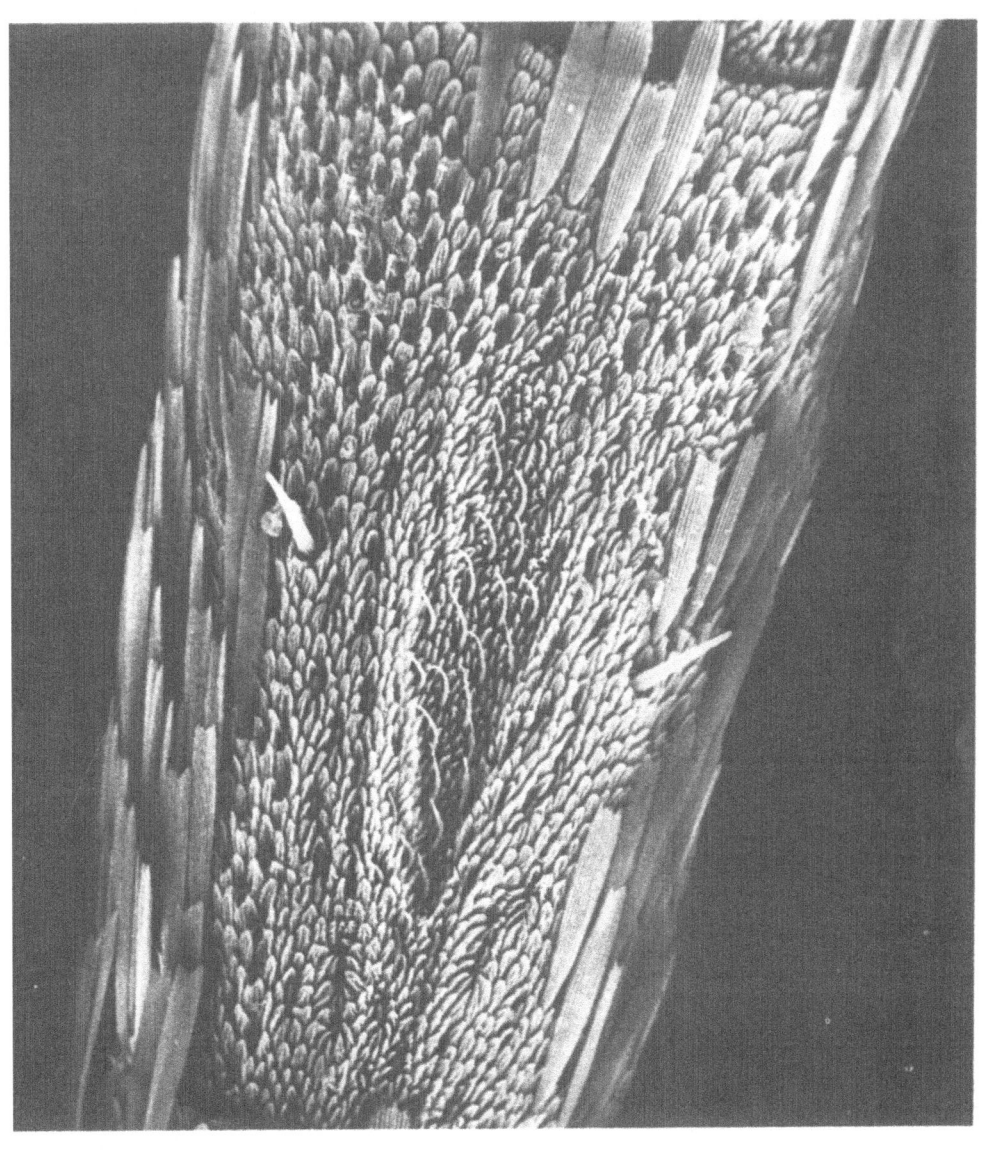

Plate VIII Female antennal segment (×400) with two types of receptors and scales (Courtesy of Louis Schoonhoven, Wageningen Agricultural University, Netherlands)

Plate IX Female antenna showing trichoid sensilla (×3000)
(Courtesy of Louis Schoonhoven, Wageningen Agricultural University, Netherlands)

Traces of tricosane and other unidentified compounds were found in *P. brassicae*. Surprisingly, a widespread ingredient of the pheromone, citral, was absent. The biological significance of these secretions was enumerated by the authors as: a) arrestant of flight activity of female; b) suppresses mate refusal posture of female; c) unknown aphrodisiac effect; d) series specific signal; e) repellant to males- f) attract males during social swarm flight; g) predator deterrent.

Tactile senses

Both the larvae and imagines are covered with hairs or setae which serve as tactile sense triggers. In the imago, the setae are particularly abundant around the thorax, but their function has not been investigated further.

In the larvae the whole body is covered with setae of different lengths. Portier (1949) published a figure of the typical tactile hair, basal cells and secretory cell, and Eastham & Eassa (1955) discussed peculiar trichia which project into the lumen of the food canal of the proboscis.

It would appear that one of the functions of the body setae is defensive, as Johansson (1951) found that the hairs on all third instar larvae and older were too large for the hymenopterous endoparasite, *Apanteles glomeratus* (Linnaeus), to oviposit. This was however, aided by the mouth secretions and body movements of the larvae which prevent the female wasp from ovipositing. The setae on first and second instar larvae do not interfere with ovipositing (cf. Chapter 7).

The larval setae may also serve to give the larva a rough texture when being handled by bird or reptile predators, which may then discard the insect. Another possibility is that the setae are used to maintain the larva's own individual distance from its neighbour when they are feeding gregariously.

In a short discussion on "the travel of larvae"(*P. brassicae*), Allan (1943) stated that larvae were not geotaxic, as they ascend and descend in search of foodplants. Further evidence presented for this is the fact that they often choose pupation sites on several planes.

Orientation: silk

As the larva moves along, it moves its head alternately from right to left and at the same time lays a silk thread. The pattern of the silk path covers an area which is much wider than the larva which made it. Darchen (1967) published some excellent photographs of silk webs. Silk is produced throughout the rest of the larval stage, even during the wandering stage of the fifth instar larva. Eliot (1944) suggested that the position of the pupation site was governed by the time when the silk glands are not able to produce any more silk, following the wandering stage. However, no experimental evidence in support of this was given.

Silk would appear to serve four functions: a) helping to keep the larva on the foodplant and not being blown off; b) allowing the larva to regain its foodplant by walking up or along its defensive silk thread extruded when the larva falls off: c) securing the pupa in position by means of a silk girdle and silk pad; and d) possibly acting as a means of communication keeping larvae together. Ochmann (1933) published a photomicrograph of the origin and part of the middle of the silk girdle of *P. brassicae* which had been artificially teased out to show how it was made of several silk threads.

Whether or not larvae use the silk thread to keep together is still open to conjecture. When a silk path is laid it is difficult to determine which direction the larva has passed, and so from human means of observation it is difficult to see how the larva can determine the correct direction to follow. Perhaps the lost larva still has a greater chance of regaining its group if it follows in either direction, rather than none at all.

Ultrastructure of cuticular setae

Ma & Schoonhoven (1973) investigated the setae on the tarsi of *P. brassicae* with scanning electron microscopy (SEM) and transmission SEM. They found two types of setae: tactile setae and bristles which they called "A-type" structures, and the smaller "B-type" trichoid sensillae. The "A-type" structures include the very small microtrichia which cover the surface of the cuticle.

The "B-type" hairs are approximately 55 μm by 5.5 μm with a pore at the distal end which is 0.6 μm in diameter. There are more "B-type" setae on the tarsi of the females (197–263 – average of four specimens) than on males. "A-type" structures were present in about equal numbers on the mid- and hind-tarsi. The "B-type" structures are innervated by five bipolar neurones which at their distal end are divided into an inner and outer segment by a ciliary structure. The cilia have nine peripheral sets of double tubules.

Auditory sense

It has not been established whether the larva or imago of *P. brassicae* is responsive to sound. However, in many other insects sound is used for communication, and detection of sound pressure waves by body setae has been established. Such a study in *P. brassicae* has not been carried out as far as the author is aware. Heath (1940) suggested that sound might be involved in courtship of *P. brassicae*, as he found a pair of insects, the male on the "back" of the female, emitting a noise which he claimed could be heard 1.5 m away. However, this does not seem to be typical.

The only claim of auditory powers in *P. brassicae* was by Baier (1930), who was supported by Eltringham (1933). Baier observed in the wild that *P. brassicae* imagines apparently responded to the human voice. When he sang to his wild insects, they reacted by flinging their heads back (stage 1), raising their last few segments in the air (stage 2) and contracting the body along the anterior-posterior direction (stage 3). The response was carried out up to 10 m away from the sound, took five seconds to effect, and varied with the time of day. Isolated larvae made no visible response. The three characteristic attitudes illustrated in Baier's (1930) thesis (cf. Portier, 1949) are very reminiscent of attitudes taken up by larvae when stimulated by tactile stimuli picked up from vibrations of the leaf. Rothschild (1980, pers. comm.) recorded that late instar larvae of *P. brassicae* respond synchronously by head waving to a bird-like whistle.

On a light hearted note, Baines (1941) observed that imagines of *P. brassicae* appeared "very alarmed" at the sound of gunfire on the battle front in France in 1940. They would "rise and dither" along with other insects such as ladybirds.

Barometric sense

Pictet (1917, 1918) investigated the effect of atmospheric pressure of the rate of eclosion of seven species of lepidoptera, including *P. brassicae*. Although other factors are involved with inducing eclosion, i.e. temperature, he found that the time taken from the first appearance of wing patterns under the cuticle of the pupae until eclosion varied between species, and that for *P. brassicae* the range was 1–3 days. Pictet (1918) mentioned that there were seven factors which affected eclosion: a) the number of individuals eclosing drops to a few when the barometer falls; b) the number of individuals eclosing is directly related to the fall in the barometric pressure, i.e. a big depression is followed by increased rate of eclosion; c) a drop in pressure is sufficient to induce the pupae to eclose; d) when the pupae are ready to hatch and the pressure then rises, eclosion is delayed for either a day or until the pressure drops; e) when the pressure drops at a time when the wing patterns are showing, the time taken for their development is shortened; f) conversely, when the pressure increases when the wing patterns are already showing, the time taken for their development increases, and g) when the pressure stays constant for a long time the pupae are not able to mature.

Tulloch (1940) suggested that imagines of *P. brassicae* were also sensitive to barometric changes. He recorded that "whites", in particular

P. brassicae, become very sensitive and restless when the weather becomes very thundery and the barometer drops. The insects do not mate and are not interested in visiting flowers.

Ultra-violet reflectance

Ultra-violet reflectance is an important aspect of sexual recognition in *P. brassicae* (Chovet, 1974, 1977) and had already been identified in sexual recognition in *P. rapae crucivora* (Obara & Hidaka, 1964, 1968). The reflecting properties of butterfly wings and the role of ultra-violet rays in the vision of insects was investigated in 50 species of lepidoptera from 12 genera by Mazokhin-Porshniakov (1957). It was found that in general more ultra-violet reflection was present in the upper wing surfaces than the lower, and that yellow and orange areas reflected ultra-violet light most strongly.

In the case of *P. brassicae* the white areas absorb much of the ultra-violet light, but small differences in the order of 3.5% reflectivity occur between the sexes. Greenish-yellow areas on the lower wing surfaces were shown to have higher reflection coefficients (see Figure 3 *in* Mazokhin-Porshniakov, 1956), so too did androconial patches.

Both fore- and hindwings of both sexes of *P. brassicae* reflect ultra-violet light between 67–90%, in the range of 420–600 nm (Chovet, 1977). This is similar to the 70% reflectance of the female wing in both *P. rapae* and *G. rhamni* (Mazokhin-Porshniakov, 1957, cf. Lozina-Lozinskii, 1953). Chovet (*loc. cit.*) found that he could attract males to similarly sized and coloured model females in the wild; 50% of all males which passed the models were attracted to them and if the model was rotated at 10 revolutions per second, corresponding almost exactly to the wing beats of the female. Attraction of males to females or to each other is therefore due to a visual cue, and may account for the observation made by Chovet that up to 50 males may be found together; this would appear to be exceptional.

It must be borne in mind when considering ultra-violet reflecting substances in the wings of butterflies that pteridine pigments also fluoresce under ultra-violet. It does not appear that anyone has considered whether *P. brassicae* can detect the ultra-violet nectar-guides of flowers.

Thermoregulation

It is a well established fact that some butterflies, such as *Colias eurytheme* Boisduval, thermoregulate themselves by aligning their wings at right-angles to the rays of the sun. This enables their muscles to heat up to the required temperature for flight.

Thermoregulation in *P. brassicae* has not been studied, but Adkin (1918) did notice that on one occasion *P. brassicae* imagines were apparently sunning themselves on elm trees for more than an hour. Similarly, Sanderman (1946) noted *P. brassicae* basking on rocks in hot sunshine at 853 m in Breconshire, Wales. Nicholson (1933) noted that *P. brassicae* was a warmth seeker, preferring the edges of fields. Whether the insects concerned in all cases were thermoregulating, or were benefiting from some other sensory stimulation was not established.

Defense mechanisms

There are a number of defensive mechanisms which rely on the visual and gustatory systems of predators which become involved in all stages of *P. brassicae*.

Ovum

Ova of *P. brassicae* are brightly coloured and possibly act as a warning signal to birds. They have only been recorded as being eaten by two species of bird: *Passer domesticus* (Linnaeus) (House Sparrow) and *Sylvia borin* (Boddaert) (Garden Warbler) (cf. Table 16.1).

Larva

Young larvae are very susceptible to attack from hymenopterous parasites; older larvae have long setae, can walk faster and have additionally the advantage of a defensive secretion which they eject from the mouth when disturbed by a predator. When in groups, the combined effect of the larvae raising their heads as a response to an alighting bird also acts as a surprise deterrent. On the leaf, larvae are aposematic but cyptic when wandering over soil (Baker, 1970).

The green defensive secretions produced by larvae contains eight protein esterases (Clements, 1967). Pictet (1918) mentioned that the secretion can irritate the skin and he gives an extraordinary story of a French peasant who almost lost the sight of an eye when the liquid accidentally got into that organ. Pictet (*loc. cit.*) believed that the irritant factor of the "blood" caused birds, especially chickens, not to take larvae in the wild. The gregarious larvae of *P. brassicae* have a peculiar evil smell, "which proceeds from them" (South, 1936; Ford, 1945), the odour of which is intensified if the larvae are trodden upon.

To some birds *P. brassicae* larvae are offensively toxic (Chapter 16). Rothschild (1964) recorded that her pet Crow (*Corvus* sp.) remembered the distastefulness of the larvae for nine months. This is probably due to the mustard oil glycosides obtained from the various foodplants, which make the larva toxic (Rodriguez & Levin, 1975) (cf. Poulton, 1890).

Pupa

Pupae are usually regarded as cryptically coloured (but also see Chapter 8). No specific organs producing toxic secretions have been identified in the pupa but when they are injected into rabbits and mice they prove lethal whether or not the larvae were reared on a normal cabbage or cabbage-free diet (Marsh & Rothschild, 1973) (see also mustard oil glycosides in relation to artificial diets in Chapters 5 and 10). Experiments show that some birds can distinguish a disagreeable smell or taste which is unassociated with a toxic substance (Marsh & Rothschild, 1973). The only other possible form of defense is a visual one, possibly associated with sound production.

Imago

The black and white colours of the imago give *P. brassicae* an aposematic colouration, advertising the fact that they are toxic. When at rest, the yellow colour of the undersides of the hindwings serves to give it a cyptic colouration, particularly on sea kale (*Crambe maritima* L.). When tasted by predators, scales of the imagines probably dislodge and cause emetic reactions in the mouth or gut. Scales can also cause severe irritation of the eye, leading to swelling of the eyelids and occlusion of the eye, as the author can substantiate.

Haemolymph reactions

The proteins contained in the haemolymph of insects are often characteristic of the species and may be identified by their electrophoretic patterns. They also give clues to important biochemical changes which take place during development. Gysels (1975) used electrophoresis to compare the effects of introducing different foreign bodies into parasitised and non-parasitised larvae and pupae of *P. brassicae*, bearing in mind the distinction between "classic immune responses", typical of vertebrates, and ordinary defense reactions found in insects. Insects tend to respond to foreign bodies by surrounding them with haemocytes.

Gysels (*loc. cit.*) showed that, when a piece of cellophane was introduced into a pupa, it "caused melanin formation, necrosis of the adjacent fat cells and connective tissues, and hypertrophy of the digestive tract epithelium: the original cell layer becoming drastically stratified". Gysel pointed out that this may be termed either a neoplasm or "melanotic tumour" according to different schools of thought. After administration of benzidin, a recognised carcinogen in vertebrates, there resulted a loss of chromatophilia elements in the generative elements, and a thickening of the testis coat and follicle wall. Gysel confirmed the presence of increasing amounts of protein in the haemolymph of older larvae (see Chapter 10).

Chemoreception

Gustation in the larva

Pioneer work on phagostimulation in insects, particularly on *P. brassicae*, was carried out by Schoonhoven (Schoonhoven & Dethier, 1966; Schoonhoven, 1967, 1969a,b, 1970, 1973, 1977). He demonstrated, using an ingenious technique of inserting glass microelectrodes into various sensory areas, that the lateral sensillum on the maxillary palp of the larva contained cells which were responsive to sugar (= sucrose), salt, mustard oil glycosides, amino acids and anthocyanin (1969a,b). One cell each was sensitive to sugar and salt and two cells controlled detection of mustard oil glycosides.

Another receptor on the median sensillum styloconicum of the maxilla was found by Ma (1969) to be sensitive to feeding inhibitors, such as quinine chloride, strychnine nitrate, quinine sulphate. No effect was found with colchicine, caffein and D-salicin at 0.01 M concentrations. The receptor was found to give highly inhibitive responses with ecdysterone and ponasterone (juvenile hormone derivatives), but this could be neutralised by adding sinalbin.

The histology of four sensilla on the labrum of the larva were described by Ma (1972) (see Table 12.1) and electrostereograms of the campaniform sensillum, papilla like organ and microtrichia may be found in his thesis. Eight sensilla were identified on the tip of the maxillary palps.

Schoonhoven (1977) pointed out that the sensitivity of the chemosensory system of an insect may vary with age, feeding history, food deprivation, adaption rate and individual variation.

Gustation in the imago

Proboscis extension is initiated by solutions containing honey or sugar, but may also be supplemented by stimulation from chemosensory stimuli received from the tarsi of the imago. Plateau (1904–6) *in* Portier (1949) brought attention to the fact that the imago may be stimulated to extend its proboscis to *Salvia horminoides* Pourr. (Wild Clary) which is slightly aromatic. Details on the structure of the proboscis may be found in the chapter on morphology and anatomy (Chapter 7).

Tarsal sensitivity of imago

Ma & Schoonhoven (1973) demonstrated that the tarsi of female *P. brassicae* were sensitive to mustard oil glycosides (MOG), and the butterflies would oviposit on broad bean (which is not a usual foodplant) when their stems were dipped into a solution of 0.001 M sinigrin. Sinigrin was just as effective as glucotropaeolin tetramethyl ammonium salt (GTA) for inducing oviposition when compared with a control. Amputation of fore- and hindlegs as well as antennae did not prevent the imago from ovipositing.

An interesting experiment that suggested that the female selects the lower (abaxial) surface of the leaf by texture on which to oviposit the majority of ova was carried out by Rothschild (unpublished results). She found that if a cabbage leaf was experimentally turned upside down so that its lower surface is now uppermost, the female will still oviposit on the original lower surface.

B-type tarsal hairs on the fourth tarsomere of the females were shown to have two types of sensitive cells, called S1 and S2, which responded differently when stimulated by solutions from an electrode. The presence of another cell (presumed) appeared to be associated with a mechano-receptor as movement of the B-type hairs caused discharge in the mechano-receptors. Cells of the tarsi responded to sodium chloride and GTA but overall, there was poor response to MOGs, due possibly to other variable factors, such as the age of the imago. One of the sensory cells inside the B-type hairs was thought by these workers to be sensitive to water and another to salt.

Ma & Schoonhoven (*loc. cit.*) thought that the B-type cells detected sugar from water or they detected the chemical principle in the complex of stimuli inducing oviposition, or both.

Eastham & Eassa (1955) reported the presence of three types of sensilla on the proboscis of *P. brassicae*: a) spines on the outside; b) trichoid sensilla sparsely distributed; c) two pairs of modified trichoid sensilla, one near the base of the proboscis, the other about half-way up its length. The latter type of sensillum is made up of an extension of the hypodermis through which the trichium protrudes into the inside of the food canal. The function of the sensilla is not known, but it was established that they did not appear to function in proboscis movement.

The presence of mustard oil glycosides in plants, as feeding stimulants for larvae of *P. brassicae*, has been known for a long time (Verschaffelt, 1910; Gautier & Riel, 1919; Dethier, 1947; Thorsteinson, 1953; an account of acceptability of foodplants is found in Chapter 4).

David & Gardiner (1966) showed that nine mustard oil glycosides, four of which are present in cabbage (Table 12.3), acted as feeding stimulants when fed to larvae in semi-synthetic diets. They found surprisingly, that the two glycosides, glucocapparin and glucotropaeolin, which do not occur in cabbage, acted as the most effective feeding stimulants. Using frass drop frequency as a measure of acceptability they showed that an increase in concentration of any of the nine glycosides produced an irregular increase in frass production. David & Gardiner (*loc. cit.*) experienced difficulty with the larvae as they demonstrated acceptance and non-acceptance of the diet and would feed at different times. The tetraacetates of three of the glycosides acted as very weak feeding stimulants. It is interesting to note that Rothschild (1980, pers. comm.) has recorded larvae of *P. brassicae* eating their own parents.

The occurrence of mustard oil glycosides in various varieties of cabbage and the different stages of *P. brassicae* has been dealt with in Chapter 10 on biochemistry.

Table 12.3 Mustard oil glycosides as feeding stimulants. (from David & Gardiner, 1966).

---

*Glucoiberin
 Glucocheirolin
*Glucoerucin
*Sinigrin
 Glucocapparin
*Progoitrin
 Glucoconringiin
 Tetraacetyl glucosides as K salts
 Acetylglucocheirolin
 Acetylsinigrin
 Acetylgluconringiin
 Glucosides as tetramethyl ammonium salts
 Glucosinalbin
 Glucotropaeolin

---

*Present in cabbage leaves.

Feeding inhibitors

It is known that certain substances have a detrimental effect on growth and development of insects. In the present decade, work has been carried out on the effect of inorganic salts, alkaloids, steroids, and terpenes on feeding inhibition of *P. brassicae*: a brief summary of this appears in Chapman (1974); Schoonhoven & Jermy (1977).

Butterworth & Morgan (1971), while working primarily on the Desert Locust, *Schistocerca gregaria*, found that the hexanotriterpenoid substance called Azadirachtin ($C_{35}H_{44}O_{16}$), which is obtaind from the Neen or Nim Tree (*Azadirachta indica* A. Joss), showed only moderate activity against fifth instar larvae of *P. brassicae*. The larvae were offered the azadirachin on filter paper. No further details of this were given.

In 1972, Ma published his extensive work on feeding inhibitors in his thesis. He tested 22 different compounds against *P. brassicae* larvae by using leaf discs. Only nine of these substances gave a strong inhibitory

Table 12.4 Feeding inhibitors used on *P. brassicae*.

Compound	Authority
Azadirachtin	Butterworth & Morgan, 1971
*Berberin hydrochloride	
Betulin	
Caffein	
Colchincine	
*Conessine	
Conline	
d-salicin	
*Ecdysterone	
*Inokosterone	
Morine	
Morphine hydrochloride	Ma, 1972
Naringin	
Pilocarpin hydrochloride	
Picrotoxin	
Ponasterone A	
*Quinine hydrochloride	
*Solanine	
*Sparteine sulphate	
*Strychnine nitrate	
*2-thio uracil	
Tomatine	
Uric acid	

*Gave best results as feeding inhibitors (cf. Ma, 1972, p. 49, Table 10).

response, and those which were most effective were of an alkaloidal or steroidal nature with a high molecular weight. Ma tested the effect of three substances especially carefully: quinine, strychnine and ecdysterone. Strychnine could apparently still be detected by the larvae which had had their maxillary organs removed. With ecdysterone, if the concentration was increased to 2×10^{-4} molar then a 50% drop in the feeding response was noticed (Ma, 1972, Figure 39, p. 53), and thereafter a slow decrease in feeding inhibition occurred.

Biting rate and receptor activity

When *P. brassicae* larvae feed, they take a very short break between every two bites, and these inter-bite periods increase gradually until a maximum is attained every 10–15 seconds (Figure 12.4) (Schoonhoven, 1972). When the maxillae are removed, the breaks do not appear. The significance of this is that possibly the breaks allow the larva to disadapt its maxillary taste receptors, ready for the next bite.

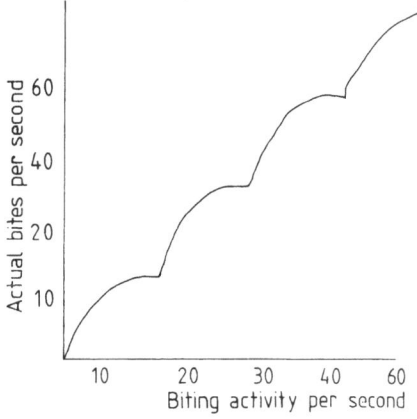

Fig. 12.4 Biting activity of *P. brassicae* larvae
(after Schoonhoven, 1972, by courtesy of Mr. B. Van Leeuwen)

References cited

Adkin, R., 1918. The abundance of white butterflies. *Entomologist* 51: 36–39.
—— 1942. *P. brassicae* and *P. rapae*. *Entomologist* 75: 91.
Ali, F.A., 1974. Structure and metamorphosis of the brain and suboesophageal ganglion of *Pieris brassicae* (L.) (Lepidoptera: Pieridae). *Trans. R. ent. Soc. Lond.* 125: 363–412.
Allan, P.B.M., 1943. The travel of larvae. *Entomologist* 76: 159–164.
Angersbach, D., 1975. The direction of incident light and its perception in the

control of pupal melanization in *Pieris brassicae. J. Insect Physiol.* 21: 1691–1696.

Baines, J.M., 1938. Butterflies stidulating. *Entomologist* 72: 220.

Baier, L., 1930. *Contribution ou physiologie de stridulation.* Dissertation, University of Friebourg.

—— 1941. Entomology under difficulty. *Entomologist* 74: 45–46.

Baker, R.R., 1970. Bird predation as a selective pressure on the immature stages of the cabbage butterflies, *Pieris rapae* and *Pieris brassicae. J. Zool. Lond.* 162: 43–59.

Barrer, P.M., 1969. *Functional studies on insect photoreceptors with special reference to lepidopteran larvae.* Ph.D. thesis, University of London (Imperial College).

Barth, R., 1938. Beobachtung über Massenwanderung des grossen Kohlweisslings (*Pieris brassicae* L.). *Anz. Schädlingsk.* 14: 60.

Baynes, E.S.A., 1961. Canary island butterflies. *Entomologist* 94: 260–262.

Beattie, A.J., 1972. Insect visitors to three species of violet (*Viola*) in England. *Entomologist's mon. Mag.* 108: 7–11.

Bedford, C.E., 1897. *Pieris brassicae* attracted by artificial flowers. *Entomologist* 30: 197–198.

Bedford, E.J., 1940. *Pieris brassicae* in the city. *Entomologist* 73: 195.

Behan, M. & Schoonhoven, L.M., 1978. Chemoreception of an oviposition deterrent associated with eggs in *Pieris brassicae. Entomologia exp. appl.* 24: 163–179.

Bergström, G. & Lundgren, L., 1973. Androconial secretion of three species of butterfly of the genus *Pieris. Zoon. Suppl.* 1: 67–75.

Brecher, L., 1924. Die Puppenpassung des Kohlweisslings, *Pieris brassicae* L. VIII: Die Farbanpassung der Puppen durch das Raupenauge. *Arch. mikrosk. Anat. EntwMech.* 102: 501.

Bünning, V.E. & Joerrens, G., 1960. Tagesperiodische antagonistische Schwankungen der Blauviolett- und Gelbrot-Empfindlichkeit als Grundlage der photoperiodischen Diapause-Induktion bei *Pieris brassicae. Z. Naturf.* 4: 205–213.

Butterworth, J.H. & Morgan, E.D., 1971. Investigation of the locust feeding inhibition of the seeds of the Neem Tree, *Azadirachta indica. J. Insect Physiol.* 969–977.

Chovet, G., 1974. *Méchanismes d'accouplement, structure et fonctionnement de l'appareil reproducteur mâle de Pieris brassicae (L.) (Lepidoptera: Pieridae).* Thèse, Diplome Doctorale 3 ème cycle. Université de Paris VI.

—— 1977. Les stimulus visuels dans le déclenchment de la parade nuptiale chez *Pieris brassicae* L. (Lepidoptera: Pieridae). Nouvelle interprétation de certains comportements grégaires. *C. r. Acad. Sci. Paris.* 284: 2127–2130.

Clements, A.M., 1967. A study of soluble esterases in *Pieris brassicae* (Lepidoptera). *J. Insect Physiol.* 13: 1021–1030.

Collenette, C.L., 1945. Orientation experiments on larvae of *Pieris brassicae* L. *Entomologist* 78: 33–36.

Dannreuther, T., 1946. Migration records, 1945. *Entomologist* 79: 97–110.

Darchen, R., 1967. Les constructions de la chenille de *Pieris brassicae* L. *Rev. comp. An.* (3) 61–76.

David, W.A.L. & Gardiner, B.O.C., 1966. Mustard oil glycosides as feeding stimulants for *Pieris brassicae* larvae in a semi-synthetic diet. *Entomologia exp. appl.* 9: 247–255.

Dethier, V.G., 1947. *Chemical insect attractants and repellants.* H.K. Lewis & Co. Ltd., London.

Dixey, F.A., 1932. The plume scales of the pierinae. *Trans. ent. Soc. Lond.* 80: 57–75.

Eastham, L.E.S. & Eassa, Y.E.E., 1955. Feeding mechanism of the butterfly, *Pieris brassicae* L. *Phil. Trans.* B. 239: 1–43.

El-Dakroury, M.S.I., 1972. *A study of the larval photoreceptors of Pieris brassicae.* Ph.D. thesis, University of London (Queen Mary College).

Eliot, N., 1944. Larval anthropomorphism. *Entomologist* 77: 22–27.

—— 1948. Some notes on the habits of butterflies. *Entomologist* 81: 64–69.

Eltringham, H., 1933. *The senses of insects.* Methuen, London.

Feltwell, J.S.E., 1977a. A choice chamber for small invertebrates. *Nat. Sci. Schools* 15: 72–74.

—— 1977b. Migration of whites. *Entomologists Mon. Mag.* 112: 88.

Ford, E.B., 1945. *Butterflies.* Collins, London.

Frantsevitch, L.I. & Pichka, V.E., 1976. The size of the binocular zone of the visual field in insects. (In Russian.) *Zh. Evol. biokhim. Fiziol.* 12: 461–465.

Frohawk, F.W., 1936. Late appearance of *Pieris brassicae* larvae. *Entomologist* 69: 63.

Gautier, C. & Riel, P., 1919. Sur l'alimentation des chenilles des genres *Pieris* et *Euchloe. C. r. Acad. Biol.* 82: 1371–1374.

Gysels, H., 1975. electrophoretical and histochemical investigations on some noxious Lepidoptera. *Acta Zool. Pathol., Antwerp* (62) 129–141.

Götz, B., 1936. Beiträge zur Analyse des Verhaltens von Schmetterlingsraupen beim Aufsuchen des Futters und des Verpuppungsplatzes. *Z. vergl. Physiol.* 22: 429–503.

Hamm, A.H., 1948. Butterfly and other visitors to Michaelmas Daisies. *Entomologist's mon. Mag.* 84: 91–93.

Heath, J., 1940. Courtship of *Pieris brassicae. Entomologist* 73: 235.

Hertz, M., 1927. Der Flug des Kohlweisslings über ein Feld. *Biol. Zbl.* 47: 569–570.

Heslop, I.R.P., 1956. White varieties of two Vanessid butterflies. *Entomologist* 89: 198–199.

Ilse, D., 1928. Über den Farbensinn der Tagfalter. *Z. vergl. Physiol.* 8: 658–692.

—— 1937. New observations on responses to colours in egg-laying butterflies. *Nature, Lond.* 140: 544–545.

Johannson, A.S., 1951. Studies on the relation between *Apanteles glomeratus* L. (Hymenoptera: Braconidae) and *Pieris brassicae* L. (Lepidoptera: Pieridae). *Norsk. ent. Tidsskr.* 8: 145–186.

Kudrna, O., 1972. On some Moroccan butterflies. *Entomologist's Rec. J. Var.* 84: 267–268.

—— 1973. On the status of *Pieris cheiranthi* Hübner (Lepidoptera: Pieridae). *Ent. Gaz.* 24: 299–304.

Littlewood, F., 1912. Collecting in Westmoreland, 1911. *Entomologist* 45: 158–161.

Longstaff, G.B., 1912. *Butterfly hunting in many lands.* Longmans, Green & Co.

Lozina-Lozinskii, L.K., 1953. The wing of a butterfly as a receptor of infrared radiation. (In Russian.) *Akad. nauk. SSSR. Dok.* 93: 369–372.

Ma, W.C., 1969. Some properties of gustation in the larva of *Pieris brassicae. Entomologia expl. appl.* 12: 584–590.

—— 1972. *Dynamics of feeding responses in Pieris brassicae L. as a function of chemosensory input: a behavioural, ultrastructural and electrophysiological study.* Thesis. Agricultural University, Wageningen, Netherlands.

Ma, W.C. & Schoonhoven, L.M., 1973. Tarsal contact chemosensory hairs of the large white butterfly, *Pieris brassicae* and their possible role in oviposition behaviour. *Entomologia expl. appl.* 16: 343–357.

Marsh, N. & Rothschild, M., 1974. Aposematic and cryptic lepidoptera tested on the mouse. *J. Zool. Lond.* 174: 89–112.

Mazokhin-Porshniakov, G.A., 1957. Reflecting properties of butterfly wings and role of ultra-violet rays in the vision of insects. *Biofizia* 2: 352–362.

Miles, P.M., 1957. Immigrant lepidoptera observed in Wales during 1954 and 1955. *Entomologist's mon. Mag.* 93: 184–188.

Minnich, D.E., 1924. The olfactory sense of the cabbage white. *J. exp. Zool.* 39: 339–356.

Muller, A., 1925. *Buddlea variabilis*, a specific bait-plant for butterflies. *Nach. PflannzenSchutz.* 5: 11.

Nicholson, C., 1933. Immigrant lepidoptera, 1932. *Entomologist* 66: 67–68.

Obara, V. & Hidaka, T., 1964. Mating behaviour of the cabbage white, *Pieris rapae crucivora*. I. The "Flutter response" of resting males to flying males. *Zool. Mag. Tokyo* 73: 131–135.

—— 1968. Recognition of the female by the male, on the basis of ultra-violet reflection in the white cabbage butterfly, *Pieris rapae crucivora* Boisduval. *Proc. Jap. Acad.* 44: 829–832.

Ochmann, A., 1933. Aus der Spinnstube einiger heimischer Raupen. *Int. ent. Z.* 27: 249–252.

Oltmer, A., 1968. Die Steuerung des Melanineinbaus in das Farbmuster der Kohlweisslingspuppe *Pieris brassicae* L. *Wilhelm. Roux. Arch. EntwMech. Org.* 160: 401–427.

Pictet, A., 1917. Influence de la pression atmosphérique sur le développement des Lépidoptères. *Arch. Sci. phys. nat.* 44: 413–454.

—— 1918. Les éclosions de papillons et la pression barometrique. *Bull. Soc. lépidopt. Genève* 4: 67–74.

Plateau, F., 1904–1906 *in* Portier (1949).

Portier, P., 1949. *Biologie des Lépidoptères*. Lechavalier, Paris.

Poulton, E.B., 1890. *The colours of animals*. Kegan, Paul, Trench, Trübner & Co. Ltd., London.

Rodriguez, E. & Levin. D.A., 1975. Biochemical parallelisms of repellents and attractants. *In: Biochemical Interaction between plants and insects*. Recent advances in phytochemistry. 10: Ed. J.W. Wallace & R.L. Mansell. 214–270.

Rothschild, M., 1964. An extension of Dr. Lincoln Brower's theory on bird predation and food specificy together with some observations on bird memory in relation to aposematic colour patterns. *Entomologist* 97: 73–78.

Rothschild, M., Gardiner, B.O.C., Valadon, L.R.G. & Mummery, R.S., 1975. The Large white butterfly, oviposition cues, control and changes in colouration. *Proc. R. ent. Soc. Lond.* 40: 13.

Rothschild, M. & Schoonhoven, L.M., 1977. Assessment of egg load by *Pieris brassicae* (Lepidoptera: Pieridae). *Nature, Lond.* 266: 352–355.

Sandeman, R.G., 1946. Autumn butterflies in Breconshire. *Entomologist* 79: 116–117.

Schlieper, C., 1928. The distribution of brilliancy in the spectrum, with reference to various insects. *Z. verg. Physiol.* 8: (2) 281–288.

Schoonhoven, L.M., 1967. Chemoreception of mustard oil glycosides in larvae of *Pieris brassicae*. *Koninkl. Ned. Akad. Wetensch. Proc. Ser. C.* 70: 556–568.

—— 1969a. Amino-acid reception in larvae of *Pieris brassicae* (Lepidoptera). *Nature, Lond.* 221: 1268.

—— 1969b. Gustation and foodplant selection in some lepidopterous larvae. *Entomologia expl. appl.* 12: 555–564.

—— 1970. Hoe Herkennen Insekten Hun Voedselplant? *Vakbl. biol.* 52: 129–135.

—— 1972. Some aspects of host selection and feeding in phytophagous insects. *Insect and Mite Nutrition*, North Holland.

—— 1973. Plant recognition by lepidopterous larvae. *Symp. R. ent. Soc. Lond. No. 6. Insect Plant Relationships.* 87–99.

—— 1977. Insect chemosensory responses to plant and animal hosts. *In: Chemical Control of Insect Behaviour.* Ed. H.H. Storey & J.J. McKelvey. John Wiley & Sons. pp. 7–14.

Schoonhoven, L.M. & Dethier, V.G., 1966. Sensory aspects of host plant discrimination by lepidopterous larvae. *Arch. Neerl. Zool.* 16: 497–530.

Schoonhoven, L.M. & Jermy, T., 1977. A behavioural and electrophysiological analysis of insect feeding deterrents. *In: Crop Protection agent – their biological evaluation.* Ed. N.R. McFarlane. pp. 133–146.

Scott, H., 1921. *Colias edusa, Pyrameis atalanta*, and *Aglais urticae* in Cambs., Guernsey, and N. France. *Entomologist* 54: 18–19.

—— 1933. Random notes from N. Oxfordshire, 1933. *Entomologist's mon. Mag.* 69: 280.

Seugé, J., 1973. Recherches du rôle des stemmates dans la perception de la photopériode et l'induction de la diapause, chez *Pieris brassicae. Bull. Soc. zool. Fr.* 98: 435–440.

SLENHS (South London Entomological and Natural History Society), 1914. March 12, Meeting. *Entomologist* 47: 158.

South, R., 1936. *Butterflies of the British Isles.* Warne, London.

Speyer, W., 1955. Kohlweisslinge–Notizen. *Z. Pflkrankh. Pflpath. Pflschutz.* 62: 552–560.

Sudah, M.I.E., 1970. *Some effects of larval age and leaf age on the feeding behaviour of Pieris brassicae L. (Lepidoptera: Pieridae).* Msc. Thesis, University of Bangor, Wales.

Thorsteinson, A.J., 1953. The chemotactic responses that determine host specificity in an oligophagous insect (*Plutella maculipennis* (Curtis): Lepidoptera). *Can. J. Zool.* 31: 52–72.

Tulloch, B., 1928. Valerian versus ivy blossom. *Entomologist's mon. Mag.* 61: 160.

—— 1940. Do butterflies get "nerves". *Entomologist* 73: 46–47.

Valletta, A., 1954. An unusual pairing between *Pieris* and *Colias* (Lepidoptera). *Entomologist* 87: 38.

Verschaffelt, H., 1910. Plantenkunde. De oorzaak der roedselkeus bij eenige plantenetende insecten. *Kon. Akad. wet., Amsterdam* 19: 595–600.

Warren, B.C.S., 1961. The androconial scales and their bearing on the question of speciation in the genus *Pieris* (Lepidoptera). *Entomol. Ts. Arg.* 82: 3–4.

Westwood, J.O., 1854. *The butterflies of Great Britain.* W.S. Orr & Co. Paternoster Row, London.

Wigglesworth, V.B., 1967. *The principles of Insect physiology.* Methuen & Co., London.

Wojtusiak, R.J., 1930. Weitere Untersuchungen über die Raumorientierung bei Kohlweisslingsraupen. *Bull. int. Acad. pol. Sci. Lett.* 1930: 631–655.

Further references

Auel, H., 1912. *Z. wiss. Insck. Biol.* 8: 258–260 (defense mechanisms).

Burkhardt, D., 1964. *Adv. Insect Physiol.* 2: 131–173 (colour discrimination).

Demoll, R., 1909. *Arch. ges. Physiol.* 129: 461–475 (pigment migration).

Goldsmith, T.H., 1964. *In: Physiology of the Insecta.* Ed. M. Rockstein. pp. 397–462 (visual system of insects).

Hocking, B., 1960. Defence research Board, Canada (bibliography of smell in insects).

Levinson, H.Z., 1976. The defensive role of alkaloids in insects. *Experimentia* 32: 408–411.

Newman, E., 1868–9. *Entomologist* 4: 130 (imagines attracted to water).

Nolte, H.W., 1949. *Der Kohlweisslinge.* Ziesem, Wittenberg (behavioural aspects of imagines).

Querci, O. & Romei, L., 1945. *Fla. ent.* 28: 20–21 (effects of solar energy).

Sanchez, D., 1916. *Trab. Lab. Invest. biol. Univ. Madr.* 14: 189–231 (retina development).

Sanchez, D.S.Y. & Sanchez, D., 1918. *Trab. Lab. Invest. biol. Univ. Madr.* 16: 213–278 (nerves of the retina).

Schneider, H., 1923. *Zool. Anz.* 56: 155–160 (hairs).

Schoonhoven, L.M., 1976. *Symp. Biol. Hung.* 16: 261–266 (chemosensory variability).

13. Economic importance

"... we forget our enmity to the 'filthy grub', which destroys so many of our useful products of our garden, when we see it metamorphosed into an ember of Innocence."

Peter Rylands writing in *The Naturalist* (1837)

Introduction

In the first part of this chapter, past and present depredations of *Pieris brassicae* have been collected together by country. It will be seen, however, that information is lacking from a number of countries and that there is great contrast between the amount published about different countries. This review of depredations by *P. brassicae* thus serves to highlight those areas where further information is needed. Helpful comments of many entomologists have been incorporated in this chapter to give a current assessment of the state of *P. brassicae* in the various countries.

Additional information regarding depredations can be inferred from the comments made by workers who have discussed distribution and migration of *P. brassicae* in their own countries; here the reader is referred to chapters 2 and 11.

The second half of the chapter assesses the current damage caused by *P. brassicae* throughout the whole of its world range on man's major crops. This is expressed in terms of financial loss and total tonnage of crops lost per annum.

A number of points must be borne in mind when considering the importance of *P. brassicae* as a pest. A female imago can lay up to 500 ova in the wild which could produce a potential 250,000 individuals in the second generation (Gardiner, 1963). However, through the attacks of naturally occurring pathogenic protozoa, viruses, bacteria, fungi, predators and parasites, a high mortality of the progeny always occurs.

One of the keys to the success of *P. brassicae* is that the larva is oligophagous and feeds on a variety of foodplants, many of which contain mustard oil glycosides which act as feeding stimulants. These foodplants comprise mostly members of the Cruciferae. Another important feature of *P. brassicae* is that in some areas it can sustain itself on wild brassicas in the first generation before the maturation of the summer cabbage crops.

Depredations by country

Western Europe

Austria. Pieris brassicae was reported to be a serious pest on cabbage particularly in August and September 1973 (Glaeser, 1974) and although it is found in the mountains it is only regarded as a pest at lower altitudes. More recent information is that *P. brassicae* is a pest in five regions of Austria in the north and west of the country: near Linz, in Niederösterreich, Eisenstadt, Burgenland and near Graz, where in all cases the insect has to be strictly controlled (Berger, 1975, pers. comm.).

Belgium. Pieris brassicae was earlier regarded as a pest (Lampere, 1907) but is currently not considered to be of economic importance (Depireux, 1975, pers. comm.).

France. There are few reports of depredations of brassicas although there must have been numerous occurrences. An old report of Marchal (1913, 1914) gives accounts of *Pieris* species involved in depredations in the *départements* of Meurthe and Moselle. Recently Brussels sprouts were reported to have been skeletonised in a relatively small vegetable patch near the Camargue (Bouches du Rhône) (Feltwell, 1977a).

Germany BRD. Great invasions of *Brassica* crops used to occur along the north German coastline which receives the southerly migrating pierid insects. The migration is supposed to commence during the last two weeks of July and the first two weeks of August. The author was in the region around Kiel during this period in 1975 but did not witness any devastation of cabbage crops by *P. brassicae* (Feltwell, 1977b). In fact the cabbage in two fields comprising eight hectares was found to be totally unfit for human consumption due to *P. rapae* larvae which were on every plant.

Today much of the region is under wheat or pasture and Mr. George

Waach of the Department of Zoology, University of Kiel-Kitzberg, who was Professor Blunck's technical assistant (Blunck was also at Kiel-Kitzberg and wrote many papers on *P. brassicae*) was able to locate exactly the region where devastation of cabbage crops occurred in the past.

There is now little cultivation of cabbage in this part of northern Germany and most of the production of cabbage is carried out in more northerly parts of Schleswig-Holstein, particularly Ditmarschen and on the island of Fehmarn. It would be interesting to study the devastations of brassicas in these regions.

Germany DDR. One of the earliest accounts is recorded for Prussia (Sachtleben & Pape, 1922). *Pieris brassicae* was a serious pest in central Germany in 1947 (Eichler, 1948). Godan (1949) reported a mass outbreak of *P. brassicae* larvae in the French sector of Berlin on red and white cabbage, cauliflower and Savoy cabbage in September 1949, where the density of larvae reached 100 per square metre. Only red cabbage was able to complete its development following this attack.

Klemm (1954) published maps of Germany DDR showing outbreaks of attack by *P. brassicae* in 1951, 1952 and 1953. In the first two years, "strong" attacks occurred in the north adjacent to the Baltic, and in 1953 strong attacks were experienced throughout the country from the Baltic to the border with Czechoslovakia, with the exception of the regions east of Berlin and the mountains in the south-west.

Great Britain and Ireland. There are several accounts of great damage to crops through the actions of *P. brassicae* larvae. Turnips in Essex were attacked (Cansdale, 1876), Brussels sprouts in East Anglia (Theobald, 1909), swedes in Wales (Walton, 1916), turnips and cabbage stripped at the Lizard, Cornwall in 1920 (Glanville-Clutterbuck, 1941) and about three hectares of kale defoliated (Anderson, 1909). Table 2.4 lists comments from those who have discussed the abundance of *P. brassicae* from the middle of the 19th century and this indicates the degree of damage to crops which has been caused. Further information regarding seasonal damage by county is available in the Ministry of Agriculture, Fisheries and Food (MAFF) *"Monthly summaries of Insect and Allied Pests occurring in England and Wales"*, which have been published since January 1948, and in the MAFF miscellaneous publications (1920, 1921, 1923) and Bulletins (1936, 1939, 1952, 1954, 1960, 1964, 1968, 1974, 1976).

Italy. *Pieris brassicae* is not generally regarded as a pest species in Italy but does attack small garden plots (Balletto, 1978, pers. comm.).

Malta. P. brassicae was recorded as a serious pest on all cultivated crucifers and its numbers were thought to be augmented by large numbers of immigrants which arrived from the north African coast (Borg, 1929, 1932). Today the insect is still very common on the island (Bonnett, 1976, pers. comm.).

The Netherlands. In The Netherlands *Pieris brassicae* is not generally regarded as a pest, but it may be more destructive in some years depending on the weather; in cold and wet years, e.g. 1974, there were not many insects, but they occur in greater abundance in warm and dry years. De Jong (1978, pers. comm.) stated that *P. brassicae* was not very injurious to crops and Lempke (1978, pers. comm.) mentioned that it was only a pest in kitchen gardens, with effective control on most *Brassica* crops with the use of insecticides.

Switzerland. As a result of a large immigration of *Pieris brassicae* in 1917 cruciferous and umbelliferous crops such as carrot and parsley throughout Switzerland were almost devastated (Pictet, 1918). Larvae were so thick in the fields that up to 983 larvae per m^2 were recorded from a dozen fields in Geneva and Yverden (France). The cause of the massive numbers of larvae was thought to be due to the lack of their principal parasite, *Apanteles glomeratus* Linnaeus). It was reported by Berne (1925) that during 1917 the Federal Government of Switzerland paid out £1,000 ($2240, at 1980 rates) for the collection and destruction of *P. brassicae* larvae.

Fennoscandia

Finland. Reported earlier as a pest on cabbage (Vappula, 1965; Linnaniemi, 1935), and subsequently recorded as numerous on cabbage in southern Finland in 1928, 1936 and 1939 (Kanervo, 1946).

Norway. Pieris brassicae was reported to be a pest on cabbage in 1916 and 1924–1925 in east and central Norway (Schøyen, 1918, 1926).

Sweden. Pieris brassicae was noted as a pest in 1912–1916 (Tullgreen, 1918) and today is regarded as a serious pest on all kinds of cabbage in southern and central Sweden (Douwes, 1975, pers. comm.).

Eastern Europe and USSR

Bulgaria. Noted as a serious pest by Karadjov (1972).

Czechoslovakia. It was estimated that *Pieris brassicae* causes as much as

10% loss of all *Brassica* vegetables in Czechoslovakia each year, and that it is a more serious pest than either *P. rapae* or *P. napi*, preferring "biotopes under human, anthropogenic and agricultural influence" (Spitzer, 1975, pers. comm.).

Poland. Cabbage was reported as destroyed in both 1923 (Garbowski, 1925) and 1924 (Woroniecka, 1924). More recently Starega (1976, pers. comm.) stated that the second generation which feeds on crucifers causes much damage to crops Przybylski (1968) made a five year study of the depredations of *P. brassicae* in south-east Poland during 1962–1967 and calculated that on average the greatest danger from *P. brassicae* was on about August 9th each year; this time coincides with the full ripening of the apple (*Malus domesticus* Barbosa var. *Inflancka*). This he described as a "phytophenological phenomenon antecedent indicator of *P. brassicae* devastation".

Rumania. Kornfeld (1935) noted *Pieris brassicae* ovipositing on *Soya* grown in the Danube delta but the larvae inflicted little injury to the crop and moved onto other foodplants. Attacks on cabbage in Rumania have been summarised by Mustaţa & Andriescu (1973). In 1949 larvae of *P. brassicae* in the north of Moldavia (NE Rumania) and the south of Bukovina nearly completely destroyed the cabbage of the region and the local population had to go as far as Iasi and Roman (120 km to the SE) to find winter cabbage (Constantineanu *et al.*, 1957, & Mustaţa, 1970). Mustaţa & Andriescu (*loc. cit.*) also found cabbage damaged in the region of Bicazul Ardelean in the valley of Bicaz (near Roman) in 1959. There were more serious outbreaks in 1848–49, 1955–56, 1962–63 and 1970–71. Manolache *et al.*, (1969) noted that outbreaks of *P. brassicae* were always more serious in the north of the country.

Turkey. Pieris brassicae was recorded as a pest by Asena (1974).

USSR. There are many accounts of damage caused by *Pieris brassicae* in the USSR, covering a long period up to the present day. No other country appears to have had such severe damage over the last 20 years.

One of the earliest reports was by Guerne (1894) who gave an account of the thousands of *Pieris brassicae* larvae wandering over a railway line at Kief (*sic*), causing a locomotive to be halted and squashing the larvae "like pâté". In 1914 Plotnikov (1914) reported that in the previous year *P. brassicae* larvae destroyed 90% of a cotton crop growing in Ferghana and Transcaucasia and injured maize and lucerne. This report is the only one known to the author which cites these atypical foodplants and must be regarded with scepticism.

The depredations wrought by *P. brassicae* in the USSR during the first two decades of this century can be considered chronologically: between 1912–16 damage was caused in Volkynia (Ksenjopolsky, 1914), Bessarabia (Vitkousky, 1915), SW of Moscow (Svjatovitch-Bielikova, 1914), Tashkent (Plotnikov, 1915), Kursk, Astrakan and Don (Spassky, 1916). From 1920–26 there were reports from Lithuania (Mostauskis, 1923), Urals (Kharitonov, 1926), Azerbaijan (Rodionov, 1924) and Astrakan (Shembel, 1922). *P. brassicae* was reported as a pest in central USSR in 1930 (Yaroslavtzev, 1931) and in Armenia in 1960 where a total of 60 cabbage pests were listed by Safaryan (1969).

Pieris brassicae is still a major pest of cruciferous crops throughout the whole of European USSR and east to the Urals and in the Caucasian Republics, where it periodically multiplies in large numbers (Osmolevski, 1964). Kadamshoev (1971) stated that the first generation of *P. brassicae* feeds on wild brassicas, while the second generation feeds on cultivated crucifers. In Transcaucasia up to five generations a year may occur.

In Lithuania *P. brassicae* is a pest in vegetable gardens (Jakimaricius, 1974) and during 1971–1973 it was responsible for a 62–96% loss of cabbage in the district of Kapsukas, Vilnius and Ignalina (Zhukauskene & Misyalyunene, 1976).

In south-west USSR both *P. brassicae* and *P. rapae* are responsible in some years for a 50% loss in the white cabbage crop (Voltikov, 1975). In the Transcaucasian Republics west of the Caspian Sea, *P. brassicae* is an important cruciferous pest (Shapiro, 1976). Damage has also been reported throughout Kazakhstan and in the Chu Valley in Kirgizia to the east of the Aral Sea (Moiseeva et al.,1975).

Yugoslavia. Severe attacks by *P. brassicae* were reported in Belgrade in 1924 by Voukassovitch (1926).

Middle East and North Africa

Iraq, Syria, Lebanon. Pieris brassicae causes serious damage to crops in these countries, but during the hot summer months their numbers are small (Talhouk, 1969).

Israel. Buxton (1923) recorded *P. brassicae* as a pest on cabbage and cauliflower and said that the insect could be caught around Jerusalem. *P. brassicae* was stated by Bodenheimer (1935) to feed on *Capparis* and *Beta* only in the neighbourhood of human settlements. Balakowsky & Mesnil (1936) however, did not list *P. brassicae* as a vegetable pest in Israel.

Libya. An earlier reference stated that *P. brassicae* was a pest on cauliflower at Benghazi (Zanon, 1919).

Indian subcontinent

India. Early reports of *Pieris brassicae* as a pest in India are by South (1891), Lefroy & Howlett (1909) and Lal (1946). Rataul (1959) reported *P. brassicae* as a pest on cabbage, cauliflower and radish in NE India. More recently Srivastara (1970) reported that both rape and mustard seed suffered from attacks of *P. brassicae*, but Lall (1963) in his book on vegetable pests in India does not list *P. brassicae*.

Far East

Afghanistan. *Pieris brassicae* was recorded as a pest on cruciferous crops on the Kabul plateau by Ahmad (1940).

China. In China *P. brassicae* is regarded as being as important a pest as it is in Great Britain (Chuang-Lung, 1976, pers. comm.). The subspecies *P. b. nepalensis* occurs in the Yunnan region of south-west China.

Nepal. The subspecies *P. b. nepalensis* is recorded as a pest of certain vegetables in the Himalayas (Chuang-Lung, 1962).

South America

Chile. Since its introduction to South America in late summer 1971, *P. brassicae* was reported as a pest on crucifers in June 1973 in provinces near to Valparaiso, chile (Gonzalez, 1972) and has now spread to many areas (Campos, 1978, pers. comm.). It is considered to represent a threat to the rape (*Brassica rapa* var. *oleracea* Linnaeus) cash crops grown in the south of the Quillota valley (Feltwell, 1978).

Estimate of world damage caused by *P. brassicae*

Crops attacked

The success of *P. brassicae* as a pest is certainly in part due to the larva being oligophagous, thus being able to sustain itself both on wild foodplants as well as man's cultivated crops. The majority of the foodplants of *P. brassicae* are members of the Cruciferae, but other frequently selected species belong to the Capparaceae, Resedaceae and the Tropaeolaceae (Chapter 4, Tables 4.2, 3, 4). There are a number of crops amongst these and other families.

The high risk crops which are most susceptible to depredations of the

Large White in Europe and USSR are cabbage, cauliflower, Brussels sprouts, rape, Kohlrabi, turnip and swede, all of which belong to the Cruciferae (Table 3.4). In the Middle East, capers and beet are liable to attack.

It is often found that more damage occurs at the periphery of the crop than at the centre (Friederichs, 1931). In fact Pollard *et al.* (1974) thought that cabbage butterflies may prefer the calm air by hedges and thus lay more ova at the margins of the fields. Even slight damage can make the plant unsightly to the consumer and in particular larval frass spoils the market value of a crop.

In a series of experiments designed to find out the minimum amount of leaf area of *Brassica napus* var. *napobrassica* (*syn.* var. *rapifora*) (Swedish Turnips) of the variety "Korund" eaten by *P. brassicae* to cause irreversible damage, Daebler *et al.* (1973) found that if 25% of the plant is eaten, 10–15% of the crop would be lost. This they calculated to be equivalent to the effect of two to three *P. brassicae* larvae from the second to fifth instar.

Brassica production

Great Britain was expected to produce about 1 m tonnes of cabbage, Brussels sprouts, cauliflower, turnips and swedes during 1976–77, with a market value of £99 m ($222 m)* (Table 13.1). During the same period imports of cabbage, cauliflower and brocolli amounted to about £4 m ($9 m).

Surprisingly, in Great Britain there are no figures available of the amount of brassicas grown in allotments where often, because *P. brassicae* is uncontrolled, damage is severe. The number of allotments in Great Britain is rapidly decreasing. In 1945 there were 1.5 m (Handiman Which, 1975), in September 1955 there were 1,004,656 (Best & Ward, 1956), in September 1973, "*The Times*" quoted 467,755 and in 1975 Handiman Which (*loc. cit.*) quoted about 333,000. Inspection of allotments today shows that brassicas are still grown on at least 5–10% of the total allotment area.

Information on the production of brassicas from gardens is also limited. Best & Ward (*loc. cit.*) stated that 374 kg of cabbage, cauliflower and Brussels sprouts can be produced from a 10 rod plot (1 rod = 5.5 sq. yards). They quoted £4.17.6d (£4.87½) as the value of the total yield of vegetables from each house with a garden in 1952, which would be equivalent to at least double that value today. According to Herwin (1977,

---

*£1 = $2.24, 6th January 1980.

Table 13.1 Brassica production in Great Britain (based on the Ministry of Agriculture, Fisheries and Food Statistics 1966/77 Forecast; * only figures available 1975/76).

Crop	Hectares	Tonnes	Value of output (£)
cabbage	24,581	468,900	40,810,000
imported	—	14,800*	1,646,000*
Brussels sprouts	13,351	141,700	25,506,000
cauliflower	14,340	193,100	21,394,000
imported (+brocolli)	—	16,500*	2,018,000*
(+Channel Isles)	—	6,000*	638,000*
turnips and swedes	5,239	124,000	7,428,000
Totals	57,511	965,000	99,440,000

Table 13.2 Cabbage and cauliflower production in Europe, USSR & Asia, 1974.

Country	Tonnes of cabbage	Tonnes of cauliflower
Europe	7,689,000	2,128,000
Asia and China	3,959,000	851,000
North America	1,504,000	151,000
South America	180,000	60,000
Europe, USSR	3,218,000	298,000
Total	16,550,000	3,488,000

(based on FAO (1975) *Production Yearbook* for 1974.)

pers. comm.) at least £0.75 m ($1.68 m) worth of crops is lost annually to the depredations of this pest in allotments and gardens.

Data from the current FAO (1975) *Production Yearbook* shows that 18 m tonnes of cabbage and cauliflower were produced in 1974 in Europe and Asia and about 2 m tonnes were produced in the Americas (Table 13.2). World figures for the production of swedes and turnips in north and south America are not available.

Crop losses

In Great Britain the Large White is now less of a problem than before 1955 when an epizootic virus naturally decimated the resident population Added to this, British farmers regularly use DDT and organophosphorus insecticides, such as mevinphos and trichlorphon, to kill cabbage caterpillars. However, in areas such as north and central Europe, USSR, India, Nepal and China, *Pieris brassicae* is certainly a major pest today.

In order to assess on a world-wide scale the damage caused by *P. brassicae* and its subspecies, it is necessary to take into account the following factors: a) attacks are often extremely localised and in small areas crop losses can be 100%; b) the strongly migratory habit of *P. brassicae* helps to disperse it over wide areas and may help it to infest areas previously free from attack; c) the larvae can develop on a wide variety of foodplants, including both cultivated and wild crucifers; d) populations are regulated by a large number of predators, hymenopterous parasites and micro-organisms such as viruses, bacteria, protozoa and fungi; and e) warm and dry years usually result in an increase in the population of this insect.

The above factors make an assessment of the overall degree of attack by this pest extremely difficult. Furthermore, there are communication difficulties in obtaining first-hand information about the degree to which *P. brassicae* attacks crops in the various countries within the world range and on the 20 or so species of crop attacked.

However, taking four brassicas most heavily attacked in Great Britain as an example (Table 4.3), a loss of 1% due to the Large White would result in a loss of approximately £1 m ($2.2 m) worth of crop in 1977. In countries such as Austria, Poland and Sweden where damage is more severe many millions of pounds worth of crop must be lost to *P. brassicae*.

Europe and USSR produced 11 m tonnes of cabbage and cauliflower in 1974 (Table 14.3). If a 4% theoretical loss of yield caused by *P. brassicae* is applied for this region, the tonnage lost would be about 0.35 m tonnes or about £35 m ($78 m) based on 1977 prices. (Forecast average of the two is nearly £100 ($224)/tonne.) In Asia and China, the production of cabbages for 1974 was about half that of Europe and for north and south America about one third. Thus one could tentatively calculate that at least £60 m ($134 m) worth of cabbage and cauliflowers are lost annually in Europe and Asia. Bearing in mind that this figure has not been corrected for inflation and also omits the other 18 species of crop attacked and the crops attacked in non-commercial areas, gardens and allotments, the figure for destruction to all crops could easily be nearer to £100 m ($224 m). A potential threat of about £10 m ($22 m) loss per annum in the Americas is posed by the presence of *P. brassicae*.

It is therefore clear that a considerable amount of potential human food is eaten each year by the larvae of *P. brassicae* and that the major areas of infestation are central Europe and Asia. The possibility of returning £110 m ($246 m) worth of food material to man each year is an incentive to look for an effective means to control the insect; even a 50% reduction in the world population of *P. brassicae* would return £30–50 m ($67–112 m) worth of crops to hungry mouths in both developed and underdeveloped countries.

References cited

Ahmad, T., 1940. A survey of the insect fauna of Afghanistan. *Indian J. Ent.* 2: 159–176.

Anderson, J., 1909. Extraordinary abundance of, and destruction by, the larvae of *Pieris brassicae*. *Entomologist* 42: 282.

Asena, N., 1974. Investigations on vegetable pests in west and south-west Anatolia (in Turkish). *Diyarbakir Bölge Zirai Mucadele Arastirma Enstitusu, Turkey.*

Balachowsky, A. & Mesnil, L., 1936. *Les insectes nuisible aux plantes cultivées.* II: 1183–1185. Etablissements Busson, Paris.

Berne, 1925. Lutte contre les ennemis des plantes cultivées. *Ann. agric. Suisse* 26: 456–457.

Best, R.H. & Ward, J.T., 1956. *The Garden Controversy.* Department of Agricultural Economics, Wye College.

Bodenheimer, F.S., 1935. *Animal life in Palestine.* L. Mayer, Jerusalem.

Borg, P., 1929. Report of the plant pathologist. *Ann. Rpt. Dept. agric. Malta* 1928–1929.

—— 1932. *Lepidoptera of Maltese Islands.* Government Printing Press.

Buxton, P.A., 1923. Applied entomology of Palestine, being a report to the Palestine government. *Bull. ent. Res.* 14: 289–339.

Cansdale, W.D., 1875–1876. Ravages of *Pieris brassicae*. *Entomologist* 8–9: 257.

Chuang-Lung, L., 1962. Results of the zoologico-botanical expedition to southwest China, 1955–1957 (Lepidoptera: Rhopalocera). (In Chinese.) *Acta ent. Sinica* 11: 172–198.

Constantineanu, M.I.I. & Mustaţa, G., 1970. Ichneumonide (Hymenoptera: Ichneumonidae) obtinute prin culturi de insecte daunatoare legumelor, in Lodova. *Stud. Com. St. Nat. Muz. Jud. Suceava* (sub tipar).

Constantineanu, M.I.I., Sucii, I., Andriescu, V., Ciochia, Y. & Pisica, C., 1957. Contributii la studiul himenopterelor parazite in albilita verzii (*Pieris brassicae* L.) din *R.P.R. Anal. St. Univ. Al. Cuza Iasi* (Sera Noua). Sect. II (St. Nat. Geogr.) 3: 1–2, 1–7.

Daebler, F., Hinz, B. & Giessmann, H.J., 1973. The effect of leaf damage on the yield of rutabaga. *Arch. Phytopath. Pflanz.* 9: 29–35.

Eichler, W., 1948. Auffällige Schädlingsvorkommen in Mitteldeutschland (1947). *Anz. Schädlingsk.* 21: 55–58.

Feltwell, J.S.E., 1977a. Skeletonisation by *Pieris brassicae*. *Entomologist's mon. Mag.* 112: 104.

—— 1977b. Migration of whites. *Entomologist's mon. Mag.* 112: 88.

—— 1978. *Pieris brassicae* (L.) in South America. *Entomologist's Rec. J. Var.* 90: 330.

Food and Agriculture Organisation of the United Nations, 1975. *Production Yearbook, 1974* 28.1, 28.2.

Friederlichs, K., 1931. Zur Ökologie des Kohlweisslings (*Pieris brassicae*). *Z. angew. Ent.* 18: 568–581.

Garbowski, L., 1925. Les maladies et les parasites animaux des plantes cultivées dans l'ouest de la Pologne en 1923. *Chor. Szkodn. Rośl.* Suppl. 1: 1–39.

Gardiner, B.O.C., 1963. Genetic and environmental variation in *Pieris brassicae*. *J. Res. Lepid.* 2: 127–136.

Glaeser, G., 1974. The occurrence of significant harmful factors in plant crops in Austria in 1973. *Pflanzenschutzberichte* 44: 113–126.

Godan, D., 1949. Massenauftreten von Kohlweisslingsraupen (*Pieris brassicae*) in Berlin. *Biol. Zentanst.* 1: 165–166.

Gonzalez, R.H., 1972. Chile. Oriental fruit moth on peach. Large cabbage butterfly. *FAO Plant Prot. Bull.* 20: 89–91.

Granville-Clutterbuck, C., 1941. Notes on lepidoptera at the Lizard in 1920. *Entomologist* 74: 121–123.

Guerne, J., 1894. Invasion des chenilles de *Pieris brassicae. Ann. Soc. ent. Fr.* 63: 241.

Handiman Which, 1975. Allotments. p. 29.

Jakimavicius, A., 1974. On the parasites of vegetable garden pests in Lithuania. (In Lithuanian.) *Lietuvos Tsr Mokslu Akad. zool. Parazit. Inst. Lithuanian SSR.* 183–188.

Kadamshoev, M., 1971. Biology of the cabbage butterfly. *Izv. Akad. nauk. Tadkz.* 2: 89–91.

Kanervo, V., 1946. Sporadic observations concerning diseases in certain species of insects. *Ann. ent. fennici* 11: 218–227.

Karadjov, S., 1972. The joint utilisation of biopreparations entobacterin (*Bacillus cereus* var. *galleriae*) and boverin (*Beaveria bassiana* Vuill.) with the egg parasite *Trichogramma evanescens* West, for the control of *Pieris brassicae* L. and *Pieris rapae* L. and the cabbage worm (*Mamestra brassicae* L.). *Acad. agric. Sci. Inst. Plant Prot. Kostinbrod. Stud. Biol. Cont. Plant Pests.* 1: 129–143.

Kharitonov, D.E., 1926. Outbreak of *Pieris brassicae* L. in the Government of Perm. *Bull. Inst. Récherches Biol. Univ. Perm* 4: 383–385.

Klemm, M., 1954. Kohlweisslingsjahr, 1954? *Nach. Deutsch. Pflanzenschutzd.* 8: 1–2.

Kornfeld, A., 1935. Schädigungen und Krankheiten der Oelbohne (*Soja*), soweit sie bisher in Europa bekannt geworden sind. *Z. Pflkrankh. PflPath. PflSchutz.* 45: 577–613.

Ksenjopolsky, A.V., 1914. Insect pests of Volhynia during period of existence of late maintenance commission (1880–1897). (In Russian.) *Zemstvo Gvt. Volhynia, Jitomir* 1914.

—— 1916. Review of pests of Volhynia. *Report of the Volhynia entomological bureau*, 1915. Jitomir. p. 24.

Lal, K.B., 1946. Crop pests of United provinces and their control. *Plant Prot. Ent. Bull. Lucknow*, p. 6.

Lall, B.S., 1963. Vegetable pests. *In: Entomology in India*, 1938–1963. Ed. N.C. Pant. Entomological Society of India. pp. 187–211.

Lampere, A., 1907. *Manuel Faune Belgique.* Lamertin, Brussels.

Lefroy, H.M. & Howlett, F.M., 1909. *Indian Insect Life.* Thacker Spink, Calcutta.

Linnaniemi, W.M., 1935. Report. Occasional plant pests in Finland, 1917–1923. *Valt. Maatalousk Julk Helsinki* 1923: 1–159.

Manolache, C.A., Savescu, G., Boguleanu, F.L., Paulian, D. & Balajm, P., 1969. *Entomologie Agricola* 563. Ed. Agro-Silica, Bucaresti.

Marchal, P., 1913. Rapport phytopathologie pour 1912. *Bull. agric. Alg. Tun.* 9: 193–199.

—— 1914. Rapport des insectes nuisables en France en 1913. *Rev. Phytopath. appl. Paris* 18–19: 9–13.

Moiseeva, N.V., Mashkina, L.G., Kalashnikova, G.I. & Ratomskaya, A.A., 1975. The effectiveness of the use of *Trichogramma* for control of certain cabbage pests in the Chu Valley, Kirgizia. *In: Arthropods of Agricultural Importance.* Ed. Protsenko, A.I.

Mostauskis, S., 1923. Entomological report for 1921–1922 (in Lithuanian). *Ent. Rpt. Dotnava* 1923: 219–240.

Mustaţa, G. & Andriescu, I., 1973. Recherches sur le complexe de parasites (Insecta) du papillon du chou (*Pieris brassicae* L.) en Moldavia (R.s. de Roumania). I. Parasites primaires. *Ecol. Ter. Genet.* 1972–1973: 191–230.

Osmolovski, G.Y., 1964. The biology of *Apanteles glomeratus* L. (Hymenoptera: Braconidae), a parasite of *Pieris brassicae*. *Entomol. Rev. (URSS)* 43: 387–388.

Pictet, A., 1918a. Les migrations de *Pieris brassicae* en Suisse en 1917. *Verh. schweiz. naturf. Ges.* 1917: 277–278.

—— 1918b. Observations biologiques sur *Pieris brassicae* en 1917. *Bull. Soc. lepidopt. Genève* 4: 53–66.

Plotnikov, V., 1914. Insects injurious to orchards, field crops and market gardens in Turkestan, with methods of fighting them. (In Russian.) *Turkestan ent. Stn. Tashkent* 1914: 122.

Pollard, E., Hooper, M.D. & Moore, N.W., 1974. *Hedges*. Collins, New Naturalist, London.

Przbylski, Z., 1968. Development of the second generation of cabbage white butterfly *Pieris brassicae* L. (Lepidoptera: Pieridae) in agricultural-climatic conditions of Rzeszow region (in Polish). *Polskie Pismo. ent.* 38: 897–906.

Rataul, H.S., 1959. Studies on the biology of cabbage butterflies. *Ind. J. Hort.* 16: 255–266.

Rodionov, Z.S., 1924. Vegetable pests in Azerbaijan. (In Russian.) *Defense des Plantes* 1: 129–131.

Rylands, P., 1837. Notes on the species and varieties of the genus *Pontia*. *Naturalist* 2: 127–130.

Sachtleben, H. & Pape, H., 1922. Krankheiten und Beschädigungen der Kolbur-pflanzen im Jahre 1920. *Mitt. biol. Reichsanst. Ld-u. Forstw.* 26: 101.

Safaryan, S.E., 1969. The injurious fauna of cabbage in Armenia. (In Russian.) *Jub. Fauna Armenia SSR. Inst. zool. Akad. nauk. Armyanskui SSR.* 1969: 50–52.

Schøyen, T.H., 1918. Report on plant diseases in Norway in 1916. *Saertryk Landtbr. Aarsb. Christiana*, 1917.

—— 1926. Report on insect pests occurring in agriculture and horticulture in 1924–1925. *Report Insect Pest Agric. Hort. Oslo* 1924–1925: 1–31.

Shapiro, V.A., 1976. *Apanteles* – a parasite of the cabbage white butterfly. (In Russian.) *Zashch. Rast.* (10) 17–18.

Shembel, S.I., 1922. Report of Astrakan station for the protection of plant pests, 1921.(In Russian.) *Report Astrakan Stn. Prot. Plant Pests* 1921: 1–15.

South, R., 1891. On the distribution in eastern Asia of certain species of lepidoptera occurring in Britain. *Entomologist* 24: 81–86.

Spassky, S., 1916. Insect pests of the province of Don. *Ann. Don Polytech. Inst. Novotch.* 5: 219–226.

Srivastava, A.S., 1970. Important insect pests of stored oil seeds in India. *Int. Pest Control* 12: 18–20.

Svjatovitch-Bielokova, A.V., 1914. Report of the working entomological bureau, Kaluga, 1913–1914. *Zemstvo Gvt. Kaluga, Kaluga*, 1914: 89–106.

Talhouk, A.M., 1969. Insects and mites injurious to crops in Middle Eastern countries. *Monographien zur angew Ent.* No. 21. Verlag Paul Parey, Hamburg & Berlin.

Theobald, F.V., 1909. Animals injurious to vegetables. *J. SE. agric. Coll. Wye* 1909: 157–164.

—— 1926. Entomological department. *SE. agric. Coll. Ann. Rpt. Res. Adv. Dept. Wye* 1926: 5–22.

Tullgreen, A., 1918. Injurious animals in Sweden, 1912–1916. *Medd. centralanst. Jorsbruksf. ent. avdel.*

397

Vappulla, N.A., 1965. Pests of cultivated plants in Finland. *Acta ent. fennica* 19: 1–239.

Vitkousky, N., 1915. *Reports of the entomological subsection.* Uprava Zemstro Gvt. Ekaterinoslan. Rev. Pests. agric. 1914. Published by Zemstvo Gvt. Ekaterinoslav.

Voltikhov, V.A., 1975. Economic evaluation of the chemical and biological methods. (In Russian.) *Zashch. Rast.* (5) 29.

Voukassovitch, P., 1926. Contribution à l'étude de *Pteromalus puparum* L. Chalcidae, parasite interne des chrysalides: sur les types aberrants chez *Pteromalus puparum. Ann. Soc. ent. Fr.* 95: 179–182.

Walton, C.L., 1916. Cabbage caterpillars. *Agric. J. Min. agric. Great Britain* 48: 243–246.

—— 1917. Some farm insects observed in Aberystwyth area, 13–16. *Ann. appl. Biol. Lond.* 4: 4–14.

Woroniecka, J., 1924. Agricultural pests observed in districts of Lublin and Kielce in 1924. *Mem. Inst. nat. Polon Econ. Rur. Pulawy* 5: 379–392.

Yaroslavtzev, G.M., 1931. Brief report on pests of field cultures in 1930 according to data of state service of dynamics and distribution of the injurious insects. (In Russian.) *Plant Prot. Leningrad* 8: 375–413.

Zanon, V., 1919. Horticulture at Benghasi, Tripoli. *Agric. Colon Florence* 13: 154–176.

Zhukauskene, Y.I. & Misyalyunene, I.S., 1976. Effect of biotoxibacillin and entobacterin-3 on cabbage pests in the Lithuanian SSR. (In Russian.) *Liet Tsr. Mokslu Akad. Darb. Ser. C. Biol. Mokslai* 4: 61–68.

Further references

Anonymous, 1948. *Plantendienst Wageningen Vlugschrift* 58: 1–4 (skeletonisation).

—— 1970. *Trop. Pest. Res. Unit, Report for 1969–70.* p. 11 (control of cotton pests).

Ashcroft, R.W., 1918. *SE. Nat. Lond.* 1918: 73–87 (allotment pests).

Averin, V.G., 1913. *Rpt. ent. Bur. Zemstvo Gvt. Charkov 1913* (pests of Charkov).

Bainbridge Fletcher, T., 1925. *Bull. ent. Res.* 16: 177–181 (migration as a factor in pest outbreaks).

Balachowsky, A.S., 1962. *Entomologie Appliquée à l'Agriculture.* Masson, Paris.

Bitzky, I.G., *Wenden* 1914: 28 (pest in kitchen gardens).

Borodin, D.N., 1915. *Rpt. Bur. Gvt. Zemstvo Poltava,* 1914 (pest in Poltava).

Briantzen, B., 1925. *La Def. Plantes* 2: 237–241 (biological control).

Campbell, J.E., 1939. *Entomologist's mon. Mag.* 75: 278 (abnormal numbers of larvae).

Canada, 1967. *Research Rpt. Res. Branch Canada Dept. Agric. for 1967*: 1–460 (resistant swede varieties).

Carpenter, G.H., 1923. *J. Dept. agric. Tech. Instr. Ireland* 23: 12–14 (pests of cabbage).

Chaure, L., 1914. *Moniteur Hort. Paris* 38: 135 (attack at Vienne, France).

Chesnokov, P.G., 1936. *Izd Vesesoyuzn akad. Skh nauk Im Lenina* (pest on crucifers).

Clarke, H.S., 1893. *Yn Lioar Manninagh* 2: 100–103 (Isle of Man).

Curtis, J., 1842. *Jl. R. agric. Soc. Eng.* 3: 306–323 (effects of various larvae on turnips).

—— 1877. *Farm Insects.* Blackie & Son, London.

D'Ambrosio, G., 1914. *Boll. Catt. Amb. agic. Brindisi* 8: 70–71 (outbreak in Brindisi).

Driest, J.Ph., 1974. *Proef. Groent. Grond. ned. Alkmaar* 1974 (pest).

Dunbar, G. & Yilmaz, N., 1973. *Ziraf Mucadele Arastirma Yilligu* 119: (pest in Samsun region).

Emchuk, E.M., 1937. *Trav. Inst. zool. Biol. acad. Sci. Ukr. Kiev* 14: 279–282 (pest in NW Ukraine).

Goriainov, A.A., 1914. *Zemstvo Gvt. Riazan Report* (pests in Riazan).

—— 1915. *Ent. Bur. Zemstvo Gvt. Riazan.* 1–138 (pest).

Goriatchkovsky, V.L., 1915. *Warsaw Hort. Soc. ann.* 1915: 64–67 (pest).

Gram, E. & Rostrup, S., 1924. *Tidjsskr. Planteavl.* 30: 361–414 (pest in Denmark).

—— 1925. *Tidjsskr. Planteavl.* 31: 353–417 (pest).

Guilbert, (–), 1939. *Les Lépidoptères ennemis de nos cultures.* Thèse Docteur, Université de Strasbourg.

Hahn, E., Petzold, D. & Ramson, A., 1973. *Nachr. PflSchutz* 5: 93–112 (crop pests).

Hukkinen, Y., 1927. *Maatalouskoelaitos (Lantbruksförsöksanstalten)* 1925: 1–164 (pest on crucifers).

—— 1931. *Maatalous, Helsingfors.* 24: 21–24 (pest in future).

Husain, M.A., 1924. *Rpt. Dept. agric. Punjab* 1: 55–90 (pest).

Isaac, P.V., 1934. *Sci. Rpt. Inst. agric. Res. Pusa* 1933: 161–166 (pest).

James, W.O. & Clapham, A.R., 1935. *The Biology of Flowers.* Clarenden Press, Oxford (proboscis length).

Kaitazov, A., 1963. *Rastch. Zasht. Sofia* 11: 20–24 (parasites and predators).

Kazanski, A.N., 1924. *Défense des Plantes* 1: 78–82 (high rate of parasitism).

Kéler, S., 1932. *Prace Wydz. Chor. Ros. Panstw. Inst. nauk. Gosp. Wiejsk. Bydgoszczy. Bromberg* 1932: 1–22 (pest).

Krainsky, S., 1914. Hort, Mkt. Gardener 1914: 329–339 (pest in market gardens).

Krasucki, A., 1929. *Mem. Inst. nat. Polon Econ. Rur. Pulaway* 10: 216–223 (pest in SE Poland).

Lefroy, H.M., 1909. *Indian Insect Life*, Thacker Spink & Co., Calcutta (pest).

Leningrad, 1929. *State Inst. exptl. agron.* 1929: 1–76 (outbreaks).

Macdougall, R.S., 1923. *Trans. Highland agric. Soc. Scotland* 1–43 (pest on various crucifers).

Maroc, 1923. *Var. Sci. Soc. Sci. nat. Maroc* 1: 22–27 (pest on vegetables).

Michailov-Doinokov, A., 1914. *Orchard Mkt. Gdn. Bachza, Astrachan* 471–472 (pest on cabbage).

—— 1915. *Orchard Mkt. Gdn. Bachza Astracan* 327–333. (damage to crops).

Monks Wood Data Cards, 1971. (three females in allotment).

—— 1975. (many accounts throughout country).

N.D.R., 1945. *Entomologist* 78: 136 (invasion at Westcliffe on Sea).

Nieuwhof, M., 1969. *Cole Crops.* Hill, London (pest on *Brassica*).

Noel, P., 1913. *Bull. Lab. Rég. ent. agric. Rouen* 1913: 15–16 (on radish).

Özer, M., 1962. *Ankara Univ. Ziraat Fak.* 1962: 13–25 (bionomics).

Rossum, G. van, Hurger, H.C. & Bund, C.F. van, 1974. *Ent. Ber.* 5: 133–135 (pest in The Netherlands).

Sacharov, N. & Shembel, S., 1915. *Rpt. ent. Stn. Myc. Branch*, 1914 (pest).

Schmidt, G., 1949. *Anz. Schädlingsk.* 22: 184 (abundance of larvae in Berlin).

Schoyen, T.H., 1915. *Rpt. Inj. Insects and Fungi of Field and Orchards in 1916, Kristiania* 37–92 (pest on cabbage).

Schreiner, J.T., 1915. *Prot. Plants Pests. Suppl. Friend of Nature, Petrograd* (pest on mustard).

Stapley, J.H., 1949. *Pests of Farm Crops.* Farmer and Stockbreeder.

Turati, E. & Zanon, V., 1922. *Atti. Soc. ital. Sci. nat.* 61: 132–178 (pest on cabbage).

Uvarov, B.P., 1913. *Rpt. Bur. ent. Stavropol* 1912: 1–32 (pest on cabbage).

Van Poeteren, N., 1919. *Versl. Meded. Phytopath. Dienst Wageningen* 1–48 (pest of cabbage).

Vereshtchagin, B., 1921. *Viata Agricola* 12: 78–89 (pest on vegetables).

—— 1920. *Furnika Sine Loco* 12–15 (pest in Kishinev).

Vielwerth, V., 1938. *Ochr. Rost. Prague* 14: 8–16 (pest on crucifers).

Vitkovsky, N., 1915. *Rpt. ent. Subsect. Uprava Zemstvo Gvt. Ekaterinoslav Rev. Pests agric.* 1914 (pest in vegetables).

Voelkel, H., 1925. *Arb. biol. Reichsanst. Land. Forstw.* 14: 97–108 (pest on cabbage).

Warburton, C., 1918. *Jl. R. agric. Soc. Lond.* 79: 258–263 (pest on cabbage).

Yaroslavtzev, G.M., 1938. *Zapadn. Obl. S-Kh. Op. Stanz.* 1927: 1–31 (pest on cabbage).

14. Parasitic control

"... I think that the ichneumon wasps prick these caterpillars with the hollow tube of their ovipositor and insert
eggs into their bodies; the maggots are hatched by the
warmth of them, and feed there until full grown; then
they gnaw through the skin, come out, and spin their
cocoons."

John Ray, 1710. *Historia Insectorum.* In Mickel, 1973

SECTION A. GENERAL INFORMATION
by M.R. Shaw, Royal Scottish Museum

Historical background

Parasites (parasitoids) associating with Lepidoptera, including *Pieris
brassicae*, attracted enough attention to be illustrated by the early
entomologist Joanne Goedart (1662) even before the nature of parasitism
was properly understood. John Ray (1710) was perhaps only one of
several naturalists who independently discovered the essence of the host-
parasite relationship in the early part of the 18th century, but it is of
particular interest here that his celebrated conclusions were inspired by
watching the braconid wasp now known as *Apanteles glomeratus* (L.)
attacking larvae of *P. brassicae*.

   As a result of observations like these it became widely realised that
parasites can destroy insect pests in a way that is beneficial to man,
although it was another 200 years before serious and large scale attempts
were made to manipulate parasites as biological control agents (see
Howard & Fiske (1911), who include a review of the few previous efforts).
Following some early successes, however, the study of host-parasite

interactions quickly blossomed into an important aspect of economic entomology, and numerous attempts to control a wide range of insect pests using parasitic Hymenoptera, and to a lesser extent Diptera, have been documented. Some of the early work was decidedly hit-and-miss and in some cases unsuitable parasites, which may even have done more harm than good, were bred in large numbers to be released among pest infestations. More recently it has become clear that extremely detailed studies of both the pest and its parasites are needed in order to choose the parasite species most likely to become established, and provide useful control, under particular field conditions. Most of what is known, in detail, of the biology of parasitic Hymenoptera has come from studies stimulated by these needs.

General biology

As a whole, the biology of parasitic Hymenoptera is a complex subject and it will be useful to present here a summary of the different ways in which hymenopterous parasites of Lepidoptera can interact with their hosts. The general biology of parasitic Hymenoptera and Diptera attacking (British) Lepidoptera is treated in greater detail by Shaw & Askew (1976).

Primary parasites attack and eventually kill the host itself, in this case *P. brassicae*, and thus effect a reduction or limitation of the host population. *Secondary parasites*, or *hyperparasites*, are parasites of primary parasites and have quite the opposite ecological effect since they constrain the population of the primary parasite. Their presence is therefore usually assumed to be harmful in most biological control contexts concerned with the protection of transient monocultures, although this is probably an oversimplification of their true ecological roles in stable multitrophic ecosystems. A few species are able to function as *facultative hyperparasites*, attacking both the host and its primary parasites. Their effect is hard to predict since it depends in part on their degree of preference, which may in turn depend on a range of variable ecological parameters, but several attempts at controlling injurious insects have certainly been spoilt by the unwitting introduction of facultative hyperparasites.

Parasites may be *ectoparasites*, which feed on the host from an external position, or *endoparasites* which feed from within the host's body. The former often attack concealed hosts, which are usually stung and either killed or permanently paralysed during the oviposition sequence, while the latter (if the larval stage is attacked) typically do not destroy or greatly impair the host until the parasite(s) become fully fed. The endoparasitised host larva thus continues in its ecological role of feeding and potentially providing prey during the period of the association. A few groups of ectoparasites attach their eggs externally on initially unimpaired exposed larvae, and these hosts similarly persist for a while in the above roles.

Parasitism may be *solitary* in which case a single parasite larva develops in or on a single host individual, or *gregarious* (several of the same species of parasite per host individual). Generally, particular parasite species are absolutely consistent in this as well as in endo- or ectoparasitism, but some facultative hyperparasites vary in one or both respects according to their actual host.

Most primary parasites attack their hosts at a precise life-history stage, and can be divided into groups reflecting these differences. It is important to appreciate that these are ecological categories and do not usually indicate phylogenetic affinities. *Egg parasites* attack, and develop to the adult stage within, the host ovum; *larval parasites* attack the host larva, also killing this stage; while *pupal parasites* attack and kill the host pupa. Some endoparasites complete their feeding in a later host stage than the one attacked, and are referred to as *egg–larval* or *larva–pupal* parasites accordingly. Most larval parasites spin cocoons near the host remains, while pupal parasites normally develop to the adult stage within the host pupa.

Secondary parasites can similarly be divided according to their habits. It is convenient to refer to those species which attack the primary parasite while it is actually feeding in its living host as *true hyperparasites*, and those which attack the primary parasite only after it has finished feeding and killed its host as *pseudohyperparasites*. Although both categories have the same overall effect, that of killing primary parasites, the division is useful because true hyperparasites are typically completely specialised to this way of life (and are generally endoparasites of endoparasites), while pseudohyperparasites comprise a wide range of usually less specialised and often ectoparasitic species, and include the facultative hyperparasites. Often pseudohyperparasites are really attacking whatever small cocoons and similar structures they find, and the cocoons of primary parasites just happen to fall into this category. Thus pseudohyperparasitism is rarely obligatory, although when primary parasite cocoons are abundant they may serve as hosts almost exclusively.

Most parasites are not absolutely specific to a single host species but instead tend to attack a group of hosts which usually bear some relation, either ecological or phylogenetic, to one another. The *host range* (or *usual hosts*) of a given parasite is a key feature of its biology, although in practice it is a nebulous term difficult to define with any precision. This is because some host species may be attacked less avidly, or with less success, than others, and furthermore it is known that such factors can vary from one population of a parasite species to another. However, it is useful to define host range in general terms as including only the species of potential hosts that the parasite is usually able to attack successfully, following a pattern of searching behaviour enabling it to encounter them regularly. Potential hosts that are attacked only by some freak of circumstance, or which only occasionally permit the parasite's full development, would thus be excluded from a useful concept of host range, as would species of otherwise suitable hosts which the parasite rarely encounters. A regular parasite of a given host must first search in the microenvironment supporting the host, second be able to recognise the host as suitable, and third be physiologically capable of full and unimpaired development upon it. These processes are dependent on a complex array of behavioural responses to physical and chemical cues which, especially in the case of endoparasites, tend to culminate in the choice of a host towards which the parasite has, to some extent specifically, evolved a biochemical compatibility. In practice host ranges often include all or most insects of a particular group delimited by a more or less severe (and variously biased) balance between host ecology and host phylogeny.

Recorded parasites of *Pieris brassicae*

By abstracting published rearing records, several authors have sought to provide definitive lists of parasitic Hymenoptera and Diptera reared from *P. brassicae*. The longest and best known of these was compiled by Thompson (1946), who incorporated an earlier list of 19 species of Hymenoptera given by Morley & Rait-Smith (1933) in his grand total of 44 Hymenoptera and seven Tachinidae (Diptera) supposedly parasitic on *P. brassicae*. More recently, Mustaţa & Andriescu (1973) have produced a list of 39 Hymenoptera and seven Tachinidae from an apparently independent literature survey, and also researched the parasite complex themselves whereby they claimed to find 21 species of Hymenoptera attacking *P. brassicae* in Rumania.

Unfortunately, lists like the above are very nearly useless in reflecting parasitism of *P. brassicae* for a variety of reasons, the most obvious being that equal weight is given to freak and regular records. However, their shortcomings run a good deal deeper than this, for the reasons enumerated below.

a) *The status of the names*

The lists based on literature surveys are simply a collection of the names that have been published in this context, and they usually include a number of species which have been entered twice or more in the same list but under different names. Some of this synonymy is now formally

404

established and fairly easy to trace, but undoubtedly much remains to be discovered since some of the groups of parasites most extensively recorded have never received a proper taxonomic revision. Also, the parasite complex of *P. brassicae* has been investigated over a wide geographical area, and this has inevitably contributed to the generation of synonymy. In other words, populations of a parasite species often maintain better contact over their geographical range than the taxonomists who study them. Some of the names listed, however, have little or no meaning even as synonyms. Sometimes incorrect or no authorship is indicated, often leaving the name ambiguous, and sometimes names have been so badly misspelt or scrambled that they can no longer be recognised with confidence. It should also be said that very many of the generic names which appear in the lists are no longer applied to the species listed.

b) *Parasite misidentifications*

Parasitic Hymenoptera are difficult to identify and the names which appear in the lists have not always been provided by competent tax-onomists. A further point is that determinations are, at their best, meaningful only at the current level of knowledge. Sometimes names correct in this sense may later turn out to have been applied to aggregates of closely related species. The application of the aggregate name can then no longer be justified and the material must be identified again in the light of the new discoveries – but mere host-parasite lists can never provide for this. It is impossible to guess how many of the listed names represent parasite misidentifications, but some insight can be gleaned from Thompson's (1953) list of 57 species of Lepidoptera recorded as hosts of *Apanteles glomeratus*. The host range of *A. glomeratus* is now known to be restricted to a few genera of Pieridae, and only eight of the listed Lepidoptera belong to this family. Most or all of the remaining 49 are now known to be attacked by *Apanteles* species quite distinct from *A. glomeratus*, but which could easily have been confused with this species in the past.

c) *Host misidentifications*

A surprisingly large number of the parasites listed certainly result from host misidentifications and have nothing whatsoever to do with the parasite complex of *P. brassicae*. Most spectacularly, these include a few parasites of aphidivorous Syrphidae (listed under the generic names *Promethes*, *Homocidus*, *Bassus* and *Diplazon*) and gall-causing Cecidomyliidae (under *Macroglenes*) recorded by Mustaţa & Andriescu (1973). Thompson's (1946) inclusion of *Melittobia* could be in the same

405

class since it is usually parasitic on bees, although it can be a pseudohyper-parasite of Lepidoptera in its much less frequent attacks on ichneumonoid cocoons. Most of the other Hymenoptera appearing in the lists are, at least potentially, regular parasites or hyperparasites of Lepidoptera. Many Lepidoptera feed on cultivated Cruciferae, however, and their larvae (or cocoons of parasites which have already killed these) must be very difficult to exclude from field collections of Cruciferae infested with *P. brassicae*. Thus it seems that parasites reared in the laboratory have often come from overlooked lepidopterous hosts and then been associated incorrectly with *P. brassicae*. It is just conceivable, however, that some parasites of other lepidoptera on Cruciferae may very occasionally develop upon *P. brassicae* although the latter is not within their usual host range. One of the least unlikely to be an example of this is *Apanteles rubecula* Marshall, a fairly common solitary endoparasite of *Pieris rapae* (L.) which certainly does not develop in *P. brassicae* at all regularly, although various authors have suggested it does so at times. Parasites which usually attack microlepidoptera have been widely included in the lists, appearing under the generic names *Angitia*, *Diadegma*, *Campoplex*, *Microbracon*, *Bracon*, *Microgaster* and *Apanteles* (some species). True hyperparasites in the genus *Stictopisthus* are also more likely to be associated with microlepidoptera than *P. brassicae*. Noctuid larvae seem to have been another frequent source of bogus records, since species listed under *Eulophus*, *Comedo*, *Microplitis*, *Apanteles* (some species), *Exetastes*, *Meloboris*, and possibly also *Sinophorus*, seem unlikely to be regular parasites of *P. brassicae*.

d) *Secondary parasitism*

There has usually been no effort made to separate primary from secondary parasites, and in some lists over half the parasites included are actually hyperparasites of one sort or another. Facultative hyperparasites are listed under the pteromalid genera *Eupteromalus*, *Habrocytus* and *Dibrachys*, while listed species of the torymid genus *Monodontomerus*, the chalcidid genus *Brachymeria*, and the ichneumonid genus *Itoplectis* also function in this way. Pseudohyperparasites belonging to the ichneumonid subfamily Phygadeuontinae (Cryptinae *auctt.*) are listed under the generic names *Hemiteles*, *Astomaspis*, *Lysibia*, *Gelis*, *Pezomachus*, *Agasthenes*, *Leptocryptus* and *Bathythrix*, although one species usually listed as *Hemiteles melanarius* Gravenhorst seems to be well-documented as a gregarious primary parasite of *P. brassicae* pupae. True hyperparasites are listed under the names *Mesochorus*, *Stictopisthus* and *Tetrastichus*; species of the latter may also be able to function as pseudohyperparasites at times.

406

Concluding remarks

Although there is a considerable amount of literature regarding parasitism of *P. brassicae*, most of the studies have been on only limited aspects. The somewhat piecemeal data resulting from these are not always easy to interpret or to correlate, especially in view of the many different languages, geographical areas, races of insects, and field or experimental conditions involved. Further, parasitism of *P. brassicae* has usually been studied in isolation rather than in conjunction with other host species, as would be necessary in order to understand the roles of the more polyphagous parasites properly. Long term and somewhat less specifically directed investigations into host-parasite interactions would no doubt help to rectify this, but unfortunately such generalised research is rarely undertaken because the likely benefits are not always clearly visible beforehand.

The fact that long lists of supposed parasites of *P. brassicae* have circulated does not, therefore, mean that this pest has an extensive parasite complex. The above treatment reduces the lists to around half a dozen species which are regular enough as primary parasites of *P. brassicae* to have any significant impact on its numbers. This rather limited fauna might possibly be expected of a strongly migratory host tending to produce transient populations, and it is significant that most of the parasites regularly attacking *P. brassicae* are equally well known as parasites of other abundant hosts. This is particularly true of *Trichogramma evanescens* Westwood (Trichogrammatidae), *Pteromalus puparum* (L.) (Pteromalidae), *Pimpla* (*Coccygomimus*) species and the non-British *Theronia atalantae* (Poda) (Ichneumonidae). Others, such as *Hyposoter* (= *Anilastus*) *ebeninus* (Gravenhorst). (Ichneumonidae) and *Apanteles glomeratus* (L.) (Braconidae), have alternative pierid hosts. As for the secondary parasites, several of the facultative hyperparasites and pseudoparasites recorded are common and fairly polyphagous species, although more host specificity might be expected of true hyperparasites in the genera *Mesochorus* (Ichneumonidae) and *Tetrastichus* (Eulophidae). Although at least four specific names have been used in conjunction with *Tetrastichus* in parasite lists of *P. brassicae*, they probably all refer to the species correctly known as *Tetrastichus galactopus* (Ratzeburg), which certainly seems to be a regular hyperparasite through *Apanteles glomatus*. *T. galactopus* has often been incorrectly identified as *T. rapo* (Walker) or *T. microgastri* (Bouché), but these two species are distinct and quite unrelated (Graham, 1961).

SECTION B. BIOLOGY OF SOME EFFECTIVE PARASITES
by J.S.E. Feltwell

Introduction

As pointed out in the preceding section, most of the species reputed to be parasites of *P. brassicae* have been included in a number of lists through inaccurate identifications, uncontrolled breeding and collecting techniques and the requoting of erroneous published material. The long lists of *P. brassicae* parasites can in fact correctly be reduced to seven parasites frequent enough to have been studied in detail, of which the most outstanding is *Apanteles glomeratus* (Linnaeus).

Most of the following section is devoted to *A. glomeratus*, although details are also presented of one egg parasite, another larval parasite and four pupal parasite species, including dipterous parasites as well as non-insect ectoparasites and non-insect endoparasites.

Egg parasite

Trichogramma evanescens Westwood

T. evanescens may be either a solitary or gregarious parasite; indeed Mustaţa & Andriescu (1973) found that one to nine adults develop in each host ovum, implying that the ovipositing female lays on average one to nine eggs per host. The parasite overwinters in the host ovum.

T. evanescens is a very effective egg parasite of *P. brassicae* in the wild (Völkel, 1925) and has already been noted to exert very high mortalities in many countries; in France up to 90% of *Pieris* ova have been found infested (Faure, 1926a,b), in USSR a 68% reduction of a *P. brassicae* population was induced experimentally (Telenga, 1955, cf. 1929), in Bulgaria 94–100% parasitism was noted (Karadjov, 1972).

In the wild the degree of parasitism may vary according to locality as Mustaţa & Andriescu (1973) found in Rumania, where percentage parasitism varied from 9–64% in six localities; or it may vary from year to year; for instance in Poland in 1966 *T. evanescens* became unusually abundant when the level of parasitism in *P. brassicae* ova on 23 August, 1966 was almost 91% (Kot & Plewka, 1968; cf. Kot, 1961). In order to achieve a mortality rate of 75% Kot (1961) recommended liberating 4000–5000 *T. evanescens* per 500–600 m² of infested cabbage.

Biological control of *P. brassicae* in the USSR using *T. evanescens* was shown to be effective as an alternative to insecticides by Moiseeva *et al.* (1975). *T. evanescens* is however, susceptible to sub-lethal doses of insecticides and herbicides in USSR (Quednau, 1956; Prokofera, 1976).

In a series of published works Meïer (1925, 1938, 1940, 1941) investigated the different races of *T. evanescens* in relation to their suitability as parasites of *P. brassicae*. He reported that nine races of *T. evanescens* had been found since 1932, and that only one, the Cabbage White Race, was suitable for control of *P. brassicae*.

Larval parasites

a)   *Apanteles glomeratus* (Linnaeus)

Introduction. One of the most important and widely known parasites of *P. brassicae* is the Braconid, *Apanteles glomeratus* (L.), which is a gregarious larval parasite (Klein, 1932; Blunck, 1944b; Speyer, 1956), and which can be reared easily in the laboratory (David & Gardiner, 1952; Soures, 1971). It seems to have been first associated with *P. brassicae* by Goedart (1662) who illustrated many of the stages, and again by Maria Graffinn (1679). A long descriptive account of the biology of *A. glomeratus* was published by Hamilton (1934 – diploma) in which are included details of mating, oviposition, hosts and resistance of hosts.

In 1660 John Ray published his first botanical work, *Catalogus Plantarum circa Cantabrigiam nascentium* in which he made some notes on the Large White Butterfly. Mickel (1973) published some of these in her review of Ray's work as a naturalist and the following quotation concerns parasitism by the "Ichneumon". This was later mentioned in Ray's 1770 *Historia Insectorum*, in which the true nature of parasitism was realised.

"I shut up ten or so of these (caterpillars) in a wooden box at the end of August 1658. They fed for a few days, and fixed themselves to the sides or lid of the box. Seven of them proved to be viviparous or verminiparous: from their backs and sides very many, from thirty to sixty apiece, wormlike animacules broke out; they were white, glabrous, footless, and under the microscope transparent. As soon as they were born they began to spin silken cocoons, finished them in a couple of hours, and in early October came out as flies, black all over with reddish legs and long antennae, and about the size of a small ant. The three or four caterpillars which did not produce maggots after a long interval changed into angular and humped chrysalids which came out in mid-April as white butterflies."

Oviposition. It was originally thought that the female *A. glomeratus* attacked the ova of *P. brassicae* (Fabre, 1908; Faure, 1925a,b; Voukassovitch, 1926) as well as the first three instars of the larva, but after

409

careful observation it was found that the parasite actually oviposited in the first instar larva (Ormerod, 1884; Gautier, 1918; Gatenby, 1918, 1919). One adult *A. glomeratus* may "sting" up to eight *P. brassicae* larvae in rapid succession, when the hosts are introduced successively into glass tubes (Hamilton, 1934).

First instar larvae are presumably more suitable for parasitism because they are much smaller and make only small movements of the head when molested, and they do not produce a green defensive fluid from their mouths as they do in the second instar (Gautier & Bonnamour, 1924; Klein, 1932). Oviposition takes between 20–40 seconds (Hamilton, 1934 – diploma, 1935). The following passage is taken from an eloquent description of oviposition by Johannson (1951); cf. Lyle, 1908, 1916, 1926; Shevirev, 1913a,b.

" ... She moves slowly, the antennae curved. Contacting the host, the movements become even slower and she seems to feel the larva carefully with the antennae. Sometimes the ovipositor can be seen moving out and in from the end of the abdomen. Starting the sting she will bend the abdomen forward. The ovipositor is now visible. The fore part of the body is raised, the wings are bent upwards, their upper sides facing each other, and kept firmly in position. During this forward movement of the abdomen she will insert the ovipositor into the host. If the host is not reached, she will take a few quick steps towards it and then insert the ovipositor. She will now stay quite quiet in this position while the sting lasts and the oviposition usually takes place. The body is often kept in a nearly vertical position with the wings at a right angle to the thorax. The antennae are often bent upwards parallel to the wings. The femurs of the legs are more at a right angle to the body than in the normal walking position. The legs are locked in this position, the femur-tibia often forming an angle, while the joint tibia-tarsus is nearly always straightened out. The legs of the metathorax are mostly kept close to the body, while the two other pairs stand out more from the body. Often she is only supported by the last pair of legs, but it may be accidental which of the six legs will contact the piece of cabbage leaf. The ovipositor is usually inserted into the ventrolateral parts of the host. When the sting is finished, the ovipositor is withdrawn, the body is straightened and wings and legs are replaced in their normal position. After the sting the female may be seen pressing the end of the abdomen against the piece of cabbage as if to clean the ovipositor."

It was pointed out by George (1927, cf. 1928) that larvae living on the following foodplants were often free from parasitic attack; *Tropaeolum majus, T. minus, Eruca sativa, Cochlearia armoracia* and *Capparis spinosa,*

410

possibly due to odours given off (or not given off) by these plants. Indeed a recent thesis by Abdulla (1978) produced some evidence that *P. brassicae* is less parasitised when feeding on Nasturtium (*Tropaeolum*) than on cabbage.

Several ova are injected into the host larva by the parasites during one "sting". By dissection of 15 freshly parasitised larvae (a small sample), Hamilton (*loc. cit.*) recorded that the average number of ova per sting was 30.5, although it ranged from 2–54. Others who have investigated the number of ova per sting are Grandori (1911) who stated that 4–12 could be laid, and Gatenby (*loc. cit.*) who quoted 16–60 ova per larva. Corunna (1934) gave a figure of about 40 ova per larva. It would be interesting to investigate whether different strains of *A. glomeratus* behave differently, and it is important to realise that parasites kept in culture may be of low performance.

The number of larvae of *Apanteles* found in the host has sometimes been quoted as much higher, for example 50–80 (Gautier, 1918), average of 35.6 (Hamilton, *loc. cit.*; Hamilton very carefully calculated his results from 179 parasitised larvae), average 44.7 (max. 100) (Moss, 1933), 142 from one larva (Bignell, 1883) and up to 160 under special conditions (Führer & Keja, 1976).

Hosts. A. glomeratus has apparently been recorded from 16 lepidopterous pests in Rumania (Lacatuscu, 1963) and from a total of 62 lepidopterous pests altogether (Tawfik, 1956); however, these lists have the shortcomings explained in the first part of the chapter.

In England and Europe *A. glomeratus* parasitises *P. brassicae* and *P. rapae* preferring the former to the latter according to Blunck (1951a) who worked in northern Germany (DDR), cf. Hamilton (1934). On the European mainland *A. glomeratus* also parasitises *Aporia crataegi* (Linnaeus) (Black Veined White) (Askew, 1971).

Effect on the host. The development of the cuticle in *P. brassicae* larvae parasitised and non-parasitised by *A. glomeratus* has been studied by Führer (1975b) who found that the parasite delayed the growth of the endocuticle. In some cases the endocuticle of parasitised larvae was half as thick as in normal larvae, but overall the parasitised larvae finished their development slightly heavier than normal larvae.

Führer & Keja (1976) showed that third instar larvae of *A. glomeratus* block the usual metabolic pathways normally present in the *P. brassicae* larvae, and regulate development to their own advantage. All the reserves of the host's body are exhausted when the parasite larvae exceed 60–80 per larva.

In the Japanese pierid, *Pieris rapae crucivora* Boisduval, dead ova of

Apanteles caused changes in haemolymph count and were encapsulated by the host (Kitano, 1974).

Larval development. While the larvae of the endoparasite *A. glomeratus* are developing inside the larvae of *P. brassicae* they are haemolymph feeders and are not thought to eat vital tissues. Some means of synchronisation may be available to *Apanteles* larvae because when they are fully grown they eat their way out of the body wall of their host almost simultaneously (Gautier, 1919, for description of emergence). Within a few minutes the larvae begin to spin lemon-yellow coloured cocoons (Frohawk, 1906), the colour of which is caused by a carotenoid pigment and can vary from almost white to deep orange (Lyle, 1916). The cocoons lie close to the dying body of the host and the cocooned larvae soon change to pupae (see colour figure of cocoons around host in Ford, 1945). Meanwhile, the parasitised host larva dies or lingers on and has been found to live for up to one month afterwards (Gatenby, 1919; Westropp, 1925).

The *A. glomeratus* larvae definitely undergo two and occasionally three larval instars according to Hamilton (*loc. cit.*) who examined some 7,891 larvae from *Pieris* hosts. Tawfik (1957 – thesis) recorded three instars, the third instar emerging from the host, forming a cocoon and pupating in it. *Apanteles* has the effect of slowing up the development of the *P. brassicae* larvae by an extra three days over the period from ova to fifth instar. Unparasitised larvae take about 16.2 days to mature, whereas parasitised larvae complete their development in 19.3 days under the same conditions (Hamilton, 1935). Corunna (1934) stated that *A. glomeratus* stays in the pupa for 7–11 days and then the imago will live for one to six days without food or 30 days with 10% sugar solution.

Eclosion from the pupa is regulated by temperature and humidity. Klein (1932) found that an increase in temperature from 12–19°C and a decrease in the relative humidity from 86–73% shortened length of the pupal stage from 24–25 days up to seven to nine days.

Gatenby (1918 – thesis) gave a very detailed account of the embryology, morphology, anatomy, behaviour and hyperparasitism of certain hymenopterous parasites including *A. glomeratus*. Tawfik (1956) studied the host-parasite specificity and early development of *A. glomeratus*.

Whether or not *A. glomeratus* overwinters as a prepupa within its cocoon or as a larva within its host is dependant upon which host species is involved, whether the species is univoltine or multivoltine, and what latitude and weather conditions prevail. Picard (1923) stated that in Belgium *A. glomeratus* are to be found inside the larvae of *P. brassicae* at the end of October. He goes on to say that any *Pieris* larvae still developing on cabbage after this period are thus liable to be free of attack.

Physiology. The biochemical constituents of the haemolymph of *A. glomeratus* show extreme similarities to those of its host; Junnikkala (1966) found that only one peptide (named Peptide I), which is normally found in *P. brassicae* haemolymph, was absent in *A. glomeratus* haemolymph. However, the two shared the same 19 free amino-acids, one amide and another peptide.

The first instar larva of *A. glomeratus* shows some degree of dependance upon the endocrine system of its host (Daniliveski, 1965), but in the last instar the larva responds to photoperiod and shows some degree of independance. This is thought to be an adaptation designed to keep the parasite in step with seasonal conditions, and allow them to exploit other hosts (cf. Askew, 1973).

In the USSR much comparative work has been done on the responses of host and parasite to photoperiod (cf. Maslennikova references). In the north of the USSR (e.g. Leningrad 60°N) diapause is stimulated in *P. brassicae* at the "normal" photoperiod experienced elsewhere of less than 15 hours per day, whereas in the south (e.g. Sukimi 43°N) *P. brassicae* goes into diapause much earlier at 12 hours per day. Experimentally, however, Maslennikova & Mustafayeva (1971) found that *A. glomeratus* would not even go into diapause between 6–16 hours at 18°C.

Effectiveness. In the wild *A. glomeratus* has been recorded from various countries where it has exerted up to 100% control of *P. brassicae* larvae (Table 14.1).

In northern Germany, Blunck (1950) thought that along with two other parasites, *Trichogramma evanescens* and *Pteromalus puparum*, *A. glomeratus* was successful in keeping the population of *P. brassicae* in check. In

Table 14.1 Percentage attack by *A. glomeratus.*

Percentage attack	Country	Authority
95	France	Gautier, 1919
5	France (Lyon)	Paillot *et al.*, 1924
1–2	France	Voukassovitch, 1926
60–100	Germany (BDR)	Blunck, 1953
53.8	Germany (BDR)	Speyer, 1956
90	Germany (DDR)	Godan, 1949
96	Great Britain	Birkett, 1944
50	Great Britain	Lyle, 1916
47.7	Netherlands	De Bruijn, 1921
13–25 (May), 30 (June)	India	Lal & Chandra, 1976
90	Rumania	Constantineanu, 1929
73	Rumania	Filipescu, 1972
91.5	Rumania	Lacatuscu, 1963

the USSR successful control of *P. brassicae* has been obtained after three years work on keeping cocoons of *A. glomeratus* overwintered at −5° to −7°C and then releasing them in the following spring (Silujanova, 1969). Berkovitch (1972) recommended that insecticides should be chosen with a view to their non-toxicity to *A. glomeratus*. However, Osmolovski (1964) thought that chemical control of cabbage pests only slightly reduced the effect of *A. glomeratus* as it is able to exist in other insects (? pierids).

A. glomeratus is also thought to be an extremely useful insect as a vector of various micro-organisms which may cause the death of *P. brassicae*. Various species of bacteria have been found associated with *A. glomeratus* in larvae of *P. brassicae*, namely, *Vibrio pieris*, *Coccobacilles* spp., *Diplococcus* spp. and *Bacillus* spp. (Paillot, 1918a,b). Paillot (1918a,b) thought that *A. glomeratus* in some way transmitted the protozoan, *Perezia legeri*, from larva to larva of the host, and was responsible for the giant cells found in their haemolymph.

Although *A. glomeratus* is principally a Palaearctic species it has been recorded from the United States where it is known as a parasite of *P. rapae*. As *P. brassicae* has now become established in Chile *A. glomeratus* may well become a useful biological control agent there.

b) *Hyposotor ebeninus* (Gravenhorst)

This parasite is a solitary larval endoparasite, which makes its cocoon inside the host larval skin considerably before the latter is fully grown. In England it is a common parasite of *Gonepteryx rhamni* (Linnaeus) (M.R. Shaw, 1978, pers. comm.).

Klaudia Lartschenko (1932) made an extensive study of the resistance

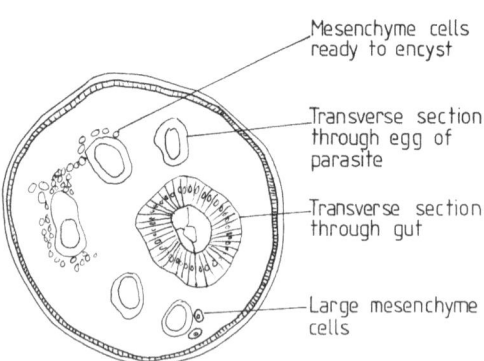

Fig. 14.1 Transverse section through the abdominal cavity of *P. brassicae* 20 hours after oviposition by *Hyposotor ebininus*
(after Lartschenko, 1932)

of *P. brassicae* (and *Margaritia sticticalis* (Linnaeus)) larvae and observed that the ova were dispersed quite randomly through the body cavity and that they could be seen near the surface. Ova were encysted by mesenchyme cells which surround them (Figure 14.1), but later stages of the parasite were not encysted.

The percentage resistance of *P. brassicae* larvae was found to vary from 47–88% but this was not found to be due to any physiological difference experienced by first or second generations of the parasite, as the percentage parasitism was the same (Lartschenko *loc. cit.*). One anomaly she found was that although her laboratory results indicated that with an increase in temperature, the resistance of the larvae increased, field trials showed that in the colder northern regions of USSR there was 100% mortality of larvae.

Pupal parasites

a) *Hemiteles melanarius* (Gravenhorst)

This is a gregarious parasite whose larva enters the shell of the stung and dead or moribund host pupa (Ciochia, 1966).

In an interesting paper Blunck & Jansen (1957) described how the larva of this parasite may be both external and internal feeders on the host pupa. The females lay their eggs on the underside of the developing wings of the pupa (see Figure *in* Blunck & Jansen, 1957; cf. Blunck, 1951a,b,c) and also in the space between the wings and the body. For a while the larvae eat through the cuticle of the pupa but soon penetrate inside the body and feed within.

Martelli (1907) and Blunck & Jansen (1957) believed that *H. melanarius* and *P. puparum* are able to exist as co-parasites within the same host, whereas Faure (1926a,b) maintained that *H. melanarius* was a secondary parasite of *P. puparum*.

b) *Pimpla instigator* (Fabricius)

This is a solitary endoparasite of the pupa of *P. brassicae* (Faure, 1924). A considerable amount of work on the biology of this solitary pupal parasite and on the cellular immune response of its host, *P. brassicae*, has been carried out by Carton (1971, 1973, 1974, 1975a,b). Carton showed by using cylindrical paper tubes (24 cm long) that *P. instigator* can detect its host by olfaction when the host is placed inside, and that "host form", that is the behaviour of the pupae when molested, played an important part in stimulating the insertion of the ovipositor.

Blunck (1953) stressed the importance of this parasite in the field in the control of *P. brassicae* pupae. Recent work has been carried out on the structure and function of the male genitalia of *P. instigator* (Hogge & King, 1975).

c) *Pimpla turionellae* (Linnaeus)

This is another solitary endoparasite of the pupa which has a very wide host range (Führer, 1975a). The larva is, however, limited in its acceptance of different host species by two physiological processes. First, the female is able to secrete substances from three accessory glands, the uterus, the poison and the lubricant gland, which help to inhibit the encapsulation reaction of the host (*Pieris*) pupae. These secretions influence the plasmocytes and granulocytes of *P. brassicae* pupae and prevent them from encapsulating the larvae. Second, the parasite larva are very susceptible to "hypersensitivity" of the host and a high mortality of larvae may thus result.

d) *Pteromalus puparum* (Linnaeus)

This is a common British and European species which is a gregarious primary endoparasite of the pupae of *P. brassicae*, (see Figure 4, Table III in Strawinski (1929), adult on *P. brassicae* pupa) and can easily be bred in the laboratory (Soures, 1971; Bouletreau, 1972). Bouletreau (1968) reared two lots of *P. puparum* larvae on semi-synthetic diet containing: a) the anterior part of *Pieris* pupae whereby he obtained 40% success; and b) the haemolymph of *Pieris* pupae whereby he obtained 70% success.

The density of *P. puparum* larvae within the body of their host can influence both morphological and physiological features. At high density the larval metamorphosis of *P. puparum* is facilitated and hastened (Bouletreau & David, 1967), their oxygen consumption is increased and their individual metabolic rate is decreased (Bouletreau, 1971a). However, food supply may be limited and their host consumed that much quicker, which will result in the inhibition of growth and smaller larvae having greater overall biomass (Bouletreau, 1971b). When there are smaller numbers of larvae per host the larvae grow for longer periods, reach higher individual weights and have a smaller biomass; they may also leave a large amount of food untouched (Bouletreau, 1971b). An increase in the density of the larvae also results in a decrease in the dry and fresh weights of the male adults (Bouletreau, 1974).

The females oviposit on freshly formed pupae when the cuticle is still soft, and have often been seen to remain by or on the backs of fully fed fifth instar larvae waiting for them to shed their skin at the pupation site.

416

Picard (1923) stated that the later the pupae of *P. brassicae* formed in the year the less is their chance of being attacked by *P. puparum*. Turner (1933) reported that as many as 224 males and seven females came out of one exit hole in a parasitised pupa.

P. puparum can inflict great damage to populations of *P. brassicae*. In the Netherlands, de Bruijn (1921) recorded that *P. puparum* attacked 35.4% of *P. brassicae* pupae; and in France Paillot (1923) recorded that during an outbreak of *P. brassicae*, *A. glomeratus* was scarce but *P. puparum* and *Chalcis femoralis* were abundant. In a mass outbreak in Germany (DDR) in 1949, Godan (1949) recorded *P. puparum* as parasitising 50% of the pupae, and in Rumania 99% of *P. brassicae* pupae were recorded as parasitised (Andriescu, 1973, cf. Constantineanu *et al.*, 1957; Constantineanu & Mustaţa, 1970; Constantineanu & Pisica, 1966).

Dipterous parasites

Some 20 species of diptera have been recorded as parasites of *Pieris brassicae*, and these belong to five families, Drosophilidae, Phoridae, Sarcophagidae, Syrphidae and Tachinidae (Table 14.2). The earliest review of dipterous parasites of *P. brassicae* appears to be that of Thompson (1946) who mentioned seven species. Seven dipterous parasites of *P. brassicae* were listed by Mustaţa and Andriescu (1973) in their recent review of *P. brassicae* parasites.

Drosophila species have been listed as parasites by a number of people but it is most likely that their presence in cultures of *P. brassicae* can be accounted for by their breeding in decaying organic matter which is often found in their cages. Bonnamour (1921) was the first to record *Drosophila* as a parasite of *P. brassicae* but Faure (1926a,b) later suggested that his results needed checking. Faure (1926a,b) believed that the *Drosophila* found in his own cultures had bred in dead *P. brassicae* larvae for three to four days and then emerged 8–10 days later.

By far the largest number of dipterous parasites of *P. brassicae* belong to the Tachinidae. Faure (1926a,b) stated that *Compsilura concinnata* Meigen was a "normal" parasite of *P. brassicae*, while *Exorista (Tachina) larvarum* Linnaeus, *Ceromasia rubrifrons* (= *Ceromasia florum*) Macquart, *Phryxe vulgaris* Fallén and *Exorista subsp. moreti* Robineau-Desvoidy were "occasional" parasites. Bisset (1934) studied tachinid parasitism of both *P. brassicae* and *P. rapae* at Imperial College, London, and successfully reared or dissected out three tachinid species including two unidentified species, of which he made drawings. He noted that the Braconid *Apanteles glomeratus* (Linnaeus) suppressed tachinid development when there was a high rate of parasitism by *A. glomeratus*.

Table 14.2 Dipterous parasites (classification according to Kloet & Hincks, 1976; Mesnil, 1944–1975).

Species	Authority
DROSOPHILIDAE	
Drosophila busckii Coquillet	Bonnamour, 1921; Thompson, 1946
Drosophila pulvilineata Villeneuve	Faure, 1926a,b
Drosophila rubrostriata Becker	Faure, 1926a,b
PHORIDAE	
Dohrniphora cornuta Bigot	Marchal & Foex, 1918
Phora chlorogastra Becker	Bonnamour, 1921; Faure, 1926a,b
SARCOPHAGIDAE	
Agria (Sarcophaga) affinis (Fallén)	Voukassovitch, 1926
SYRPHIDAE	
Xanthandris comtus Harris	Portier, 1926; Lyon, 1969
TACHINIDAE	
Ceromasia rubrifrons Macquart	Faure, 1926a,b; Mustaţa & Andriescu, 1973
Ceromasia rutila Meigen	Thompson, 1946
Compsilura concinnata Meigen	Martelli, 1907; Blunck, 1922
Masicera serriventris Rondani	Picard, 1922a,b; Faure, 1926a,b; Mustaţa & Andriescu, 1973
Drino (Carcelia) lota Meigen	Thompson, 1946
Epicampocera succincta Meigen	Bisset, 1934
Eurithia (Ernestia) consonbrina Meigen	Serebroskii *et al.*, 1944
Exorista larvarum Linnaeus	Marchal & Foex, 1920; Faure, 1926a,b; Mustaţa & Andriescu, 1973
Exorista (Tachina) subsp. *moreti* Robineau-Desvoidy	Martelli, 1907; Faure, 1926a,b
Thicholyga segregata Rondani	Mustaţa & Andriescu, 1973
Exorista (Tachinid) sp.	Ghosh, 1914
Phryxe nemea Meigen	Bisset, 1934
Phryxe (Exorista, Voria) vulgaris Fallén	Bordeaux, 1917; Faure, 1926a,b; Bisset, 1934; Mustaţa & Andriescu, 1973; Thompson, 1946; Marchal & Foex, 1920
Phryxe (Voria) trepida Meigen	Thompson, 1946
Unknown species A, B	Bisset, 1934

In France, Marchal & Foex (1920) believed that a *Voria* sp. is a natural enemy of *P. brassicae*, along with *A. glomeratus* and *Pteromalus larvarum*. A study of the syrphid *Xanthandrus comtus* Harris in the south of France showed that it has several generations a year, each one in a different host, that being in *P. brassicae* during November and December (Lyon, 1969). A complete review of the parasites of *P. brassicae* in Rumania was published by Mustaţa & Andriescu (1973), who also collated the data of

Martelli (1907), Picard (1922a,b) and Faure (1926a,b) as well as from other sources.

Ectoparasites

In his book *Mites of Moths and Butterflies* Treat (1975) stated that 171 larvae of the acarine *Thrombidium mediteraneum* (Berlese) had been reported by Robaux (1971) on *Pieris brassicae*. Another parasite, *Pyemotes herfsi* (Oudemans), was apparently found in a pupa of *P. brassicae* which had been parasitised by the hymenopteran *Pteromalis puparum* (Herfs in Treat, *loc. cit.*). According to Robaux (*loc. cit.*) larvae of these acarine ectoparasites rarely stay on *P. brassicae* for more than 24 hours.

Endoparasites

In 1974 Stanuszek recorded the nematode *Neoplectana feltiae pieridarum* N. from *P. brassicae*. Naturally occurring bacteria in the alimentary canal of *P. brassicae* have been dealt with in chapter 15.

References cited

Abdulla, I., 1978. *Studies on host discrimination, mating and sex determination in Apanteles glomeratus (L.) and Pteromalus puparum (L.); hymenopterous parasitoids of Pieris species.* Ph.D. thesis, University of London (Wye College) Unpublished.

Andriescu, I., 1973. Importance economique des chalcids en Roumanie. *Ecol. Ter. Gen.* 1972–1973: 155–190.

Askew, R.R., 1971. *Parasitic Insects.* Heinemann Educational Books. London. pp. 316.

Berkovitch, I., 1972. Seeking pest solutions at Silwood Park. *Int. Pest Control* 14: 24–25.

Bignell, G.C., 1883. Extra-ordinary number of *Apanteles glomeratus* infesting *Pieris brassicae*. *Entomologist* 16: 263.

Birkett, N.L., 1944. The natural control of *Pieris brassicae* by its Braconid parasite, *Apanteles glomeratus*. *Entomologist* 77: 13–14.

Bisset, G.A., 1934. *Tachinids parasitising Pieris rapae and Pieris brassicae at Imperial College Field Station, Slough, Berks with a note on the extent of parasitism in Pieris rapae.* Diploma. University of London (Imperial College).

Blunck, H., 1922. Über den Massenwechsel des grossen Kohlweisslings bei Hamburg. *Mitt. biol. Zent. Anst. Berl.* 21: 182–184.

—— 1944a. Zur Kenntnis der Hyperparasiten von *Pieris (Autographa) brassicae* L. Pt. 1. *Mesochorus pectoralis* Ratz. und seine Bedeutung für den Massenwechsel des Kohlweisslings. *Z. angew. Ent.* 30: 418–491.

—— 1944b. Zur Kenntnis der Hyperparasiten von *Pieris brassicae* L. Relative Einflüsse auf die Populationsdichte von *Apanteles glomeratus* und *Pieris brassicae* im Vergleich zu andern Begrenzungsfaktoren des Massenwechsels. *Z. angew. Ent.* 30: 465–483.

—— 1950. Zur Kenntnis des Massenwechsels von *Pieris brassicae* L. mit besonderer Berücksichtigung des Dürrejahres 1947. *Z. angew. Ent.* 32: 141–171.

—— 1951a. Zur Kenntnis der Hyperparasiten von *Pieris brassicae* L. 3. Beitrag. *Hemiteles simillimus* Taschb. nov. var. *sulcatus*. Kennzeichen und Verhalten der Vollkerfe. *Z. angew. Ent.* 32: 335–405.

—— 1951b. Zur Kenntnis der Hyperparasiten von *Pieris brassicae* L. 4. *Gelis* cf. *transfusa* Forst. *Z. angew. Ent.* 33: 217–267.

—— 1951c. Parasiten und Hyperparasiten von *Pieris brassicae* L. *Z. PflKrankh. PflPath. PflSchutz* 58: 25–54.

—— 1952a. Zur Kenntnis der Hyperparasiten von *Pieris brassicae* L. 5. *Hemiteles simillimus sulcatus*, die Metamorphose. *Z. angew. Ent.* 33: 421–459.

—— 1952b. Zur Kenntnis der Hyperparasiten von *Pieris brassicae* L. 6. *Gelis corruptor* Forst. und *Gelis faunus* Forst. *Beitr. Ent. Berl.* 2: 94–109.

—— 1953. *Tierische Schädlinge an Nutzpflanzen*. 1. Teil. Lepidopteren und Trichopteren. Berlin.

Blunck, H. & Jansen, M., 1957. Zur Kenntnis von *Hemiteles melanarius* Grav. (Ichneumonidae). Ein Fall des Übergangs von Ekto- zum Endoparasitismus. *Z. PflKrankh. PflPath. PflSchutz* 64: 600–606.

Bonnamour, S., 1921. Note sur deux diptères parasites nouveaux de la piéride du chou (*Drosophila rubrostriata* Beck. et *Phora chlorogasta* Beck.). *Bull. Soc. ent. Fr.* (15): 217–219.

Bordeaux, 1917. Ravages des chenilles de *Pieris brassicae* avec les cruciferes. *Bull. Soc. Étude vulg. zool. agric. Bordeaux* 16: 108–110.

Bouletreau, M., 1968. Développement et croissance larvaires et conditions semiartificielles et artificielles chez un hyménoptère entomophage: *Pteromalus puparum* L. (Chalc.). *Entomophaga* 17: 265–273.

—— 1971a. Métabolisme respiratoire de *Pteromalus puparum* L. (Hym. Chalc.) au cours de développement et influence de la densité de population larvaire. *Ann. Zool. ecol. anim.* 3: 195–207.

—— 1971b. Croissance larvaire et utilisation de l'hôte chez *Pteromalus puparum* (Hym. Chalc.): Influence de la densité de population. *Ann. Zool. ecol. anim.* 3: 305–318.

—— 1972. Premiers résultats de l'élevage des larvas d'un Hyménoptère chalcidien (*Pteromalus puparum* L.) sur hemolymphe des lépidoptères. *Entomophaga* 13: 217–222.

—— 1974. Influence de la densité de population larvaire sur quelques caractères biometrique des adultes. Études chez un hyménoptère parasite *Pteromalus puparum*. *Ann. Zool. ecol. anim.* 6: 200–204.

Bouletreau, M. & David, J., 1967. Influence of the density of the larval population on the size of the adults, the duration of development and the frequency of diapause in *Pteromalus puparum*. *Entomophaga* 12: 187–197.

Carton, Y., 1971. Biologie de *Pimpla instigator* F. 1793 (Ich. Pimplinae) I. Mode de perception de l'hôte. *Entomphaga* 16: 285–296.

—— 1973. La biologie de *Pimpla instigator* (Ich. Pimplinae). *Entomophaga* 18: 25–39.

—— 1974. Sur la biologie de *Pimpla instigator* (Ich. Pimplinae) 3. Analysis experimental sur le recognition de l'hôte pupal, *Pieris brassicae*. *Ent. exp. appl. Amst.* 17: 265–278.

—— 1975a. Reactions haemolytiques dans pupae des lépidoptères. *C. r. hebd. Séanc. Acad. agric. Fr.* 281: 579–582.

—— 1975b. Colloquium on immunity in insects. Angers, 30th Sept. to 1st Oct. 1975. *Annls. Parasit. Huma. comp.* 52: 53–99.

Ciochia, V., 1966. Sur la biologie de l'ichneumonidae *Hemiteles melanarius* Grav. *Anal. St. Univ. Al. 1. Cuza Iasi (Seria noua) Sect.* II (St. Natur-Geogr.) 13: 355–364.

Constantineanu, I.M., 1929. *Contribution à l'étude des Ichneumonides en Roumania.* Thèse dans les Annal. St. Univ. Jassy. 15: 389–642.

Constantineanu, I.M. & Pisica, C., 1966. Ichneumonides de la Roumanie obtenus par les cultures et leurs hôtes. *Anal. St. Univ. Al. 1 Cuza* (Ser. noua) Sect. II. 12: 205–215.

Constantineanu, M.I.I., Suciu, I., Andriescu, V., Ciochia, V. & Pisica, C., 1957. Contributii la studiul himenopterelor parazite in albilita verzii (*Pieris brassicae* L.) din R.P.R. *Anal. St. Univ. Al. Cuza Iasi* (Sera noua). Sect. II. 3: 1–7.

Constantineanu, M.I. & Mustaţa, G., 1970. Ichneumonide (Hym. Ichn.) obtinute prin culturi de insecte daunatoare legumelor in Moldava. *Stud. Com. St. Nat. Muz. Jud. Suceava.*

Corunna, 1934. Mem. work. Phytopath. Stn. Corunna. *Publ. Estac. Fitopat. Agric. Galicia* 1935: 1–79.

Danilevskii, A.S., 1965. *Photoperiodism and Seasonal Development of Insects.* English Translation of 1961 edition. Oliver & Boyd, Edinburgh & London.

David, W.A.L. & Gardiner, B.O.C., 1952. Laboratory breeding of *Pieris brassicae* L. and *Apanteles glomeratus* L. *Proc. R. ent. Soc. Lond.* 27: 54–56.

De Bruijn, J.F., 1921. Wat kant er in den herfst van de Witjesrupsen terecht. *De Levende Natuur* 25: 210–215.

Fabre, J.H., 1908. La chenille de chou. *Rev. Quest. Sci. Louvain* 14: 349–374.

Faure, J.C., 1924. Sur la determination de la ponte chez *Pimpla instigator* (Ichneumonidae) parasite de *Pieris brassicae* L. *Feuille Nat.* 45: 153–157.

—— 1925a. Sur la multiplicité des parasites de l'*Apanteles glomeratus* L. *C. r. Séanc. Soc. Biol.* 93: 524–526.

—— 1925b. Contribution à l'étude des hyménoptères parasitiques. *Rev. path. veg. ent. Agric.* 12: 293–305.

—— 1926a. *Contribution à l'étude d'un complexe biologique: La Piéride du chou (Pieris brassicae L.) et ses parasites hyménoptères.* Thèse. Université de Lyons (Faculté de Science). pp. 222.

—— 1926b. Sur la spécificité relative des insectes parasites polyphages. *C. r. hebd. Séanc. Acad. Sci. Paris* 182: 243–245.

Faure, J.C. & Zolotarewsky, B., 1925. Contribution à l'étude biologique de *Dibrachys boucheanus* Ratz. *Rev. path. veg. ent. Agri.* 12: 144–161.

Ferrière, C. & Faure, J.C., 1924. Sur *Trichogramma evansecens* Westw. parasite des oeufs de *Pieris brassicae* L. *Rev. path. veg. ent. agri.* 11: 104–118.

—— 1925. Contribution à l'étude des Chalcidiens parasites de *Apanteles glomeratus. Ann. Épiphyt.* 11: 221–234.

Filipescu, I.C., 1972. Contributiuni la studiul sistematic, biologic, ecologic si economic a Fam. Braconidae (Hymenoptera) parazite in insecte daunatoare agriculturii Teza, Univ. Al. I Cuza.

Fitton, M.G., 1976. The western Palaearctic Ichneumonidae of British Authors. *Bull. Br. Mus. nat. Hist. (Ent.)* 32: 301–373.

Ford, E.B., 1945. *Butterflies.* Collins, London.

Frechin, E., 1959. Mes captures d'hymenoptères en 1957 et 1958. *Bull. Soc. ent. nord. Fr.* (104) 3.

Frohawk, F.W., 1906. Life history of *Aporia crataegi*. *Entomologist* 39: 132–138.

Führer, E., 1972. Abnorme Glykogenspeicherung in Larven von *Pieris brassicae* (Lepidoptera: Pieridae) als Folge des Parasitismus von *Apanteles glomeratus Z. angew. Ent.* 70: 370–374.

—— 1975a. Über die physiologische Spezifität des polyphagen Puppenparasiten *Pimpla turionella* L. (Hym. Ichneumonidae) und ihre ökologischen Folgen. *Sond. Centr. Ges. Forst.* 92: 218–227.

—— 1975b. Der Einfluss einiger Endoparasiten auf die Entwicklung der Kutikula ihrer Wirte. *Z. angew. Ent.* 77: 274–279.

Führer, E. & Keja, T.D., 1976. Physiologische Wechselbeziehungen zwischen *Pieris brassicae* und dem endoparasiten *Apanteles glomeratus*, der Einfluss der Parasitierung auf Wachstum und Körpergewicht des Wirtes. *Ent. exp. appl. Amst.* 19: 287–300.

Gatenby, J.B., 1918. Notes on the bionomics, embryology, and anatomy of certain Hymenoptera Parasites, especially of *Microgaster connexus* (Nees). *J. Zool. Linn. Soc.* 33: 287–416.

—— 1919. Note on *Apanteles glomeratus*, a braconid parasite of the larva of *Pieris brassicae. Entomologist's mon. Mag.* 55: 19–26.

Gautier, C., 1918. Études physiologiques et parasitologiques sur les lépidoptères nuisibles. *C. r. Soc. Biol.* 81: 44–46, 197–199, 801–803, 1152–1155.

—— 1919. Recherches physiologiques et parasitologiques sur les larves des lépidoptères nuisibles. *C. r. Soc. Biol.* 82: 72–721, 1000–1002, 1369–1371.

Gautier, C. & Bonnamour, S., 1924. Recherches sur *Tetrastichus rapo* Walker. *Rev. path. veg. ent. Agric.* 11: 246–253.

George, L., 1927. Observations sur la biologie de deux Hymenopteres entomophages. *Bull. Soc. Hist. nat. Afr. N. Algiers* 18: 55–71.

—— 1928. Sur la biologie de l'*Apanteles glomeratus* L. *Bull. Soc. Hist. nat. Afr. N. Algiers* 19: 104–112.

Ghosh, C.C., 1914. Life history of Indian insects. Lepidoptera. *Mem. Dept. Agric. India Pusa* 5: 1–72.

Godan, D., 1949. Massenauftreten von Kohlweisslingsraupen (*Pieris brassicae*) in Berlin. *Biol. Zentanst.* 1: 165–166.

Goedart, J.B., 1662. *Metamorphosis et Historia naturalis Insectorum.* Middlesborough. (A copy is in the British Museum (Natural History) library.)

Graffinn, M.G., 1679. *Der Raupen wunderbare Berivandelung und Sonderbare Blumen-nahrung.* Graffen, Frankfurt.

Graham, M.W.R. de V., 1961. The genus *Aprostocetus* Westwood, sensu lato (Hym. Eulophidae); notes on the synonymy of European species. *Entomologist's mon. Mag.* 97: 34–64.

—— 1969. Pteromalidae of north west Europe. *Bull. Br. nat. hist. ent.* Suppl. 16.

Grandori, R., 1911. Contributo all'embriologia e alla biologia dell' *Apanteles glomeratus* (L.) Reinh., immenottero parasita del bruco di *Pieris brassicae* L. *Redia* 7: 363–428.

Hamilton, A.G., 1934. *The biology of Apanteles glomeratus (Braconidae) parasitising Pieris brassicae and Pieris rapae* L. Diploma. University of London (Imperial College).

—— 1935. Miscellaneous observations on the biology of *Apanteles glomeratus* (Braconidae). *Entomologist's mon. Mag.* 71: 262–270.

Hogge, M.A.F. & King, P.E., 1975. The structure and function of the male genitalia of *Pimpla instigator* F. (Hym. Ichneumonidae). *Entomologist's mon. Mag.* 111: 193–200.

Howard, L.O. & Fiske, W.F., 1911. The importation into the US of the parasites of the gipsy moth and brown tail moth. *US Dept. agric. Bull.* 91: 344 pp.

Imms, A.D., 1925. *General Textbook of Entomology.* Methuen, London.

Johannson, A.S., 1951. Studies on the relation between *Apanteles glomeratus* (Hym. Brac.) and *Pieris brassicae* L. (Lep. Pieridae). *Norsk. ent. Tidsskr.* 8: 145–186.

Junnikkala, E., 1966. Effect of braconid parasitization on the nitrogen metabolism of *Pieris brassicae* L. *Ann. Acad. Sci. fenn.* 100: 1–83.

Karadjov, S., 1972. The joint utilisation of biopreparations entobacterin (*Bacillus cereus* var. *galleriae*) and boverin (*Beauveria bassiana* Vuill.) with the egg parasite *Trichogramma evanescens* West. for the control of *Pieris brassicae* L. and *Pieris rapae* L. and the cabbage worm (*Mamestra brassicae* L.). *Acad. agric. Sci. Inst. Plant Prot. Kostinbrod. Stud. Biol. Cont. pl. Pests* Vol. 1. 129–143.

Kitano, H., 1974. Effects of the parasitization of a braconid, *Apanteles*, on the blood of its host, *Pieris*. *J. Insect Physiol.* 20: 315–327.

Klein, H.Z., 1932. Studien zur Oekologie und Epidemiologie der Kohlweisslinge. II. Zur Bionomie von *Pieris brassicae* L. und deren Parasit *Microgaster glomeratus* L. *Z. wiss. InsektBiol.* 26: 192–199.

Kloet, G.S. & Hincks, W.D., 1972. *Check list of British Hymenoptera.* British Museum (Natural History).

Kot, J., 1961. Experiments on the use of *Trichogramma evansecens* Westw. in pest control of the vegetable garden pests *Plutella maculipennis* Curt., *Pieris brassicae* L. and *Pieris rapae* L. *Ekol. Polska* B 5: 83–88.

Kot, J. & Plewka, T., 1968. The parasitism of eggs of the large white cabbage butterfly, *Pieris brassicae* and the small white cabbage butterfly, *Pieris rapae* by *Trichogramma evanescens* in natural conditions. (In Polish.) *Polskie Pismo ent.* 38: 619–625.

Lacatuscu, M., 1963. *Studiul Braconidelor din R.P.R. (Sistematic, morfologic, biologic, zoogeographic).* Teza de doctorat, Minist. Inv. Univ. Bucuresti.

Lal, O.P. & Chandra, J., 1976. Some parasites of cabbage worm *Pieris brassicae* L. (Lepidoptera: Pieridae) from the Kulu Valley, Himachal Pradesh (India). *Curr. Sci.* 45: 766–767.

Lartschenko, K., 1932. Definite reaction to parasites of larvae of *L. sticticalis* and *Pieris brassicae*. *Z. Parasitenk.* 5: 679–707.

Lyle, G.T., 1908. Oviposition of a hyperparasite (Chalcid) of *Pieris brassicae*. *Entomologist* 61: 249.

Lyle, G.T., 1916. Contribution to our knowledge of the British Braconidae. *Entomologist* 49: 121–125.

—— 1926. *Pieris brassicae* and its parasite *Apanteles glomeratus*. *Entomologist* 59: 302–303.

Lyon, J.P., 1969. Contribution to study of the biology of *X. comptus*. *Ann. Épiphyt.* 19: 683–693.

Marchal, P. & Foex, E., 1920. Rapport phytopathologique pour les années 1919–20. *Ann. Épiphyt.* 7: 1–87.

Marshall, T.A., 1885. Monograph of British Braconidae Part I. *Trans. R. ent. Soc. Lond.* 1–280.

Martelli, G., 1907. Contribuzioni alla Biologia della *Pieris brassicae* L. e di alcuni suoi parassiti ed iperparassiti. *Boll. Lab. zool. gen. Agrar R. Scuola. Sup. Agric. Portici* 1: 170–224.

Maslennikova, V.A. & Mustafayeva, T.M., 1971. An analysis of photoperiodic adaptations of *Apanteles glomeratus* L. (Hymenoptera, Braconidae) and *Pieris brassicae* L. (Lepidoptera: Pieridae). *Ent. Rev.* 50: 281–284.

423

Matheson, R., 1907. The life history of *Apanteles glomeratus*. *Can. Ent.* 39: 205–207.

Meïer, N.F., 1925. Study of some larvae in relation to their parasites. (In Polish.) *Ann. State Inst. Expt. Agron.* 3: 260–265.

—— 1938. Biological methods of controlling injurious insects and results of application in USSR (with additional material on races of *Trichogramma*. (In Russian.) *Zool. Zh. Moscow* 17: 905–932.

—— 1940. Species and races of the genus *Trichogramma* Westw. (In Russian.) *Bull. Plant Prot. Leningrad* 1940: 70–77.

—— 1941. Ecology and results of utilisation for the control of injurious insects. (In Russian.) *Sel'khozgiz Moscow* 1941: 1–175.

Mesnil, L., 1944–1975. *Tachinidae. Fliegen palaearkt. Reg.* 9–10: 1–1435.

Mickel, C.E., 1973. John Ray: Indefatigable student of nature. *A. Rev. Ent.* 18: 1–16.

Moiseeva, N.V., Mashkina, L.G., Kalashnikova, G.I. & Ratomskaya, A.A., 1975. The effectiveness of the use of *Trichogramma* for control of certain pests in the Chu Valley, Kirgizia. *In: Arthropods of Agricultural Importance*. Ed. A.I. Protsenko. Emt. issled Kirgizii Vypusk. Kirgiz SSR.

Mokrzecki, Z., 1934. Chalcidoid parasites and hyperparasites of forest pests. *Polskie. Pismo ent.* 12: 143–144.

Morley, C., 1936. Notes on Braconidae XV: Microgasterinae. *Entomologist* 1936: 39–42.

Morley, C. & Rait-Smith, W., 1933. The hymenopterous parasites of British Lepidoptera. *Trans. ent. Soc. Lond.* 81: 133–183.

Moss, J.E., 1933. The natural control of the cabbage caterpillars, *Pieris* spp. *J. Anim. Ecol.* 2: 210–231.

Mustaţa, G. & Andriescu, I., 1973. Recherches sur le complexe de parasites (Insecta) du Papillon du chou (*Pieris brassicae* L.) en Moldavie (R.s. de Roumanie) I. Parasites primaires. Lucrările Staţiunii "Stejarul". *Ecol. Ter. gen.* 1972–73: 191–230.

Nonell Comas, J., 1922. Entomophagous insects. Use of coccinelid *N. cardinalis* against Australia Scale *I. purchasi. Rev. Inst. Agric. Catalan S. Isidro* 72: 211–213.

Ormerod, E.A., 1884. *Notes and observations of injurious insects and common crop pests, 1884*. Simpkin, Marshall, London.

Osmolovski, G.Y., 1964. The biology of *Apanteles glomeratus* L. (Hymenoptera: Braconidae), a parasite of *Pieris brassicae*. *Ent. Rev. URSS* 43: 387–388.

Paillot, A., 1918a. Un nouveau parasite coccobacilli du cockchafer. *C. r. hebd. Séanc. Acad. Sci. Paris* 167: 1046–1048.

—— 1918b. Deux nouveaux parasites microsporidiens de larva de *Pieris brassicae*. *C. r. Soc. Biol. Paris* 81: 66–68.

—— 1923. Sur *Chalcis femorata* Panz. nouveau parasite hyménoptère des chrysalides de *Pieris brassicae* L. *Rev. path. veg. ent. Agric.* 10: 342–345.

Paillot, A., Ferriere, C. & Faure, J.C., 1924. Note préliminaire sur les parasites des *Apanteles* hôtes de *Pieris brassicae* L. dans la région de Lyon 1923. *Rev. path. veg. ent. Agric.* 11: 78–85.

Picard, F., 1921. Sur la biologie de *Tetrastichus rapo* Walk. (Hym. Chal.). *Bull. Soc. ent. Fr.* 1921: 206–208.

—— 1922a. Contribution à l'étude des parasites de *Pieris brassicae* L. *Bull. Biol. Fr. Belg.* 56: 54–130.

—— 1922b. Note sur la biologie de *Melittoba acasta* Walk. Bull. *Soc. ent. Fr. Paris* 1922: 301–304.

—— 1923. L'hibernation des chenilles de *Pieris brassicae* L. *Bull. Biol. Fr. Belg.* 57: 98–106.

Portier, P., 1949. *La Biologie des Lépidoptères.* Lechavalier, Paris.

Prokofera, N.A., 1976. *Trichogramma* in cabbage fields. (In Russian.) *Zasch. Rast.* No. 9. 19–20.

Quednau, W., 1956. Die biologischen Kriterien zur Unterscheidung von *Trichogramma* Arten. *Z. PflKrankh. PflPath. PflSchutz* 63: 333–344.

Ray, J., 1710. *Historia Insectorum.* pp. 114.

Richards, O.W., 1940. Biology of *Pieris rapae* with special reference to factors controlling abundance. *J. Anim. Ecol.* 9: 243–288.

Robaux, P., 1971. *Recherches sur le développement et la biologie des Acariens, Thrombidiidae.* Thèse. Université de Paris.

Schwarz, R., 1948. *Butterflies Diurnal.* Vesmir, Prague.

Serebroskii, A.S., Tupikov, V. & Khvostova, V.B., 1944. *Ernestia consobrina* Meig. a parasite of the larvae of Noctuids which attack vegetables. (In Russian.) *Proc. Lenin. Acad. Agric. Sci. USSR* 9: 16–19.

Shaw, M.R. & Askew, R.R., 1976. Parasites. *In: The Moths and Butterflies of Great Britain and Ireland.* Ed. J. Heath. Vol. 1. Blackwell, Oxford. pp. 24–56.

Shenfelt, R.D., 1972. *Hymenopterorum Catalogus* (Nova edition) Ed. J. van der Vecht and R.D. Shenfelt.

Shevirev, I.J., 1913a. Regulation of sex of their offspring by female Ichneumonidae. (In Russian.) *Bull. lab. biol. St. Petersburg* 111: 24–30.

—— 1913b. Oviposition in Ichneumon flies. *Jl. R. Microscop. Soc. Lond.* 385.

Silujanova, O.E., 1969. Control of cabbage white butterflies. *Zashch. Rast.* 14: 51.

Soures, B., 1971. La lutte biologique et ses avantages. *Pepinieristes Horticulteurs Maraichers* 119: 47–48.

Speyer, W., 1956. *Pieris brassicae* L. in den Dünen der Nordseeinsel Amrum. *Z. PflKrankh. PflPath. PflSchutz* 63: 12–14.

Stanuszek, S., 1974. *Neoplectana feltiae pieridarum* N. ecotype (Nematoda: Rhaoditoidea, Steinernematidae) a parasite of *Pieris brassicae* L. and *Mamestra brassicae* L. in Poland. Morphology and biology. *"Nematodes and cultivated and natural environment"* Zesz. Probl. Post. nauk. Roln. 154: 361–393.

Strawinsky, K., 1929. Cabbage butterfly. Biology and control. *Polskie Pismo ent.* 8: 227–248.

Talhouk, A.M., 1969. *Insects and mites injurious to crops in Middle Eastern countries.* Monographien zur Angew. Entomologye No. 21. Verlag Paul Parey, Hamburg & Berlin.

Tawfik, M.F.S., 1956. *On the host-specificity and early development of Apanteles glomeratus (Brac. Hym.)* Ph.D. thesis. University of Dublin (Trinity College).

Telenga, N.A., 1929. Hymenopterous parasites of F. Ichneumonidae, reared at Kuban Plant Protection Station, 1927. *Plant Prot. Leningrad* 6: 225–226.

—— 1955. Fauna SSSR. Pereponciatokrillie Tom. V. vip. 4. Sem. Braconidae, Podsem. Microgasterinae, posem, Agathinae. Zool. Inst. Akad. *SSSR Nov. Ser. N. 61. Izd. Acad. nauk. SSSR Moskva-Leningrad.* pp. 311.

Thompson, W.R., 1946. *A catalogue of parasites and predators of insect pests.* Section 1. Part 8. Imperial Agricultural Bureaux, Belleville.

—— 1953. *A catalogue of the parasites and predators of insect pests.* Section 2. Part 2. Commonwealth Agricultural Bureaux, Ottawa.

Torka, V., 1927. *Angita rufipes* Grav. ein Parasit der Kohlweisslingsraupe. *Anz. Schädlingsk.* 3: 97.

Treat, A.E., 1975. *Mites of Moths and Butterflies.* Comstock Publishing Association. Cornell University Press.

425

Turner, J., 1933. South London Entomological and Natural History Society. Meeting. July 27th. *Entomologist*.

Verma, G.C., Singh, S. & Bindra, O.S., 1976. Record of a hyperparasitoid, *Catolaccus crassiceps* (Masi) (Hymenoptera: Pteromalidae) from India. *Entomologist's Newsletter* 6: 44.

Völkel, H., 1925. Ueber die praktische Bedeutung der Schlupferwespe *Trichogramma evanscens* Westw. *Arb. biol. Bundanst Forst.* 14: 97–108.

Voukassovitch, P., 1926. Observations biologique sur les parasites de la Pieride du chou (*Pieris brassicae* L.). *Rev. zool. agric. appl. Bordeaux* 25: 81–90, 103–108, 113–121, 134–140.

Westropp, M.S.D., 1925. *Pieris brassicae* in January. *Entomologist* 58: 39.

Wilbert, H., 1960. *Apanteles pieridis*, a parasite of *Aporia crataegi*. *Entomophaga* 5: 183–211.

Wilkes, B., 1747–49. *The English Moths and Butterflies, together with the plants, flowers, and fruits whereon they feed, and are usually found.* Benjamin Wilkes, Fleetstreet, London.

Wilkinson, D.S., 1928. Revision of the Indo-Australian species of the genus *Apanteles*. Pt. 1. *Bull. ent. Res.* 19: 79–146.

—— 1939. Two species of *Apanteles* (Hym. Brac.) not previously recognised from the western Palaearctic region. *Bull. ent. Res.* 30: 77–84.

Further references

Adaskievici, B.P., 1972. *In: Min. Sel'sk. Hoz. Mold. S.S.R. Naucin. Isled. Inst. Oros.* 3–29 (parasites in Rumania).

Alder, S., 1918. *Z. wiss. InsektBiol.* 14: 182–186 (biology of *Apanteles*).

—— 1920. *Aus. Natur.* 16: 236–243 (biology of *Apanteles*).

Allen, W.W., 1958. *Hilgardia* 27: 515–541 (biology of *Apanteles*).

Bendel-Janssen, M., 1962. *Z. PflKrankh. PflPath. PflSchutz* 69: 526–529 (parasitism of *Apanteles*).

Benedek, P., 1972. *Acta Phytopath. Acad. Sci. Hung.* 7: 445–452 (effect of parasitism on host stability).

Benitez Morera, A., 1945. *Bol. Pat. Veg. ent. agric. Madrid* 13: 463–466 (*P. instigator* on *P. brassicae*).

Birkett, N.L., 1944. *Entomologist* 73: 13–14 (control by *A. glomeratus*).

Bisset, G.A., 1938. *Parasitology* 30: 111–122 (tachinids).

Bjegovic, P., 1962. *Arh. Poljopr. nauk Belgrade* 15: 125–131 (tachinid).

Blunck, H., 1951a. *Z. angew. Ent.* 32: 335–405 (hyperparasites).

—— 1951b. *Z. angew. Ent.* 33: 217–267 (hyperparasites).

Bree, W.T., 1832. *London's Mag. Nat. Hist.* 5: 105 (*A. glomeratus*, colour of larvae).

Ceballos Jimenez, P., 1961. *Bol. Serv. Plag. For. Madrid* 4: 97–101 (reproduction of *P. instigator* in captivity).

Chandra, G. & Gupta, V.K., 1977. *Ichneumonologia orientalis*. Part VII. The Tribes of Lissonofini and Banchini. Oriental Insects Monograph No. 7. (*Exetastes illusor*).

Creighton, C.S., McFadden, T.L., Cuthbert, R.B. & Onsager, J.A., 1972. *J. econ. Ent.* 65: 1399–1402 (control with *B. thuringiensis*).

Crutchley, G.W., 1922. *Entomologist* 55: 245–246 (biology of *A. glomeratus*).

Debach, P., 1974. *Biological Control by Natural Enemies*. Cambridge University Press (early details of parasitism by *A. glomeratus*).

Dewitz, J., 1912. *Bull. ent. Res.* 3: 343–354 (experiments on cocooning of *A. glomeratus*).

Doutt, R.L., 1959. *A. Rev. Ent.* 4: 161–182 (biology of parasites).

Dowden, P.B., 1935. *J. agric. Res. Washington* 50: 495–523 (*B. intermedia*).

Farwick, S., 1947. *Zur Kenntnis der Hyperparasiten von Pieris brassicae L. Über einige Chalcididen die Parasiten von Apanteles glomeratus*. Dissertation. University of Bonn. pp. 108.

Faure, J.C., 1926. *Congres. Nat. sur la lutte contre les ennemis des cultures Lyon.* 1–375 (methods of destroying *P. brassicae*).

Führer, E., 1975. *Z. angew. Ent.* 77: 274–279 (biochemical effects of parasite).

Grandori, R., 1911. *Redia* 7: 363–428 (embryology of *A. glomeratus*).

Harvey, J.H., 1944. *Lond. Nat.* 1943: 12–19 (parasites).

Janssen, M., 1960. *Z. PflKrankh. PflPath. PflSchutz* 67: 19–24 (biology of *A. glomeratus*).

Jegen, G., 1918. *Landw. Jb. Schweiz.* 32: 524–550 (parasites).

Kot, J., Krukierek, T., Plewka, T., 1975. *In: Studies on crop-field ecosystem*. Ed. L. Ryszkowski, pp. 173–182 (resistance of *T. evanescens* to insecticides).

Krukierek, T., Plewka, T. & Kot, J., 1975. *In: Studies on crop-field ecosystem*. Ed. L. Rzszkowski. pp. 183–196 (susceptibility of parasites to insecticides).

Kurir, A., 1946. *Zentbl. Ges. Forst. Holzw.* 70: 81–88 (*P. puparum*).

Kutin, A., 1917. *Z. PflKrankh. PflPath. PflSchutz* 26: 452–454 (*A. glomeratus*).

Lyle, G.T., 1922. *Entomologist* 55: 281 (defensive fluid of *P. brassicae*).

Marchall, P., 1927. *C. r. Acad. Sci. Paris* 185: 489–493 (*T. evanescens*).

Martin, F., 1910. *Mem. Soc. Vulg. Sci. Nat. Deux-Sevres* 1: 84–90 (larval parasite).

Miles, P.M., 1950. *Entomologist's mon. Mag.* 86: 141 (*Drosophila* on larvae).

Moiseeva, T.S., 1960. *Trudy. Vses. Inst. Zashch. Rast.* 14: 51–56 (specialisation of *A. glomeratus*).

Mokrzecki, Z., 1934. *Polskie Pismo Ent.* 12: 143–144 (parasites and hyperparasites).

Muggeridge, J., 1939. *N. Z. J. Agric.* 58: 305–307 (control of *P. rapae* with *A. glomeratus*).

Muller Hale, D.E.W., 1966. *Der deutsche Gartenbau* 13: 218–219 (parasites).

Nikol'skaya, M., 1934. *Bull. ent. Res.* 25: 129–143 (chalcids reared in USSR).

Nixon, G.E.L., 1973. *Bull. ent. Res.* 63: 169–228 (Revision of NW European *Apanteles*).

Olenev, N.O., 1925. *Defense des Plantes* 2: 377–378 (biology of *A. glomeratus*).

Parker, F.D., 1970. *Agric. Res. Wash.* 18: 8–9 (build up of parasites).

Pavlichek, K.I., 1962. *Zashch. Rast.* (10) 52 (*Trichogramma*).

Rabbm, R.L. & Thurston, R., 1969. *Annl. ent. Soc. Am.* 62: 125–128 (diapause in *Apanteles*).

Richter, V., 1960. *Ent. Z.* 70: 114–116 (parasites).

Ritzema, J., 1924. *Tijdschr. Plantenz.* 30: 65–67 (life history of *A. glomeratus*).

Sakharov, N.L., 1938. *Soc. Grain Fmg. Saratov* 1938: 164–168 (races of *Trichogramma*).

Schoonhoven, L.M., 1962. *Symp. Diapause in Relation to Insect pest control*. London 1962. 617–621 (discussion on parasite host systems).

Seurat, L.G., 1899. *Ann. Sci. nat. zool.* 10: 1–159 (parasites).

Shapiro, V.A., 1970. *Byull. Vses. Nauch. Issled. Inst. Zash. Rast.* 1968: 3–7 (host parasite relationships).

Shapiro, V.A. & Khotyanovitch, A.V., 1962. *Proc. Symp. use of Biophys. in Field of Plant Protection. Leningrad. Vsesoyuz. Inst. Zashch. Rast.* 1961. 67–69 (developmental characteristics of *A. glomeratus*).

Steinberg, D.M., 1961. *Dokl. Akad. nauk. SSSR* 138: 1477–1480 (host/parasite relationships).

SLENHS (South London Entomological and Natural History Society), 1946. *Entomologist's mon. Mag.* 82: 120 (*P. puparum* exhibited).

Spanjer, W., Grosu, L. & Piek, T., 1977. *Toxicon* 15: 413–422 (effect of homogenate from *Microbracon hebetor* on *P. brassicae*).

Stellwaag, F., 1929. *Verh. dt. Ges. angew. Ent.* 31: 15–32 (biological control).

Suster, P., 1931. *Annal. Sci. Univ. Jassy* 16: 57–249 (tachinids).

Taschenberg, E.L., 1865. *Naturgeschichte der wirbellosen Tiere.* Leipzig. (*H. melanarius*).

Thorpe, W.H., 1930. *Bull. Ent. Res.* 21: 387–412 (parasite on *P. brassicae*).

Tuhan, N.C., Bindra, O.S. & Grewal, C., 1977. *Ind. J. Ent.* 37: 208–209 (*P. brassicae* tried experimentally as host for *T. israeli*).

Weiseenberg, R., 1908. *Sitzungsber Ges. naturf. Freunde Berlin* (biology of *A. glomeratus*).

Wilbert, H., 1958. *Z. PfKrankh. PflPath. PflSchutz* 11: 661–673 (*A. glomeratus*).

—— 1959. *Beitr. Ent. Berl.* 9: 874–898 (*A. glomeratus*).

Willers, D., 1974. *Untersuchungen über die physiologische Eignung verschiedener Schmetterlingspuppen als Wirte für den Puppenparasiten Pimpla turionella* L. Diplomarbeiten. University of Göttingen.

15. Pathogenic control

General introduction

Several workers have been so closely associated with pathogenic control of insects that their names immediately come to mind in any discussion on this subject; they have also published much on *P. brassicae*. The best known team on this subject is David and Gardiner of the Glasshouse Crops Research Institute, Littlehampton and of the Agricultural Research Council Unit of Invertebrate Physiology at Cambridge; they have published more than 50 works on viral control of *P. brassicae* since 1962. In France early work on bacteria and viruses was done by Paillot (26[+] refs) and more recently Burgerjon 15[+] refs), Lecadet (7[+] refs) and Martouret (5[+] refs).

A general account of pathogenic control of insects, including details about means of control, application and interpretation of results may be found in Grison (1956a,b) and Hurpin (1970). The last author said that, of the 1000 entomopathogenic organisms so far identified, only five had come to any practical application; two of these, *Bacillus thuringiensis* Berliner and the fungus *Beauveria bassiana* (Balsamo) Vuilemin, are the only ones he mentioned which are effective against *P. brassicae*. Both of these have also been selected for commercial or experimental production. Hurpin did not mention other viruses such as the granulosis and polyhedrosis viruses, as well as protozoa and other species of fungi, which are pathogens of *P. brassicae* both in the wild and in the laboratory.

Grison & Silvestre de Sacy's (1956) paper gives a good account of the diseases which affect *P. brassicae* cultures and it includes photographs of larvae suffering from muscardine fungus attack, as well as a series of photographs showing the progressive effects of granulosis virus attack after infection.

There are advantages in using pathogens as biological control agents as they are sometimes cheaper and quicker to produce, do not leave toxic residues, and are non-toxic to man. However, caution should be exercised in using these micro-organisms as several are not host specific and may have secondary effects on beneficial insects including parasites.

Since completing this chapter an important review of the granulosis virus of *P. brassicae* has been published by David (1978). It includes details on general properties of the virus, effect of chemical and physical agents, bioassay, the relationships of the virus with its host, its epizootiology, classification and use in biological control.

SECTION A. BACTERIAL CONTROL
by J.S.E. Feltwell and H.D. Burges (Glasshouse Crops Research Institute)

a) *Bacillus thuringiensis* Berliner

Historical

Between 1870 and 1970, 90 species of bacteria had been described from insects (Falcon, 1971) and one – *Bacillus thuringiensis* Berliner – the most useful of these pathogenic bacteria – was first described as early as 1915 (Sandoz, 1974a,b).

Early work on bacterial control of *P. brassicae* in field and laboratory resulted in the naming of a profusion of bacterial species (Table 15.1) but none of these is recognised in present day bacteriology. However, *B. cazaubon* Toumanoff which was shown to give 100% mortality in two days was most probably *B. thuringiensis*.

In the 1950s there was a controversy concerning the correct nomenclature of *B. thuringiensis*: should it be called *B. thuringiensis* or *B. cereus* (Frankland & Frankland) Toumanoff (1956, 1959) used the name *B. cereus*, even up to his death, but modern authors virtually all use *B. thuringiensis*.

Table 15.1 Bacteria associated with *P. brassicae* up to 1944.

Bacterium	Authority
Bacillus agrotidis (from *Agrotis segetum*)	Pospelov, 1929
Bacillus cazaubon	Metalnikov, 1930
Bacillus hoplosternus non-liquifaciens-γ *Bacillus melolonthae non-liquifaciens-γ* }	Paillot, 1920
Bacillus pieris liquefaciens	Sweetman, 1936
Bacillus pieris non-liquefaciens-α	Paillot, 1920
Bacillus pirenei	Metalnikov, 1930; Pospelov, 1936
Metalnikov's *Bacillus-pirenei* type (?) (from *Galleria mellonella*)	Pospelov, 1936, 1939, 1944

Much work was carried out on *B. thuringiensis* in the 1950s. Rapidly, the number of strains in use by interested workers rose dramatically and some degree of standardisation was called for (Burges, 1967a,b,c). Today, 375 isolates of *B. thuringiensis* have been assembled at the United States Department of Agriculture.

A large proportion of the Lepidoptera are susceptible to the spore-crystal complex of one or other of the *B. thuringiensis* strains, although the degree of susceptibility of lepidopterous species varies greatly. *P. brassicae* falls into the most susceptible twenty or so of about 500 species that have been tested across the world. It is also susceptible to the beta-exotoxin. *B. thuringiensis* was reported to act more rapidly on larvae of *P. brassicae* than those of *Evergestis forficalis* (Linnaeus) (Pyralidae) (Garden Pebble Moth) and *Thaumetopoea pityocampa* (Schiffermüller) (Notodontidae) (Pine Processionary moth) (Grison, 1956a,b).

For further information on bacterial control of insects, the reader is referred to the very complete work on *Microbial Control of Insects and Mites* edited by Burges and Hussey (1971a,b), and its sequel *Microbial Control of Pests and Plant Diseases* edited by Burges (1981), in which there are several important chapters on *B. thuringiensis* (cf. Franz, Cooksey, Bond *et al.*, Burgerjon & Martouret, Burges, Burges & Hussey, Norris).

Structure of B. thuringiensis

A fully grown, sporulating, bacterial cell of *B. thuringiensis* has at one end a spore, which has a two-layered thick wall in an endosporium and at the other end a crystal of protein, as shown in Plate X.

The centre of a spore contains all types of compounds necessary to sustain life, i.e. carbohydrates, fats and proteins. When the spore is ripe, the cell wall decomposes, liberating the spore and the crystal of the original bacterial cell.

The crystal is bipyramidal in shape; thus if cut in the longitudinal plane it looks diamond-shaped. Biochemically the crystal is made up of a large protein, a protoxin which is inert when in the crystal form. However, on solution by gut enzymes, a smaller toxic molecule is released. This has an unknown molecular weight and is toxic only to lepidopterous larvae, and larvae of mosquitoes and blackflies which is a very valuable degree of specificity.

Serotypes

Twenty-one H-serotypes of *B. thuringiensis* have been described (Table 15.2). These are based virtually entirely on reaction to specific H-antisera in the definitive analysis. This classification agrees very closely with classifications based on biochemical characters (De Barjac & Bonnefoi, 1973) and

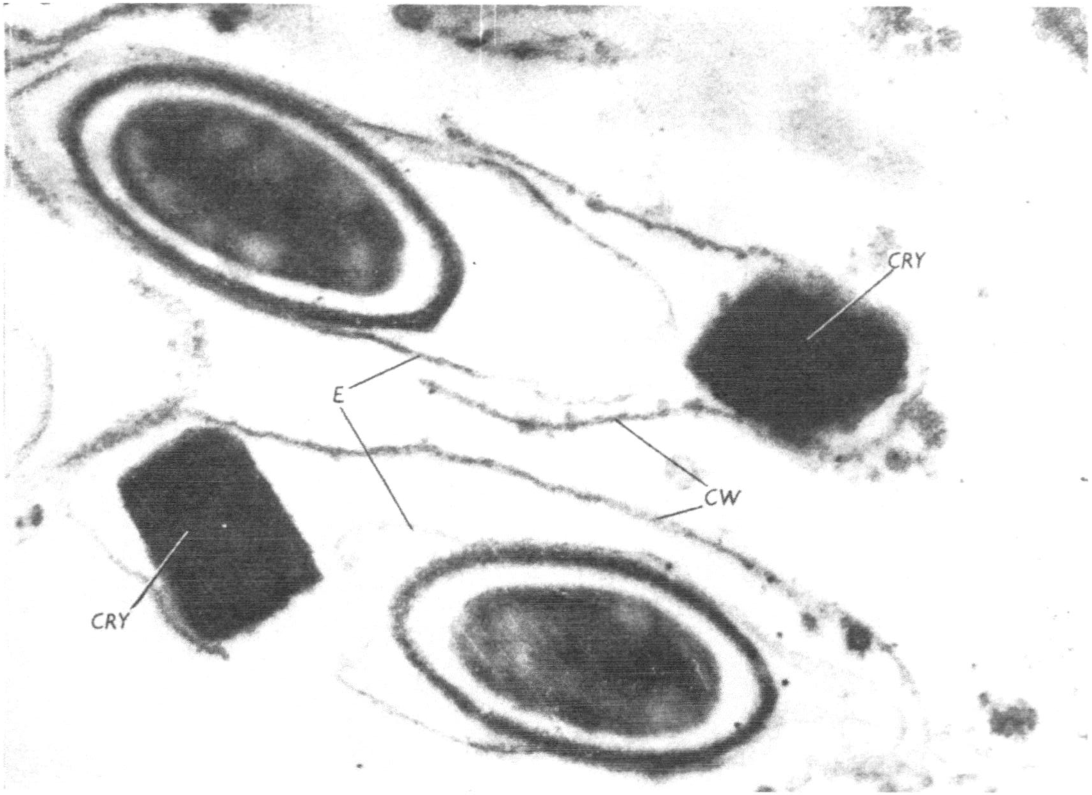

Plate X Bacillus thuringiensis Berliner sporulating
(Courtesy of H.D. Burges of the Glasshouse Crops Research Institute, and J.R.Norris of the Meat Research Institute)

Table 15.2 H-serotypes of *Bacillus thuringiensis.*

H-sero-type	Variety	Workers who used *P. brassicae*
I	*Berliner* (= *thuringiensis*)	Burgerjon, 1959; ARC, 1962; Martouret, 1962b; van der Geest, pers. comm. Geest, pers. comm.
II	*finitimus*	van der Geest, pers. comm.
IIIa	*alesti* (= *alesti anduze*)	Toumanoff & Grison, 1954; Vago *et al.*, 1961; Galowalia *et al.*, 1973; van der Geest, pers. comm.
IIIaIIIb	*kurstaki*	Lemoigne *et al.*, 1956; Biliotti *et al.*, 1956; Martouret, 1962b; van der Geest, pers. comm.
IVaIVb	*dendrolimus*	van der Geest, pers. comm.
IVaIVc	*kenyae*	van der Geest, pers. comm.
IVaIVb	*sotto*	Galowalia, 1970; van der Geest, pers. comm.
VaVb	*galleriae*	Burges *et al.*, 1975; Galowalia *et al.*, 1973; van der Geest, pers. comm.
VaVc	*canadensis*	van der Geest, pers. comm.
VI	*entomocidus*	Galowalia *et al.*, 1973; Chilingaryan *et al.*, 1969; van der Geest, pers. comm.
VI	*subtoxicus*	Galowalia *et al.*, 1973; van der Geest, pers. comm.
VII	*aizawai*	Galowalia *et al.*, 1973; van der Geest, pers. comm.
VIIIaVIIIb	*morrisoni*	Galowalia *et al.*, 1973; van der Geest, pers. comm.
VIIIaVIIIc	*ostriniae*	
IX	*tolworthi*	Galowalia *et al.*, 1973; van der Geest, pers. comm.
X	*darmstadiensis*	van der Geest, pers. comm.
XI	*toumanoffi*	van der Geest, pers. comm.
XII	*thomsoni*	⎫
XIII	*pakistani*	⎬ not yet tested
XIV	*israelensis*	⎪
XV	*indianae*	⎭

on vegetative cell esterases (Norris, 1964). Some confusion is encountered in the literature by the use of synonymous names and this is aggravated by the use of names from "in house" lists, but as there are so many strains and serotypes of *B. thuringiensis*, such a state of affairs is inevitable.

The most important serotypes used in commercial products are IIIaIIIb, I and VaVb. The strain *anduze* (IIIa) was discovered in a silkworm rearing room in the Cévennes (Delaporte & Beguin, 1955; Grison, 1956a,b; Lemoigne *et al.*, 1956).

It is worth recording that until recently 11 of the serotypes of *B. thuringiensis* had not been tested on *P. brassicae*, although many of them, except the last three have been available for a long time. Van der Geest (unpublished) has, however, recently tested all except the newest serotypes on *P. brassicae*.

Bacteriology

B. thuringiensis contains two toxins which are lethal to *P. brassicae*, the crystalline delta endotoxin (Norris, 1971; Cooksey, 1971; Burges *et al.*, 1975) and the heat stable exotoxin (Bond *et al.*, 1971; Maas Geesteranus, 1965a,b). The crystal of toxic protein, or δ-endotoxin, is found only in sporulation, is destroyed by heat, insoluble in water, in a pro-toxin form in the intact crystal, from which toxin(s) are released by the appropriate gut enzymes of susceptible Lepidoptera. The heat stable exotoxin, or β-exotoxin, is heat stable at autoclave temperatures, soluble in water, formed at the log phase of vegetative bacterial multiplication, is an adenine nucleotide, kills by interrupting moulting and does not specifically attack the gut. The β-exotoxin is absent from all modern commercial products because it has not been registered for safety clearance.

The protein crystal of *B. thuringiensis* (Martouret, 1962a) is solubilised by proteases released from the gut of *P. brassicae* (Norris, 1964; Lecadet & Dedonder, 1965; Cooksey, 1968). Morrison (1967) recorded that *P. brassicae* proteases digested 93 isolates of *B. thuringiensis*. De Barjac & Bonnefoi (1968, 1973) tested 12 H-serotypes for activity of DNAse, RNAse and argininealdehydrolase. For the products of hydrolysis of the crystal and the molecular weight of the bacterial proteins, the reader is referred to Lecadet (1965), Lecadet & Dedonder (1967) and Lüthy & Trumpi (1977).

Exotoxins have been identified in *B. thuringiensis* serotype I (Burgerjon & De Barjac, 1962), serotype VIII and IX (De Barjac *et al.*, 1966) and serotypes IVa, IVc (Burgerjon & De Barjac, 1967).

Burgerjon & De Barjac (1965) tested the effectiveness of the exotoxin on third instar larvae and showed that 100% mortality could be achieved using the supernatant taken from centrifuging a broth culture of *B. thuringiensis*. Galowalia tested eight varieties of *B. thuringiensis* endotoxin on *P. brassicae* (Table 15.2).

Some endotoxin was found in the spore wall, and the crystal is thought to have evolved as overproduction of spore-wall protein (Somerville & Pockett, 1975). Those interested in pursuing the effects of the bacterial toxin on the mesenteron are referred to the papers of Lhoste & Martouret (1968) and Lhoste *et al.* (1965).

Mode of action

The effect of the spore-crystal complex on *P. brassicae* varies according to concentration and *B. thuringiensis* strain. At ingestion of high con-

434

centrations of active strains, the crystals dissolve in the mid gut, causing a fall in pH, paralysis of the mouthparts and gut musculature with the cessation of feeding. Intense secretory activity of the epithelial cells is initiated very quickly (White, 1968) and the gut wall is made more permeable (Benz, 1962, 1966). Cytoplasm is budded off the epithelial cells which soon disintegrate entirely and separate from the basement membrane to form a slurry of broken down cell material between the basement membrane and the peritrophic membrane. This, together with the lowering of the pH, creates favourable conditions for the spores to germinate and bacteria to multiply. Often, normal bacterial species of the gut flora multiply faster than *B. thuringiensis* in these conditions. Disruption of this magnitude is usually lethal in a few hours. The rapidly reproducing bacteria produce enzymes that disrupt the basement membrane allowing invasion of the body cavity, the contents of which are even more favourable for bacterial multiplication; gross septicaemia results (Martouret, 1962a,b; Isakova, 1964a,b; Martouret *et al.*, 1965). When the bacteria have increased to such an extent that they use up one or more vital nutrients in the insect body, or foul the body contents with their waste products, sporulation ensues and the cadaver becomes virtually a bag of spores, usually black in colour. The cuticle may disintegrate releasing a black liquid, or the cadaver may drop to the ground and dry up if the ground is dry. At lower doses, or with less active strains, death may take several days, the cause of death being penetration of bacteria into the body cavity and septicaemia. Larvae may recover from low doses, damage to the epithelium being less and repairable. On resumption of feeding, more bacteria may be ingested, the sequel depending on dose. A succession of small doses may lead to eventual death by weakening, to small pupae in which septicaemia may develop, or to small imagines which may be infertile or may lay few ova, fertile or infertile, and have shortened life (Burgerjon & Biache, 1967a,b).

The thermostable beta exotoxin kills larvae at moulting. Survivors may be deformed, as may surviving pupae or resulting imagines. The mouthparts are the organs most usually deformed (Burgerjon *et al.*, 1969); Tipton *et al.* (unpublished) *in* Bond *et al.*, 1971).

Production

Methods have been described for culturing and harvesting preparations of spores and crystal toxins (Cantwell, 1964) and preparing a viable powder of *B. thuringiensis* which can be kept in a bottle at laboratory temperature (Beguin & Martouret, 1957) for up to 12 years without losing its virulence (Hurpin, 1973a,b).

Commercial production resulted in much research on formulation. Two types of formulation were produced. One was liquid suspension, stabilised to prevent death of spores and deterioration of crystals (Biliotti *et al.*, 1956). This has excellent properties of mixing with water and remaining in suspension. However, it required cool storage because its shelf life at high temperatures is limited and so it has largely been surpassed by the second type of formulation, the wettable powder (Martouret & Milaire, 1963). This has a long shelf life but is more difficult to mix and soon settles in spray tanks unless continually agitated.

Bioassay

The only satisfactory way of measuring the activity of a *B. thuringiensis* preparation is by bioassay with susceptible insects. This is now used for the standardisation of products and their activities are expressed in *Trichoplusia ni* (Hübner) units (USA) or *Anagasta kuehniella* Zeller units (Europe) according to the species of moth selected for the bioassay. A vital feature in conducting these bioassays is complete standardisation of method and the use of larvae of even size (Burgerjon & Biache, 1967a,b). A product, E61, designated as an international standard is bioassayed in parallel with the test products and all units are expressed in comparative terms against this international standard (Burges, 1967a,b,c; Dulmage, 1975).

The activity of *B. thuringiensis* against *P. brassicae* can be measured accurately by bioassay in the laboratory and several techniques have been described. Burgerjon (1967) sprayed suspended bacteria onto leaves (cf. Biliotti *et al.*, 1956; Martouret & Milaire, 1963). Rogers *et al.* (1966) incorporated them into artificial agar diets and Galowalia *et al.* (1973) treated the tip of a triangular piece of leaf, ensuring that the whole tip was eaten by the larva. Bohm (1961), however, did not have any success in dusting or spraying larvae in the laboratory, which is not surprising because the bacteria attack the larvae only when ingested.

With regard to the relative toxicities of the various strains of *B. thuringiensis*, Galowalia *et al.* (1973) recorded that *morrisoni, entomocidus* and *galleriae* were rather the same, *tolworthi* was a third as toxic as *galleriae*, and *aizaiwai* was half as toxic as *tolworthi*. The serotypes *alesti* and *sotto* were found to have a weak reaction so that no accurate results could be obtained. Van der Geest (1978, pers. comm.) stated that in his experience very effective action against *P. brassicae* was found for the serotypes *kurstaki, alesti* and *thuringiensis*, moderate action for *galleriae, kenyae, aizawi* and *darmstadiensis*, while no effect was experienced with *finitimus*.

Burgerjon & Martouret (1971) also compared the effectiveness of *entomocidus, berliner* and *dendrolimus*, and Aizawa (1971) brought attention to the fact that the difference in virulence of *alesti* and *berliner* was in the crystals and their digestion by the gut enzymes. Norris (pers. comm. *in* Aizawa, 1971) increased the virulence of a strain of *B. thuringiensis* by selecting from UV-irradiated bacteria.

Comparisons of different doses of the purified crystal of *B. thuringiensis* were also carried out by Martouret (1962a) who showed that a concentration of 0.5 μg/insect would give 96% mortality in 97 hours compared to only 10% mortality at 5 μg/insect (see both his graphs).

Dulmage, who has assembled all the strains of *B. thuringiensis* together in America, has released six of the best strains to industry as a result of the work of an International Program. Franz (1961) recommended that fourth instar *P. brassicae* larvae should always be tested for the sake of standardisation.

Martouret (1965) made extensive trials on the relative adhesiveness of different preparations of *B. thuringiensis* when subjected to known amounts of simulated rainfall in a specially prepared treatment tower (see his figure for details). He tested the supernatant from cultures containing heat stable toxin, and spore-crystal mixtures either as an aqueous solution or as a powder. Spore-crystal preparations formulated as wettable powders were found to be very resistant.

Van der Laan & Wassink (1969) found little eperimental evidence that *B. thuringiensis* could be transmitted from infected to healthy larvae in the laboratory. When infected larvae were reared together with healthy larvae in the same containers, only 1.7% of the healthy third instar larvae and 2-day old larvae became infected. Infection of healthy larvae is not likely to occur until dead bodies break down and release spores and crystals. In the field, cross infection would depend on larval density and sequence of generations, but infestations on crops can never be allowed to reach a density at which spread of *B. thuringiensis* would be noticeable.

Field trials

Initial work on the effect of *B. thuringiensis* (cited as *B. cereus*) strains on *P. brassicae* was carried out by Toumanoff & Grison (1954) in the Institut Pasteur in Paris. They recorded 100% mortality using the variety *alesti* on third and fifth instar larvae and did preliminary experiments on the effects of temperature, age of larvae and persistence of this bacterium (Table 15.3). They stated that as early as 1878, the control of insects had been envisaged.

Published recommendations for treatment in the field quoted high mortalities after several days and in some cases several weeks. The larvae caused little damage to crops in this time because the crystal toxin paralyses the gut and greatly reduces feeding.

Table 15.3 Bacillus thuringiensis (cited as *B. cereus*) used on *P. brassicae.*

H-serotype	Authority
Bacillus cereus	Isakova, 1958, 1963, 1964a,b; Isakova & Moiseeva, 1967, 1968
alesti B_3 *cazaubon* *cazaubon* No. 2 Toumanoff *galechiae* Toumanoff *galleriae*	Toumanoff & Grison, 1954; Toumanoff, 1956; Leskova, 1960

Concentration of the *Bacillus* is a very important factor (Martouret, 1962a,b; Burgerjon & Biache, 1967a,b). Ingestion of *B. thuringiensis* crystals at 2.50 μg/insect gives 100% mortality in just over two days, whereas at 0.10 μg/insect, 10% mortality is achieved only after 10 days. Concentrations recommended in the literature over the years have progressively decreased, reflecting the improvement of products (Biliotti *et al.*, 1956; Martouret & Milaire, 1963; Vankova, 1962). Manufacturers' recommendations for modern products should be followed closely.

Bacterial preparations

A number of commercial bacterial preparations are produced in western Europe and America for control of cabbage pests (Table 15.4) and these have been successfully used on *P. brassicae* larvae on crops in Britain, France, Switzerland, Japan & USSR.

Modern bacterial products such as the three Dipel, Thuricide, and Bactospeine are very effective and give good results using a drenching spray of 0.1%. The amount used per hectare depends on the size of the crop. A mature crop of Brussels sprouts, for instance, would need about 200–300 gallons (908–1362 litres) per hectare. These three products, and some of the Russian ones are available on sale now. Manufacturers' information for Thuricide is typical of them all.

Thuricide is made by Sandoz (Switzerland) and contains 1.5–2.0% each of protein crystals (delta-endotoxin), spores, sporangium material and fermentation solids. This wettable powder has a shelf life of 2–3 years provided it is kept at 21–24°C in a sealed container (Sandoz Commercial Literature, 1974b). It is standardised at 30×10^9 spores/g and 16,000 units of activity (*Trichoplusia ni*).

Thuricide is effective, when used at the recommended dose of 0.5–1.0 kg/ha ($\frac{1}{2}$–1 lb/acre), against 137 lepidopterous pests, including *P.*

438

Table 15.4 Bacterial preparations tested on *P. brassicae.*

Preparation	Authority
Bakthane	Bohm, 1961
Biotrol BTB, Thuricide	Lipa, 1962
Thuricide WP	Maas Geesteranus, 1963;
H	Herfs & Krieg, 1963
Thuricide 25B	Maas Geesteranus, 1963
Thuricide HPSC	Maas Geesteranus, 1963;
	Varma *et al.*, 1974
Thuricide HP-1	Sandoz, 1974a,b.
Biotrol BTB	Maas Geesteranus *et al.*, 1967
E_{61}, Plantibac, Thuricide, Baturin	Burgerjon, 1962
Biospor 2802	Maas Geesteranus, 1963
Cela, L-69 Minoc	Maas Geesteranus, 1963
Dipel WP	Varma *et al.*, 1974
Entobacterin	Lebeder, 1970
Entobacterin-3	Misalyunene, 1976
Entobacterin-3	Zhukauskene & Misalyunene, 1976;
	Isakova & Mogilevskaya, 1975;
	Bulbulshoev & Archipov, 1976;
	Voskresenskaya, 1977
Toxibacterin	Voskresenskaya, 1977

brassicae, P. rapae and *P. napi* (Sandoz, 1974a,b). The spore and delta-endotoxin must be ingested to take effect, then they cause cessation of feeding and larval death due to toxicosis, bacteraemia or starvation in 3–5 days. Thuricide is non-phytotoxic, non-persistent in the wild, non-toxic to mammals and beneficial insects, and does not leave a harmful residue.

Many earlier products were produced for trials but were never marketed commercially. Table 15.4 summarises literature about these and also about earlier versions of Dipel, Thuricide, Biotrol and Bactospeine. Biotrol BTB, which has been used on *P. brassicae* (Table 15.4) was manufactured from *B. thuringiensis* Berliner serotype 1 by Nutrilite Products Inc., Lakeview, California (Nutrilite, 1977).

Effect on parasites and other beneficial insects

The effect of spore-crystal preparations of *B. thuringiensis* on *P. brassicae* and braconid parasites was investigated by Marchal-Segault (1974). Biache (1975) recorded that the pupae from larvae of *P. brassicae* which had survived treatment with *B. thuringiensis* were able to support the development of *Pimpla instigator* Fabricius (Hymenoptera) larvae. Isakova (1958) found that *B. thuringiensis* (cited as *B. cereus*) var. *galleriae* extracted from *Galleria mellonella* (Linnaeus) (Greater Wax Moth) killed 85–90% of *P. brassicae* larvae, but was not harmful to its parasites *A. glomeratus* and *Pteromalus puparum*.

B. thuringiensis is non-toxic to bees as was demonstrated by Lecompte & Martouret (1959) who sprayed *B. thuringiensis* containing several times the normal amount of crystal required to kill *P. brassicae* onto rape and found that bees were not affected when foraging in the crop. Recently *B. thuringiensis* has been incorporated into foundation wax to control wax moth in bee comb. without any effect on the bees.

b) Miscellaneous bacteria

Bacillus cereus (Frankland & Frankland)

Ratcliffe & Gagen (1976) and Gagen & Ratcliffe (1976) studied the defence reaction of host cells against *B. cereus* in *P. brassicae*. By injecting specially killed *B. cereus* (NCTC 2599) into larvae (instar not stated), the defence reaction was found to follow a particular pattern: a) 5 minutes after infection clumps of granular haemocytes, plasmocytes and bacteria were found attached to the internal surfaces of the insect; b) the clumps broke down and appeared as melanic acellular substance, and c) nodules were formed by encapsulation by plasmocytes.

Other bacteria

Paillot (1920) recorded 10 types of bacteria in larvae of *P. brassicae* which had been parasitised by *A. glomeratus*; these were *Coccobacilli* (6 spp.), *Diplococcus* (1 sp.), *Bacillus* (2 spp.) and a *Vibrio* sp.

The haemolymph of a number of insects was found by Stephens (1963) to exhibit some antibacterial activity only against those organisms that were non-pathogenic to the test insects. However, that of *P. brassicae* did not show any signs of activity against *Pseudomonas aeruginosa* (Schroeter), a bacterium which is normally associated with decomposition.

Ratcliffe (1975) investigated the effect of three bacteria, *Staphylococcus aureus* (Strain 0, Hospital No. 174/8), *Escherichia coli* (K 12) and *B. thuringiensis* on monolayer cultures of haemocytes of *P. brassicae*. He observed that the bacteria were not ingested but formed a "sticky" layer around the cells and this he thought may be important in nodule and capsule formation. (See also Fast's chapter in Burges 1981.)

Some free-living bacteria inhabit the gut of *P. brassicae* and play an important part in septicaemia after *B. thuringiensis* crystals increase the permeability of the gut wall (see Mode of Action). Some, like *Enterobacter aerogenes* (Kruse) Hormaeche & Edwards (= *Aerobacter aerogenes* (Kruse) Beijerinck), can upset experiments (Martouret *et al.*, 1965) and grow rapidly inside artificially infected *P. brassicae*. Isakova (1964a,b) noted the presence of bacteria belonging to the Enterobacteriaceae in the gut of *P. brassicae*.

Galowalia (1970 – thesis), studied *Streptococcus faecalis* in larvae on *P. brassicae* and found a complex relationship between this pathogen and its host (Rogers 1967 – thesis). Earlier, White (1969) had thought quite independently that *Streptococcus faecalis*, found in the gut of insects, might play a part in the pathogenic process, after the gut epithelium had been damaged by *B. thuringiensis*.

A recent study of the intestinal flora of *P. brassicae* showed that *Pseudomonas aeruginosa*, *Streptococcus faecalis*, *Micrococcus* sp. and *Sarcina* sp. were present; more occurred in *P. brassicae* than either *Phytometra gamma* or *Mamestra* (= *Barathra*) *brassicae* (Isakova & Mogilevskaya, 1975). These authors believed that the composition of the gut flora seemed to have some effect in disease development.

SECTION B. VIRUS CONTROL

Introduction

Early work up to 1950. The effects of viruses on insects have been known for a long time; in fact in 1679 Maria Graffinn in Germany described what appeared to be a virus attack on *P. brassicae* larvae.

Paillot (1926a,b,c) in France first detected granules in *P. brassicae* larvae and described two nuclear diseases which were very contagious and caused lesions in the larva. From then on up to the 1950s nothing was apparently seen of granulosis virus and in 1965 Claude Rivers at the Unit of Invertebrate Virology at Oxford posed the question – how did the virus maintain itself during this period without presumably being witnessed or documented?

1950–1970. In the 1950s scientists were saying that there was no suitable virus to control *P. brassicae* larvae and in fact they had been trying several different types but without success (Smith, 1959).

The year of 1955 is significant as it was in this year that laboratory cultures of *P. brassicae* in several locations were curiously afflicted with a novel virus disease. The natural populations in the wild in England were rapidly reduced also. This was blamed on contaminated immigrants from the continent, for there had been a large influx that year, and the virus had been reported earlier in France in 1954 (Rivers, 1978, pers. comm.). An additional problem was that cabbage in the wild was infected with virus which was consequently brought into the laboratory as food for *P. brassicae* stocks. It was curious too, to find that the Russians had apparently not heard of granulosis virus disease at a conference in Russia in the late 1950s and that the disease was apparently absent from Germany at that time. This would tend to suggest that little mixing of populations takes place although *P. brassicae* is a migratory species (Rivers, 1978, pers. comm.).

Work on the virus diseases of *P. brassicae* then started by David & Gardiner at the Agricultural Research Council unit in Cambridge (England), their interest being initiated when they found that wild caught stock from Oxfordshire and Staffordshire appeared very susceptible to the virus disease when reared alongside their own virus resistant strain.

1970 to present day. Today the field of viruses has seen great progress (cf. David, 1975a,b, Table I), as David (*loc. cit.*) stated that no less than 544 viruses have been classified from the lepidoptera alone, while Hughes (1957) (cf. Matignoni & Langston, 1960) listed 667 insect species with virus disease. Smith (1963) published a list of 80 lepidopterous species which had been recorded with cytoplasmic polyhedosis.

In Great Britain and Ireland, *P. brassicae* is widely distributed in the southern part of the mainland (Plate II) but in recent years it has become rather scarce in places. The epizootic outbreak of the naturally-occurring virus in 1955 is thought to have exerted its effect on the whole population of *P. brassicae* in subsequent years and this may account for its present decline. Reports before the Second World War frequently report thousands of "whites" seen in cabbage fields. In part these estimations are confused by the much greater abundance of the small white, *P. rapae*.

Useful books and reviews on insect viruses include Bergold (1953), Smith (1976), Smirnov (1976) and Jacques (1977) who gives an interesting discussion on the effects of sunlight, temperature, substrate, soil, humidity and chemicals on the stability of entomopathogenic viruses drawn, with regard to *P. brassicae*, on much of David & Gardiner's work.

Classification of associated viruses

Two types of virus have been reported in *P. brassicae*, granulosis virus (GV) and cytoplasmic polyhedrosis virus (CPV), but most work has

442

concentrated on the granulosis virus. More recently the generic name "Baculovirus" (BV) has been used to describe those pathogens which cause both granulosis and nuclear polyhedrosis. Crozier & Meynadier (1972) found that the proteins of BV inclusion bodies of *P. brassicae* were typical of those of insect viruses. (A certain amount of work has been carried out on the effect of Tipula Iridescence virus (TIV) in *P. brassicae*.) The granulosis virus was initially termed the "pseudograsserie" by Paillot (1918a). Bergoldiavirus as used by Vago & Adger (1961) is an unfamiliar form of classification for granulosis virus and appears to originate from Kelsey (1958) who tentatively named a New Zealand strain of the virus *Bergoldia virulenta* Tanada.

The Cambridge Stocks

David & Gardiner during their work at Cambridge (1950–1960) developed three distinct cultures of *P. brassicae*.

a) *Virus free "Cambridge Stock"*. This was established in 1957 and by 1962 had been successfully bred through 40 generations in 5 years (David, 1962, David & Gardiner, 1965a). Typically this culture had to be periodically freed of virus infection by using rigorous physical and chemical methods such as i) destroying virus on eggshell (cf. Vago, 1954); ii) using virus free food, iii) selective breeding to eliminate transovarian transmissions, iv) testing for hardiness by varying extrinsic factors.

b) *Virus resistant "Cambridge Stock"*. This was established in 1951 and in 14 years was successfully bred through about 126 generations (David & Gardiner, 1965a). It was also referred to as the Cambridge Stock by these workers and thus must be distinguished from the former stock. By selective breeding David & Gardiner (*loc. cit.*) were able to produce a stock which had the genotypic ability to resist viral infection.

c) *P. brassicae cheiranthi stock*. This was established in about 1965 from specimens supplied from the Canary Islands and was successfully bred for 30 generations. Typical of this culture is the facility to transmit granulosis virus from one generation to the next.

The importance of these three cultures was that other insect viruses could be tested on them, and that studies on stress factors and latency could be investigated. Furthermore it is worth pointing out that many research centres in Europe received specimens from these stocks thus providing a constancy of parent characteristics. This is particularly important in view of the fact that different stocks of *P. brassicae* vary in their susceptibility to the virus (Sidor, 1959).

Granulosis virus (GV)

a) *Physical properties*

Vago (1959) and Vago & Croissant (1959) investigated the structure of GV and cytoplasmic polyhedrosis virus (CPV) in *P. brassicae* larvae which were simultaneously suffering from both viruses, using specimens which had been collected in the wild in southern France (Gard *département*). Using the electron microscope they showed the presence of rod shaped viruses which were 270–300 mμ long by 75–80 mμ wide. Virions which they extracted from haemolymph were spherical, being 50–60 mμ in diameter and were either found scattered or in groups. An electron micrograph of viruses of *P. brassicae* was also published in 1955 by Vago *et al.* who demonstrated small rod-like structures.

David & Gardiner (1967c) investigated a wide range of temperature changes on the activity of *P. brassicae* GV. Samples of dried virus were sealed in shell vials and polythene tubes and placed in waterbaths, refrigerators, deep freezes and kept in the glasshouse in the light and dark. Virus free larvae were used to test the subjected virus suspensions for activity. These workers found that the virus was inactivated at 70°C for 10 minutes increasing to 24 hours at 40°C. At −20°C the virus could be stored indefinitely. Two other points are worth noting: first, that dried virus keeps better (at least up to 4 years) if stored in the dark, and second, that a crude dried extract of virus when stored in the dark (up to 10 years) will lose its activity. However, dried virus loses more activity than suspensions at temperatures up to 30°C (David *et al.*, 1971).

In other studies David *et al.* (1971) found that highly purified dried virus lost a significant amount of activity in two days at 20°C and thereafter a loss increased with temperature. However, large fluctuations in relative humidity did not have any effect on virus activity. They pointed out that this has particular significance as in the wild the virus ends up as a dry film and is affected by fluctuations in the temperature and humidity.

GV capsules (or granules) are about 350 × 200 nm in size with a virion (or virus particle) 225 × 65 nm which enclosed a nucleocapsid 250 × 35 nm (Figure 15.1, note the synonymous names) (Brown *et al.*, 1977). The nucleocapsid is larger because it is bent within the envelope.

The inclusion body protein contains a polypeptide of 27,500 molecular weight, while the virion contains 12 polypeptides, eight of which are associated with the nucleocapsid. The virus contains DNA which has a molecular weight of 69.8 and 74.6 million determined by different methods (Brown, *loc. cit.*). Croizier & Meynadier (1973) found that the proteins of GV and CPV are very similar. Virus polyhedra are made up of a crystal network of virus particles in inclusion body protein (Figure

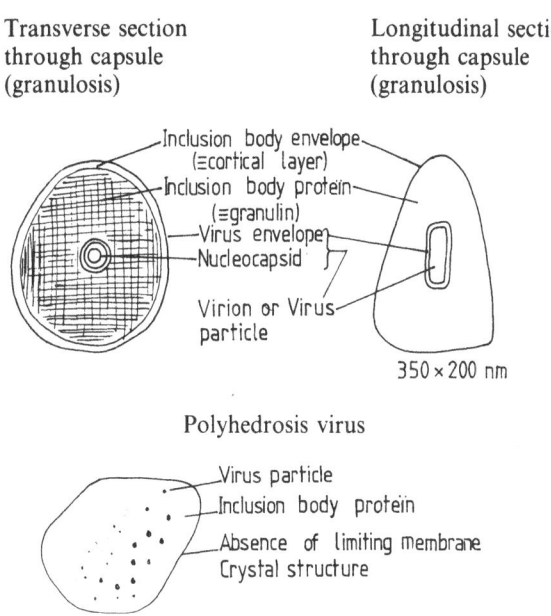

Transverse section
through capsule
(granulosis)

Longitudinal section
through capsule
(granulosis)

Inclusion body envelope
(≡cortical layer)
Inclusion body protein
(≡granulin)
Virus envelope
Nucleocapsid
Virion or Virus particle

350 × 200 nm

Polyhedrosis virus

Virus particle
Inclusion body protein
Absence of limiting membrane
Crystal structure

Fig. 15.1 Granulosis and polyhedrosis virus inclusion bodies

15.1). CPVs are very much the same morphologically from different insect hosts (Payne, 1978, pers. comm.).

b) *Chemical properties*

The GV of *P. brassicae* comprises three parts when dissolved in alkali: i) inclusion body protein, ii) enveloped virus particles, iii) virus particle envelope (Longworth *et al.*, 1972). From the inclusion body proteins two proteins were extracted by gel chromatography. Protein A was present only in the inclusion body proteins; Protein B was also found associated with the virus particle envelopes and the virus particle. It must be remembered that Croizier & Meynadier (1972) found two proteins (T & C) in inclusion bodies of BVs.

In a series of serological experiments conducted at the Insect Pathology Unit, Oxford, using their own Oxford stock, Cunningham (1968) used antisera prepared in rabbits to compare the GV and nuclear polyhedrosis virus (NPV) of seven lepidopterous species including *P. brassicae*. By using the complement filtration test backed up with light and electron microscopy he demonstrated that GV from *P. brassicae* was serologically indistinguishable from GV from *P. rapae* and *P. napi*, and that it was distantly related to NPV.

a) *Physiological effect on host.* GV infection causes a large number of inclusion bodies in the haemolymph, while CPV causes hypertrophy of the gut and accumulation of polyhedral bodies (1–3 μm diameter). When the disease is advanced polyhedral bodies penetrate the gut wall, and the basement membrane which "appear injured" (Vago, 1959).

When the larva has been killed by the virus its black corpse hangs down in a characteristic "V" shape and this has been photographed and illustrated by several workers (Martouret, 1959a,b; Smith, 1967; Hurpin, 1974). Smith & Rivers (1956) described the sequence of events after the larvae are infected. First, they cease feeding and remain motionless, become pallid especially in the thoracic region and then hang up and fall into the "V" position. At this time granules may be found in the nucleus, hypodermis and fat body.

The action of the virus on larvae is relatively slow (David, 1965a); younger larvae die off more quickly than older larvae, indeed Biliotti *et al.* (1956) stated that when larvae were experimentally infected they showed no signs of the disease for three days, and only started dying off 10 days after treatment (cf. Nurlybaera, 1975 – effect of host).

Ripa (1978) and Ripa *et al.* (1976) showed that gut fluids secreted by the larva degrade the capsules and virions almost completely and that inactivation of the virus occurs in the mid-gut of *P. brassicae*. Microscopic investigation of the peritrophic membrane, which is known to act as a barrier to pathogens, showed that it originated from a ring of mid-gut cells which arise at the hatching of the larva (Ripa, *loc. cit.*).

b) *Foodsource variables.* Only two constituents, sucrose and casein, out of 11 tested, were found by David *et al.* (1972a,b) to increase the incidence of virus significantly when they were omitted from semi-synthetic diet medium. The other constituents which did not have any effect in their absence were calcium, magnesium, potassium, mixed minor elements, choline chloride, sugar, casein, ascorbic acid, mixed vitamins, aureomycin and methyl p-hydroxy benzoate (both preservatives). However, these workers did not offer an explanation of the results.

In further work on the effect of sucrose lack on the larvae of *P. brassicae* David & Taylor (1977) suggested that the gut becomes more permeable to the GV, and so this may account for the increased mortality. They suggested that the mortality was not due to any physiological lack of sugar.

c) *Susceptibility of different stocks.* David & Gardiner (1965a) showed that completely different stocks of *P. brassicae* differed in their susceptibility to

GV. They compared the reaction of *P. b. cheiranthi* and *P. b. brassicae* from Cambridge to GV, and found that *cheiranthi* was significantly more susceptible, i.e. that the Cambridge stock was virus resistant. Sidor (1959) also noted that susceptibility varied in different stocks.

Ripa (1978) tested the susceptibility of three stocks of *P. brassicae* and *P. b. cheiranthi* to GV and found that *cheiranthi* was most susceptible, followed by a French, a Dutch stock and the virus free strain (cf. David & Gardiner, 1974). The dosage-mortality responses of the four stocks are given in Figure 15.2.

d) *Transovarian transmission*. It is not now considered that GV virus can be transmitted transovarially, indeed when David (1975a,b) and David & Taylor (1974) tested GV with a virus-free stock of *P. brassicae* they could not find any evidence that the virus could enter the micropyle (cf. David, 1975a,b and David *et al.*, 1973). It is more likely that succeeding generations are infected from contamination of the ova as well as from the soil (Jacques, 1977).

Persistence in the environment

This subject has been dealt with by the Cambridge workers who investigated the effects of sunlight including ultra-violet radiation on GV (David & Magnus, 1967; David *et al.*, 1968b), as well as the effects of rainfall, heat and cold, and retention by different soils (David & Gardiner,

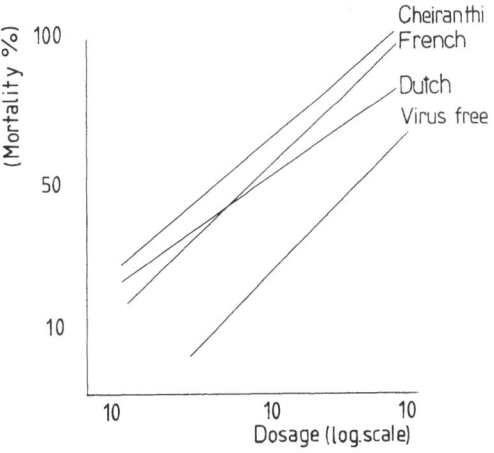

Fig. 15.2 Dosage-mortality response of fourth instar larvae of four stocks of *P. brassicae*
(after Ripa, 1978)

1966; David, 1967 – short review). David & Gardiner (*loc. cit.*) also investigated how well the virus remained on the leaves of cabbage when rinsed in water or scrubbed in detergent.

a) *Photodegradation.* The note communicated by David & Magnus (1967) presented evidence which suggested that the GV of *P. brassicae* could be activated by strong visible light immediately after it had been inactivated by exposure to ultra violet radiation at 250 or 260 nm. However, further information along this line has not been forthcoming.

When GV was experimentally applied to the upper surface of cabbage leaves and exposed to sunlight (no shade) it was found that it was inactivated sufficiently that after three hours it gave significantly less kill of larvae than non-exposed virus (David *et al.*, 1968). They pointed out that wavelengths of about 2600 Å are most efficient for inactivation of viruses but that above 3000 Å it is relatively slow. However, nothing less than 2915 Å reaches the earth from the sun. The lesson to be gained from this work is that when the virus is applied in the wild some means of protecting the virus is necessary.

b) *Leaching.* Experiments conducted using artificial rain and mechanically scrubbing virus infected cabbage leaves failed to remove the virus from the leaf surface (David & Gardiner, 1966). Potted cabbage plants were exposed to 20 inches of rain per hour and received in total 100 inches of artificial rain, and in the other experiments leaves were dipped in a wetting agent and then scrubbed with an artist's brush. Viruses extracted from the leaves of both experiments had the same fatal effect on virus-free larvae. David & Gardiner thought that perhaps the virus was held in the waxy coating of the leaf but this requires verification.

c) *Persistence in soil.* The type of soil does not appear to have any real effect in leaching away the virus, which stays mostly in the upper surfaces, (David & Gardiner, 1967b). These workers tested the persistence of GV in soil samples by mixing ground up corpses of *P. brassicae* larvae into soil and sand samples contained in clay pots and burying these in a greenhouse. The pots were sprayed weekly for two years and the soil samples tested for virus activity in the usual way on virus free larvae.

Biological control

General considerations. David (1965a) found by experiment in the laboratory that the virus was not destroyed at $-36°C$ for 5 weeks indicating that it would survive winter conditions and that when mixed with soil and sand for 14 months it maintained its virulence and that the virus remained

448

effective up to 4 months after spraying cabbage leaves. However, Payne *et al.*, (1978, pers. comm.) have found that the 99.99% infectivity of virus is destroyed in 2 days after spraying unprotected purified virus.

Fluctuations in population density was also found by David (*loc. cit.*) not to have any effect on the susceptibility of the larvae to the virus. However, a change in the foodplant was found to cause a high incidence of infection in both virus free and virus resistant stocks as well as the *cheiranthi* stock.

There are certain disadvantages in the use of GV in biological control. First, that the virus acts slowly, more so in old larvae and continued defoliation would occur (actually the larvae do stop feeding just before death); second, for an epizootic to occur a few larvae would have to survive until the next generation, thus damage to crops would continue; and third, not all surviving larvae do transmit the GV to others (David, *loc. cit.*). In the wild, success has been had with GV for the control of pierids in Europe (Biliotti *et al.*, 1956) and also in Japan (Ito & Sakiyama, 1975).

The GV of *P. brassicae* exhibits some degree of wider specificity to other members of the Pieridae as it is possible to infect both *P. rapae* and *P. napi*. Kelsey (1958) experimentally tested the effectiveness of English GV (supplied by K.M. Smith, UK) on *P. rapae* in the wild in New Zealand and found that it gave 100% control of the larvae in 11 days. K.M. Smith had said that it could potentially kill larvae in the laboratory in 4 days – but Kelsey kept the virus for 16 months before using it. In comparison a New Zeland strain of the virus, which was tentatively named as *Bergoldia virulenta* Tanada, gave control of larvae in 9 days (Kelsey, *loc. cit.*).

Granulosis virus of *P. brassicae* especially imported into the island of Okinawa from Japan was found to play a successful part in controlling larvae of *Pieris rapae crucivora* Boisduval, however the problem from this pest reoccurred each year with the arrival of immigrants from the mainland (Ito & Sakiyama, 1975). Viruses are commercially available to control some pests, but not yet for *P. brassicae* (Burges, 1981).

Commercial production

It has been estimated that 25 GV infected larvae in an advanced state of the disease disintegrated in five gallons (22.75 l) of water would be sufficient for one acre (0.41 hectares) (Smith, 1959). This is hardly surprising as it is known that the liquid which oozes from the GV corpse contains "millions of capsules" each containing single virus rods.

GV outbreaks have been induced in some *P. brassicae* cultures by feeding etiolated heart leaves of an old cabbage (Grison & Silvestre de Sacy, 1956), or increasing temperature (Gardiner, 1978, pers. comm.).

Vago & Atger (1961) stated that GV (Bergoldia virus) could be produced in quantity by feeding larvae orally at the end of the feeding period and harvesting the pupae.

David (1965a) stated that in the USA the cost of production per infected larva was one penny, and that the cost of commercial production in the UK would not be "prohibitive".

However, commercial production of viruses using larvae is not yet a viable proposition in England because of the labour involved. Also, considerable testing of the virus has to be rigorously carried out to ascertain its non-toxicity to mammals and its specificity to the pest concerned. Only two commercially produced viruses are available in the USA, and they came out during 1976–1978; *Helicorerpa (= Heliothis) zea* Boddie NPV and *Orgyia* NPV are both grown in the insect (FAO, 1977). On a small scale, larvae of *P. brassicae* may be used to manufacture viruses, as at the Glasshouse Crops Research Institute, but this is only for virus screening purposes. Safety precautions in the use of baculoviruses for insect control have been written up in the book by Summers *et al.*, 1975.

A virus preparation called Virin-Ex has been used on *P. brassicae* in USSR with success (Spaar, 1973).

Application

Granulosis virus is very readily denatured *in vivo* by light, therefore it must be administered directly onto the larvae or immediately onto the leaves where they are. It would be preferable to spray the undersides of the leaves but this is not a commercially viable proposition, or to coat the viruses in the form of a pellet.

In a series of experiments David *et al.* (1971) found that varying the following criteria did not have any significant effect on the activity of GV: repeated freezing, thawing virus stock between assays, varying the number of droplets applied to the leaves over the range 40–80, spreading partially evaporated droplets or exposing the droplets during evaporation to radiation from fluorescent and infrared lamps.

Granulosis virus in a laboratory culture

In many cultures virus attack is responsible for a very small percentage of deaths and this does not usually call for any means of control. Should the conditions of the culture change however, such as a significant drop in temperature, or a supply of a different or inferior foodplant, then an outbreak of virus disease may often occur. When cultures are maintained over several years many workers have found that the insects become more susceptible to virus as well as fungal and bacterial attack. In these cases revitalisation of the culture can be achieved by introducing ova from another source.

The virus is extremely virulent and is not easy to remove from a once infected laboratory. Fumigation with formaldehyde is not always effective even when the infected room and apparatus are left under fumigation for a month (author's experience). Sterilisation of apparatus by boiling in water for 10 minutes would be sure to kill the virus.

Tipula Iridescent Virus (TIV)

Successful infection of *P. brassicae* larvae with TIV (from *Tipula paludosa* Meigen, the Daddy Long Legs) was first reported by Smith & Rivers (1959) who stated that infection could be induced either by injection or ingestion (cf. NERC, 1970). However, the injection method appears to be the most effective as the larvae are inclined to regurgitate virus treated leaves (Smith *et al.*, 1961).

The TIV virus can be used to experimentally control a variety of insect species and Smith *et al.* (1961) reported successful use of TIV on 7 spp. of Diptera, 11 spp. of Lepidoptera and three spp. of Coleoptera. They found that TIV has a characteristic shape, particular optical properties and is produced in very large amounts in the insect, thus aiding identification. Apparently TIV extracted from infected *P. brassicae* larvae is almost indistinguishable from TIV from *T. paludosa* (Stobbart *in* Smith *et al.*, 1961).

When TIV first affects *P. brassicae* larvae the fat body is attacked, followed by the cuticle, muscles, wing buds, legs and finally the head is invaded. Smith & Hills (1959) demonstrated by formalin fixation combined with electron microscopy that where an infected area of tissue meets an uninfected area of fat body large numbers of hexagonal shells, $1,200-1,400 \text{ Å}$ in diameter, are encountered. The TIV pellets extracted from different hosts all give different irridescent colours, that of *P. brassicae* is green.

The disadvantages of using TIV to control *P. brassicae* are that when innoculation occurs in the second and third instar, larvae do not die until after pupation and that the effect on the prepupae and pupae is less drastic when compared to the effects of the virus on *Lymantra dispar* (Linnaeus). Excellent photographs of prepupae and pupae distorted by TIV, and pupae with malformed wing shields can be seen in Smith *et al.* (1961).

Oliveira & Ponsen (1966) at Wageningen studied the viral antigens in the haemolymph of TIV infected *P. brassicae* larvae, by means of micro-precipitation tests and fluorescent antibody techniques using specimens derived from the Cambridge stock via Philips Duphar. Fifth instar larvae were injected with TIV suspensions, and it was found that the first small centres of antigen multiplication occurred in the cytoplasm of the proleucocytes and amoebocytes (both haemocytes) three days after infection. Later the haemocytes were found to increase in size causing some of them to burst, due to the multiplication of TIV inside.

Cytoplasmic Polyhedrosis Virus (CPV)

Effect on host. CPV causes hypertrophy of the cytoplasm of the intestinal walls with the appearance of a number of polyhedral bodies (see Vago & Croissant, 1959, Figure 1). These polyhedra multiply, fill their host cells and then burst out during advanced stages of the disease.

On tests of virulence of CPV Vago & Croissant (*loc. cit.*) showed that oral administration of polyhedral bodies (suspensions of 1,000,000 PIB cm^3) to third instar larvae caused 70–80% mortality in 12 days.

Experimental infection with other viruses

A number of specific insect viruses have been used on *P. brassicae* but they have all been unsuccessful. Generally insect viruses when used *in vivo* are genus specific but when used in cell culture they may attack other non-related species.

A non-occluded virus of *Jugonia coenia* (Hübner) (N. American Buckeye Butterfly: Nymphalidae) which was sent to England and cultivated in larvae of another Nymphalid *Aglais urticae* (Linnaeus) (Small Tortoiseshell) was found to have no effect on first instar larvae of *P. brassicae*. Krieg (1957a,b) also experimented with a virus (called *Borrelina aporiae* – a name formerly used for NPV) which he extracted from another pierid *Aporia crataegi* (Linnaeus) (Black-veined white) but this also had no effect on *P. brassicae* larvae. Vago (1953) also extracted a virus (*Borrelina pithyocampae* sp. n. from *Thaumetopoea pithyocampa* (Schiffermüller) and this also had no effect when injected into *P. brassicae* larvae.

Cell cultures of *P. brassicae* have been used successfully to culture TIV viruses of the coleopteran *Sericesthis pruinosa* (Kelly, *pers. comm.* in Kalmakoff *et al.*, 1974), and the GV of *P. brassicae* has been successfully grown in cell cultures of fibroblasts from the ovarial tubes of *Lymantria dispar* (Linnaeus) (Gypsy moth) (Vago & Begoin, 1963; cf. Pospelov & Noreiko, 1929).

P. brassicae as an insect vector

Initial work by Martini (1953, 1956) on the possible transmission of Cauliflower Mosaic Virus (CMV) from one plant to another by *P.*

brassicae larvae showed that the chance of this happening was < 1:10,000.

Using another virus Proesler (1971) showed that *P. brassicae* larvae could "occasionally" transmit Turnip Yellow Mosaic virus (TYMV) to other turnip plants, although it is not at all clear from his paper what exactly is meant by "occasionally". Proesler (*loc. cit.*) infected larvae by spraying the virus over their bodies and securing them by their "afterfeet" (= prolegs), thus allowing them contact only with their mouthparts and forelegs with the virus free plants. Martini (1958) also showed that *P. brassicae* larvae (first and third instars) acted as insect vectors for TYMV.

There would appear to be little significance to the fact that *P. brassicae* larvae act as insect vectors of certain plant viruses, as their behaviour restricts them generally to a small area. Unless the source of foodplant runs out *P. brassicae* larvae rarely travel beyond one or two plants, although the wandering fifth instar larva may travel several metres and thus present a greater menace.

SECTION C. FUNGAL CONTROL

According to information found in the literature Fresenius (1858) was the first person to describe the fungus *Entomophthora sphaerosperma* Fresenius from *P. brassicae* (cf. Steinhaus, 1949). Since then 10 other fungi have been used against *P. brassicae* (Table 15.5).

One of the earliest fungi to be used against *P. brassicae* was *Beauveria bassiana* (= *Botrytis bassiana*), the white Muscardine fungus (Arnaud, 1923, 1927; Pospelov, 1938, 1944), which was effective using different measures of application. At 20–22°C and 54–85% RH the fungus was

Table 15.5 Fungi used on *P. brassicae*.

Fungi	Authority
Beauveria (Botrytis) bassiana (Balsamo) Vuillemin	Arnaud, 1927
Beauveria densa (Link) Picard	Arnaud, 1927
Beauveria globulifera (Spegazzini) Picard	Arnaud, 1927
Cordyceps militaris (Fries) Link	Müller-Kögler, 1965
Entomophthora punctata Garbowski	Garbowski, 1927
Entomophthora sphaerosperma Fresenius	Garbowski, 1927
Paecilomyces farinosus (Dickson *ex* Fries) Brown & Smith	
(= *Spicaria farinosa* (Fries) Vuillemin)	Arnaud, 1927
Paecilemyces fumoso-roseus (Wize) Brown & Smith	
(= *Spicaria aphodii* Vuillemin)	
(= *Spicaria fumoso-rosea* (Wize) Vasil'evskii)	Vasil'evskii, 1929
Sorosporella uvella (Krassilstschik) Giard	Paillot, 1943
Tarichium gammae Weiser	Koval, 1969

found to cause 80–100% mortality of *P. brassicae* larvae. Later Pospelov (1944) found that *B. bassiana* was most effective in the field between an optimum temperature of 20–28°C killing all the larvae within six days.

Weiser's (1977) book *Atlas of Insect Diseases* gives a very good photograph of *B. bassiana* invading the cuticle of *P. brassicae* in which there were germinating conidia with hyphae. Some resistance is put up by *P. brassicae* tissues as the lymphocytes group together (more than 20) to form giant cells which destroy the hyphae. Burges & Hussey (1971a) mentioned that a factory in the USSR was then being built near Kiev which was expected to produce 250 tons (246 tonnes) of *B. bassiana*, sufficient to treat 5 million acres (2 million ha) to control the Colorado beetle.

Experiments with the two *Spicaria* species showed that the pupae of *P. brassicae* were resistant, while larvae were susceptible to the fungi in infected soil or when the spores were dusted over them (Vassijevski, 1929). There was high mortality of the larvae which varied with temperature and humidity but 4% mortality of ova.

In the wild *E. sphaerosperma* has been reported to control *P. brassicae* quite naturally. However, Woroniecka (1928) found that even the combined effect of *E. sphaerosperma* and *Apanteles glomeratus* could not control the ravages of the Large White in Poland in 1928. In the Ukranian region of Poltava another fungus *Tarichium gammae* Weiser was used against *P. brassicae* (Koval, 1969).

In Finland, Kanervo (1946) noted that *E. sphaerosperma* caused "very considerable destruction" of *P. brassicae* particularly in the autumn during the years 1928, 1936 and 1939. Often laboratory experiments with wild collected *P. brassicae* larvae were spoilt by high mortalities due to this fungus.

Unsuccessful attempts were made by Müller-Kögler (1965) to infect larvae and pupae of *P. brassicae* with a fungus isolated from *Tipula paludosa* Meigen.

For details about the combination of both fungus and chemicals into proprietary products such as Entobackterin, the reader is referred to Section B on bacterial control earlier in this chapter. Fungi may be used to control some Lepidoptera, but probably not *P. brassicae* (Burges, 1981).

SECTION D. PROTOZOAN CONTROL

Paillot first described protozoans in *P. brassicae* from a natural outbreak of diseased larvae at Lyon (France) in 1917. He recorded their presence in adipose tissue and "certain cells of the blood", in the Malpighian tubules, silk glands and wall of the digestive glands (Paillot, 1918b, 1924a, 1924c).

Microsporidium mesnili (Paillot) has been recorded from *P. brassicae* by both Paillot and Blunck (Table 15.6).

Experiments carried out by Fantham (1939) to infect *P. brassicae* larvae with two *Nosema* species pathogenic to *Cactoblastis* sp. (Lepidoptera) were inconclusive (Table 15.6). However, *Nosema bombycis* Naegli was found to be successful in infecting *P. brassicae* (Veber, 1959). Low incidence of *Perezia mesnili* Paillot in *P. brassicae* in the wild near Kiel, West Germany was reported by Blunck (1952).

The control of *P. brassicae* using protozoa may also have detrimental effects on one of its parasites, *Apanteles glomeratus* (L.) (Issi & Maslennikova, 1964, 1966; Laird, 1973). Issi & Maslennikova (*loc. cit.*) showed that larvae fed with 4–5,000 spores of *Nosema* passed these onto the endoparasite. *P. brassicae* larvae infected with *Perezia polyvora* are less susceptible to sublethal doses of DDT but they are more susceptible to bacterial preparations (Issi, 1965). Protozoa are not likely to be practical for *P. brassicae* control (Burges, 1981).

The action of *Perezia mesnili* on *P. brassicae* has been illustrated by Weiser (1977) in various photographs. In four different plates the distribution of this protozoan is shown in the gut wall, the fat body, the Malpighian tubules and in the ovaries where the spores can be transmitted onto the ova during oviposition.

Other pathogens

Only two species have been described here, *Spirochete pieridis*, which was recorded as inducing septicaemia of the larval body tissues of *P. brassicae* (Paillot, 1940), and *Rickettsiella melolonthae* which was shown to remain latent in *P. brassicae* and appear in the next generation (Hurpin, 1971).

Table 15.6 Protozoa used on *P. brassicae*.

Species	Authority
Microsporidium mesnili (Paillot)	
Thelohania mesnili Paillot	Paillot, 1924a; Blunck, 1952
Nosema apis Zander	Fantham, 1939
Nosema bombycis Naegeli	Veber, 1959
Nosema cactoblastis Fantham	Fantham, 1939
Nosema cactorum Fantham	Fantham, 1939
Perezia mesnili Paillot	Paillot, 1918b, 1924a,b;
Nosema polyvora Blunck	Blunck *et al.*, 1959
Nosema mesnili Paillot	Weiser, 1977
Perezia legeri Paillot	
Perezia pieris Paillot	

This species can be propagated in insect explants, including tissue or dispersed cell substrates of *P. brassicae.*

References cited

Aizawa, K., 1971. Strain improvements and preservation of virulence of pathogens. *In: Microbial control of Insects and mites.* Ed. H.D. Burges & N.W. Hussey. Academic Press, London. pp. 591–621.

Agricultural Research Council, 1962. Pest infestation research report. 1962: 13.

Arnaud, G., 1923. La lutte control sur les insectes par les champignons parasitiques. *C. r. Acad. agric.* 9: 863–867.

Arnaud, M., 1927. Recherches preliminaires champignons entomophytes. *Ann. Épiphyt.* 13: 1–30.

Balachowsky, A. & Mesnil, L., 1936. Les insectes nuisables aux plantes. II. *Establissements Busson, Paris.* 1183–1185.

Beguin, S. & Martouret, D., 1957. Essais de traitement microbiologique par poudrage. *Int. cong. Crop Prot. Proc.* 4: 885–887.

Benz, G., 1962. A toxic principle in the digestive fluids of *Pieris brassicae* (Linnaeus). *J. Insect Physiol.* 4: 492–495.

—— 1966. Die Pufferkapazität, Blut und Verdauungssaft von *Pieris brassicae* und der Einfluss von Anoxie und *Bacillus thuringiensis* endotoxin auf die Permeabilität der Darmwand. *J. Insect Physiol.* 12: 137–151.

Bergold, G.H., 1953. Insect viruses. *Adv. Insect Physiol.* 1: 91–139.

Biache, G., 1975. Effects of *Bacillus thuringiensis* on *Pimpla instigator* (Ichneumonidae: Pimplidae). *Ann. Soc. ent. Fr.* 11: 609–617.

Biliotti, E., Grison, P. & Martouret, P., 1956. L'utilisation d'une maladie à virus comme méthode de lutte biologique contre *Pieris brassicae* L. *Entomophaga* 1: 35–44.

Blunck, H., 1950. Zur Kenntnis des Massenwechsels von *Pieris brassicae* L. mit besonderer Berücksichtigung des Dürrejahres 1947. *Z. angew. Ent.* 32: 141–171.

—— 1952. Über die bei *Pieris brassicae* L., ihren Parasiten und Hyperparasiten schmarotzenden Mikrosporidien. *Int. Cong. ent. Trans.* 9: 432–438.

Blunck, H., Krieg, R. & Scholtyseck, E., 1959. Weitere Untersuchungen über die Mikrosporidien von Pieriden und deren Parasiten und Hyperparasiten. *Z. PflKrankh. PflPath. PflSchutz* 66: 129–142.

Böhm, O., 1961. Erfahrungen mit Bakterienpräparaten zur Schädlingsbekämpfung im Gemüsebau. *Pflanzenart, Vienna* 14: 106–108.

Bond, R.P.M., Boyce, C.B.C., Rogoff, M.H. & Shieh, T.R., 1971. The thermostable exotoxin of *Bacillus thuringiensis. In: Microbial control of insects and mites.* Ed. H.D. Burges & N.W. Hussey. Academic Press, London. pp. 275–302.

Brown, D.A., Bud, H.M. & Kelly, D.C., 1977. Biophysical properties of the structural components of a granulosis virus isolated for the cabbage white butterfly (*Pieris brassicae*). *Virology* 81: 317–327.

Bulbulshoev, T. & Archipov, G.E., 1976. Effectiveness of Entobacterin. (In Russian.)

Burgerjon, A., 1959. Titrage et définition d'une unité biologique pour les préparations de *Bacillus thuringensis* Berliner. *Entomophaga* 4: 201–206.

—— 1962. Principles thermostables dans les preparations industrielles a base de *B. thuringiensis. Coll. Int. Pathol. insectes Paris* 1962: 227–237.

Burgerjon, A. & Biache, G., 1967a. Divers effects spéciaux et symptômes tératologiques de la toxine thermostable de *Bacillus thuringiensis* en fonction de l'âge physiologique des insectes. *Annls. Soc. ent fr.* 3: 929–952.

—— 1967b. Effects teratologiques chez les nymphs et les adultes d'insectes dont les larves ont ingéres des doses sublethales de toxine thermostable de *Bacillus thuringiensis* Berliner. *C. r. Acad. Sci. Paris* 264: 2423–2425.

Burgerjon, A., Biache, G. & Cals, P., 1969. Teratology of the Colorado Beetle, *Leptinotarsa decemliniata*, as provoked by larval administration of the thermostable toxin of *Bacillus thuringiensis. J. Invert. Pathol.* 14: 274–278.

Burgerjon, A. & De Barjac, H., 1960. Nouvelles données sur le rôle de la toxine soluble thermostable produite par *Bacillus thuringiensis* Berliner. *C. r. Acad. Sci. Paris* 251: 911–912.

—— 1962. Tests on the insecticidal role of the thermostable toxin produced by *Bacillus thuringiensis. XI Int. Cong. ent. Vienna* 1960: 835–839.

—— 1967. Another serotype (4, 4a, 4c) of *Bacillus thuringiensis* which produces thermostable toxin. *J. Invert. Pathol.* Notes: 574–577.

Burgerjon, A. & Martouret, D., 1971. Determination and significance of the host spectrum of *Bacillus thuringiensis. In: Microbial control of insects and mites.* Ed. H.D. Burges and N.W. Hussey. Academic Press, London. pp. 305–322.

Burges, H.D., 1967a. Standardization of *Bacillus thuringiensis* products: Homology of the standard. *Nature, Lond.* 215: 664–665.

—— 1967b. The standardization of products based on Bacillus thuringiensis. *Proc. Int. Coll. Ins. Path. Micr. Control. Wageningen,* 1966. pp. 306–314.

—— 1967c. The standardization of *Bacillus thuringiensis*: tests on three candidate reference materials. *Proc. Int. Con. Ins. Path. Micr. Control Wageningen,* 1966. pp. 314–337.

—— 1971. Possibilities of pest resistance to microbial control agents. *In: Microbial control of insects and mites.* Ed. H.D. Burges & N.W. Hussey. Academic Press, London. pp. 445–456.

—— 1975. Control of insects by *Bacillus thuringiensis.* 5th. Brit. Insecticides and Fungal Conference. p. 405.

—— 1976. The standardization of products based on *Bacillus thuringiensis. Meded. Rijks. Land. Gent* 31: 536–665.

Burges, H.D., (ed.) (1981). *Microbial control of pests and plant diseases 1978–1980.* Academic Press, London.

Burges, H.D., Hillyer, S. & Chanter, D.O., 1975. Effect of ultraviolet and gamma rays on the activity of δ-endotoxin protein crystals of *Bacillus thuringiensis. J. Invert. Pathol.* 25: 5–9.

Burges, H.D. & Hussey, N.W., 1971a. *Microbial control of insects and mites.* Academic Press, London.

—— 1971b. Past achievements and future prospects. *In: Microbial control of insects and mites.* Ed. H.D. Burges & H.W. Hussey. Academic Press, London. pp. 687–709.

Cantwell, G.F., Heimpel, A.M. & Thompson, N.J., 1964. The production of an endotoxin by various crystal forming bacteria related to *Bacillus thuringiensis* var. *thuringiensis* Berliner. *J. Insect Pathol.* 6: 466–480.

Chilingaryan, V.A., Ormanyan, Z.K. & Kazaryan, B.K., 1969. Microbial preparations for the control of injurious insects in agriculture. *Jub. Fauna Armenia SSR Inst. zool. Akad. nauk. Armyanskoi SSR* 1969: 54–57.

Cooksey, K., 1968. Purification of a protein from *Bacillus thuringiensis* toxic to larvae of lepidoptera. *Biochem. J.* 106: 445–454.

Cooksey, K.E., 1971. The protein crystal of *Bacillus thuringiensis*: Biochemistry and mode of action. *In: Microbial control of insects and mites*. Ed. H.D. Burges & N.W. Hussey. pp. 247–274.

Croizier, G. & Meynadier, G., 1972. Les protéines des corps d'inclusion des Baculovirus. I. Étude de leur solubilisation. *Entomophaga* 17: 231–239.

—— 1973. Les proteines des corps d'inclusion des Baculovirus III. Étude comparée de la granulose de *Pieris brassicae* et de la polyhedrose de *Bombyx mori*. *Entomophaga* 18: 259–269, 431–437.

Cunningham, J.C., 1968. Serological and morphological identification of some nuclearpolyhedrosis and granulosis viruses. *J. Invert. Pathol.* 11: 132–141.

David, W.A.L., 1962. *Pieris brassicae* and its granulosis virus disease. *Proc. XI Int. Cong. ent. Vienna* 1960: 777–780.

—— 1965a. The granulosis virus of *Pieris brassicae* L. in relation to natural mutation and biological control. *Ann. appl. Biol.* 56: 331–334.

—— 1965b. Observations on cultures of *Pieris brassicae* L. susceptible and resistant to granulosis virus disease. *XII Int. Cong. Ent. London*. 1964: 744–745.

—— 1967. Factors influencing the persistence of the virus of *Pieris brassicae* in the environment. *In: Insect Pathology and Microbial Control*. Proceedings of the International Colloquium, Wageningen, September 1966. Ed. Van der Laan, P.A. pp. 174–178.

—— 1968. The effect of heat, cold and prolonged storage on a granulosis virus of *Pieris brassicae*. *J. Invert. Path.* 9: 555–562.

—— 1969. The effect (inactivation) of ultraviolet radiation of known wavelengths on a granulosis virus of *Pieris brassicae*. *J. Invert. Path.* 14: 336–342.

—— 1971. The effect of ultra-violet irradiation of known wavelength on a granulosis virus of *Pieris brassicae* L. *All Union Ent. Soc.* 2: 59.

—— 1975a. The status of viruses pathogenic for insects and mites. *A. Rev. Ent.* 20: 97–117.

—— 1975b. The large white butterfly, *Pieris brassicae* and its granulosis virus. *Annual report of the Glasshouse Crops Research Institute for 1975*. p. 101–102.

—— 1978. The granulosis virus of *Pieris brassicae* (L.) and its relationship with its host. *Adv. Insect Res.* 22: 111–161.

David, W.A.L., Ellaby, S. & Taylor, G. 1969. Formaldehyde as an antiviral agent against a granulosis virus of *Pieris brassicae* L. *J. Invert. Path.* 14: 96–101.

—— 1971. The stability of a purified granulosis virus of the European Cabbage worm, *Pieris brassicae* in dry deposits of intact capsules. *J. Invert. Path.* 17: 228–233.

—— 1972a. The fumigant action of formaldehyde incorporated in a semi-synthetic diet on the granulosis virus of *Pieris brassicae* and its evaporation from the diet. *J. Invert. Path.* 19: 76–82.

—— 1972b. The effect of reducing the content of certain ingredients in a semisynthetic diet on the incidence of granulosis virus disease in *Pieris brassicae*. *J. Invert. Path.* 20: 332–340.

—— 1973. The large white butterfly, *Pieris brassicae* and its granulosis virus. *Annual Report of the Glasshouse Crops Research Institute for 1973*. pp. 97–98.

David, W.A.L., Ellaby, S.J. & Taylor, G., 1972. Viruses. *Annual Report of the Glasshouse Crops Research Institute for 1972*. pp. 85–86.

David, W.A.L. & Gardiner, B.O.C., 1960. A *Pieris brassicae* (Linnaeus) culture resistant to a granulosis. *J. Insect Path.* 2: 106–114.

—— 1965a. Resistance of *Pieris brassicae* (Linnaeus) to granulosis virus and the virulence of the virus from different host races. *J. Invert. Path.* 7: 285–290.

—— 1965b. The incidence of granulosis deaths in susceptible and resistant *Pieris brassicae* (Linnaeus) larvae following changes of population density, food and temperature. *J. Invert. Path.* 7: 347–355.

—— 1966a. Persistence of a granulosis virus of *Pieris brassicae* on cabbage leaves. *J. Invert. Path.* 8: 180–183.

—— 1966b. Breeding *Pieris brassicae* apparently free from granulosis virus. *J. Invert. Path.* 8: 325–333.

—— 1967a. The persistence of a granulosis virus of *Pieris brassicae* in soil and sand. *J. Invert. Path.* 9: 342–347.

—— 1967b. The effect of heat, cold and prolonged storage on a granulosis virus of *Pieris brassicae* (L.). *J. Invert. Path.* 9: 555–562.

David, W.A.L., Gardiner, B.O.C. & Clothier, S.E., 1968a. Laboratory breeding of *Pieris brassicae* transmitting a granulosis virus. *J. Invert. Path.* 12: 238–244.

David, W.A.L., Gardiner, B.O.C. & Woolner, M., 1968b. The effect of sunlight on a purified granulosis virus of *Pieris brassicae* applied to cabbage leaves. *J. Invert. Path.* 11: 496–501.

David, W.A.L. & Magnus, I.A., 1967. Preliminary observations on the *in vitro* photoreactivation of an insect (*Pieris brassicae*) virus inactivated with ultraviolet. *J. Invert. Path.* 9: 266–268.

David, W.A.L. & Taylor, C., 1974. The large white butterfly and its granulosis virus. *Annual report of the Glasshouse Crops Research Station for 1974.* p. 97.

David, W.A.L. & Taylor, C.E., 1977. The effect of sucrose content of diets on subsceptibility to granulosis virus disease in *Pieris brassicae*. *J. Invert. Path.* 27: 117–118.

De Barjac, A. & Bonnefoi, A., 1973. Mise en point sur la classification des *Bacillus thuringiensis*. *Entomophaga* 18: 5–17.

De Barjac, A., Burgerjon, A. & Bonnefoi, A., 1966. The production of heat-stable toxin by nine serotypes of *Bacillus thuringiensis*. *J. Invert. Path.* 8: 537–533.

De Barjac, H. & Bonnefoi, A., 1968. A classification of strains of *Bacillus thuringiensis* with a key to their identification. *J. Invert. Path.* 11: 335–347.

Delaporte, B. & Beguin, S., 1955. Étude d'une souche de *Bacillus*, pathogène pour certains insectes, identifiable à *Bacillus thuringiensis* Berliner. *Ann. Inst. Pasteur, Paris* 89: 632–643.

Dulmage, H.T., 1975. The standardisation of formulations of the δ-endotoxins produced by *Bacillus thuringiensis*. *J. Invert. Path.* 25: 279–281.

Fantham, H.B., 1939. *Nosema cactoblastis* sp. n. and *Nosema cactorum* sp. n. Microsporidian parasites of species of *Cactoblastis* (Lepidoptera) destructive to prickly pear. *Proc. zool. Soc. Lond.* 108: 689–705.

Food and Agricultural Organisation of the United Nations, 1977. FAO plant production and protection papers, No. 6. Pest resistance to pesticides and crop loss assessment. I. Report of the first session of the FAO panel of experts. Washington, D.C., 16–19 August, 1976. p. 27.

Franz, J.M., 1961. Biological control of pest insects in Europe. *A. Rev. Ent.* 6: 183–200.

—— 1971. Influence of environmental and modern trends in crop management of microbial control. *In: Microbial control of insects and mites.* Ed. N.D. Burges & N.W. Hussey. Academic Press, London. pp. 407–440.

Fresenius, G., 1858. Über die Pilzgattung *Entomophthora*. *Abhandlg. Senkenberg. Gesell.* 2: 201–210.

Gagen, S.J. & Ratcliffe, N.A., 1976. Studies on the *in vivo* cellular reactions and fate of injected bacteria in *Galleria mellonella* and *Pieris brassicae* larvae. *J. Invert. Path.* 28: 17–24.

Galowalia, M.M.S., 1970. *A study of the pathological effects of the crystalline endotoxin on several varieties of Bacillus thuringiensis in the larvae of Pieris brassicae.* Thesis. University of Leeds. pp. 184.

Galowalia, M.M.S., Gibson, N.H.E. & Wolf, J., 1973. The comparative potencies of the crystalline endotoxin of eight varieties of *Bacillus thuringiensis* to larvae of *Pieris brassicae. J. Invert. Path.* 21: 301–308.

Garbowski, L., 1927. Sur les entomophthorées. 2. *Entomophthora (Tarichium) punctata* sp. n. sur *Phytomoma variabilis* Hbst. et *E. sphaerosperma* Fresn. sur *Pieris brassicae. Prace. Wydz. Chorob. Roslin. Panstw. Inst. nauk-Roln.* Bydgoszczy 1927: 1–24, 25–44.

Graffinn, M.G., 1679. *Der Raupen Wunderbare Berivandelung und Sonderbare Blumen-Nahrung.* Graffen, Frankfurt.

Grison, P., 1956a. Organisation de la lutte biologique en France et resultats obtenus dans l'utilisation des agents pathogènes. *Proc. 10th. Int. Cong. Ent. Montreal.* 4: 675–679.

—— 1956b. Quelques aspects de la lutte microbiologique contre les insectes ravageurs des cultures. *Ann. Épiphyt.* 4: 543–562.

Grison, P. & Silvestre de Sacy, R., 1956. L'élevage de *Pieris brassicae* L. pour les essais de traitements microbiologiques. *Ann. Épiphyt.* 4: 663–676.

Herfs, W. & Krieg, A., 1963. Untersuchungen zur Beurteilung der Wirksamkeit industrieller Präparate von *Bacillus thuringiensis* Berliner, für die Bekämpfung des Kohlweisslings (*Pieris brassicae* (L.)). *Nachrbl. Dt. Pflschutzdienst. Braunschw.* 15: 49–54.

Hughes, K.M., 1957. An annotated list and bibliography of insects reported to have virus diseases. *Hilgardia* 26: 597–629.

Hurpin, B., 1970. La lutte microbiologique contre les insectes ravageurs en agriculture. *Ann. zool. ecol. anim.* 2: 639–652.

—— 1971. Specificity of *Rickettsiella melonthae* and its pathogenicity to some vertebrates. *Ann. Soc. ent. Fr.* 7: 439–469.

—— 1973a. Lutte microbiologique et lutte integrée. *Proc. FAO conf. Ecology in relation to plant pest control.* pp. 207–247.

—— 1973b. La specificité des microporganismes entomopathogènes et son rôle en lutte biologique. *Ann. zool. ecol. anim.* 5: 283–304.

—— 1974. L'utilisation de microorganismes entopathogènes dans la lutte contre les ravageurs agricoles et forestiers. *Phytoma. Defense des Cultures.* 263: 13–22.

—— 1975. Perspectives et contraintes de la lutte microbiologique. *Z. angew. Ent.* 77: 377–386.

Isakova, N.P., 1958. The effect of a spore-producing bacterium of the type of *Bacillus cereus* Fr. on some injurious insects. (In Russian.) *Ent. URSS.* 37: 846–855.

—— 1963. Pathogenese der Insektenkrankheit, welche durch *Bacteria cereus* var. *galleriae* verursacht ist. *Sbornik. Mikrob. Moskva. Acad. nauk. SSSR.* 49–56.

—— 1964a. Le rôle du micro-flora de l'intestine des insectes au développement d'un infection causée de *Bacillus thuringiensis* var. *galleria. Entomophaga* 1964: 175–178.

—— 1964b. On the mechanism of the action of entomopathogenic bacteria *Bacillus cereus* var. *galleria* on insects. (In Russian.) *Zashch. Rast.* 21: 89–94.

—— 1965. Entomophagene und bakterielle Infektionen. *Zashch. Rast.* 10: 51.

Isakova, N.P. & Mogilevskaya, A.B., 1975. Intestinal flora of the cabbage white butterfly (*Pieris brassicae* L.), the gamma moth (*Phytometra gamma* L.) and the cabbage moth (*Barathra brassica* L.) and some factors influencing its composition. (In Russian.) *Zashch. Rast.* 42: 80–84.

Isakova, N.P. & Moiseeva, T.S., 1967. Decrease in immunity of insects under the influence of bacterial infection transmitted by hymenopterous parasites. *Proc. Int. cong. Insect Path. Micro. Control Wageningen* 1966: 281–282.

—— 1968. The decline in insect immunity to hymenopterous parasites under the influence of Entobacterin. (In Russian.) *Zashch. Rast.* 31: 367–370.

Issi, I.V., 1965. The effect of microsporidiosis on the susceptibility of cabbage white butterfly to insecticides. (In Russian.) *Zashch. Rast.* 24: 170–174.

Issi, I.V. & Maslennikova, V.A., 1964. The effect of microsporidisis on the diapause and survival of the parasite *Apanteles glomeratus* (Hymenoptera: Braconidae) and the cabbage white butterfly (*Pieris brassicae*) (Lepidoptera: Pieridae). (In Russian.) *Ent. Obozr.* 43: 112–117.

—— 1966. The role of *Apanteles glomeratus* L. in the transmission of *Nosema polyvora* Blunck. *Ent. Obozr.* 45: 494–499.

Ito, Y. & Sakiyama, M., 1975. Population dynamics of *Pieris rapae crucivora* Boisduval (Lepidoptera: Pieridae), an introduced insect pest in Okinawa. III. Results of the introduction of *Apanteles glomeratus* and a cabbage butterfly granulosis virus. *Jap. J. appl. ent. zool.* 19: 285–289.

Jacques, R.P., 1977. Stability of entomopathogenic viruses. *Misc. Publ. entom. Soc. Am.* 10: 99–116.

Kalmakoff, J., Moore, S. & Pottinger, R.P., 1972. An iridescent virus for the Grass frub, *Costelytra zealandica*: a serological study. *J. Invert. Path.* 20: 70–76.

Kanervo, V., 1946. Sporadic observations concerning diseases in certain species of insects. (In Finnish.) *Ann. ent. fennici* 11: 218–227.

Kelsey, J.M., 1958. Control of *Pieris rapae* by granulosis virus. *N. Z. Jl agric. Res.* 1: 778–782.

Koval, E.Z., 1969. Rare entomophthoraceous fungi in the Ukrainian SSR. *Ukr. Bot. Z.* 26: 117–118.

Krieg, A., 1957a. La lutte possible de piéride du chou par application artificiel du bacteria. *Z. PflKrankh. PflPath. PflSchutz* 64: 321–327.

—— 1957b. Eine Polyedrose von *Aporia crataegi* (Lepidoptera). *Z. PflKrankh. PflPath. PflSchutz* 64: 657–662.

—— 1965. Aux titres biologiques du pathogènes d'insect particulièrement *Bacillus thuringiensis*. *Entomophaga* 10: 3–20.

—— 1956. Ein Symposium über Insektenpathologie in Darmstadt, 1956. *Entomophaga* 1: 98.

Laird, M., 1973. Environmental impact of insect control by micro-organisms. *Ann. N.Y. Acad. Sci.* 217: 218–226.

Lebeder, G.I., 1970. Utilisation des methodes biologiques de lutte contre les insectes nuisibles et les nauvaises herbes en union sovietique. Coll. Franco-Sovietique sur l'utilisation des Entomophages. *INRA. Ann. zool. ecol. anim.*

Lecadet, M.M., 1965. *Isolement et caractérisation de deux protéases des chenilles de Pieris brassicae L. et étude de leur action sur l'inclusion parasporale de Bacillus thuringiensis*. Thèse Doctorat d'état, Fac. Sci. Paris. pp. 168.

Lecadet, M. & Martouret, D., 1962a. Étude composition hydrolyse enzymatique des cristaux des souches *Bacillus thuringiensis* serotype I. Berliner et *Bacillus thuringiensis* serotype III Anduze. *Coll. Int. Path. Insectes Paris*. 205–212.

—— 1962b. La toxine figuree de *Bacillus thuringiensis*. Production enzymatique de substances soluble toxiques par injection. *C. r. Acad. Sci. Paris* 254: 2457–2459.

—— 1964. A comparative study of the enzymatic hydrolysis of crystals of *Bacillus thuringiensis* serotype I Berliner and *Bacillus thuringiensis* serotype III Anduze. *Entomophaga Mem. Hors. Ser.* 2 1964: 205–212.

—— 1965. The enzymatic hydrolysis of *Bacillus thuringiensis* Berliner crystals and the liberation of toxin fractions of bacterial origin by the chyle of *Pieris brassicae*. *J. Invert. Path.* 7: 105–108.

Lecadet, M.M. & Dedonder, R., 1967. Enzymatic hydroysis of the crystals of *Bacillus thuringiensis* by the proteases of *Pieris brassicae* I. Preparation and fractionation of the lysates. *J. Invert. Path.* 9: 310–321.

Lecadet, M.M. & Martouret, D., 1967. Enzymatic hydrolysis of the crystals of *Bacillus thuringiensis* by the proteases of *Pieris brassicae* II. Toxicity of the different fractions of the hydrolysate for larvae of *Pieris brassicae*. *J. Invert. Path.* 9: 322–330.

Lecompte, J. & Martouret, D., 1959. *Annls. Abeille* 2: 171–175.

Lemoigne, M., Bonnefoi, A., Beguin, S., Grison, P., Martouret, D., Schenk, A. & Vago, C., 1956. Essais d'utilisation de *Bacillus thuringiensis* Berliner contre *Pieris brassicae* Linnaeus. *Entomophaga* 1: 19–34.

Leskova, A.T., 1960. A bacterial preparation against cabbage pests. (In Russian.) *Zashch. Rast.* 5: 31–32.

—— 1967. Susceptibility of insects to entobacterin. *Proc. Int. Coll. Ins. Path. Micr. Con. Wageningen 1966.* 214–215.

Lhoste, J. & Martouret, D., 1968. Études des lesions intestinales consecutives a l'intoxication de *Pieris brassicae* L. par *Bacillus thuringiensis*. *XII Cong. Int. Ent. Moscow, 1968. All Union Ent. Soc. Leningrad* 2: 80–81.

Lhoste, J., Martouret, D. & Roche, A., 1965. The processes of poisoning by the toxin of the crystals of *Bacillus thuringiensis*; action on the mesenteron of *Pieris brassicae*. *XII Int. Cong. Ent. London, 1964.* 737.

Lipa, J.J., 1962. Control of some lepidopterous pests of crucifers with two commercial microbial insecticides (Biotrol 25W & Thuricide WP) containing *Bacillus thuringiensis*. *Biul. Inst. Ochr. Rósl.* 16: 235–256.

Longworth, J.F., Robertson, J.S. & Payne, C.C., 1972. The purification and properties of inclusion body protein of the granulosis virus of *Pieris brassicae*. *J. Invert. Path.* 19: 42–50.

Lüthy, P. & Trumpi, B., 1977. Enzymatic degradation of the crystalline endotoxin of *Bacillus thuringiensis*. *Inst. Micr. Swiss Fed. Inst. Tech.* 138–139.

Maas Geesteranus, H.P., 1960. Bacterieziekten van planten en dieren project 1-12-4. *Inst. Plantenz. Onderzoek* 1960: 80–82.

—— 1963. Bacterieziekten by planten en insekten (project 1-12-4). *Inst. Plantenz. Onderz.* 1963: 52–53.

—— 1964. Ontwikkeling van een methode ter bepaling van de toxiciteit van bacterien voor insekten (project 1-12-7). *Inst. Plantenz. Onderzoek.* 1964: 53–54.

—— 1965a. Titre determination of preparations of *Bacillus thuringiensis* Berliner. *Entomophaga* 10: 27.

—— 1965b. Ont van een methodiek om de toxische stoffen in preparaten van *Bacillus thuringiensis* te bepalen (project 1-12-7). *Inst. Plantenz. Onderzoek.* 1965. 49–50.

—— 1967. Ont van een methode om de toxische samenstelling van preparaten van *Bacillus thuringiensis* te bepalen (project 1-12-7). *Inst. Plantenz. Onderzoek.* 1967. 45–46.

Maas Geesteranus, H.P., Noordink, J.P.W. & Van der Anker, C.A., 1967. A bioassay to characterise strains and preparations of *Bacillus thuringiensis*. *In: Insect Pathology and Microbial Control.* Ed. Van der Laan. N. Holland Publishing Co., Amsterdam. pp. 302–306.

Marchal-Segault, D., 1972. *Contribution à l'étude d'interactions entre le complexe spores-cristaux de Bacillus thuringiensis Berliner, deux hyménoptères braconides entomophages, et leurs hôtes Pieris brassicae L. (Pieridae) et Anagasta kuehniella Zeller (Pyralidae).* Thèse Doctorat 3ème cycle. Université de Paris VI. pp. 58.

—— 1974. Sensibilite des hymenopteres braconides *Apanteles glomeratus* et *Phanerotoma flavitestacea* au complexe spores cristaux de *Bacillus thuringiensis.* *Ann. zool. ecol. anim.* 6: 521–528.

Martigoni, M.E. & Langston, R.L., 1960. Supplement to an annotated list and bibliography of insects reported to have virus diseases. *Hilgardia* 30: 1–40.

Martini, C., 1956. Eine Herkunft des Blumenkohlmosaikvirus (Cauliflower mosaic virus) aus der Umgebung von Bonn. *Z. PflKrankh. PflPath. PflSchutz* 63: 577–583.

—— 1958. The transmission of turnip viruses by biting insects and aphids. *Proc. Conf. on Potato virus diseases. 3rd. Lisse Wageningen,* 1957. 106–113. Veenman & Zonen, Wageningen, Netherlands.

Martouret, D., 1958. Application diverses et normes d'utilisation de *Bacillus thuringiensis* souche Anduze. *2nd. Coll. de la C.I.L.B. sur la pathologie des insectes, Paris* 1958.

—— 1959a. Applications diverses et normes d'utilisation de *Bacillus thuringiensis* souche Anduze. *Entomophaga* 4: 211–220.

—— 1959b. Les conditions d'utilisation des preparations à base de *Bacillus thuringiensis* contre les larves de lépidoptères. *Rev. zool. agric. appl.* (1–3): 1–11.

—— 1961. Les toxines de *Bacillus thuringiensis* et leur processus d'action chez les larves de lépidoptères. 13th Symposium de Phytopharmacie et de phytiatrie Gan. 1961. *Medel. Landbouwhog. Gent* 26: 1116–1126.

—— 1962a. Études préliminaires sur le mode d'action de *Bacillus thuringiensis* var. *thuringiensis* Berliner vis à vis de *Pieris brassicae* Linnaeus. *Verh. XI Int. Kongr. Ent. Vienna, 1960* 2: 849–855.

—— 1962b. Intoxication chez *Pieris brassicae* L. par une fraction enzymatique de la toxine des cristaux des souches *Bacillus thuringiensis* serotype I. Berliner et *Bacillus thuringiensis* serotype III Anduze. *Coll. Int. Pathol. Ins. Micr. Con. Paris,* 1962. 213–220.

—— 1969. La lutte microbiologique contre les insectes nuisibles en URSS. *Rev. zool. agric. path.* 68: 5–12.

Martouret, D. & Lhoste, J., 1962. Etude comparee de l'hydrolyse enzymatique des crystaux de souches *Bacillus thuringiensis* serotype I Berliner et *Bacillus thuringiensis* serotype III Anduze. *Coll. Int. Path. Insectes, Paris,* 1962. 205–212.

Martouret, D., Lhoste, J. & Roche, A., 1965. Action sur le mésentéron de *Pieris brassicae* L. de la toxine de l'inclusion parasporale de *Bacillus thuringiensis* Berliner. *Entomophaga* 10: 349–365.

Martouret, D. & Milaire, H., 1963. Expérimentation de produits bactériens à base de *Bacillus thuringiensis* à l'échelon agricole. *Phytiat. Phytopharm.* 12: 71–80.

Metalnikov, S., 1930. Utilisation microbes dans la lutte contre *Lymantria* et autre insectes nuisibles. *C. r. Soc. Biol.* 105: 535–537.

Misyalyunene, I.S., 1976. Changes in the morphology and proportions on various types of cells of the haemolymph in larvae of the cabbage white butterfly, during infection with entobacterin. (In Russian.) *Tsitologiya* 18: 1220–1225.

—— 1976b. Effect of entobacterin-3 on large white cabbageworm of different ages in the Lithuanian SSSR. (In Russian.) *Liet. Tsr. Mokslu. Acad. Darb. Ser. C. Biol. Mokslai.* 1: 51–56.

Müller-Kögler, E., 1965. *Cordyceps militaris* (Fr.). Observations and tests in connection with a discovery of it on *Tipula paludosa*. *Z. angew. Ent.* 55: 409–418.

NERC, 1970. Report of council for year 1 April 1969 to 31 March 1970. *HMSO Report 1970*. 63–64.

Norris, J.R., 1964. The classification of *Bacillus thuringiensis*. *J. appl. Bact.* 27: 439–447.

—— 1971. The protein crystal of *Bacillus thuringiensis*: biosythesis and physical structure. *In: Microbial control of insects and mites*. Ed. H.D. Burges & N.W. Hussey. Academic Press, London.

Nurlybaera, R., 1975. Natural regulators of the numbers of pests. (In Russian.) *Zashch. Rast.* (5), 30–31.

Nutrilite, 1977. Biotrol. Biological insect control product. *Nutrilite Products Inc., Lakeview, California.*

Oliveira, A.R. & Ponsen, M.B., 1966. The development of a viral antigen in the hemocytes of *Pieris brassicae* inoculated with Tipula iridescent virus. *Neth. J. Plant Path.* 72: 259–264.

Paillot, A., 1918a. Two new microsporidian parasites of larvae of *Pieris brassicae*. *C. r. Séanc. Soc. Biol.* 81: 66–68.

—— 1918b. *Perezia legeri*, a new microsporidian parasite in *Pieris brassicae*. *C. r. Séanc. Soc. Biol.* 81: 187–189.

—— 1920. La phagocytose chez les insectes. *C. r. Séanc. Soc. Biol.* 83: 425–426.

—— 1924a. Sur *Thelohania mesnili*, microsporidie nouvelle, parasite des chenilles de *Pieris brassicae* L. *C. r. Séanc. Soc. Biol.* 90: 501–503.

—— 1924b. Sur la transmission des maladies à microsporidies chez les insectes. *C. r. Séanc. Soc. Biol.* 90: 504–506.

—— 1924c. Sur *Perezia pieris* Microsporidie nouvelle par de *Pieris brassicae* L. *C. r. Séanc. Soc. Biol.* 90: 1255–1257.

—— 1924d. Sur une nouvelle maladie des chenilles de *Pieris brassicae* L. et sur les maladies du noyau chez les insectes. *C. r. hebd. Séanc. Acad. Sci., Paris* 179: 1353–1356.

—— 1926a. Sur une nouvelle maladie du noyau ou grasserie des chenilles de *Pieris brassicae* et un nouveau groupe de microorganismes parasites. *C. r. hebd. Séanc. Acad. Sci., Paris* 182: 180–182.

—— 1926b. Sur un vibrion parasite des chenilles de *Pieris brassicae* L. *C. r. Séanc. Soc. Biol.* 94: 67: 69.

—— 1926c. Contribution à l'étude des maladies à virus filtrant chez les insectes. Un nouveau groupe de parasites ultramicrobiens: les Borellina. *Ann. Inst. Pasteur* 40: 314–352.

—— 1929. Contribution à l'étude des microsporidiens parasites de *Pieris brassicae* L. *Arch. anat. micr. Paris* 25: 212–230.

—— 1940. Existence d'une septicemie à spirochetes *S. pieridis* sp. n. chez les chenilles de *Pieris brassicae* en France. *C. r. Acad. Sci., Paris* 210: 615–616.

—— 1943. On nouveau mycose des larves de *P. brassicae*. *C. r. Acad. Sci.* 217: 383–384.

Pospelov, V.P., 1929. Symbiotic micro-organisms and their relation to diseases of insects. *Plant Prot.* 6: 13–20.

—— 1936. Results of work accomplished by the laboratory for insect diseases in the development of a system for the microbiological control of harmful insects. *Summ. Sci. Res. Wk. Inst. Pl. Prot. Leningrad, 1935*. pp. 318–321.

—— 1938. Methods of infecting insects with entomophagous fungi. *Summ. Sci. Res. Wk. Inst. Pl. Prot. Leningrad, 1936*. pp. 64–67.

—— 1939. Results of utilisation of fungous, bacterial and virus disease as a control measure against agricultural pests. *Leningrad Acad. agric. Sci.* 1940: 125–129.

—— 1944. Microbiological method of controlling agricultural pests. (In Russian.) *Proc. Lenin. Acad. agric. Sci. USSR* 9: 3–8.

Pospelov, V.P. & Noreiko, E.S., 1929. Wilt disease (Polyederkrankheit) of larvae and the yeast *Debaryomyces tyrocola* Kon and its virus. (In Russian.) *Izv. Prikl. ent.* 4: 167–183.

Proeseler, G., 1971. Übermittlung des Rübemosaischesvirus durch Insekten mit scharfem Mundstück. *Arch. Pflschutz.* 7: 391–397.

Ratcliffe, N.A. & Gagen, S.J., 1976. Cellular defense reactions of insect haemocytes in vivo: nodule formation and development in *Galleria mellonella* and *Pieris brassicae* larvae. *J. Invert. Path.* 28: 373–382.

Ratcliffe, N.A., 1975. Spherule cell-test particle interactions in monolayer cultures of *Pieris brassicae* hemocytes. *J. Invert. Pathol.* 26: 217–223.

Ripa, R., 1978. *Studies of the susceptibility of Pieris brassicae (L.) to a granulosis virus.* Ph.D. thesis, University of London (Imperial College).

Ripa, R., David, W.A.L. & Atkey, P.T., 1976. Inactivation of granulosis virus in the gut of *Pieris brassicae. Annual report of the Glasshouse Crops Research Institute for 1976.* pp. 106–107.

Rivers, C.F., 1959. Virus resistance in larvae of *Pieris brassicae. Trans. Ist. Int. Conf. Ins. Path. Biol. Con. Bratislavia Slov. Akad. Vied. 1959*: 205–210.

—— 1965. The natural and artificial dispersion of pathogens. *In: Insect pathology and microbial control.* N. Holland, Amsterdam. pp. 252–263.

Rogers, A.H., 1967. *The bioassay of crystals of Bacillus thuringiensis using larvae of Pieris brassicae (L.)* Thesis. University of Leeds. 145 pp.

Rogers, A.H., White, A.G., Wolf, J. & Gibson, N.H.E., 1966. The quantitative administration of bacterial pathogens to insect larvae. *Lab. Pract.* 15: 427–430.

Sandoz, 1974a. Thuricide HP. International recommendations for use. March 1974. *Sandoz Ltd.*

—— 1974b. Thuricide HP. General data. November 1974. *Sandoz Ltd.*

Sidor, C., 1959. Susceptibility of larvae of the large white butterfly (*Pieris brassicae* L.) to two virus diseases. *Ann. appl. Biol.* 47: 109–113.

Smirnov, O.V., 1976. Mixed virus infections in insects. (In Russian.) *Rev. ent. URSS* 55: 712–719.

Smith, K.M., 1959. The use of viruses in the biological control of insect pests. *Outlook on Agric.* 2: 178–184.

—— 1960. Viruses as insecticides. *Agric. vet. chem.* 1: 132–134, 136–137.

—— 1963. The cytoplasmic virus disease. *In: Insect pathology.* Ed. E.A. Steinhaus. Academic Press, London & New York. pp. 457–497.

—— 1967. *Insect virology.* Academic Press, London & New York.

—— 1976. *Virus–insect relationships.* Longman, London & New York.

Smith, K.M. & Hills, G.J., 1959. Further studies on the electron microscopy of Tipula iridescent virus. *J. molec. Biol.* 1: 277–280.

Smith, K.M., Hills, G.J. & Rivers, C.F., 1961. Studies on the cross-inoculation of Tipula iridescent virus. *Virology* 13: 233–241.

Smith, K.M. & Rivers, C.F., 1956. Some viruses affecting insects of economic importance. *Parasitology* 46: 235–242.

—— 1959. Cross-inoculation studies with the Tipula iridescent virus. *Virology* 9: 140–141.

Somerville, H.J. & Jones, M.L., 1972. DNA competion studies within the *Bacillus cereus* group of Bacilli. *J. gen. Microbiol.* 73: 257–265.

465

Somerville, H.J. & Pockett, H.V., 1975. An insect toxin from spores of *Bacillus thuringiensis* and *Bacillus cereus*. *J. gen. Microbiol.* 87: 359–369.

Spaar, D., 1973. Information from the socialist country. Microbiological preparations for plant protection in the USSR. *Nachrbl. Pflanzenschutzdienst DDR* 27: 47–48.

Steinhaus, E.A., 1949. *Principles of insect pathology*. McGraw-Hill Book Company, New York. pp. 757.

—— 1957a. New records of insect–virus diseases. *Hilgardia* 26: 417–430.

—— 1957b. List of insects and their susceptibility to *Bacillus thuringiensis* Berliner and closely related bacteria. *Lab. Insect Path.* Berkeley (4) 1–24.

—— 1963. *Insect pathology*. Vol. 1. Academic Press, New York & London.

Stephens, J.M., 1963. Bacterial activity of haemolymph of some normal insects. *J. Insect Path.* 5: 61–65.

Sweetman, H.L., 1936. *The biological control of insects*. Constable & Co. Ltd., London.

Toumanoff, C., 1956. Virulence experimentale d'une souche banale de *Bacillus cereus* Frank. et Frank. pour les chenilles de *Galleria melonella* L. et *Pieris brassicae* L. *Ann. Inst. Pasteur* 90: 660–665.

—— 1959. Observations concernant le rôle probable d'un predateur dans la transmission d'un bacille aux cheniles de *Pieris brassicae*. *Ann. Inst. Past.* 96: 108–110.

Toumanoff, C. & Grison, P., 1954. Études préliminaires sur l'utilisation des bactéries et champignons entomophages contre les insectes nuisables. *C. r. Acad. agric. Fr.* 40: 277–280.

Toumanoff, C. & Malmahche, M., 1956. Virulence experimentale d'une souche banale de *Bacillus cereus* F. et F. pour les chenilles de *Galleria mellonella* L. et *Pieris brassicae* L. *Ann. Inst. Pasteur* 90: 660–665.

Touzeau, J., 1962. Observations sur la lutte en plein champ contre la pièride du chou (*Pieris brassicae*). *C. r. Activ. exp. Def. Cult. Tunisie*, 1962: 43–47.

Vago, C., 1953. La polyedrie de *Thaumetopoea pityocampa*. *Ann. Épiphyt.* 3: 319–332.

—— 1954. Action du permanganate de potassium comme désinfectant des oeufs d'insectes vis à vis des viroses. *Bull. Soc. zool. Fr.* 79: 138–141.

—— 1959. On the pathogenesis of simultaneous virus infections in insects. *J. Insect Path.* 1: 75–79.

—— 1963. Predispositions and interrelations in insect diseases. *In: Insect pathology*. Ed. E.A. Steinhaus. Academic Press, New York & London. pp. 339–379.

Vago, C. & Atger, P., 1961. Multiplication massive des virus d'insectes pendant la mue nymphale. *Entomophaga* 6: 53–56.

Vago, C. & Bergoin, M., 1963. Développement des virus à corps d'inclusion du lépidoptère *Lymantria dispar* en cultures cellulaires. *Entomophaga* 8: 253–261.

Vago, C. & Croissant, O., 1959. Sur une virose cytoplasmique de *Pieris brassicae* L. Lépidoptère. *Experimentia* 15: 102–103.

Vago, C., Lepine, P. & Croissant, O., 1955. Mise en evidence du virus de la "granulose" (pseudograsserie) de *Pieris brassicae* L. *Ann. Inst. Pasteur* 89: 458–460.

Vago, C., Martouret, D. & Heitor, F., 1961. Conservation de virus et de bactéries entomopathogènes sous forme de comprimes. *Entomophaga* 6: 185–189.

Van der Laan, P.A. & Wassink, H.J.M., 1969. On the capacity of *Bacillus thuringiensis* to spread in insect populations. *Neth. J. Pl. Path.* 75: 105–108.

Vankova, J., 1962. The application of bacterial preparations of *Bacillus thur-*

ingiensis against some pests of agricultural crops, I. Its use against cabbage white butterfly, (*Pieris brassicae*). (In Russian.) *Rostlinná Vyroba* 8: 571–576.

Varma, G.C., Bindram, O.S. & Darshan, S., 1974. Comparative efficacy of *Bacillus thuringiensis* formulations against *Pieris brassicae* L. *Curr. Sci.* 43: 734–735.

Vasil'evskii, N.I., 1929. Pink muscardine and its casual agents, *S. aphodii* & *S. fumosorosca*. (In Russian.) *Morbi Plantarum Leningrad* 18: 113–148.

Verber, J., 1959. Comparative histopathology of *Nosema bombycis* in different hosts. (In German.) *Trans. 1st. Int. Conf. Ins. Path. Biol. Con. Bratislavia Slov. Aakad. Vied,* 1959. 301–313.

Voskresenkaya, V.N., 1977. Toxobakterin and entobakterin on cabbage. (In Russian.) *Zashch. Rast.* (4) 54.

Weiser, J., 1977. *An atlas of insect diseases.* 2nd edition. Dr. W. Junk, The Hague.

White, A.G., 1968. *The histopathological effects and the site of action of Bacillus thuringiensis in the larvae of Pieris brassicae and its bioassay against the larvae of Galleria mallonella.* Thesis, University of Leeds. pp. 110.

—— 1969. Bacterial insecticide against caterpillars. *Proc. R. ent. Soc. London* 34: 25–27.

Woroniecka, J., 1928. Observations on pests of cultivated plants that appeared in district of Lublin and in part of the district of Kiele 1926–27. (In Russian.) *Mem. Inst. Nat. Polon. Econ. Rur. Pulaway* 9: 216–251.

Zhukauskene, Y.I. & Misyalyunene, I.S., 1976. Effect of biotoxibacillin and entobacterin-3 on cabbage pest in the Lithuanian SSR. (In Russian.) *Liet. Tsr. Mokslu. Akad. Darb. Ser. C. Biol. Mokslai.* 4: 61–68.

Further references

Arkhipov, G.E., 1976. *Zashch. Rast.* (10) 19 (effectiveness of entobacterin).

Arnott, H.J. & Smith, K.M., 1969. *J. Invert. Path.* 13: 345–350 (branched rods of granulosis).

Atger, P., Dussausoy, G. & Bourguignon, S., 1965. *Rev. path. veg. ent. agric.* 43: 191–194 (*Aerobacter* sp.).

Blunck, H., 1956. *Int. Cong. zool. Proc.* 14: 344–345 (microsporidians).

—— 1958. *Proc. X. Int. Cong. Ent.* 4: 703–710 (microsporidia).

Boczkowska, M., 1932. *Rocz. nauk. Roln. Lesn.* 27: 137–156 (effect of *Entomopthora*).

Boehm, H., 1974. *Pflanzenarzt.* 27: 22 (biological insecticide).

Böhm, O., 1961. *Pflanzenarzt.* 14: 106–108 (experiments with *Bacillus*).

Bonnefoi, A., Burgerjon, A. & Grison, P., 1958. *C. r. Acad. Sci. Paris* 247: 1418–1420 (experiments with *B. thuringiensis*).

Bourguignon, S., Dusaussoy, G. & Atger, P., 1964. (*Aerobacter* preparation.) *Rev. path. veg. ent. Agric.* 43: 191–194.

Breed, R.S. & Petraitis, 1954. *Int. Bull. bact. Nom. Taxonomy* 4: 189–214 (taxonomy of viruses).

Burgerjon, A., 1957. *Entomophaga* 2: 129–135 (application of pathogenic organisms).

—— 1964. *Ann. Épiphyt.* 15: 73–84 (adhesiveness of *B. thuringiensis* preparations).

—— 1974. *Ann. Parasit. Humaine Comp.* 47: 835–844 (effect of *B. thuringiensis* on physiology).

Burgerjon, A. & Grison, P., 1959. *Entomophaga* 4: 201–209 (effect of different sources of *B. thuringiensis*).

Burgerjon, A. & Yamvrais, C., 1960a. *XI Int. Cong. Ent. Vienna* 1960: 842–844 (titration of *B. thuringiensis*).

— 1960. *C. r. Acad. Sci., Paris* 249: 2871–2872 (titration of *B. thuringiensis* preparation).

Burges, H.D., 1964. *World Crops* 16: 70–76 (control of insects with bacteria).

Cunningham, J.C., 1968. *J. Invert. Path.* 11: 132–141 (identification of some nuclear polyhedrosis and granulosis viruses).

De Barjac, H., Burgerjon, A. & Bonnefois, A. 1968. *J. Invert. Path.* 12: 465–467 (detoxification of *B. thuringiensis* crystals).

Dedonder, R. & Lecadet, M., 1964. *Entomophaga mem. Hors. Ser.* 2. 1964: 197–203 (effect of gut enzymes on *B. thuringiensis* crystals).

Dulmage, H.T., 1973. *Ann. N.Y. Acad. Sci.* 217: 187–199 (standardisation of microbial preparations).

Dulmage, H.T. & Rhodes, R.A., 1971. *In: Microbial control of insects and mites.* Ed. H.D. Burges & H.W. Hussey. Academic Press, London. p. 507 (production of pathogens in artificial media).

Eremeeva, A.M., 1925. *Bolezni Rastenii* 14: 100–103 (epidemic of *E. sphaerosperma* in 1919).

Ferdinandsen, C. & Rostrup, S., 1918. *Tidsskr. Plant Copenhagen* 26: 683–733 (fungus diseases of pests).

— 1919. *Tidsskr. Plant. Copenhagen* 27: 399–450 (fungus diseases of pests).

— 1921. *Tidsskr. Plant. Copenhagen* 27: 667–759 (fungus diseases of pests).

Fresenius, G., 1856. *Botan. Z.* 14: 882 (*E. sphaerosperma* account).

Hostounsky, Z., 1970. *Acta ent. Bohemoslov* 67: 1–5 (microsporidians used on *P. brassicae*).

Huger, A., 1963. *In: Insect Pathology.* Ed. E.A. Steiner. pp. 531–575 (granulosis of insects).

Jaques, R.P., 1977a. *Res. Stn. agric. Can. Harrow Ontario* 175–178 (granulosis of *P. brassicae*).

— 1977b. *Misc. Publ. Entomol. Soc. Am.* 10: 99–116 (stability of entomopathogenic viruses).

Kelsey, J.M., 1957. *N.Z. J. Sci. Technol.* 38: 644–646 (*P. brassicae* granulosis used on *P. rapae*).

Kowalska, T., 1971. *Biul. Inst. Ochr. Roslin* (48) 123–130 (preparations of *B. thuringiensis*).

Lipa, J.J., Prusszynski, S. & Bartkowski, J., 1970. *Buch. Inst. Ochr. Rost. Proznon* 347–354 (use of biopreparations for protection of brassica crops).

Masera, E., 1955. *Int. Ser. Tech. Con. Acts.* 33: 4 (vector of prebrine).

Metalnikov, S. & Metalnikov, S.S., 1935. *Ann. Inst. Pasteur* 55: 709–760 (bacterial methods used against *P. brassicae*).

Nizi, G., 1963. *Note Appunti Sper. ent. agri. Fasc. Perugia* 1963: 95–118 (different preparations of *B. thuringiensis*).

Noordink, J.P.W., 1964. *Inst. Plantenz. Onderzoek* 1964: 140–144 (toxicity of *B. thuringiensis*).

— 1966. *Inst. Plantenz. Onderzoek* 1966: 47 (feeding *B. thuringiensis*).

Noordink, J.P.W., Maas Geesteranus, H.P. & Van der Anker, C.A., 1967. *In: Insect Pathology and Microbial Control.* Ed. Van der Laan, P.A. North Holland Publishing Company, Amsterdam.

Nurylaeva, R.N., Bokhvalov, S.A. & Vorobeyva, H.H., 1974. *Hayka* 88–92 (polyhedral virus).

Paillot, A., 1915. *Ann. Épiphyt.* 2: 188–232 (use of micro-organisms).

— 1918. *C. r. Soc. Biol.* 80: 66 (microsporidians).

—— 1919. *C. r. hebd. Séanc. Acad. Sci.* 169: 740–742.

—— 1925. *C. r. Acad. Sci. Paris* 180: 1797–1799 (cytological changes).

—— 1927. *C. r. Acad. Sci. Fr.* 185: 673–675 (use of protozoans).

—— 1933. *L'infection chez les insectes.* Imprimier de Trevoux, G. Patissier.

—— 1943. *C. r. Acad. Sci. Paris* 217: 383–384 (a new mycosis).

Pendleton, I.R., & Morrison, R.B., 1967. *J. appl. Bacteriol.* 30: 402–405 (action of proteases on *B. thuringiensis* toxin).

Petch, T., 1948. *Trans. Birt. Mycol. Soc.* 31: 286–304 (*E. sphaerosperma*).

Ponsen, M.B., 1967. *Proc. int. Coll. Ins. Path. Micr. Control, Wageningen.* 1966. pp. 86–89 (development of virus antigen).

Rautapää, J., 1967. *Annls. agric. Fenn.* 6: 103–105 (toxicity of *B. thuringiensis*).

Rivers, C.F., 1964. *Discovery,* September (virus pesticides).

Rivers, C.F. & Longworth, J.F., 1972. *J. Invert. Pathol.* 20: 369–370 (effect of lepidopteran virus on *P. brassicae*).

Russo, L.F., 1968. *Boll. Lab. ent. Agrar. F. Silvestro Portici* 26: 263–270 (*B. cereus*).

Scherrer, P.S. & Somerville, H.J., 1977. *Eur. J. Biochem.* 72: 479–490 (toxicity of outer layers of spore of *B. thuringiensis*).

Shchepetilnikova, V.A. & Fedorintchik, N.S., 1964. *Entomophaga* 1964: 511–514 (use of micro-organisms and parasites).

Shchepetilnikova, V.A., Kapustina, O.V., Molchanova, V.A. & Shchichenkov, P.I., 1968. *Trudy Vses. Inst. Zashch. Rast.* 31: 86–98 (use of fungi).

Shvetzova, O.I. & Zurabova, E.R., 1967. *Proc. Int. Coll. Ins. Path. Micr. Con. Wageningen* 1966: 216–217 (alteration of properties of *B. thuringiensis*).

Smirnov, O.V., 1976. *Rev. ent. URSS* 55: 712–719 (mixed virus infections in insects).

Smith, K.M., 1960. *Rpt. 7th. Comm. Entomol. Conf. London.* 1960. 111–118 (*P. brassicae* larvae eat cadavers).

Spaar, D., 1973. *Nachrbl. Pflanzenschutzdienst DDR* 27: 47–48 (viral preparation used on *P. brassicae*).

Steinhaus, E.A., 1952. *Yearbook of Agriculture.* 388–394 (infectious diseases of insects).

—— 1949. *Principles of Insect Pathology.* McGraw-Hill Book Company, New York. pp. 757.

—— 1957. *Hilgardia* 26: 417–430 (virus diseases).

—— 1953. *Ann. N.Y. Acad. Sci.* 56: 517–537 (taxonomy of insect viruses).

Trump, B., Kwanda, N. & Luthy, P., 1974. *Path. Microbiol.* 41: 171–173 (degradation of endotoxin of *B. thuringiensis*).

Vago, C., 1952. *6th. Int. Cong. Path. comp. Madrid* 1: 121–133 (invertebrate diseases).

Vasiljevic, L., 1954. *Plant Prot. Beograd* 26: 118–120 (*B. thuringiensis*).

Veremchuk, G.V. & Issi, I.V., 1970. *Parazitologiya* 4: 3–7 (development of microsporidia).

Weiser, J., 1971. *All Union ent. Soc. Leningrad* 2: 107–109 (host specificity of protozoa).

Yamvrais, C., 1962. *Entomophaga* 7: 101–159 (mode of action of *B. thuringiensis*).

Zaitzeva, A.Y., 1938. *Summ. Sci. Res. Wk. Inst. Plant. Prot.* 1936 (3) 69–73 (effectiveness of *Bacillus*).

16. Predators

Introduction

Outlined in this chapter are accounts of 26 species of bird predators, seven species of hymenoptera, seven species of hemiptera, three species each of coleoptera and diptera, and other miscellaneous insect species belonging to five other insect orders, as well as several unidentified arachnid species, three species of mammal and one species each of a reptile, an amphibian and an insectivorous plant.

It is difficult to obtain a clear picture of bird predation on *P. brassicae* in nature, but there are sufficient records to suggest that it is taken regularly as food by some species. This is true at all stages of its life cycle, though the number of imagines which are captured can only be assessed by a few lucky observations and the wing debris seen in the area of heavy attacks.

It has been suggested (Marsh & Rothschild, 1974) that *P. brassicae* is a member of a worldwide Müllerian mimicry complex and there is little doubt that its conspicuous black and white colouration indicates that it is an aposematic species. It is equally certain that no defense is absolute and it is to be expected that various predators will be adequately equipped to prey upon this butterfly.

Considering how exceedingly common this species is, and how relatively simple field observations would be, it is astonishing that no serious attempt has been made to carry out a study in nature. It will be seen that a wide variety of predators has been recorded capturing *P. brassicae*, but many are known from only a single record. It is also worth noting that some authors (Stidson, 1933) claim that, for example, *Musicapa striata* (Pallas) avoid imagines of *P. brassicae*, while Witherby (1938) noted that Whites are taken by this bird. Also there is much scattered evidence to suggest that certain species of birds, in particular the Corvidae, have a strong aversion to feeding upon *P. brassicae*. There is a big opportunity here for further investigation.

470

Bird predators

An interesting account by John Westwood (1854) tells us that at least one bird species has been used very effectively as a biological control agent. He describes how Adrian Haworth had a perfect means of controlling *P. brassicae* in his garden.

"Seagulls with their wings cut are of infinite service. I had one 8 years, which was last killed by accident, that lived entirely all the while upon the insects, slugs and worms which he found in the garden."

There are at least 26 species of birds, representing 15 families, which have been recorded capturing or preying upon *P. brassicae*; the Thrush family (Turdidae) is particularly well represented (Table 16.1). Most of the records refer to the capture of larvae and imagines rather than to ova and pupae. It is generally believed that birds exert quite a reasonable degree of control over the numbers of *P. brassicae* in nature (Friederichs, 1931; Hale Carpenter, 1940; Berkovitch, 1972).

It has been shown (Aplin *et al.*, 1975) that *P. brassicae* as a larva sequesters and stores mustard oil glycosides from its food plant which are carried through to the imago via the pupa. It is not known towards which predator the effects of these substances are directed, but it is assumed they must perform a protective function. These authors also demonstrated the presence of another toxic substance (possibly a protein) which is present in larva, pupa, and imago. All these stages are lethal when injected intraperitoneally into the laboratory mouse, and this toxic factor is present in specimens reared on artificial diet lacking singrin; via the oral route they had no effect on a captive magpie. However, the same bird would not eat *P. brassicae* pupae reared on cabbage and experience of the latter specimens conferred absolute protection on the former which were then rejected on sight. It is therefore evident that some form of chemical protection is obtained for the larvae and pupae from the ingestion of cabbage. Other authors have noted that, reared on *Tropaeolum*, the pupae are also rejected by captive birds.

One of the chief bird predators of *P. brassicae* is *Passer domesticus* (Linnaeus) which has been seen to eat ova, larvae and imagines (SLENHS, 1909: Table 16.1). It is often a common sight to see fledgling sparrows trying to catch "whites". Musy (1918) pointed out that in France bird life should be protected in winter as a scarcity of sparrows might result in an increase in the numbers of *P. brassicae*. Baker (1970) found that older larvae of *P. brassicae* are attacked by different predators; first and second instar larvae are more often found by the ground searching *Turdus ericetorum*. A more recent account of predation, presumed to be by the sparrow, was given by Measures (1976) who found a collection of butterfly wings below a *Buddleja* sp. bush.

Table 16.1 Bird predators.

Specific names	Common name	Stage eaten O = ova, L = larva, P = pupa, I = imago	Authority
General statements about "birds" or "small birds ..."			Anonymous, 1917; Pocock, 1911; Frohawk, 1922 1941; Moss, 1933; Roer, 1957; Baker, 1970
COLUMBIDAE			
Columba sp.	Pigeon	L	Rothschild (unpublished)
CUCULIDAE			
Cuculus canoris Linnaeus	Cuckoo	L$_5$	Hale Carpenter, 1940; Parslow pers. comm. via Skelton, pers. comm.
EMBERIZIDAE			
Emberiza citrinella Gengler	Yellowhammer	I	Collenette, 1935; Gÿory & Reichart, 1965
FRINGILLIDAE			
Fringilla coelebs Linnaeus	Chaffinch	I	Collenette, 1935
HIRUNDINIDAE			
Hirundo rustica Linnaeus	Swallow	I	Marshall, 1909; Witherby, *et al.*, 1938
LARIDAE			
Larus sp.	Seagull	L	Westwood, 1854
MELEAGRIDIDAE			
Melaegris gallopara	Turkey	?	D'Ambrosia, 1914
	Chicken	L	Seitz, 1906; Roer, pers. comm.
	Fowls	?	Roebuck, 1925
MOTACILLIDAE			
Anthus pratensis (Linnaeus)	Meadow Pipit	I	Collenette, 1935
MUSCICAPIDAE			
Musicapa sp.	Spotted Flycatcher	I	Marshall, 1909; Williams *et al.*, 1942
PARIDAE			
Parus caeruleus			Witherby *et al.*,

Table 16.1 continued

Specific names	Common name	Stage eaten O = ova, L = larva, P = pupa, I = imago	Authority
PARIDAE — continued			
Linnaeus	Blue Tit	$L_{1,2,3}$	1938
		P	Baker, 1970
Parus major Linnaeus	Great Tit	$L_{4,5}$	Frohawk, 1940; Gÿory & Reichart, 1965
		P	Baker, 1970
PHASIANIDAE			
Phasianus colchicus			
Linnaeus	Pheasant	I	Collenette, 1935
PLOCEIDAE			
Passer domesticus			Marshall, 1909;
(Linnaeus)	House Sparrow	$O, L_{1,2,3}$ I	Rattray, 1913; Musy, 1918; Measures, 1976
STURNIDAE			
Sturnus vulgaris			
Linnaeus	Starling	I	Collenette, 1935
SYLVIIDAE			
Sylvia borin (Boddaert)	Garden Warbler	O	Witherby *et al.*, 1938
Sylvia communis			Taylor,
Latham	Whitethroat	I	pers. comm.
TURDIDAE			
Erithacus robecula			Marshall, 1909;
(Linnaeus)	Robin	I	Witherby *et al.*, 1938
Oenanthe oenanthe			
(Linnaeus)	Wheatear	L	Menegaux, 1920
Phoenicurus ochruros			Gÿory &
(Gmelin)	Black Redstart	L	Reichart, 1965
Saxiola rubetra			Gÿory &
(Linnaeus)	Whinchat	L	Reichart, 1965
Turdus sp.	Thrushes	L	Marshall, 1909; Gardiner, pers. comm.
Turdus ericetorum			
Turta		P	Baker, 1970
Turdus philomelos			Witherby *et al.*,
Brehm.	Song Thrush	$L_{4,5,}$ I	1938
Turdus merula			
Linnaeus	Blackbird	?	Ormerod, 1880

Table 16.1 continued

Specific names	Common name	Stage eaten O=ova, L=larva, P=pupa, I=imago	Authority
TURDIDAE — *continued*			
Turdus musicus			
Linnaeus	European Thrush	L	Gardiner, 1974
Turdus mustelinus	—	?	Marshall, 1909
NEGATIVE RESPONSES			
HIRUNDINIDAE			
Delichon sp.	Swallows	I	Lane, 1957a
Riparia sp.	Martins	I	Lane, 1957a
KITTACINCLIDAE			
Kittacincla malabarica			
Gmelin	Shama (female)	L,P,I	Lane, 1957b
MUSCICAPIDAE			
Musicapa striata			Stidson, 1933;
(Pallas)	Spotted Flycatcher	I	Witherby *et al.*, 1938
CORVIDAE			
Corvus sp.	Crow	L	Rothschild, 1964
Pica pica Linnaeus	Magpie	L,P	Rothschild (unpublished)

When pest proportions of *P. brassicae* occur, it is likely that there will be more bird–insect encounters which will result in more accounts of different bird predators. In Hungary, Gÿory & Reichart (1965) found that *Parus major* Linnaeus, *Phoenicurus ochrurus* (Gmelin) and *Emberiza citrinella* Gengler contained parts of *P. brassicae* in their guts at such times. On the other hand, in the laboratory Rothschild & Lane (1960) found that *P. major* gave alarm calls on touching and later on seeing "various whites".

The larvae of *P. brassicae* are aposematic and display various characteristics common to species with these attributes. They are colonial in habit, aggregating on the surface of leaves, their colour contrasts with the foliage on which they are feeding, they emit a foul odour, and in the third and fourth stages jerk their heads in unison in response to a shrill sound. Several authors, however, have pointed out that if these larvae fall to the ground they are well concealed against this type of background, and illustrate the principle of crypsis-from-a-distance, although they are aposematic at close quarters. Frohawk (1922) commented that although the larvae of *P. brassicae* were not highly coloured, they were nevertheless aposematic, and lacked bird predators. Evidence is available (Table 16.16) to show that they are in nature taken by a variety of birds and other predators.

When macerations of fourth and fifth instar larvae were injected interperitoneally into mice, Marsh & Rothschild (*loc. cit.*) found the fifth instar larva more toxic than the fourth, but less so than the pupal stage. Moss (1933) experimented with pupae placed in natural and unnatural positions and found that they were eaten by wild birds. Baker (1970) however, only partially corroborated this observation.

Hymenopterous predators and scavengers

Wasps have often been reported to readily attack, devour and scavenge living or dead larvae and imagines of *P. brassicae* (Table 16.2, cf. Gardiner, 1952. Fowler (1897) and Baker (1931) both recount how they saw a wasp suck fluids from a larva (halfgrown) and fly off with the carcass. Control of *Pieris* larvae by ants in a cabbage field in Italy was

Table 16.2 Hymenopterous predators.

Specific name	Common name	Stage attacked (as for 16.1)	Authority
FORMICOIDEA			
Myrmica laevinodis Nyl.	Ant sp.	—	Speyer, 1956
—	Argentine ant sp.	L	Querci, 1957
Myrmica rubra (Linnaeus)	Black ant	I (dead)	Gardiner pers. comm.
VESPOIDEA			
Wasps – unspecified	—	—	Nicholson, 1931; Baker, 1931; Gardiner, 1952; Feltwell (unpublished)
Polistes sp.	Hunting wasp	L_5	Anonymous, 1976
Polistes jokahamae Radosybowsky	Hunting wasp	L_5	Anonymous, 1976
Vespa crabo Linnaeus	Hornet	—	Easton, 1946
Vespula germanica (Fabricius)	German wasp	L	Blunck, 1935, 1950
Vespula vulgaris (Linnaeus)	Common wasp	L	Fowler, 1897; Blunck, 1935; SLENHS, 1951a,b

claimed by Querci (1957). He transported ants nests enclosed in pots to an infested field and found that the larvae of *P. brassicae* subsequently disappeared. The ants did not eat the ova but preferred the young larvae. Querci (*loc. cit.*) stated that in the Formia district near Naples invasion of the area by these ants successfully reduced the insect population and he recommended the use of movable ants nests for insect control.

Anonymous (1976) reported that in south-east China hunting wasps (*Polistes jokahamae* Radosykowsky and another *Polistes* sp.) preferred late instar larvae of *Pieris* sp (species not stated but probably *P. rapae crucivora* Boisduval), as well as three other insect species.

Hemipterous predators

In the wild some Pentatomid and Reduviid bugs exert natural control over *P. brassicae* (Table 16.3). For instance during a bad outbreak of *P. brassicae* in south-west France, Anonymous (1917) noted that *Aptus mirmicoides* (Costa) ate ova and larvae. Faure (1923) also noted that another *Nabis* species, *N. ferus* Linnaeus ate *P. brassicae* larvae in the wild and in the laboratory. Larvae three or four times their own size were successfully paralysed with one puncture of their body and their juices sucked out. Faure (*loc. cit.*) also mentioned as a footnote that another hemipteron, *Zircona coerulea* Linnaeus, attacked *P. brassicae* larvae.

Table 16.3 Hemipterous predators.

Specific name	Stage attacked (as for 16.1)	Authority
PENTANOMIDAE		
Picromerus bidens Linnaeus	L	Mayné & Breny, 1940; Blunck, 1950, 1953
Podiscus maculiventris (Say)	P,I	Sazonova *et al.*, 1976
Printhaeus sanguinipes Fabricius	L (young or old)	Temple, 1939
Troilus luridus Fabricius	L	Temple, 1939; Blunck, 1953
Zicrona coerulea Linnaeus	(L)?	Faure, 1923
REDUVIIDAE		
Nabis ferus Linnaeus	L (any age)	Faure, 1923
Nabis lativentris Boheman ? = *Aptus lativentris* (Boheman) now *A. mirmicoides* (Costa)	O, L	Anonymous, 1917

Temple (1939) noted that both *Printhaeus sanguinipes* Fabricius and *Troilus luridus* Fabricius were effective predators in the wild on a variety of insects. They are able to paralyse "large" *P. brassicae* larvae and may spend up to 26 hours sucking out the fluids. *T. luridus* Fabricius can also be fed on the larvae of *P. brassicae* as well as other insects in the laboratory.

A pentatomid bug *Picromerus bidens* Linnaeus apparently prefers slow-moving, firm insect bodies with smooth thin cuticle and accepted *P. brassicae* larvae; however, when they were fed larvae injected with polyhedsis virus, this caused a high mortality (Mayné & Breny, 1940).

A predatory bug *Podiscus maculiventris* (Say), which was introduced into the USSR in 1974–5, was recorded to eat pupae and imagines of *P. brassicae* (Sazonova *et al.*, 1976).

Coleopterous predators

In India, adults of *Broscus punctatus* Klugenburg were recorded as predators of *P. brassicae* larvae (Fletcher, 1919) and Singh *et al.* (1976) recorded that *Coccinella septempunctata* Linnaeus punctured and "sucked out" the body fluids of first instar larvae of *P. brassicae* both in the laboratory and in the wild (Table 16.4).

It is interesting to note that Dempster (1969) recorded five coleopterous predators of *P. rapae* which caused mortality in young larvae.

Other insect predators

An insect of some importance in control is *Chrysopa perla* Linnaeus (Lacewing) which has been recorded eating 10–11 *P. brassicae* larvae each day (Burakova, 1929). Other insects, such as earwigs and predatory flies,

Table 16.4 Coleopterous predators.

Specific name	Common name	Stage attacked (as for 16.1)	Authority
Broscus punctatus Klugenburg	—	L	Fletcher, 1919
Coccinella sp.	Ladybird	I	Blunck, 1935; Blyth, 1945
Coccinella septempunctata Linnaeus	7-spot ladybird	L_1	Singh *et al.*, 1976

Table 16.5 Other insect predators.

Specific name	Common name	Stage attacked (as for 16.1)	Authority
ORTHOPTERA			
Locusta viridissima	Great green		SLENHS, 1918b;
Linnaeus	bush cricket	$L_{1,2}$,I	Zeuner, 1940
PHASMIDA			
Mantis religiosa	Praying Mantis	I	
Linnaeus	(immature)	(caught only)	SLENHS, 1910
DERMAPTERA			
Euborella moesta			
Geve	Earwig	L	Berland, 1929
Forficula auricularia		L	Pussard, 1925;
Linnaeus	Earwig	(small L_3)	Blunck, 1950, 1953
NEUROPTERA			
Chrysopa carnea			
Stephens	Golden Eye	O	Kowalska, 1968
Chrysopa perla			
Linnaeus	Lacewing	L	Burakova, 1929
LEPIDOPTERA			
Pieris rapae			
(Linnaeus)			
(larvae)	Small White	L	Buckstone, 1938
DIPTERA			
Drosophila bucksii			
Coquillett		L	Miles, 1950
Exorista vulgaris			
Fallén		L?	Anonymous, 1917
Xanthandrus comtus			
Harris	Syrphid (Larva)	L?	Poutiers, 1926
—	Tachinid	L?	Ghosh, 1914
NEGATIVE RESPONSES			
Odonata			
Unspecified	Dragonfly sp.	I	
		(not eaten)	Killington, 1927

might deserve more attention as a few examples have been recorded for these orders. A Syrphid larva was seen to attack *P. brassicae*, and during two outbreaks of *P. brassicae* in India and France, Tachinids have been seen to eat *P. brassicae* (Table 16.5).

Two accounts of *Locusta viridissima* Linnaeus eating *P. brassicae* have been given, the first by South London Entomological and Natural History Society (SLENHS) (1918b), who fed it a *P. brassicae* larva (age not stated) which it ate "readily"; the second by Zeuner (1940) who noted that it ate young larvae, the older ones being "poisonous."

Vertebrate and other miscellaneous predators and scavengers

a) *Mammals.* Roer (1975, pers. comm.) made an interesting observation about *Mus musculus* Linnaeus, indicating that it could distinguish between parasitised and unparasitised pupae, and selectively ate only parasitised larvae. Cats are often seen catching and eating butterflies; Mosley (1926) noted that for two years a "young kitten" ate *P. brassicae* imagines with relish, and Eliot (1938) recorded a tom cat in the French Riviera which chewed and swallowed an imago (cf. Walker, 1938, Table 16.6). On the toxicity of *P. brassicae* to mammals, Worden (1941)

Table 16.6 Vertebrate and other predators.

Specific name	Common name	Stage attacked (as for 16.1)	Authority
MAMMALIA			
Apodemus sylvaticus			Roer, 1975,
(Linnaeus)	Field mouse	P	pers. comm.
Felis (domestic)	Cat	I	Eliot, 1938
Mus musculus			
Linnaeus	House mouse	I	Moss, 1933
REPTILIA			
Unidentified sp.	African lizards	I	van der Geest, 1975, pers. comm.
AMPHIBIA			
Bufo calamita			
Laurenti	Natterjack toad		Speyer, 1956
ARACHNIDA			
Unidentified sp.	Spider		Michael, 1947
Achaearanea			
tepidariorum			Rothschild,
(C.L. Koch)	—	I	pers. comm.
—	Common		Rothschild,
	Bathroom Spider	I	pers. comm.
—	Large		
	unidentified sp.	I	Blunck, 1935
INSECTIVOROUS			
PLANT			
Drosera sp.	Sundew	I	Williams, 1958; Oliver *in* Heslop-Harrison, 1978

noted that *P. brassicae* was a so-called "cryptotoxic form" which is said to cause stomatitis, colic and paralysis of the hind limbs.

b) *Reptiles.* The only reptiles which are known to feed on *P. brassicae* are a pair of 20 cm long African lizards (species not identified) which were being kept (1975) in the Laboratory of Experimental Entomology at Amsterdam University, and which had been living very healthily on a diet of *P. brassicae* imagines for the previous six years (Van der Geest, 1975, pers. comm.).

c) *Amphibians.* Speyer (1956) recorded *Bufo calamita* Laurenti feeding on *P. brassicae* in the dunes on the north sea coast of Germany.

d) *Arachnids.* It is not an unfamiliar sight to find *P. brassicae* imagines caught on spiders' webs and devoured by the spiders, particularly when these insects are reared in the laboratory; at least two species have been identified so far (Table 16.6).

e) *Plant.* There is a remarkable account of the insectivorous plant *Drosera* sp. on a two acre (0.82 ha) island off Essex, trapping an estimated six million immigrant *P. brassicae* imagines as they alighted (Williams, 1958).

f) *Mollusc.* Heslop-Harrison (1950) noted that the snail, *Helix aspersa* Müller ate a dead imago of *P. brassicae*.

References cited

Anonymous, 1917. Ravages of *Pieris* caterpillars on crucifers. *Bull. Sec. étude vulg. zool. agric. Bordeaux* 16: 108–110.
—— 1976. A preliminary study on the bionomics of hunting wasps and their utilisation in cotton insect control. *Acta ent. Sinica* 19: 303–308.
Aplin, R.T., Arcy Ward, D. & Rothschild, M., 1975. Examination of the large white and small white butterflies (*Pieris* spp.) for the presence of mustard oils and mustard oil glycosides. *J. Ent.* A. 50: 73–78.
Baker, H.W., 1931. Wasps killing larvae. *Entomologist* 64: 36.
Baker, R.R., 1970. Bird predation as a selection pressure on the immature stages of the cabbage butterflies, *Pieris rapae* and *Pieris brassicae. J. Zool. Lond.* 162: 43–59.
Berkovitch, I., 1972. Seeking pest solutions at Silwood Park. *Int. Pest Control* 14: 24–25.
Berland, L., 1929. Les forficules sont elles carnivores? *Bull. Soc. ent. Fr.* 1929: 289–290.
Blunck, H., 1935. Methods for breeding *Pieris brassicae. Arb. Physiol. angew. Ent. Berlin* 2: 78–87.
—— 1950. Zur Kenntnis des Massenwechsels von *Pieris brassicae* L. mit

besonderer Berücksichtigung des Dürrejahres 1947. *Z. angew. Ent.* 32: 141–171.

—— 1953. *Tierische Schädlinge an Nutzpflanzen. Lepidoptera und Trichoptera.* 8: 510.

Blyth, N., 1945. Ladybirds (Coccinellidae) destroying cabbage white (*Pieris brassicae*). *Suffolk nat. hist. Trans.* 5: 214.

Buckstone, A.A.W., 1938. *Pieris rapae* a cannibal. *Entomologist* 71: 34.

Burakova, L., 1929. Quantitative analysis of over-ground fauna. Population of cabbage. (In Russian.) *Trav. Soc. nat. Leningrad* 59: 83–94.

Collenette, C.O., 1935. Notes confirming attacks by British birds on butterflies. *Proc. zool. Soc. Lond.* 1935: 201–207.

Cox, W.E., 1940. Attacks of birds on *Pieris brassicae* L. (Lepidoptera). *Entomologist's mon. Mag.* 76: 161.

D'Ambrosio, G., 1914. An attack of caterpillars. *Boll. Catt. Amb. agric. Brindisi* 8: 70–71.

Dempster, J.P., 1969. Some effects of weed control on the numbers of the small cabbage white (*Pieris rapae* L.) on Brussel sprouts. *J. appl. Ecol.* 6: 339–345.

Easton, N.T., 1946. *Pieris brassicae* L. attacked by a hornet. *Entomologist's Rec. J. Var.* 58: 72.

Eliot, N., 1938. Winter and spring riviera butterflies. 1937–1938. *Entomologist* 72: 31–36.

Faure, J.C., 1923. Note sur un hemiptère predateur. *Rev. path. veg. ent. agric.* 10: 253–254.

Fletcher, T.B., 1919. Second hundred notes on Indian insects. *Agric. Res. inst. Pusa Bull.* 1919: 1–102.

Fowler, J.H., 1897. Lepidoptera in 1896. Notes from Ringwood. *Entomologist* 30: 107–112.

Friederichs, K., 1931. Zur Ökologie des Kohlweisslings (*Pieris brassicae*). *Z. angew. Ent.* 18: 568–581.

Frohawk, F.W., 1922. Destruction of *Papilio machaon* larvae by cuckoos. *Entomologist* 55: 280–281.

—— 1940. Destruction of *Pieris brassicae* by birds. *Entomologist* 73: 137–138.

—— 1941. Immigration of *Pieris brassicae. Entomologist* 74: 5.

Gardiner, B.O.C., 1952. Wasps attacking *Pieris brassicae* L. *Entomologist's Rec. J. Var.* 64: 355.

—— 1974. Observations on green pupae in *Papilio machaon* and *Pieris brassicae* L. *Wilhelm. Roux. Arch. EntwMech. Org.* 176: 13–22.

Ghosh, C.C., 1914. Life history of Indian insects. Lepidoptera. *Mem. Dept. agric. India Pusa* 5: 1–72.

Györy, J. & Reichart, G., 1965. Madártáplálkozás-Vissgálatok Jelentösebb erdöés Mezögazdaságikártevóktomeges Megjelenése Idején. *Aquila Budapest* 72: 67–98.

Hale Carpenter, G.D., 1940. Extensive destruction of *Pieris brassicae* (L.) by birds. *Entomologist's mon. Mag.* 76: 224–229.

Heslop-Harrison, J.W., 1950. *Pieris brassicae* L. eaten by a snail. *Entomologist's mon. Mag.* 86: 78.

Heslop-Harrison, Y., 1978. Carnivorous plants. *Sci. Am.* 238: 104–115.

Killington, F.J., 1927. Notes on a swarm of *Aeschna mixta*, Latr., (Odonata) in Hampshire, 1926. *Entomologist* 60: 244–245.

Kowalska, T., 1968. A report of studies on the biology of the golden-eyed lacewing (*C. carnea* Stephens = *C. vulgaris* Schneider). *Prace. nauk. Inst. Ochr. Roslin* 10: 145–157.

Lane, C., 1957a. Swallows and house martins taking moths near an MV trap. *Entomologist* 90: 297.

—— 1957b. Preliminary note on insects eaten and rejected by a tame Shama (*K. malabarica* Gm.) with the suggestion that in certain species of butterfly and moths females are less palatable than males. *Entomologist's mon. Mag.* 93: 172–179.

Marsh, N. & Rothschild, M., 1974. Aposematic and cryptic lepidoptera tested on the mouse. *J. Zool. Lond.* 174: 89–122.

Marshall, G., 1909. Birds as a factor in the production of mimetic resemblances among butterflies. *Trans. ent. Soc. Lond.* 329–383.

Mayné, R. & Breny, R., 1940. Predateurs et parasites du doryphore. *Bull. Inst. agron. Gembloux* 9: 61–80.

Mayné, R. & Breny, R., 1940. Predateurs et parasites du doryphore. *Bull. Inst. agron. Gembloux* 9: 61–80.

Measures, D.G., 1976. *Bright wings of summer*. Cassell, London.

Menegaux, A., 1920. Oiseaux utiles. *J. agric. Prat. Paris* 34: 174–176.

Michael, P., 1947. Lepidopterous prey of a large arachnid. *Entomologist* 80: 195–196.

Miles, P.M., 1950. *Drosophila busckii* Coq: (Diptera: Drosophilidae) in Wales. *Entomologist's mon. Mag.* 86: 141.

Mosley, C., 1926. Pierids eaten by a cat. *Entomologist* 59: 254.

Moss, J.E., 1933. The natural control of the cabbage caterpillars *Pieris* spp. *J. anim. Ecol.* 2: 210–231.

Musy, M., 1918. Les chenilles de chou, leur ennemis et leur moyens de les combattre. *Bull. Soc. Fribourg Sci. nat.* 24: 120–122.

Nicholson, C., 1931. Wasps killing larvae. *Entomologist* 64: 139–140.

Ormerod, E.A., 1880. *Notes of observations of injurious insects, report for 1880.* 25–27. West Newman & Co. London.

Pictet, A., 1918. Observations biologiques sur *Pieris brassicae* en 1917. *Bull. Soc. lépidopt. Genève* 4: 53–66.

Pocock, R.I., 1911. On the palatability of some British insects with notes on the significance of mimetic resemblances. Notes on the experiments by Poulton, E.B. *Z. Soc. Lond.* 11: 809–868.

Poutiers, R., 1926. Observations sur deux insectes vivants aux depens de *Pieris brassicae, Chalcis femorata* Ranz. et *Xanthandrus comtus* Harr. *Rev. path. veg. ent. agric.* 13: 31–32.

Pussard, R., 1925. A propos du regime alimentaire du perce-oveille *Forficula auricularia* L. (Dermaptera). *Bull. Soc. sci. nat. Rouen* 1925: 7–13.

Querci, O., 1957. Ants as a control against other insects. *Entomologist's Rec. J. Var.* 69: 101.

Rattray, R.H., 1913. Bird eating butterflies. *Entomologist* 46: 334.

Roebuck, A., 1925. Use of poultry against farm pests. *Eggs* 11: 206–207, 210–212, 418–419; 12: 3–7.

Roer, H., 1957. Tagesschmetterlinge als Vorzugsnahrung einiger Singvögel. *J. Orn.* 98: 416–420.

Rothschild, M., 1964. An extension of Dr. Lincoln Browers' theory on bird predation and food specificity, together with some observations on bird memory in relation to aposematic colour patterns. *Entomologist* 97: 73–78.

Rothschild, M. & Lane, C., 1960. Warning and alarm signals by birds seizing aposematic insects. *Ibis* 102: 328–330.

Sazonova, R.A., Shagov, E.M. & Stradimova, 1976. *Podiscus* – a predator of the

American white butterfly and the Colorado beetle. (In Russian.) *Zashch. Rast.* (8) 52.

Seitz, A., 1906. *Macrolepidoptera of the World.* Alfred Kernen Verlag, Stuttgart.

Singh, D., Ramzan, M. & Sandhu, G.S., 1976. Some observations on the feeding behaviour of adult *Coccinella septempunctata. Sci. & Cult.* 42: 178–179.

SLENHS, 1909. Meeting. *Entomologist* 42: 262.

—— 1910. Meeting, November 25th. *Entomologist* 43: 43–44.

—— 1918a. Meeting, September 13th, 1917. *Entomologist* 51: 22.

—— 1918b. Meeting, September 13th, 1917. *Entomologist's mon. Mag.* 54: 19–20.

—— 1951a. Meeting, April 25th, 1951. *Entomologist* 84: 167–168.

—— 1951b. Meeting April 25th, 1951. *Entomologist's mon. Mag.* 87: 191–192.

Speyer, W., 1956. *Pieris brassicae* L. in den Dünen der Nordseeinsel Amrum. *Z. PflKrankh. PflPath. PflSchutz* 63: 12–14.

Stidson, S.T., 1933. *Muscipa striata,* the spotted flycatcher, catching and eating *Vanessa urticae. Entomologist* 66: 92.

Tempel, W., 1939. A mass occurrence of Asopinae. *Arb. Physiol. angew. Ent. Berlin* 6: 51–56.

Walker, J.J., 1938. A precocious specimen of *Pieris brassicae. Entomologist's mon. Mag.* 76: 89.

Westwood, J.O., 1854. *The butterflies of Great Britain.* W.S. Orr & Co., Paternoster Row, London.

Williams, C.B., 1958. *Insect Migration.* Collins, New Naturalist Series, London.

Williams, C.B., Cockbill, G.F., Gibbs, M.E. & Downes, J.A., 1942. Studies on the migration of lepidoptera. *Trans. R. ent. Soc. Lond.* 92: 101–280.

Witherby, H.F., Jourdain, F.R.C., Ticehurst, N.F. & Tucker, B.W., 1938. *The Handbook of British Birds.* 5 vols. Suppl. addenda & additions, Witherby, London.

Worden, A.N., 1941. The presence of a viable noctuid larva (*Antitype flavicincta* F.) in the subcutaneaous tissues of a dog. *Entomologist's mon. Mag.* 77: 94–96.

Zeuner, F.E., 1940. *Phugiolopsis henryi* n.g. n.sp., a new Tettigonid and other Saltatoria (Orth.) from Kew. *J. Soc. Brit. ent. Kings Somborne* 2: 76–84.

17. Chemical control

Introduction

The chemical methods used for control of *Pieris brassicae* differ only slightly from those methods used on many other phytophagous insect pests and in fact they reflect the history of pesticide usage on insects. *P. brassicae* has always been widely available as a test insect and numerous insecticides have been tested on this species. In fact most work refers to the screening of insecticides in the laboratory and there is probably much that could be done in the field with compounds which have shown good results in the laboratory.

Up to 1942, with the discovery of synthetic insecticides, farmers used both simple inorganic and organic compounds or concoctions of deterrent plants to kill larvae or dissuade (with varying degrees of success) *P. brassicae* imagines from ovipositing. Many of these "anecdotal" compounds were, however, relatively ineffective or unacceptably toxic by today's standards. The value of plant extracts must not be overrated as no-one has yet produced a commercially useful repellant or anti-feedant compound; and in fact the desirability of using plant extracts may be questioned as in cases such as *Hyoscamus niger* Linnaeus (Hen-Bane), very toxic alkaloids are present which are more toxic than many modern insecticides.

These early attempts at pest control were rapidly overtaken in popularity by the use of the organophosphorus insecticides (OPs) and organochlorines (OCs), the latter including both cyclodienes and DDT derivatives. Carbamates, thiocyanate derivatives and phenylurea derivatives have also been used to control *P. brassicae*.

Biochemists have experimented with the administration of hormones for the control of *P. brassicae* and, as this particular field has progressed quickly through juvenile hormone analogues, juvenile hormone mimics and insect growth regulators, the biochemists have systematically tested many different compounds on *P. brassicae*.

Many of the present day organochlorine and organophosphorus insecticides have both contact and stomach poison properties. They can act via the cuticle or can take effect through the gut wall after ingestion. The many different and unsolved ways in which insecticides work in the insect body have been discussed by Corbett (1974). The toxicity of insecticides to mammals, their persistence in the environment and their phytotoxicity to plants, are all aspects of which the scientist has to be aware today.

As over 100 insecticides, excluding plant-derived compounds, have been included here, it has not been possible to draw up a table giving their concentration, mode of application and their effectiveness; instead the insecticides have been classified according to past and present methods of control in the field, and those compounds which have been tested on *P. brassicae* solely as a laboratory insect. Much of the earlier information describing methods of controlling insects including *P. brassicae*, is of historical value as this kind of information has not been assembled together before. Readers will be able to look up further information in the references provided, particularly with regard to more recent synthetic insecticides. Further details about recommended rates and usage of the latest insecticides can be obtained from many of the manufacturing companies.

SECTION A. CONTROL PRE-1945

a) Anecdotal methods

Those who have been interested in preventing *Pieris brassicae* from destroying their cabbage crops have used several traditional methods. These include dissuading the imago from seeing or smelling and ovipositing on the crop, destroying the ova once on the leaves, collecting or spraying the larvae and catching the imagines. Few have recommended collecting the pupae.

Distracting the female from ovipositing is a ploy used by those who have tried to mask the smell of the cabbage by using other plants which possess particularly offensive smells (at least to man). This is discussed in the next section. Here mention may be made of a few bizarre methods which have been recommended. Pliny, who lived 2000 years ago, apparently had great success in dissuading *P. brassicae* from ovipositing on cabbage by hanging up a bleached horse's head (preferably a mare's) in the cabbage patch (Fabre, 1912). Fabre (*loc. cit.*) said that the local people in southern France, near Montpellier, put eggshells on sticks in their fields and the female was attracted to these and laid her ova on them (eggshells tend to fall off sticks easily in the wind and do not appear to attract the attention of passing Whites in the author's experience). Theobald (1909) also reported that pieces of brimstone and jam pots on sticks were effective in keeping butterflies away in hot weather.

Even today, a walk through any allotment in Great Britain reveals a motley selectiôn of devices to dissuade imagines from landing on cabbage, including various coloured strips of plastic, silver paper and even scarecrows, although some of these are almost invariably used for avoiding pigeon damage. Most of these devices offer visual and auditory stimuli to the insect pests, although there are also mechanical barriers such as cages, string and cotton, many of doubtful effectiveness.

By far the best means of controlling *P. brassicae* on a small scale is to crush the ova or collect the larvae (Ormerod, 1879; Korolkov, 1914; Blanchard, 1917, 1921; Strawinsky, 1928; Anonymous, 1969a,b) but this cannot be achieved over large areas because of labour and time problems. Some control over *P. brassicae* has been accomplished in cases where there have been helpers in sufficient numbers, for instance Walker (1918) described how schoolchildren collected 6000 white butterflies in one parish in the Oxford district and those who participated received "head" money. In small cabbage patches it is relatively easy to check the undersides of the leaves every four to five days for batches of ova and to squash them. Collecting larvae is a little more tedious and has, according to Gautier (1918), the disadvantage that the parasitic insects are destroyed at the same time.

b) Nauseous smelling plants

Long before man invented insecticides he relied on the nauseous effects of certain plant species in dissuading imagines from ovipositing or killing larvae. Some 34 species belonging to nine plant families have been recorded in the literature as having been tested for their antagonistic effect on *P. brassicae* (Table 17.1). Many of the methods are recommended for small plots such as back gardens and allotments, but their degree of effectiveness must be reviewed with suspicion. The olfactory cues involved are undoubtedly complex and would certainly be worthy of further investigation.

Notable among the list are many plants, including herbs, which are strongly smelling to man; these include *Artemisia absinthium* Linnaeus (Wormwood), *Foeniculum vulgare* Miller (Fennel), *Helleborus foetidus* Linnaeus (Stinking Hellebore), *Lycopersicon esculentum* Miller (Tomato) and *Mentha* sp. (Mint). It was apparently discovered by accident that *Peucedanum graveolens* Bentham & Hooker (Dill) growing as a weed

Table 17.1 Plants which have been tested for their antagonistic action on *P. brassicae.*

Specific name	Common name	Authority
CANNABACEAE		
Cannabis sativa Linnaeus	Hemp	Marshal & Foex, 1918; Linsbauer, 1919; Blanchard, 1921; Ziarkiewicz & Anasiewice, 1961; Philbrick *et al.*, 1972
CAPRIFOLIACEAE		
Sambucus canadensis Burman	Elderberry	Eckstein, 1920
Sambucus nigra Linnaeus	Elderberry	Linsbauer, 1919; Lundgren, 1975
COMPOSITAE		
Artemisia abrotanum Linnaeus	Southernwood	Philbrick *et al.*, 1972
Artemisia absinthium Linnaeus	Wormwood	Gorianov, 1916; Philbrick *et al.*, 1972; Lundgren, 1975
Artemisia dracunculus Linnaeus	Tarragon	Philbrick *et al.*, 1972
Helianthus tuberosus Linnaeus	Jerusalem Artichoke	Marshal & Foex, 1918; Blanchard, 1921
Matricaria recutita Linnaeus	Wild Camomile	Philbrick *et al.*, 1972
LABIATAE		
Hyssopus officinalis Linnaeus	Hyssop	} Schreiber, 1915a,b
Lavandula officinalis Chaix, *in* Villars	Lavender	
Mentha sp.	Mint	} Philbrick *et al.*, 1972
Mentha piperita Linnaeus	Peppermint	
Origanum vulgare Linnaeus	Wild Marjoram	
Rosmarinus officinalis Linnaeus	Rosemary	Schreiber, 1915
Salvia officinalis Linnaeus	Sage	Philbrick *et al.*, 1972; Lundgren, 1975
Thymus vulgaris Linnaeus	Thyme	Philbrick *et al.*, 1972; Lundgren, 1975
LILIACEAE		
Allium cepa Linnaeus	Onion	Lundgren, 1975
Allium schoenoprasum Linnaeus	Chive	
Veratrum album Linnaeus	Chive	} Schreiber, 1915
Veratrum nigra Linnaeus	Chive	
RANUNCULACEAE		
Helleborus foetidus Linnaeus	Stinking Hellebore	Schreiber, 1915
PAPILIONACEAE		
Sarothamnus scoparius Linnaeus Wimmer *ex* Koch	Broom	Ormerod, 1880; Blanchard, 1917, 1921

Table 17.1 continued

Specific name	Common name	Authority
SOLANACEAE		
Datura stramonium Linnaeus	Thorn-Apple	Philbrick *et al.*, 1972
Hyoscyamus albus Linnaeus	—	Sprenger, 1915
Hyoscyamus majus Miller	—	} Sprenger, 1912
Hyoscyamus niger Linnaeus	Hen Bane	
Lycopersicon esculentum Miller	Tomato	Schreiber, 1915a,b; 1916; Musy, 1916–18; Gorianov, 1916; Philbrick *et al.*, 1972
Nicotiana tabacum Linnaeus	Tobacco	Gorianov, 1916
Petunia violacea Lindley	Petunia	Sprenger, 1915
Solanum lycopersicum Linnaeus	Tomato	Lundgren, 1975
UMBELLIFERAE		
Anthriscus caucalis Bieberstein	Bur Chervil	
Foeniculum vulgare Miller	Fennel	
Ligusticum scoticum Linnaeus	Lovage	Philbrick *et al.*, 1972
Petroselinum crispum (Miller) Nyman	Parsley	
Peucedanum graveolens Bentham & Hooker	Dill	Schreiber, 1915

beside cabbage kept *P. brassicae* larvae away. Equally interesting is the observation by Godan (1949) who saw a mass of wandering larvae avoiding *Robinia pseudoacacia* Linnaeus (False Acacia) which was in their path.

Schreiber (1915a,b, 1916) deterred oviposition of *P. brassicae* by laying tomato leaves and stalks around cabbages in the field. Musy (1916–18) found that tomatoes grown next to cabbages resulted in less ovipositing and Schreiber (1915a,b) also found that concoctions made up of *Helleborus foetidus* Linnaeus and *L. esculentum* were equally effective. The same effect by laying twigs of thyme and planting marjoram and sage next to cabbages was quoted by Philbrick *et al.* (1972), and also for *Sarothamnus scoparius* L. Wimmer *ex* Koch (Broom) by Ormerod (1880), *Sambucus nigra* Linnaeus (Elderberry) by Lindsbauer (1919) and Eckstein (1920), and *Helianthus tuberosus* L. (Jerusalem Artichoke) (Marchal & Foex, 1918).

After three years of field trials (1960–62) in Poland (Poznan) Turlowski (1963) found that only tomatoes grown in double rows between cabbage were partially effective in controlling ovipositing by *P. brassicae*. Similar results were achieved in the USSR (Don district, Vostrikov, 1915) and Switzerland (Musy, 1916–1918). Other plants which have a strong or seemingly unpleasant smell, such as *Allium* sp. (Onion), *Apium graveolens* L. (Celery), *Coriandrum sativum* L. (Coriander) and *Mentha* sp. (Mint) had no detectable effect on *P. brassicae*. More recently evidence in support of the hypothesis that many of these plants give off a deterrent odour has been established for four species: *Solanum lycopersicum* L., *Thymus vulgaris* L., *Salvia officinalis* L. and *Artemisia absinthium* L. (Lundgren, 1975). *Artemisia abrotanum* L. and *Sambucus nigra* L. did not show any repellent effect in the dual-choice chamber apparatus designed for this experiment.

The growing of hemp (*Cannabis sativa* L.) adjacent to cabbage, either in alternate rows or surrounding the small plots of cabbage, was a method used in the early part of the 20th century in many countries for deterring *P. brassicae* imagines from ovipositing. Certainly the inter-planting of hemp and cabbages in gardens was practised in the Netherlands and was "a well known fact" amongst the peasants (Jentink, 1913). Prince D'Arenburg (1913) also recommended the interplanting of hemp or *Genista* sp. (Broom) with cabbage and said that if imagines of *P. brassicae* are placed in a cage with both cabbage and hemp, they collected on the other side of the cage, i.e. away from the deterrent odours of the hemp. In a similar manner, Blanchard (1921) in France, recommended planting rows of hemp and *Helianthus tuberosus* L. (Jerusalem Artichoke), but suggested that their greater height might physically obscure the cabbage from the butterflies. In Austria, Linsbauer (1919) thought that the active principle in hemp could be used as a defensive spray. However, in experiments carried out in Poland during 1956–1958, Ziarkiewicz & Anasiewicz (1961) could not demonstrate any protective effect of hemp.

Whether or not hemp deters ovipositing females because it grows much higher than cabbage and the butterflies cannot smell and see the crop, or whether nauseous odours deter the insect directly has not been explained sufficiently. Laboratory experiments (under licence) designed to test the effect of *C. sativa* L. on ovipositing *P. brassicae* were rather inconclusive (Rothschild, 1978, pers. comm.). However, the use of hemp and other plants mentioned in Table 17.1 would be socially less acceptable today than many of the more dangerous synthetic insecticides.

c) Insecticides derived from plants

It is popularly believed that as plant-derived insecticides are naturally produced, they are more socially acceptable than synthetic insecticides which are dangerous. This is true for some groups such as pyrethrins which are non-toxic to man, but not true for others such as nicotine and rotenone which are relatively toxic to man if misused, i.e. nicotine is

Table 17.2 Natural insecticides used on *P. brassicae.*

Compound	Authority
Derris	Cottier, 1936
Derris resin W216 from *D. elliptica*	Potter *et al.*, 1946
Derris	Martin *et al.*, 1943
Derris	Potter & Tattersfield, 1943
Derris	Oldham, 1950
Rotenone	Martin *et al.*, 1943;
	Potter & Tattersfield, 1943;
	Potter *et al.*, 1946;
	Ugolini, 1959
Nicotine	Cottier, 1936;
	Potter & Tattersfield, 1943;
	Potter *et al.*, 1946;
	Oldham, 1950
Nicotine salts:	
citrate	⎫
hydrofluoride	⎬ David & Gardiner, 1953b
metaphosphate	
orthophosphate	⎭
nicotine sulphate	Cottier, 1936;
	Verma *et al.*, 1970
Hellebore	Ormerod, 1881
Tannin from Quebracho tree	Potter, 1951
Pyrethrum	Paillot & Faure, 1923;
	Feytaud, 1924; Cottier, 1936;
	Voskresenskaya, 1938;
	Potter & Tattersfield, 1943;
	Potter *et al.*, 1946;
	Oldham, 1950; Tittanen, 1964a,b
Chrysanthemum balsamita Willdenow	Mardzhanyan, 1941
Chrysanthemum carneum Bieberstein	
Caucasian pyrethrum	Feytaud, 1924
Chrysanthemum chyliophyllum Fischer & Meyer	Mardzhanyan, 1941
Chrysanthemum cineratiaefolium Trevieranus	
Dalmation pyrethrum	Isachenko & Goritzkaya, 1931
Chrysanthemum roseum Linnaeus	Feytaud, 1924
Chrysanthemum szowitssi Boissier	Mardzhanyan, 1941
Chrysanthemum tamrutense Sosnowsky	
ex Grassheim	Mardzhanyan, 1941
Anacyclus pyrethrum De Candolle	
African pyrethrum	Feytaud, 1924
Anacyclus offiniarum Hayne	
German pyrethrum – root only	Feytaud, 1924

dangerous if inhaled or imbibed as a concentrate, rotenone is only available as the product derris, which is dilute enough to be safe in practice. However, now that the structure of some of these "natural" insecticides has been elucidated they are manufactured synthetically.

It was the custom for centuries to grow certain combinations of plants together because it was known that they deterred insects. Cut stems of some of these plants and extracts of them were either laid or sprayed over cabbages as explained in the preceding section. The extracted principles were soon found to be nicotine from tobacco, pyrethrum from chrysanthemums and rotenone (= derris) from *Derris* root.

Natural pyrethrins were widely used for the control of *P. brassicae*, particularly by the Russians (Table 17.2). However, pyrethrins are of limited use in agriculture today as they are photolabile, relatively ineffective and much too expensive to produce, although in the laboratory and in certain field experiments they appeared to be highly effective. Early work on extracted pyrethrum showed that toxicity of even the same species of *Chrysanthemum* grown in different localities in the USSR varied considerably (Isachenko & Goritzkaya, 1931), for instance *C. cineratiaefolium* Treviranus grown in Transcaucasia was more effective than if it were grown in Leningrad or Moscow. An application rate of $3-5$ mg/cm^2 of pyrethrin from *C. tamrutense* Sosnowsky *ex* Grassheim gave 100% mortality of *P. brassicae* larvae after 48 hours (Mardzhanyan, 1941).

Generally all the extracted compounds or whole plants tested on *P. brassicae* mentioned in Table 17.2 were claimed to be very effective, giving mortalities up to 100% on larvae and were equally effective as a dust or as a spray.

d) Early chemical methods

Some of the early chemical methods for the control of cabbage caterpillars recommended the use of such compounds as Paris green (sodium acetoarsenate) and Bordeaux mixture (containing copper sulphate (1.8 kg), quicklime (1.8 kg) and water (227 l)) which were apparently effective against *P. brassicae* (Table 17.3). In Denmark Ferdinandsen *et al.* (1919) recommended Dufour's mixture made up from 2.7 kg of soap with 1.35 kg insecticide powder and 20 gallons of water. As can be seen from Table 17.3 certain inorganics have been used since 1945, for instance arsenates, barium chloride and sodium aluminofluride, although in some cases these compounds are fairly toxic to man.

Eleanor Ormerod in her many publications on the observations of injurious insects (Ormerod, 1878, 1879) gave accounts of several of the early methods of control of cabbage caterpillars: sprinkle with fine salt,

Table 17.3 Inorganic compounds used on *P. brassicae.*

Compound	Authority
Arsenates	
arsenates	Smith, 1896; Paillot, 1928; Voskresenskaya, 1936a,b, 1939; Pilat, 1935; Ugolini, 1959
calcium arsenate	Cottier, 1936; Fulmek, 1929
copper acetoarsenate	Strawinsky, 1928; Fulmek, 1929
lead arsenate	Fulmek, 1929; Cottier, 1936
sodium arsenate	Pilat, 1935
zinc and iron arsenate	Fulmek, 1929
Others	
barium chloride	Arkhangelski, 1925a,b
barium fluosilicate	Cottier, 1936
Bordeaux mixture	Pilat, 1935
calcium oxide (= lime)	Smith, 1891
gas lime, hot lime, fine powdered caustic lime, caustic soot	Ormerod, 1881
"cyanhyolic acid"	Portier, 1949
hydrogen cyanide	Torzeau, 1962; Yana, 1962
nickel sulphate	Ormerod, 1887
sodium aluminofluoride	Avidov & Harpas, 1969
sodium chloride	Ormerod, 1878, 1887; Cottier, 1936
sodium fluorosilicate	Pilat, 1935
sodium fluoride	Pilat, 1935

fine powdered caustic lime or caustic soot (Ormerod, 1878); "gas" lime, hot lime, salt water or even soapsuds (Ormerod, 1881, 1887) (Table 17.3). In France, Feytaud (1918) suggested applying wood ash and lime when the "dew was falling".

At the South-eastern Agricultural College at Wye (now Wye College, University of London) the Director, Frederich Theobald (1928) recommended salt and Fuller's earth applied to wet brocolli, or alcoholic solution of pyrethrum, as effective measures. He believed that salt was of particular value and that a 2% salt solution in a gallon of water was best but it did tend to scorch nasturtiums.

Paris green was a favourite compound to use against larvae earlier this century. In Poland, Strawinsky (1928) recommended 0.45 kg of Paris green with 0.90–1.35 kg of lime made up to 455 litres, sufficient to give 50–80% mortality of *P. brassicae* larvae. In Austria Fulmek (1929) drew attention to the variability of the amount of arsenic in commercially prepared insecticides and drew up a list of about 17 arsenite and arsenate salts according to their effectiveness against *P. brassicae* larvae. In the USSR, laboratory experiments showed that poisons such as sodium arsenate and Paris green accelerated the passage of food through the gut of *P. brassicae* (Voskresenskaya, 1936a,b). Also in the USSR, Schreiner (1914) recommended using acid chromate of lead with added molasses for use against *P. brassicae* larvae. He reported that it was apparently slow in acting but achieved remarkable results. In Italy, arsenates were also being used as well as bacterial preparations made up from *Bacillus thuringiensis* (Ugolini, 1959) (See also Chapter 15).

Pieris rapae, following its introduction into the USA, was also being controlled with Paris green and other arsenates; it was pointed out that cabbages treated with arsenates would only be poisonous to humans if about a dozen plants were eaten (Smith, 1891). Smith (*loc. cit.*) also stated that the simplest method of control was to use fresh slaked or dry hydrated lime, sifted and dusted on cabbage when it was wet. In New Zealand successful control of *P. rapae* was achieved using lead arsenate at all concentrations; dusts were found less effective (Cottier, 1936).

SECTION B. CONTROL POST-1945 IN THE FIELD

Since the manufacture of organic insecticides, several different groups of chemicals have been used against *P. brassicae* in the field; these include DDT and its derivatives, carbamates, γ-HCH, cyclodienes and organophosphates.

DDT has been widely used in the field (Table 17.4) and is still recommended for use today (Table 17.8). It has been used either as a spray or dust against the ova and larvae of *P. brassicae* with varying degrees of success when used at many different concentrations (Potter & Perkins, 1946; Frey, 1950; Thakur, 1966). Potter & Perkins (*loc. cit.*) recorded that 5% of DDT in kaolin or as a 0.2% w/v solution in water gave excellent control of larvae on kale. DDT acts primarily on the insect nervous system, although there is a report of it having an effect on the haemolymph (Singh & Atwal, 1969).

In comparison with the number of insecticides which have been tested on *P. brassicae* in the laboratory, which are dealt with in the next section, few have undergone trials in the field (Table 17.5). Most of these are organophosphates, the majority of which have had fairly favourable results. Touzeau (1962) tested five different insecticides in the field on *P. brassicae* larvae and found that all those except malathion gave good results at the concentrations tested.

In Germany, Brouwer (1970) tested the effectiveness of several sprays containing 10 insecticides and found that the best ones for controlling *P. brassicae* larvae on Brussels sprouts were bromophosethyl mixed with demeton-s-methyl (2 applications), thiometon (2 applications) and mevinphos (1 application). In India, Deshmukh & Adarsh (1971) tested 11

Table 17.4 Insecticides used on *P. brassicae* in the field (see also recommended insecticides in Table 17.8). Key: C=carbamates; CD=cyclodienes; DN=dinitrophenols; M=miscellaneous; O=simple organic compounds; OP=organophosphates; OC=organochlorines; T=thiocyanate derivative; OB=organobromide. NB. The names of the insecticides quoted here and in subsequent tables are in accordance with Martin & Worthing's (1977) *Pesticide Manual*, with their trade names, where these have been used in the literature quoted in brackets. Those insecticides not quoted in Martin & Worthing are indicated with an asterisk.

Compound	Code	Authority
bromophos-ethyl	OP	Brouwer, 1970
chlorfenvinphos	OP	Finlayson *et al.*, 1975
demeton-s-methyl	OP	Brouwer, 1970
DDT	OC	Buckhurst, 1947; Oldham, 1950; Thakur, 1966; Deshukh & Adarsh, 1971
endrin	CD	Sleesman, 1957
fensulfothion	OP	Finlayson *et al.*, 1975
isofenphos	OP	Finlayson *et al.*, 1975
malathion	OP	Touzeau, 1962
mevinphos (phosdrin)	OP	Sleesman, 1957; Touzeau, 1962; Brouwer, 1970
orthodibrome	OB	Touzeau, 1962
parathion	OP	Touzeau, 1962
propachlor	M	Finlayson *et al.*, 1975
thiometon	OP	Brouwer, 1970

insecticides in the laboratory but only recommended DDT and endosulphan for control in the field.

In the USA, mevinphos was found to be as effective in controlling *P. rapae* as endrin when used at a concentration of 0.23 kg per 0.41 ha. However, there was the hazard that earthworms were destroyed in the soil. In Great Britain, Finlayson *et al.* (1975) also investigated the effect of insecticides as well as herbicides on weeds, insects including *P. brassicae*, and earthworms in a "mini cauliflower crop".

SECTION C. *P. BRASSICAE* AS A LABORATORY INSECT FOR INSECTICIDE USE

Pieris brassicae has been chosen for experimental work such as in insecticide screening in the laboratory for a number of reasons; it is easy to rear all the year and it has a relatively large larva for manipulation. It is not surprising therefore that many different groups of compounds have been tried out on *P. brassicae* in the laboratory, not primarily because it is a pest, but merely as a convenient species on which to compare results.

In recent years scientists have studied the effectiveness of a group of compounds collectively known as insect growth regulators. This expression was coined to widen the category, from the original one of juvenile hormone mimics, to include compounds of structure similar to juvenile hormones but not necessarily acting in the same way and also to include other growth disruptants, such as phenylurea derivatives.

a) Organic insecticides

A host of some 50 organic insecticides, including carbamates, organochlorine and organophosphate insecticides, and also many of their derivatives, are reported as having been tested on *P. brassicae* in the laboratory (Table 17.5). However, this does not include those insecticides and formulations which are synthesised on a regular basis by the major chemical companies in their competitive search for successful compounds. The table therefore comprises information from published work.

There is much more information contained in many of the references in Table 17.5 than is possible to tabulate here, thus many of the publications give accounts of the effectiveness of several compounds, for instance Del Rivero *et al.* (1969a,b) tested 15 compounds. Martin *et al.* (1943) tested 16

Table 17.5 Organic insecticides used on *P. brassicae* in the laboratory. Symbols as for Table 17.4.

Compound	Code	Authority
aldoxine carbamates, thio series	C	Felton, 1968
aldrin	CD	Verma *et al.*, 1970
azinphos-methyl	OP	Vlasveld, 1975
2-(2-butoxyethyl) ethyl thiocyanate (butyl-"Carbitol" thiocyanate	T	Martin *et al.*, 1943
bromophos-ethyl (Nexagan)	OP	Del Rivero *et al.*, 1969a,b
*bis-dimethylamine fluorophosphoroxide	OP	Bennett & Martin, 1948; Bennett, 1949
*bis-dimethylaminophosphorus anhydride	OP	David & Gardiner, 1953a,b
*bis-(dimethylamino) fluorophosphine oxide	OP	David & Gardiner, 1953a,b
camphechlor (Toxaphene)	OC	Yana, 1962; Thakur, 1966
carbaryl (Sevin)	C	Yana, 1962; Touzeau, 1962; Alkmaar, 1975; Mulder & Sweenen, 1967; Deshmukh & Adarsh, 1971; Vlasveld, 1975
carbofuran	C	Deshmukh & Adarsh, 1971; Finlayson *et al.*, 1975

Table 17.5 continued

Compound	Code	Authority
DDT (Gerasol) and other trade names	OC	Martin *et al.*, 1943
		Potter & Perkins, 1946;
		Potter *et al.*, 1946;
		Frey, 1950; Yana, 1962;
		Del Rivero *et al.*, 1969;
		Singh & Atwahl, 1969a,b
		Verma *et al.*, 1970;
		Mulder & Sweenen, 1973
diacetate of αα-bis (hydroxyphenyl) βββ-tri-chlorethane	OC	Martin *et al.*, 1943
DFDT (Gix), a fluro derivative	OC	Speyer, 1950
analogues in which other halogens substituted for the para-chlorine atoms showed greatest potency with fluroine than bromine or iodine	OC	Martin *et al.*, 1943
Tritox (mixture of DDT and BHC and methyl-DDT)	OC	Blazejewska, 1964
*diethyl p-nitro phenylphosphate (E600)	OP	David & Gardiner, 1953a,b
demetron (Systox)	OP	David & Gardiner, 1954;
		David, 1959;
		David *et al.*, 1960
diazinon	OP	Deshmukh & Adarsh, 1971;
		Del Rivero *et al.*, 1969a,b;
		Wahla *et al.*, 1976;
		Verma *et al.*, 1970
dichlorvos	OP	Vlasveld, 1975
dicrotophos	OP	Vlasveld, 1975
dimethoate (Fosfamid)	OP	Mulder & Sweenen, 1967
DNOC (4,6-dinitro-o-cresol)	DN	Martin *et al.*, 1943
*dodecyl nitrile	O	Martin *et al.*, 1943
*dodecylthiocyanate	T	Martin *et al.*, 1943
endosulphan	CD	Del Rivero *et al.*, 1969a,b;
		Deshmukh & Adarsh, 1971;
		Alkmaar, 1972
endrin	CD	Walker & Turner, 1962;
		Singh & Atwal, 1969;
		Verma *et al.*, 1970;
		Speyer, 1950
fenitrothion	OP	Del Rivero *et al.*, 1969a,b;
		Deshmukh & Adarsh, 1971;
		Vlasveld, 1975
fenthion (Lebaycide)	OP	Yana, 1962
gerasol (see DDT)		
HCH (gamma-BHC, Lindane)	OC	Frey, 1950; Yana, 1962;
		Turenen, 1975; Thakur,
		1966; Herfs & Krieg, 1963
malathion	OP	Verma *et al.*, 1970;
		Yana, 1962

Table 17.5 continued

Compound	Code	Authority
menazon	OP	Deshmukh & Adarsh, 1971
mephosfolan (-Cytrolane)	OP	Deshmukh & Adarsh, 1971
methiocarb	C	Del Riviero *et al.*, 1969a,b
mevinphos (Phosdrin)	OP	Yana, 1962; Touzeau, 1962; Verma *et al.*, 1970; Vlasveld, 1975; Alkmaar, 1974
*N-monothylylamise de Pacide o,o,dimethyl dithophosphoryl acetique	OP	Yana, 1962
monocrotophos	OP	Deshmukh & Adarsh, 1971
naled (dibrone)	OP	Yana, 1962
parathion	OP	David & Aldridge, 1957; Yana, 1962; Singh & Atwahl, 1969; Verma *et al.*, 1970; cf. Clark *et al.*, 1976
*phenothiazine	M	Martin *et al.*, 1943b
phenyl-N-methylcarbamate	C	Meltzerm & Welle, 1969
phosalone	OP	Del Rivero *et al.*, 1969a,b
phosmet (Imidan)	OP	Del Rivero *et al.*, 1969a,b
phosphamidon	OP	Verma *et al.*, 1970
*phosphoric acid ester	OP	Frey, 1950
*new phosphate insecticides; Bayer 5691, 5621, I-674538	OP	Del Rivero *et al.*, 1969a,b
*polychlorethyl derivatives	OC	Martin *et al.*, 1943b
schradan	OP	David & Gardiner, 1959
sodium fluoroacetate	O	David & Gardiner, 1953a, 1959
sodium fluoroacetamide	O	David *et al.*, 1958
TEPP	OP	Potter, 1955, 1957; Potter *et al.*, 1957; Daalen *et al.*, 1972 Post *et al.*, 1973; Post & Mulder, 1974
tetrachlorvinphos (Gardona)	OP	Del Rivero *et al.*, 1969a. Vlasveld, 1970; Deshmukh & Adarsh, 1971; Levchenko *et al.*, 1974; Novozhilov & Evstigneeva, 1977
trichlorphon (Dipterex)	OP	Del Rivero *et al.*, 1969; Sundukov, 1969, 1971
*thiocyanates (aliphatic)	O	Martin *et al.*, 1943b
*thiodiphenylamine	M	Martin *et al.*, 1943b

compounds. Mulder & Sweenen's (1973) work with dimethoate gave sufficiently encouraging results that further tests on a larger scale were thought useful. There has also been much work on the physiological way in which the various insecticides work, how they are dealt with by the gut, how toxic they are, where they are absorbed, the effect on the central nervous system and/or the haemolymph (e.g. David & Gardiner, 1953a,b; Singh & Atwal, 1969; Novozhilov & Evstigneeva, 1977).

b) Phenylurea derivatives

A number of phenylurea derivatives have been tested on *Pieris brassicae in vitro* but the only one which is commercially available is Dimilin (Philips-Duphar, Netherlands) (Table 17.6). For further information on the effects of Dimilin on *P. brassicae*, a long account can be seen in Philips-Duphar (1975). A small fraction of all derivatives are very active *in vitro* and are very good stomach poisons which appear to act by upsetting chitin synthesis. Many of the derivatives listed in Table 17.6 are the subject of further investigation; some showed interesting larvicidal properties on *P. brassicae* (Wellinga *et al.*, 1973, 1977).

Table 17.6 Urea derivatives used on *P. brassicae*.

Compound	Comments	Authority
diflubenzuron (Dimilin)	good at 100 ppm, inhibits cuticle synthesis	Wellinga *et al.*, 1973; Mulder & Sweenen, 1973; Mulder & Gijswift, 1973; Philips-Duphar, 1975
1-(2,6-dichlorobenzoyl)-3-(3,4-dichlorophenyl) urea	good internal properties but does not penetrate cuticle	Daalen *et al.*, 1972
several 1-(2,6-disubstituted benzoyl-3-phenylurea derivatives including DU 19111	cuticular chitin synthesis inhibitor; causes lowering of respiratory rate and upset of pentose cycle	Post, 1974; Post & Mulder, 1974; Moreau *et al.*, 1975
1-(2,6-dichlorobenzoyl)-3-phenylureas. 57 were listed	evaluated	Wellinga *et al.*, 1973, 1977
1-(4-chorphenyl)-3-(2,6-dichlorobenzyl urea	larva failed to moult and ruptures	Wellinga *et al.*, 1973a,b Mulder & Sweenen, 1973
(PH 60-38)	new cuticle	Mulder & Gijswift, 1973

Table 17.7 Juvenile hormone analogues used on *P. brassicae*.

Compounds	Authority
Farnesenic acid	Benz, 1970, 1971
Juvenoid RO 20–3600 (isomeric mixture of 6,7-epoxy-3,7-dimethyl-1-(3,4-(methylene-dioxide)-phenoxy)-2-nonene	Benz, 1974
derivatives of aryl-pentenoic acid and aroxyl-butenoic acid	Scheurer *et al.*, 1975
ethyl 3,7-dimethyl-9-cyclohexyl-2-4-nonadrenoate and its 2-methylcyclohexyl homologue	Sehnal *et al.*, 1976

c) Juvenile hormone analogues

This is a field of control which is very much in its infancy (Table 17.7). Early in the 1970s, Benz (1971, 1973, 1974) used topical application of juvenile hormone (JH) analogues to induce reversal of spinning behaviour in the larva, which also underwent an extra moult, and developed leg-like structures as modified palps on the head. Sehnal *et al.* (1976) tested the effects of a juvenoid containing the cyclohexane moiety on 16 insect species from seven orders, which included *P. brassicae*, and believed that this was one of the most effective juvenoids available. Experiments carried out by Scheurer *et al.* (1975) using aryl-pentenoic acid and aroxyl-butenoic acid indicated that these were not toxic to the principal hymenopterous parasite of *P. brassicae*, *Apanteles glomeratus* L. They thus suggested that these insect growth regulators could be used during the long term to reduce pest populations.

d) Plant extracts: Azadirachtin

Azadirachtin is a highly oxidised triterpenoid compound, the structure of which is still being investigated (Butterworth & Morgan, 1971; cf. Zanno *et al.*, 1975). It can be extracted from *Azadirachta indica* A. Joss (Neem tree) (Butterworth & Morgan, *loc. cit.*) and is probably also present in another member of the Meliaceae, *Melia azedarach* L. (Atwal & Pajni, 1964). These last two workers investigated the effect of applying extracts of ground dried drupes of *M. azedarach* in different solvents on third instar larvae of *P. brassicae*. Larvae dusted with powder of the plant extract gave 60% mortality in 96 hours, while a higher percentage mortality (78.3%) was obtained over the same period using a 10% emulsion in ethanol. Although Butterworth & Morgan tested the effect of azadirachtin as a feeding inhibitor primarily on locusts, they tested the reaction of *P. brassicae* larvae by way of comparison and found that the compound showed only moderate activity.

SECTION D. CURRENT CHEMICAL CONTROL OF *P. BRASSICAE* INFESTATIONS

In some countries certain organic and inorganic compounds which were reported earlier are still in active use today (cf. Table 17.4). However, in countries like Great Britain and Ireland recommendations have been published by the Ministry of Agriculture, Fisheries and Food for the use of certain insecticides (MAFF, 1976). Their leaflet on *The Control of Cabbage Caterpillars* recommends the use of eight different insecticides and gives instructions as to formulation and application (Table 17.8).

The same compounds are approved for use in France (Le Ministre Pleni-potentaire, London, 1977, pers. comm.). In the Netherlands, endosulphan is recommended (Wageningen, 1971). In Germany, a single treatment with one of the various preparations is usually sufficient for the control of *P. brassicae* larvae which cause much damage (Anonymous, 1969a,b). *P. brassicae* presumably has to be strictly controlled in those countries where severe damage occurs even in the 1970s, such as Austria, Bulgaria, Czechoslovakia, Poland, Rumania and southern and central Sweden.

In the USSR trichlorphon is used (Zaitseva & Ponomareva, 1976) but much of the current chemical control measures make use of bacterial and fungal formations based on *Bacillus thuringiensis* and *Beauveria bassiana*, of which at least the former acts in a chemical manner via a proteinaceous toxin. The recent published work from the USSR would seem to indicate that these are the main methods of chemical control (see Chapter 15), but they are also known to be keen to buy insecticides from western countries.

Table 17.8 Recommended insecticides for control of cabbage caterpillars (from MAFF, 1976).

Insecticide	Amount of formulation per hectare
DDT (25% emulsifiable concentrate)	2.8–4.2 litres (2–3 pints)
derris (liquid containing 5% rotenone)	1.4 litres (1 pint)
azinphos-methyl (22% emulsifiable concentrate)	1.4 litres (1 pint)
iodofenphos (50% wettable powder)	1.1 kg (1 lb)
mevinphos (99% emulsifiable concentrate)	280 ml (4 fl.oz.)
tetrachlorvinphos (75% wettable powder)	1.1 kg (1 lb)
triazophos (40% emulsifiable concentrate)	850 ml (12 fl.oz.)
trichlorphon (80% soluble powder)	1.1–1.7 kg (1–1½ lb)

Toxicity to parasites and other invertebrates

One of the disadvantages of chemical control is that the useful parasitic insects may be destroyed at the same time as the pest species. Indeed, with reference to *P. brassicae* Berkovitch (1972) recommended that an insecticide should be selected with a view to its non-toxicity to *Apanteles glomeratus*. However, earlier Osmolovski (1964) thought that chemical control of the cabbage pests would only slightly reduce the effect of *A. glomeratus* as it can maintain itself on other lepidopterous hosts. More recently in the USSR, Moiseeva *et al.* (1975) demonstrated that it was just as possible to control *P. brassicae* in the wild by using the egg parasite *Trichogramma evanescens* Westwood as by using insecticides. Alternatively, an insect growth regulator which is non-toxic to *A. glomeratus* was found suitable for controlling *P. brassicae* (Scheurer *et al.*, 1975).

In 1958 Champ at Imperial College, London carried out research into the effects of sub-lethal doses of insecticides, testing 16 different kinds, on the incidence of parasitism in *P. brassicae*. Larvae already parasitised were shown to be more susceptible to insecticides than those not parasitised. He published several photographs showing the effects of these insecticides on the various stages of *P. brassicae* and *A. glomeratus*.

Beneficial insects and arachnids, such as ladybird species, lacewings, predatory mites, parasitic wasps and ground beetles may also be significantly affected either by many of the insecticides used on *P. brassicae* or indirectly by elimination of their prey. Some of the reasonably persistent insecticides are not necessarily environmentally unacceptable.

References cited

Alkmaar Report, 1972. Insektenbestriydiff in sprintkool. *Bericht* No. 1824.
—— 1974. Growing of cauliflower. *Bericht* No. 10., 2,4,8,9,11,15.
Anonymous, 1969a. Agreport pests and diseases 38. I. Cabbage, *Brassica oleracea*. *Agr. vet. chem.* 10: 13–14.
—— 1969b. Schädigen bewirkt bei der fressenden Raupen. *Prakt. Schädlbekämpf.* 21: 113–114.
Arkhangelski, P., 1925a. Note on *Pieris brassicae*. *Defense des Plantes* 1: 239.
—— 1925b. Control of cabbage worms. (In Russian.) *Reg. Plant Prot. Stn. N. Caucasus* B. 1–7.
Atwal, A.S. & Pajni, H.R., 1964. Preliminary studies on the insecticidal properties of drupes of *Melia azedarach* against caterpillars of *Pieris brassicae* L. (Lepidoptera: Pieridae). *Indian J. Ent.* 26: 221–227.
Avidov, Z. & Harpaz, I., 1969. *Plant pests of Israel*. Israel University Press.
Bennett, S.H., 1949. Preliminary experiments with systemic insecticides. *Ann. appl. Biol.* 36: 160–163.

Bennett, S.H. & Martin, H., 1948. Qualitative examination of insecticidal properties. Progress Report, 1947. *Rep. agric. hort. Res. Stn. Bristol* 1947: 147–156.

Benz, G., 1970. Stimulation of oogenesis in *Pieris brassicae* by the juvenile hormone derivative farnesenic acid ethyl ether. *Experimentia* 26: 1012.

—— 1971. Failure to demonstrate sterilans effect of juvenile hormone mimetics in *Pieris brassicae* and *Galleria mellonella*. *Experimentia* 27: 581–582.

—— 1973. Reversal of spinning behaviour in last instar larvae of *Pieris brassicae* treated with juvenile hormone derivatives. *Experimentia* 29: 1437–1438.

—— 1974. Homoeotic transdetermination caused by juvenoid in larvae of *Pieris brassicae* L. *Experimentia* 30: 1264–1265.

Berkovitch, I., 1972. Seeking pest solutions at Silwood Park. *Int. pest Control* 14: 24–25.

Blanchard, E., 1917. Dégâts causés par les chenilles du chou dans le *département* de la Loire et regions avoisantes. *Vie agric. Rur. Paris* 7: 419–420.

—— 1921. Contre la pièride du chou. *Vie agric. Rur. Paris* 19: 26–27.

Blazejewska, A., 1964. The effect of tritox dust on the alimentary canal of larvae of the cabbage white butterfly, *Pieris brassicae*. (In Polish.) *Polskie Pismo Ent.* B. 1964: 25–34.

Brouwer, W.M.T.J. de, 1970. Protection of brussels sprouts against insects. *Gewasberbeschherming* 1: 104–110.

Buckhurst, A.A., 1947. White butterflies. *Amat. Gard.* 64: 17.

Butterworth, J.H. & Morgan, E.D., 1971. Investigation of the locust feeding inhibition of the seeds of the Neem Tree, *Azadirachta indica*. *J. Insect Physiol.* 17: 969–977.

Champ, B.R., 1958. *The effect of sub-lethal dosages of insecticides on parasitism in Pieris brassicae.* Ph.D. thesis, University of London (Imperial College).

Clark, A.G., Cropp, P.L., Smith, J.N., Speir, T.W. & Tan, B.J., 1976. Photometric determination of methyl parathion GSH S-methyl transferase. *Pest. Biochem. Physiol.* 6: 126–131.

Corbett, J.R., 1974. *The biochemical mode of action of pesticides.* Academic Press. New York & London.

Cottier, W., 1936. The use of insecticides in the control of the white butterfly. *N.Z. J. agric.* 52: 24–29.

Daalen, J.J. van, Meltzer, J., Mulder, R. & Wellinga, K., 1972. Ein auswählendes Schädlingsbekämpfungsmittel mit einer neuartigen Form der Wirkung. *Naturwissenschaften* 59: 312–313.

D'Arenberg, P., 1913. A possible method of preventing the attack of *Pieris brassicae* on cabbages. *Bull. Soc. nat. acclimat.* 1913: 18.

David, W.A.L., 1959. The systematic insecticidal action of paraoxon on the eggs of *Pieris brassicae* (L.). *J. Insect Physiol.* 3: 14–27.

David, W.A.L. & Aldridge, W.N., 1957. Insecticidal material in leaves of plants growing in soil treated with Parathion. *Ann. appl. Biol.* 45: 332–346.

David, W.A.L. & Gardiner, B.O.C., 1953a. Systemic insecticidal action of nicotine and certain other organic bases. *Ann. appl. Biol.* 40: 91–105.

—— 1953b. The systemic insecticidal action of sodium fluoacetate and of three phosporous compounds on the eggs and larvae of *Pieris brassicae* L. *Ann. appl. Biol.* 40: 403–417.

—— 1954. The action of the systemic insecticide systox on *Pieris brassicae* L. and *Phaedon cochleariae* F. *Bull. ent. Res.* 45: 693–702.

—— 1959. The action of the systemic insecticide fluoroacetamide on certain aphids and on *Pieris brassicae* (L.). *Bull. ent. Res.* 50: 25–38.

David, W.A.L., Gardiner, B.O.C., Chapman, C. & Phillips, M.A.M., 1958. Fluoroacetamide as a systemic insecticide. *Nature, Lond.* 181: 1810–1811.

David, W.A.L., Metcalf, R.L. & Winton, M., 1960. The systemic insecticidal properties of certain carbamates. *J. econ. Ent.* 53: 1021–1025.

Del Rivero, J.M., Tuset, J.J., Roig, F.J. & Lafuente, M., 1969a. La eficacia de varios insecticidas, solos o combinados contre las orugas de *Pieris brassicae*. *An. Inst. nac. invest. Agron.* 18: 145–155.

Del Rivero, J.M., Tuset, J.J., Roig, F.J., Miquel, E. & Lafuente, M., 1969a. La eficacia de varios insecticidas contra *Pieris brassicae* y pruebas de fitotoxicidad en el campo. *Boln. Patol. veg. ent. Agric.* 31: 85–91.

Deshmukh, S.N. & Adarsh, H.S., 1971. Comparative toxicity of some common formulations of common insecticides to caterpillars of *Pieris brassicae* (Cabbage butterfly). *J. Res. Punjab agric. Univ.* 8: 339–342.

Eckstein, K., 1920. Zur Bekämpfung der Kohlweisslinge. *Naturwissenschaften* 18: 234–235.

Fabre, J.H., 1912. *The Life of the Caterpillar.* Translated by A.T. de Mattos. Hodder & Stoughton.

Felton, J.C., 1968. Insecticidal activity of some oxime carbamates. *J. Sci. Fd. agric. Suppl.* 19: 32–37.

Ferdinansen, C., Lind, J. & Rostrup, S., 1919. Report of insect pests and diseases of orchards in 1916–1917. *Tijdschr. Planteavl. Copenhagen* 26: 297–334.

Feytaud, J., 1918. Note sur la pièride du chou. *Bull. Soc. étude vulg. zool. agric. Bordeaux* 17: 33–38.

—— 1924. La pyrèthre. *Rev. zool. agric. appl.* 23: 238–243.

Finlayson, D.G., Campbell, C.J., Roberts, H.A. & Vancouver, B.C., 1975. Herbicides and insecticides: their compatibility and effects on weeds, insects and earthworms in the minicauliflower crop. *Ann. appl. Biol.* 79: 95–108.

Frey, W., 1950. The use of newer contact insecticides for the control of larvae of the cabbage white butterfly. *NachrBl. dt. PflschutzDienst. Braunschw.* 2: 168–170.

Fulmek, L., 1929. Giftigkeitsunterschiede gebräuchlicher Arsenmittel. *Fortschr. Landw. Vienna* 4: 209–212.

Gautier, C., 1918. Physiological and parasitological studies on injurious lepidoptera: on some facts relating to Pierid larvae. *C. r. Soc. Biol. Paris* 81: 197–199, 801–803.

Godan, D., 1949. Massenauftreten von Kohlweisslingsraupen (*Pieris brassicae*) in Berlin. *Biol. Zentanst.* 1: 165–166.

Gorianov, A., 1916. Experiments with some vegetable and mineral insecticides. (In Russian.) *Protection of Plants from Pests, supplement to Friend of Nature, Petrograd* 1916: 1–28.

Herfs, W. & Krieg, A., 1963. Studies for assessing the efficiency of commercial preparations of *Bacillus thuringiensis* Berliner for the control of the cabbage white butterfly, (*Pieris brassicae* L.). *NachrBl. dt. PflschutzDienst. Braunschw.* 15: 49–54.

Isachenko, V.B. & Goritzkaya, O.V., 1931. Some data of toxicological analysis of the pyrethrum. (In Russian.) *Bull. Plant. Prot. Leningrad* 3: 165–174.

Jentink, F.A., 1913. Hemp as a deterrent of *Pieris brassicae*. *Review of Applied Entomology* A: 191. Cited *in* Review of Applied Entomology as letter to D'Arenberg.

Korolkov, D.M., 1914. Insects injurious to orchards and market-gardens and remedies against them. (In Russian.) *Orchard & Market Garden, Moscow* (5) 235–241.

Levchenko, E.I., Bolotryi, A.V. & Yurkova, E.F., 1974. Gardona. *Zashch. Rast.* (8) 43–44.

Linsbauer, L., 1919. Zur Bekämpfung der Kohlweisslinge. *Naturwissenschaften* 17: 147–149.

Lundgren, L., 1975. Natural plant chemicals acting as oviposition deterrents on cabbage butterflies (*Pieris brassicae, Pieris rapae, Pieris napi*). *Zool. Scr.* 4: 253–258.

Ministry of Agriculture, Fisheries & Food (MAFF), 1976. *Cabbage caterpillars.* Advisory leaflet No. 29. Her Majesty's Stationery Office, London.

Marchal, P. & Foex, E., 1918. Rapport phytopathologique pour les années 1916 et 1917. *Ann. Épiphyt.* 5: 1–35.

—— 1920. Rapport phytopathologique pour les arnées 1919 et 1920. *Ann. Épiphyt.* 7: 1–87.

Mardzhanyan, G.M., 1941. Question of the toxic characters of different *Pyrethrum* species (in Russian). *Proc. Lenin Acad. Agri. Sci. U.S.S.R.* 6: 26–29.

Martin, H., Stringer, A. & Wain, R.L., 1943. Qualitative examination of insecticidal properties. Progress Report. *Rept. agric. Res. Stn. Bristol* 1943: 62–76.

Martin, H. & Wain, R.L., 1945. Qualitative examination of insecticidal properties. Progress Report. *Rept. agric. Res. Stn. Bristol* 1944: 121–140.

Martin, H. & Worthing, C.R., 1977. *Pesticide Manual* (5th edition). British Crop Protection Council.

Meltzerm, J. & Welle, H.B.A., 1969. Insecticidal activity of substituted Phenyl N-Methylcarbamates. *Entomologia expl. appl.* 12: 169–182.

Moiseeva, N.V., Mashkina, L.G., Kalashnikova, G.I. & Ratomskaya, A.A., 1975. The effectiveness of the use of *Trichogramma* for control of certain cabbage pests in the Chu Valley, Kirgizia. *In: Arthropods of Agricultural Importance.* Ed. Protsenko, A.I. Ent. Issled. Kirgizia Vypusk. Kirgiz. SSR.

Moreau, R., Castex, C. & Lamy, M., 1975. Examen préliminaire de quelques aspects des effects métaboliques d'un nouvel insecticide de synthese chez deux insectes nuisibles: *Pieris brassicae* L. et *Thaumetopoea pityocampa* S. (Lépidoptères). *Ann. zool. ecol. Anim.* 7: 161–170.

Mulder, R. & Gijswijt, M.J., 1973. The laboratory culture evaluation of two promising new insecticides which interfere with cuticle deposition. *Pest. Sci.* 4: 737–745.

Mulder, R. & Sweenen, A.A., 1973. Small scale field experiments with PH-60-38 and PH 60-40, insecticides inhibiting chitin synthesis. *Proc. Brit. Insect Fung. Conf.* 7: 729–735.

Musy, M., 1916–18, Les chenilles du chou, leurs ennemis et les moyens de les combattre. *Bull. Soc. Fribourg. Sci. nat. Fribourg.* 24: 120–122.

Novozhilov, K.V. & Evstigneeva, T.A., 1977. The toxicity of Gardona for caterpillars of the cabbage butterfly. (In Russian.) *Khim. Sel. Khoz.* 15: 73–76.

Oldham, C., 1950. *Vegetable growers guide.* Lockwood, London.

Ormerod, E.A., 1878 & 1879. *Notes of observations of injurious insects, report 1878.* West Newman & Co. London.

—— 1880. *Notes of observations of injurious insects, report 1880.* West Newman & Co. London. pp. 25–27.

—— 1881. *Report of observations of injurious insects during 1881.* Swan Sonnenschein, London. pp. 5–8.

—— 1887. *Notes of observations of injurious insects and common crop pests in 1887.* pp. 87–89.

Osmolosvski, G.Y., 1964. The biology of *Apanteles glomeratus* L. (Hymenoptera: Braconidae), a parasite of *Pieris brassicae*. (In Russian.) *Ent. Rev.* (URSS) 43: 387–388.

Paillot, A., 1928. Sur les properties insecticides de l'arsenite de calcium. *C. r. Acad. agric. Fr.* 14: 12–13, 502–506.

Paillot, A. & Faure, J.C., 1923. Cultive du pyrethre et utilisation sur place de la recolte. *C. r. Acad. agric. Fr.* 9: 806–809.

Philbrick, H., Gregg, R.B. & Hatfield, A.W., 1972. *Companion Plants*. Robinson & Watkins.

Philips-Duphar, Bv., 1975. *Dimilin. Experimental insecticide*. Technical Information.

Pilat, M., 1935. Action of insecticides on the intestinal tube of insects. *Bull. ent. Res.* 26: 165–180.

Portier, P., 1949. *Biologie des Lépidoptères*. Lechevalier, Paris.

Post, L.C., Jong, B.T. & Vincent, W.R., 1973. A new insecticide inhibits chitin synthesis. *Naturwissenschaften* 60: 431–432.

Post, L.C. & Mulder, R., 1974. Insecticidal properties and mode of action of 1-(2,6-dihalogenbenzoyl)-3-Phenylureas. *Abstr. papers Am. chem. Soc. 167th meeting*.

Potter, C., 1951. Insecticides and fungicides department. *Rothamsted Experimental Station, Harpenden, Report for 1950*. 101–111.

—— 1955. Insecticides and fungicides department. *Rothamsted Experimental Station, Harpenden, Report for 1954*. 103–119.

—— 1956. Insecticides and fungicides department. *Rothamsted Experimental Station, Report for 1955*. 112–130.

—— 1957. Insecticides and fungicides department. *Rothamsted Experimental Station, Harpenden, Report for 1956*. 127–148.

Potter, C., Lord, K.A., Kenten, J., Slakeld, E.H. & Holdbrook, D.V., 1957. Embryonic development and esterase activity of eggs of *Pieris brassicae* in relation to TEPP poisoning. *Ann. appl. Biol.* 45: 361–375.

Potter, C. & Perkins, J.F., 1946. Control of brassica pests by DDT. *Agric.* 53: 109–113.

Potter, C. & Tattersfield, F., 1943. Ovicidal properties of certain insecticides of plant origin (Nicotine, pyrethrins, derris products). *Bull. ent. Res.* 34: 225–244.

Potter, C., Tattersfield, F. & Gillham, E.M., 1946. Laboratory comparison of toxicity as a contact poison of DDT with nicotine, derris products and the pyrethrins. *Bull. ent. Res.* 37: 469–496.

Scheurer, R., Fluck, V. & Ruzette, M.A., 1975. Experiments with insect growth regulators (IGRs) on lepidopterous pests and some of their parasitoids. *Mitt. Schweiz. ent. Ges.* 48: 315–321.

Schreiber, A.F., 1915a. Vegetable insecticides. (In Russian.) *Horticulturist. Rostov on Don* 903–912.

—— 1915. Control of *Pieris brassicae. Orchard Market Garden, Moscow* 140–142.

—— 1916. Deterrents to insects. (In Russian.) *Horticulturist, Rostov on Don* 15: 50–51.

Schreiner, J., 1914. Provincial entomological conference Zemstvo of Charkov. 50–52.

Sehnal, F., Romanuk, M. & Streinz, L., 1976. Potent juvenoids with cyclohexane moiety in the molecule. *Acta ent. Bohemsoslov* 73: 1–12.

Singh, K. & Atwal, A.S., 1969. Haemocytopenia in insects caused by insecticides. *J. Res. Punjab agric. Univ.* 6: 927–932.

Sleesman, J.P., 1957. New materials to control DDT-resistant cabbage worms and loopers. *Ohio Fm. Home Res.* 42: 64–65.

Smith, J.B., 1891. *Insect life* 3: 218.

—— 1896. *Economic entomologist. Farmer and Fruit Grower.* Lippencott Co. Philadelphia, USA.

Speyer, W., 1950. Do modern contact insecticides have an ovicidal action? *NachrBl. dt. PflSchutdienst. Berl.* 2: 2–3.

Sprenger, (-)., 1915. *In* Schreiber, A.F., Vegetable insecticides. (In Russian.) *Horticulturalist, Rostov on Don.* 903–912.

Strawinski, K., 1928. Caterpillars on cabbage and their control by chemical measures. (In Russian.) *Stacja Doswiadcz Ochrony Roslin w Zgierzu.* 1–9.

Sundukov, O.V., 1969. The action of organophosphorus insecticides on the central nervous system of lepidopterous larvae (Lepidoptera). (In Russian.) *Ent. Obozr.* 48: 71–80.

—— 1971. The mechanism of pathogenesis in larvae of *Pieris brassicae* poisoned with chlorofos (Trichlorphon). *All Union ent. Soc. Leningrad* 2: 281–282.

—— 1972. Features of the insecticidal action of organophosphorus compounds. (In Russian.) *Ent. Obozr.* 51: 561–572.

Thakur, M.R., 1966. Common insect pests of temperate vegetable crops with special reference to Upper Kulu valley. *Punjab Hort. J.* 6: 188–192.

Theobald, F.V., 1909. Animals injurious to vegetables. *Jl. S.E. agric. Coll.* Wye 1909: 157–164.

Theobald, F.W., 1928. The large cabbage white butterfly (*Pontia brassicae*) and a simple method of control. *Jl. S.E. agric. Coll. Wye* 1928: 75–78.

Tittanen, K., 1964a. The effect of pyrethrum preparations on certain common horticultural pests. *Ann. agric. fenn.* 3: 272–274.

—— 1964b. The possibility of using pyrethrum preparations in the control of plant pests. *Pyrethrum Post* 7: 27–32.

Touzeau, J., 1962. Observations sur la lutte en plein champ contre la Pieride du chou (*Pieris brassicae*). *C. r. act. exp. Def. Cult. Tunisie* 1962: 43–47.

Turenen, S., 1975. Effect of gamma-BHC on lipid digestion and utilisation in *Pieris brassicae* (Lepidoptera: Pieridae) reared on an artificial diet. *Ann. zool. fenn.* 12: 275–279.

Turowski, W., 1963. The influence of plant repellents on the appearance of the white butterfly (*Pieris brassicae* L.). (In Polish.) *B. Inst. Ochrony Rosl.* (19) 239–247.

Ugolini, A., 1959. La cavolaia. *Inf. Fitopat.* 9: 254–256.

Verma, A.N., Sharma, P.D., Saramma, P.U., 1970. Relative toxicity of some contact insecticides against cabbage caterpillar of *Pieris brassicae* L. *J. Res. Punjab agric. Univ.* 6: 197–199.

Vlasveld, W.P.N., 1975. Gids voor Ziekten Onkruidbestrijding in land Tuinbouw. Wageningen. 179.

Voskresenskaya, A.K., 1936a. Poison penetration through intestinal wall. *Zashch. Rast.* 7: 25–36.

—— 1936b. Reaction of throwing out the poison being the cause of resistance of insects to arsenical compounds. *Summ. Sci. Res. Wk. Inst. Plant Prot. Leningrad* 1936: 380–383.

—— 1938. Can pyrethrum be used as a stomach insecticide? (In Russian.) *Summ. Sci. Res. Wk. Inst. Plant Prot.* 1936. 93–95.

—— 1939. Resistance to arsenicals in insects. *Bull. Plant Prot. USSR* 19: 132–144.

Vostrikov, P., 1915. Tomatoes as insecticides. The importance of Solanaceae in control of pests of agriculture. (In Russian.) *Husbandry on the Don, Novotcherkassk* 10: 9–12.

Wageningen, 1971. Control of insects on Brussels sprouts (in Dutch). *Plant Protection Service no. 1797.*

Wahla, M.A., Gibbs, R., Gibbs, G. & Ford, J.B., 1976. Diazinon poisoning in the large white butterfly larvae and the influence of sesamex and piperonyl butoxide. *Pest. Sci.* 7: 367–371.

Walker, J.J., 1918. The butterflies of the Oxford district. *Entomologist's mon. Mag.* 54: 246–250.

Walker, P.T. & Turner, M.L., 1962. The toxicity of endrin granules to larvae of *Pieris brassicae*, including the effect of particle size and humidity. *Rept. Trop. Pestic. Res. Unit Porton* 219: 6.

Wellinga, K., Grosscurt, A.C., Hes, R. van, 1977. 1-phenylcarbamoyl-2-pyrazolines, a new class of insecticides. I. Synthesis and insecticidal properties of 3-phenyl-1-phenylcarbamoyl-2-pyrazolines. *J. Agric. Food Chem.* 25: 987–992.

Wellinga, K., Mulder, R. & Daalen, J.J. van, 1973. Synthesis and laboratory evaluation of 1-(2,6-disubstituted benzoyl)-3-phenylureas, a new class of insecticides, I. 1-(2,6-Dichlorobenzoyl)-3-phenylureas. *J. Agric. Food Chem.* 21: 348–354, 993–998.

Yana, A., 1962. Essais de lutte chimique contre le pieride du chou en laboratoire. *C. r. act. exp. Def. Cult. Tunsie* 1962: 36–43.

Zaitseva, V.G. & Ponomareva, E.A., 1976. A test of the economic effectiveness of plant protection. (In Russian.) *Zashch. Rast.* (4) 34–36.

Zanno, P.R., Miura, I., Nakanishi, K. & Elder, D.L., 1975. Structure of the insect phagorepellent azardirachtin; application of PRFT/CWD carbon 13 nuclear magnetic resonance. *J. Am. Chem. Soc.* 97: 1975–1977.

Ziarkiewicz, T. & Anasiewicz, A., 1961. Investigation on influence of hemp on the incidence of *Pieris brassicae* on cabbage. *Roczn. nauk. Roln Warsaw* 83: 641–649.

Further references

Arkhangelski, P., 1925. *Defense des Plantes* 1: 239 (Control in USSR).

Austin, M.D., 1929. *Jl. S.E. agric. Coll. Wye* 1929: 124–135 (use of pyrethrum).

Austin, M. & Theobald, F.V., 1928. *Jl. S.E. agric. Coll. Wye* 1928: 59–67 (pyrethrum experiments).

Bell, C.H., 1977. *J. Stored Prod. Res.* 13: 119–127 (methyl bromide).

Board of Agriculture & Fisheries, 1918. *Miscellaneous publication no. 21* (dust with slaked lime or spray with salt solution).

Bogdanov-Katkov, N.N., 1919. *Nat. Comm. agric. Moscow Petrograd Kiev* 1919: 1–3 (control measures).

—— 1921. *Petrograd Kitchen Garden.* 1921: 47–78 (review of control measures).

—— 1922. *All Russ. Union agric. Co-op. Mosc. & Petrograd* 1–20 (control measures).

Bovien, P. & Staple, C., 1940. *Tidsskr. Planteavl, Copenhagen* 45: 39–83 (derris experiments).

Brown, A.W.A., 1951. *Insect control by chemicals.* Wiley, New York (many references to *P. rapae*, a few to *P. brassicae*).

Coulon, J., 1963. *Phytiat. Phytopharm., Paris* 12: 45–49 (ethyl parathion, methyl paration).

De Bussy, L.P., Van der Laan, P.A. & Jacobi, E.F., 1935. *Tijdschr. Plziekt, Wageningen* 41: 33–50 (derris dust mixed with kaolin).

Dig for Victory, 1941. *Leaflet Nos. 1, 5. 16* (derris recommended).

Dındon, P.J., 1914. *Horticulturist, Rostov on Don* (12) 893–896 (control in USSR).

Dodonov, B.A., 1936. *Bull. Plant Prot. Leningrad* 1936: 55–77 (experiments with arsenic and fluorine derivatives).

Edwards, C.A. & Heath, G.W., 1964. *Principles in agricultural entomology.* Chapman & Hall, London (recommended dosages of DDT and phosdrin).

Falck, R., 1920. *Z. angew. Ent.* 7: 37–47 (resinol solution).

Fryer, J.C.F., Stenton, R., Tattersfield, F. & Roach, W.A., 1923. *A. appl. Biol.* 10: 18–34 (derris).

Ginsburg, J.M. & Granett, P., 1934. *J. econ. Ent.* 27: 393 (derris root powder).

Gustafsson, M., 1958. *Medd. Växtskyddsans. Stockholm* 11: 81–130 (*P. brassicae* used to test significance of droplet size in spraying).

Güvener, A., Yücel, I. & Tuzun, H., 1973. *Ziraf Mucadele Arastirma Yilligi* 120 (pesticide residues of those pesticides used against *P. brassicae*).

Hartley, G.S. & West, T.F., 1969. *Chemicals for pest control.* Pergamon Press, London.

Janische, R., 1926. *NachrBl. dt. PflSchutzdienst. Berl.* 6: 18–20 (21 compounds tested by new method).

Jones, F.G.W. & Jones, M.G., 1964. *Pests of field crops.* Arnold, London (rotenone, pyrethrum recommended).

Laska, P., 1973. *Bull. Vysk. Ustav. Zel. Olomouc* (16–17) 85–93 (effectiveness of certain insecticides).

Lesne, P., 1917. *J. agric. Prat. Paris* 30: 410–411 (spray soap or turpentine).

Lind, J., Rostrup, S. & Kolpin Ravn, F., 1915. *Beret. Stat. Forsgsv. Plantek.* Kobenhavn 1915 (control by arsenical sprays).

Lindblom, A., 1928. *Meddel. Centralanst. Försöks Jordbruks. Lanthruksent Avdel.* 53: 1–33 (nicotine, soap).

Macleod, W.S., 1949. *Review of literature on systemic insecticides.* Division of entomology, Department of agriculture, Canada (Pestox-III).

Meltzer, J., 1972. *Neth. J. Pl. Path.* 78: 77–88 (MYC 8005 – an acaricide with strong inhibitory properties on *P. brassicae*).

Morozov, S.F., 1935. *Plant Prot. Leningrad* 1935: 38–58 (effect of insecticides on cuticle penetration).

Nikitin, I.V., 1939. *Lenin. Acad. agric. Sci.* 1940: 140–143 (pyrethrum).

Potter, C., Lord, K.A. & Solly, R., 1960. *Int. Cong. Crop. Prot.* 4: 1169–1172 (action of organophosphoros compounds on insect ova).

Prochaska, R.G., Cuthbert, F.P. & Reid, W.J., 1964. *J. econ. Ent.* (Zectrand and Bayer 44646 show good results).

Rakushev, F.N., 1914. *Horticulturalist, Rostov-on-Don* (10) 801–804 (carbolic acid, potash soap recommended).

Razauska, E., 1967. *Kartofel' Ovoshchi* (8) 45 (DDT).

Ripper, W.E., Greenslade, R.M. & Hartley, G.S., 1950. *Bull. ent. Res.* 40: 481–501 (bis (bis dimethylamine phosphorous) anhydride).

Rothe, G., 1937. *NachrBl. dt. PflschDienst. Berlin* 15: 46–48 (mineral oils).

Schreiner, J., 1914. *Prov. ent. Conf. Zemstvo of Charkov* 50–52 (acid chromate of lead).

Sekun, N.P., 1968. *Ent. Obozr.* 47: 726–730 (physiological changes during intestinal poisoning).

Sharma, P.B. & Chopra, S.L., 1970. *J. Res. Punjab Agric. Univ. Ludhiana* 7: 365–368 (thiometon).

Slade, R., 1945. *Chem. Trade J. Lond.* 116: 279–281 (gammexane, highly toxic).

Stapel, C. & Peterson, H.I., 1944. *Tidsskr. Planteavl.* 48: 631–654 (gerasol).

Strawinsky, K., 1928. *Publ. Stan. Prot. Zgierz* (2) (control of cabbage larvae).

Swingle, M.C., 1934. *J. econ. Ent.* 27: 1101–1102 (pyrethrins).

Switzerland, 1922. *Schweiz. Z. Obst Weinbau Frauenfeld* 31: 219–220 (control).

Szczepanska, K. & Kowalska, T., 1969. *Prace nauk. Inst. Ochrony Roslin* 11: 175–179 (DDT, fenitrothion).

Szwejda, J., 1973. *Biul. Warsywniczy* 14: 319–332 (several insecticides tested).

Tarasova, K.L., 1936. *Izr. Kurs. Prikl. zool. Leningrad* 6: 2–14 (anabasine).

Thalenhorst, W., 1937. *Z. angew. Ent.* 23: 615–652 (derris).

Theobald, F.W., 1927. *Ann. Rpt. Res. Adv. Dept. SE agric. Coll. Wye* 1–16 (pyrethrum).

Turunen, S., 1972. *Ann. Acad. Sci. Fenn. A.* 193: 1–7 (effect of chlorinated insecticide on fatty acids of fat body).

—— 1977. *Ent. expl. appl.* 21: 254–260 (effect of lindane).

Way, M.J., Smith, P.M. & Hopkins, B., 1951. *Bull. ent. Res.* 331–354 (insects as test subjects).

Yule, W.N., 1964. *Ann. appl. Biol.* 53: 15–28 (pyrethrum).

18. Integrated control

As the name implies integrated control means the use of all systems of control, i.e. biological and cultural, with the careful use of chemical control methods when these are selective and not detrimental to biological methods (Greathead, 1976; van Emden, 1974a,b; Way, 1977). However, for perennial crops integrated control measures can be implemented much more easily and with a greater chance of success than with annual crops, such as cabbage, which of necessity therefore can only be controlled by limited integrated methods. Nevertheless, where large monocultures of cabbage are grown certain measures such as ploughing, rolling, time of seeding and, less often, interplanting can be practised to combat the wide variety of cabbage pests, including *P. brassicae*.

Many countries are concerned about the control of cabbage pests and in some cases have carried out special campaigns, particularly Germany (Kollar, 1846; Kallenbach, 1948; Born, 1957), India (Saini *et al.*, 1968), Netherlands, (Oostendorp, 1954), Portugal (Urquijo Landaluze, 1938), Switzerland (Ferrière, 1922) and the USSR (Leningrad, 1938).

To plan a successful integrated control programme it is essential to understand the biology and ecology of the species under attack. Not much work has been carried out in this direction in relation to *P. brassicae*, but much can be gleaned from independent work on various aspects of the ova, larva, pupa and imago of *P. brassicae* (cf. Chapters 3,4,5).

Ova of *P. brassicae* are laid in groups of up to 200 on the underside of cabbage leaves. Here they are susceptible to the parasitic hymenopteran *Trichogramma evanescens* Westwood (Chapter 14). In the USSR Dub (1977) achieved 70–80% parasitism of *P. brassicae* in a cabbage field by introducing ova parasitised with *T. evanescens*. However, with the ova of *P. brassicae* being on the undersides of leaves, chemical solutions applied from above will have little effect upon them. Contact poisons and compounds applied as an aerosol spray would have a greater chance of success. On a small scale by far the best method of control is to collect or crush ova manually.

Larvae are easier to see than ova, are exposed to the upper surface of leaves when half-grown and are naturally afflicted by a diverse collection of parasites, hyperparasites, predators and naturally occurring pathogenic organisms; thus it is not surprising that much work has concentrated on the control of the larvae of *P. brassicae*, even though this is the most destructive stage. Chansigaud (1964) went some way in trying to understand the biology and ecology of *P. brassicae* with a view to devising control stratagies. He studied the effects of temperature on the distribution of *P. brassicae* larvae on the foodplants and their speed of development, the natural disappearance of different groups of larvae by predation and parasitism of larvae by *Apanteles glomeratus*. He showed that the most effective time to treat the larvae was during the second instar when they are exposed on the leaf and fairly well developed.

Timing of attack on larvae is therefore of importance in the control of *P. brassicae* and it is necessary to be aware of regional differences in bionomics, for instance, the number of broods; Ghirlanda (1921) noted that there were up to four broods per year in northern Italy, with only two broods in the south.

Anecdotal methods of chemical control were described in the previous chapter but there are other biological methods which deserve mention, as they take into account other relationships. Querci (1957) suggested that ant colonies should be prepared in buckets which are then transplanted to fields infested with *P. brassicae* larvae where apparently they inflict great loss on the pest. Engerbund (1914) recommended that nest boxes for tits should be set up as these birds eat a large number of larvae.

Quite a large number of bird predators (25 species) feed on all stages of *P. brassicae* and their effect in reducing overall populations of *P. brassicae* is probably considerable; but there have been problems in assessing the degree of predation in the field. Birds are also not practical to use in the field. Other predators such as various Hemiptera and Coleoptera species have only recently been tested in the field. Results have been encouraging and this approach appears to merit further investigation (Chapter 16).

The most important factor in the natural control of *P. brassicae* according to Chansigaud (1964) is from birds and insects, notably from the Coleoptera, Diptera and Hymenoptera. Bird predators were responsible for depleting experimental groups of 100, 300 and 500 larvae by 59, 69 and 68% respectively. When he studied parasitism of larvae, Chansigaud (*loc. cit.*) found that the number of parasitised larvae in any group varied, i.e. using the same groups of larvae as above, he found that there were 9–14, 18–94 and 54–160 parasitised in any sample, thus making assessment of parasite success difficult. A smaller percentage of control (4.4%) was recorded for pathogenic organisms. Control by parasites and pathogens however, were most effective during the period from fifth instar to pupation.

Combination of biological methods with and without chemical methods has proved effective on the few occasions it has been practised; for instance, a mixture of the fungus *Beauveria bassiana* and the bacterium *Bacillus thuringiensis* gave up to 98% mortality of larvae (Kalyuga, 1968) (Chapter 15). However, there are some strains of *B. thuringiensis* resistant to *P. brassicae* (Empson, 1968). Moiseeva *et al.* (1975) found that the use of the parasite *Trichogramma evanescens* was perfectly adequate for the control of *P. brassicae* in the USSR, instead of using insecticides. Successful integrated control of *P. brassicae* and *P. rapae* in Bulgaria by using Entobacterin (*B. cereus* var. *galleriae*) and Boverin (*Beauveria bassiana* Vuill.), together with the ova parasite *Trichogramma evanescens*, was accomplished by Karadjov (1972).

It has been stated that as mixtures of *B. thuringiensis* and low levels of DDT have been effective in controlling the moth *Agrotis ipsilon* (Hüfnagel) these measures may be usefully applied to *P. brassicae* and other cruciferous pests (Burges, 1971). However, two points which must be considered are that when two or more pests occur on the same plant and their susceptibility to the same pathogen varies, then the combination of two pathogens is required and has been seen to be successful (Franz, 1971). The other point is that *P. brassicae* larvae which are already infected with protozoa are less susceptible to Entrobactin-3 (Issi, 1965). Ripper *et al.* (1948) recommended using DDT coated with degraded cellulose to make it less harmful to predators.

There seems to be potential in other less well known microorganisms and parasites. The long list of primary parasites of *P. brassicae* is overlooked by the effective control inflicted by *A. glomeratus*, but many of these could be economically useful if they were produced in the laboratory in sufficient quantities, released and their effectiveness measured. Similarly there are several other viruses worthy of further attention, other than the two main ones, the granulosis virus and the cytoplasmic polyhedosis virus; also fungi other than *Beauveria bassiana* and *Entomophthora sphaerosperma* might provide useful species for integrated control.

References cited

Born, M., 1957. Wann muss der Kohlweissling (*Pieris brassicae*) bekämpft werden? *Mitschurin Bewegung* 6: 690–691.

Burges, H.D., 1971. Control of insects by *Bacillus thuringiensis*. *5th Brit. Insect Fung. Conf.* 405.

Chansigaud, J., 1964. Observations préliminaires sur les essais d'infestations artificielles avec *Pieris brassicae* L. au stade larvaire dans les conditions naturelles. *Revue zool. agric. Talence* 63: 55–61.

Dub, P.A., 1977. *Trichogramma* in the centre of a town. (In Russian.) *Zashch. Rast.* (5) 59.

Empson, D.W., 1968. Papers presented at a conference of advances in entomology. 11–12th April, 1967. *J. appl. Ecol.* 5: 489–516.

Engerbund, (–)., 1914. A propos des nichoirs artificiels. *Bull. Soc. étude vulg. zool. agric. Bordeaux* 13: 16–17.

Ferriere, C., 1922. Les problèmes moderne de la lutte contre les insectes et leur application en suisse. *Berne Edn. Ernest Bir. Cher.* 1–36.

Franz, J.M., 1971. Influence of environmental and modern trends in crop management of microbial control. *In: Microbial control of insects and mites.* Ed. H.D. Burges and N.W. Hussey. Academic Press, London.

Ghirlanda, C., 1921. Il Bruco verde dei caroli (*Pieris brassicae* L.). *Revta Oss. Fitopat. Turin* 11: 1–4.

Greathead, D.J., 1976. A review of biological control in western and southern Europe. Tech. Comm. No. 7., Commonwealth Institute of Entomology.

Issi, I.V., 1965. The effect of microsporidiosis on the susceptibility of cabbage white butterfly to insecticides. *Trudý vses. nauchnóissléd. Inst. Zashch. Rast.* (24) 170–174.

Kallenbach, F., 1948. Kohlweissling (*Pieris brassicae*) und Kohleule (*Mamestra brassicae*) rechtzeitig vernichten! *Land. wald. Gart.* 3: 168.

Kalyuga, M.V., 1968. The combined use of microbiological preparations for the control of pests. (In Russian.) *Ent. Obozr.* 47: 450–453.

Karadjov, S., 1972. The joint utilisation of biopreparations entobacterin (*Bacillus cereus* var. *galleriae*) and boverin (*Beauveria bassiana* Vuill.) with the egg parasite *Trichogramma evanescens* Westw. for the control of *Pieris brassicae* L. and *Pieris rapae* L. and the cabbage worm (*Mamestra brassicae* L.) (in Bulgarian). *Acad. agric. sci. inst. plant Prot. Kostinbrod. Stud. Biol. Contr. pl. Pests* 1: 129–143.

Kollar, V., 1846. Wie schützt man die Kraut- oder Kohlgärten vor ihrem verderblichsten Feinde, dem Kohlweisslinge *Papilio brassicae? Neuer Wirtsch Kalender, Wien* 9: 19–23.

Leningrad, 1938. Summary of a scientific working institute of plant protection, 1936 (II). Virus and bacteria disorders of plants: biological, chemical and mechanical methods of plant protection. (In Russian.) *State Publ. Off. Lit. Coll. Co-op. Farming Selkhozgiz, Leningrad* 1–111.

Moiseeva, N.V., Mashkina, L.G., Kalashnikova, G.I. & Ratomskaya, A.A., 1975. The effectiveness of the use of *Trichogramma* for control of certain pests in the Chu Valley, Kirgizia. *In: Arthropods of Agricultural Importance.* Ed. A.J. Protosenko. Ent. issled Kirgizii, Vypusk, Kirgiz. SSR.

Oostendorp, F.L., 1954. The cabbage worm (*Pieris brassicae*) and its control. (In Dutch.) *Floralia* 74: 202–203.

Querci, O., 1957. Ants as a control against other insects. *Entomologist's Rec. J. Var.* 69: 101.

Ripper, W.E., Greenslade, R.M., Heath, J. & Barker, K. 1948. Coated DDT to protect parasites and predators. *Nature, Lond.*, 161: 484.

Saini, S.S., Sharma, B.R. & Sharma, P.P., 1968. Cabbage seed production – a revolution in the economy of Kalpa Valley of Himachal Pradesh. *Himachal Hort.* 9: 3–6.

Urquijo Landaluze, P., 1938. Posibilidades de lucha biologica contra las arugas de la col. *Boln. patol. veg. Ent. agric.* 8: 171–178.

Van Emden, H.F., 1974a. *Pest control and its ecology.* Arnold, Studies in Biology.

—— 1974b. *Ecological aspects of insect control in crops.* Arnold, London.
Way, M.J., 1977. Integrated control – practical realities. *Outlook on Agriculture* 9: 127–135.

Botanical specific name index
(including bacteria, fungi and viruses)

515

517

Zoological specific name index

Subject index

Individually-named amino-acids, carotenoids, enzymes, lipids, pteridines and so on are excluded from this index. These may be searched for under their separate headings.

Potential research students may wish to know that topics requiring further investigation are to be found on pages 4, 27, 160, 163, 164, 165, 168, 186, 257, 293, 358, 470, 486.